Climate-smart production of coffee

Improving social and environmental sustainability

It is widely recognised that agriculture is a significant contributor to global warming and climate change. Agriculture needs to reduce its environmental impact and adapt to current climate change whilst still feeding a growing population, i.e. become more 'climate-smart'. Burleigh Dodds Science Publishing is playing its part in achieving this by bringing together key research on making the production of the world's most important crops and livestock products more sustainable.

Based on extensive research, our publications specifically target the challenge of climate-smart agriculture. In this way we are using 'smart publishing' to help achieve climate-smart agriculture.

Burleigh Dodds Science Publishing is an independent and innovative publisher delivering high quality customer-focused agricultural science content in both print and online formats for the academic and research communities. Our aim is to build a foundation of knowledge on which researchers can build to meet the challenge of climate-smart agriculture.

For more information about Burleigh Dodds Science Publishing simply call us on +44 (0) 1223 839365, email info@bdspublishing.com or alternatively please visit our website at www.bdspublishing.com.

Related titles:

Achieving sustainable cultivation of coffee: Breeding and quality traits
Print (ISBN 978-1-78676-152-1) Online (ISBN 978-1-78676-154-5; 978-1-78676-155-2)

Agroforestry for sustainable agriculture
Print (ISBN 978-1-78676-220-7) Online (ISBN 978-1-78676-222-1; 978-1-78676-223-8)

The sustainable intensification of smallholder farming systems
Print (ISBN 978-1-78676-430-0) Online (ISBN 978-1-78676-432-4; 978-1-78676-433-1)

Achieving sustainable management of tropical forests
Print (ISBN 978-1-78676-248-1) Online (ISBN 978-1-78676-250-4; 978-1-78676-251-1)

Chapters are available individually from our online bookshop:
https://shop.bdspublishing.com

climate•SMART•publishing
IN AGRICULTURAL SCIENCE

BURLEIGH DODDS SERIES IN AGRICULTURAL SCIENCE
NUMBER 111

Climate-smart production of coffee

Improving social and environmental sustainability

Edited by Professor Reinhold Muschler, CATIE, Costa Rica

burleigh dodds
SCIENCE PUBLISHING

Published by Burleigh Dodds Science Publishing Limited
82 High Street, Sawston, Cambridge CB22 3HJ, UK
www.bdspublishing.com

Burleigh Dodds Science Publishing, 1518 Walnut Street, Suite 900, Philadelphia, PA 19102-3406, USA

First published 2022 by Burleigh Dodds Science Publishing Limited

Library of Congress Control Number: 2022934197

British Library Cataloguing in Publication Data
A catalogue record for this book is available from the British Library

ISBN 978-1-78676-483-6 (print)
ISBN 978-1-78676-486-7 (PDF)
ISBN 978-1-78676-485-0 (ePub)
ISSN 2059-6936 (print)
ISSN 2059-6944 (online)

DOI 10.19103/AS.2021.0096

Typeset by Deanta Global Publishing Services, Dublin, Ireland

Contents

Part 2 Sustainable pest and disease management

Series list

Title	Series number
Achieving sustainable cultivation of maize - Vol 1 From improved varieties to local applications Edited by: Dr Dave Watson, CGIAR Maize Research Program Manager, CIMMYT, Mexico	001
Achieving sustainable cultivation of maize - Vol 2 Cultivation techniques, pest and disease control Edited by: Dr Dave Watson, CGIAR Maize Research Program Manager, CIMMYT, Mexico	002
Achieving sustainable cultivation of rice - Vol 1 Breeding for higher yield and quality Edited by: Prof. Takuji Sasaki, Tokyo University of Agriculture, Japan	003
Achieving sustainable cultivation of rice - Vol 2 Cultivation, pest and disease management Edited by: Prof. Takuji Sasaki, Tokyo University of Agriculture, Japan	004
Achieving sustainable cultivation of wheat - Vol 1 Breeding, quality traits, pests and diseases Edited by: Prof. Peter Langridge, The University of Adelaide, Australia	005
Achieving sustainable cultivation of wheat - Vol 2 Cultivation techniques Edited by: Prof. Peter Langridge, The University of Adelaide, Australia	006
Achieving sustainable cultivation of tomatoes Edited by: Dr Autar Mattoo, USDA-ARS, USA and Prof. Avtar Handa, Purdue University, USA	007
Achieving sustainable production of milk - Vol 1 Milk composition, genetics and breeding Edited by: Dr Nico van Belzen, International Dairy Federation (IDF), Belgium	008
Achieving sustainable production of milk - Vol 2 Safety, quality and sustainability Edited by: Dr Nico van Belzen, International Dairy Federation (IDF), Belgium	009
Achieving sustainable production of milk - Vol 3 Dairy herd management and welfare Edited by: Prof. John Webster, University of Bristol, UK	010
Ensuring safety and quality in the production of beef - Vol 1 Safety Edited by: Prof. Gary Acuff, Texas A&M University, USA and Prof. James Dickson, Iowa State University, USA	011
Ensuring safety and quality in the production of beef - Vol 2 Quality Edited by: Prof. Michael Dikeman, Kansas State University, USA	012
Achieving sustainable production of poultry meat - Vol 1 Safety, quality and sustainability Edited by: Prof. Steven C. Ricke, University of Arkansas, USA	013
Achieving sustainable production of poultry meat - Vol 2 Breeding and nutrition Edited by: Prof. Todd Applegate, University of Georgia, USA	014

Introduction

Coffee cultivation faces a number of crucial challenges, including increasing biotic and abiotic stresses related to climate change, concern about its environmental impact and the vulnerability of many smallholder coffee farmers. This volume addresses the need for more resilient and sustainable methods of cultivation which produce high-quality products at a reduced cost with minimum environmental impact.

The book is split into two parts: Part 1 chapters focus on improving social and environmental sustainability of coffee, focusing on areas such as coffee's global importance, its importance for smallholder farmers and the environmental and social impact assessment of coffee. Chapters also examine the use of specialty coffees as drivers of change, the meaning of 'fair' in fair-trade coffee, advances in Arabica coffee breeding and the optimisation of postharvest practices in coffee cultivation. Chapters in Part 2 address sustainable pest and disease management, drawing specific attention to insect pests and diseases affecting coffee such as coffee leaf rust and coffee wilt disease. Chapters on integrated management of nematode, soil-borne insect and fungal pests are also included and a chapter on integrated weed management of coffee is also provided.

Part 1 Improving social and environmental sustainability

The book opens with a chapter that focuses on global coffee production and sustainability. Chapter 1 begins by examining trends in coffee production, particularly focusing on factors affecting the low share of global revenue received by smallholders responsible for a large share of production. The chapter also reviews key market developments such as the degree of concentration, developments in specialty and differentiated coffees as well as Arabica and Robusta coffees. A section on creating an enabling environment to make coffee farming more sustainable is also provided, which is then followed by an overview of the case of Brazil in improving coffee grower incomes as well as ongoing issues in improving livelihoods for smallholders in creating a more sustainable coffee sector.

The subject of Chapter 2 is the coffee sector and the role of smallholders in coffee cultivation. The chapter first provides an overview of the main challenges faced by small-scale coffee farmers, which is then followed by a discussion on the challenges faced by smallholder farmers, focusing specifically on market trends. A section on the role of women and young people in the coffee growing sector is also included and the impact of the COVID-19 pandemic on small and medium-scale coffee farmers is also examined. The chapter also looks at potential solutions for improving the livelihoods of smallholders, including women and young people and examines diversification in the coffee sector. The environmental services provided by coffee smallholders are also touched upon.

Chapter 3 concentrates on assessing and managing the environmental and social impact of coffee production. The chapter begins by examining the complexity of coffee value chains. It reviews ways of measuring and improving the sustainability of coffee production systems, including good agricultural practices (GAP), life cycle assessment (LCA), ways of calculating the carbon footprint of coffee as well as methods for assessing social impacts. It also discusses ways of improving sustainability in the coffee value chain, particularly by understanding and optimising coffee production as an agroecosystem. The chapter reviews techniques to support agroecosystem function and resilience such as shade, soil health monitoring and biodiversity management.

The next chapter examines the role and importance of specialty coffees as drivers of change. Chapter 4 first looks at the emergence of the specialty coffee sector and the issues that arise in terms of defining these coffees in terms of their sensory quality attributes such as unique flavour, geographical origin and methods of production. The chapter also reviews factors affecting pricing and the role of geographical indicators for specialty coffees. The importance of good agricultural practices in standardising, improving and validating the quality attributes that underpin speciality coffees is also discussed.

Chapter 5 discusses fair-trade coffee and what 'fair' means in the coffee sector. The chapter begins by defining what 'fair' means and how 'fair' wages and incomes are established, then moves on to address some of the major Voluntary Sustainability Standards (VSS) that are used to help calculate 'fair' prices for coffee producers. This is then followed by an overview of two alternative approaches to calculating prices, the first being the 'Living Income Reference Price Formula' and the second the 'Living Income/Fair Price Approach'. This is a bottom-up approach for defining what a 'fair' price is from the farmer's perspective.

The subject of Chapter 6 is advances in Arabica coffee breeding, focusing specifically on developing and selecting the right varieties. The chapter first draws attention to the reproductive behaviour of Arabica coffee, focusing on characteristics such as allotetraploidy, autogamy and homozygosity. This is then followed by an overview of Arabica genetic resources such as varieties derived from Typica and Bourbon types as well as their mutants, varieties derived from crosses with the Timor Hybrid, introgressed lines derived from interspecific hybrids, wild and semi-domesticated genotypes and other plant species or organisms. The chapter also address coffee germplasm banks, breeding methods in Arabica coffee such as conventional crossing, intraspecific hybrids, coffee tissues culture, the use of androsterile parents and genome modifications. This is followed by a review of the main traits that are selected as key for breeding Arabica coffee varieties. The chapter also includes a case study for improved Arabica varieties which focuses on CENICAFE's breeding programme.

The final chapter of Part 1 addresses optimising post-harvest practices in coffee cultivation. Chapter 7 begins by first focusing on coffee types and their processing, it breaks down the components of a coffee cherry and highlights the three final products that can be produced depending on the processing method: natural coffee, pulped natural or honey coffee and washed coffee. The chapter then moves on to review the main processing steps for these three different types of coffee, such as harvesting, cleaning and separation, pulping, fermentation and mechanical removal of mucilage. Sections on drying processes, smallholder farmers, the environmental impacts of processing and the storage of dry coffee are also provided. Cleaning processes, hulling and polishing, size grading, gravity separation and micro lot processing are also discussion. The chapter draws attention to colour sorting, coffee storage in dry mills, blending and bulking, weighing and packing for shipment as well as processing lines.

Part 2 Sustainable pest and disease management

Part 2 of the book begins with a chapter that focuses on insect pests affecting coffee as well as the importance of understanding agroecosystems and the alternative methods for insect pest control. Chapter 8 reviews some of the most important pests affecting coffee plantations worldwide. In each case, the chapter discusses the damage each insect pests causes, their biology and ecology and how this understanding can be used to develop alternative more environmentally-friendly and sustainable methods of pest management. The chapter also highlights some of the natural enemies for key insect pests in coffee agroecosystems as well as the various impacts climate change can have. A case study on improving integrated pest management of coffee berry borer in commercial farms in Hawaii is also included, focusing specifically on coffee farms from the Ka'u district.

Chapter 9 provides an overview of the diseases that can affect coffee and the measures that can be put in place in order to control their spread and effect. It draws specific attention to diseases that can affect adult plants, such as coffee leaf rust, black rot and anthracnose. Sections on the main root diseases that can affect coffee plants are also provided, such as brown root and red root disease, black root disease and santavery root disease. The chapter also examines the presence of coffee trunk canker and berry blotch, which is then followed by an overview of the various diseases that can affect coffee seedlings. Diseases of minor importance are also mentioned, they are defined as 'minor diseases' due to their occurrence on a smaller scale than to those previously mentioned.

Flowering parasites of coffee are also mentioned, before the chapter concludes with an overview of why developing effective and economic disease control methods is important for coffee plantations.

The next chapter of Part 2 focuses on the ecological perspectives of coffee leaf rust. Chapter 10 first provides a brief overview of the research which has stemmed from a more technocentric approach to understanding and subsequently control coffee leaf rust. The chapter highlights key areas which are of potential ecological and evolutionary interest within this thread of research. Next, the chapter synthesises and discusses the sparsely available information about the community ecology and natural enemies of coffee leaf rust around the world. The chapter then presents two case studies about coffee which the authors feel represent a more ecological (i.e. intelligible) approach to understanding the pathogen.

Chapter 11 examines coffee wilt disease, it first provides an overview of the history and impact of the disease when it was first becoming prevalent in coffee cultivation. The chapter highlights how, through systematic sanitation and the establishment of breeding programmes in countries it affected, the disease did decline for a time. This is then followed by a brief discussion on other potential hosts for the disease. The chapter draws attention to the symptoms that can occur as a result of the disease and also the fungus' ecology. A section on historical management recommendations is provided, followed by an overview of the disease's re-emergence in the late 20th century. The chapter also provides a case study on CABI's Regional Coffee Wilt Programme (RCWP) which occurred in 2000-2007.

The subject of Chapter 12 is integrated management of nematodes of coffee. The chapter reviews the main soil-based nematodes affecting coffee plants, such as the root-knot nematode (*Meloidogyne* spp.), *Meloidogyne exigua* and *Meloidogyne paranaensis* and *M. incognita*. The chapter also draws attention to minor species of *Meloidogyne* and root lesion nematodes. In each case, the chapter discusses pest distribution and range, symptoms and damage caused, diagnosis, life cycle and control measures. These include preventative measures to limit movement of infected material, use of resistant varieties, cultural control measures such as use of rotations as well as biological control using parasitic microorganisms.

Expanding on topics previously discussed in Chapter 12, Chapter 13 moves on to address integrated management of soil-borne insects and fungal pests of coffee. It begins by assessing insect pests such as termites, leaf-cutting ants, mealybugs, cicadas, root flies and white grubs. In each case, the chapter includes an overview, discussion of pest life cycle, damage caused and management options. The chapter then discusses key soil-borne fungal diseases: *Rhizoctoniosis*, *Rosellinia* root rot, *Ceratocystis* canker stain and coffee wilt disease. In each case it covers disease symptoms, factors affecting incidence, epidemiology and disease management.

The final chapter of the book examines integrated weed management in coffee production. Chapter 14 discusses the effects of weed species on coffee plantations and the main integrated weed management practices that are applied in the crop, focusing on nonherbicidal methods. Emphasis is given to the width of strip weeding required to prevent competition in young coffee plantations without suppressing natural (or introduced) vegetation in the inter-rows. Sustainable weed control practices used worldwide such as cover crops, grazing with ovines, and electrical discharge are discussed. Finally, some major aspects of weed control in Robusta (Conilon) coffee are also presented.

Acknowledgements

We wish to acknowledge the following for their help in reviewing particular chapters:

- Chapter 2: Professor Ed Price, Texas A&M University, USA and Dr Charles Agwanda, CAB International, Africa Regional Centre, Kenya
- Chapter 3: Dr Peter Läderach, CGIAR, Senegal
- Chapter 4: Dr Thomas Oberthur, African Plant Nutrition Institute, Morocco and Dr. Esteban Escamilla Prado, Professor - Researcher, Chapingo Autonomous University, Mexico
- Chapter 5: Dr Eberhard Krain, Independent Consultant, Agricultural and Rural Development, Bremen, Germany
- Chapter 6: MSc. William Solano, CATIE, Costa Rica
- Chapter 8: Dr Juan F. Barrera, El Colegio de la Frontera Sur, Mexico
- Chapter 10: Professor Laercio Zambolim, University of Viçosa, Brazil
- Chapter 11: Dr Luc Villain, IRD, France
- Chapter 13: Dr Cezar Araujo-Junior, IAPAR, Brazil

Part 1

Improving social and environmental sustainability

Chapter 1

Global coffee production and sustainability

Carlos H. J. Brando, P&A Marketing, Brazil

1 Introduction

Coffee is one of the most consumed beverages in the world, and one of the most widely traded agricultural commodities, with the sector having an overall value of over US$200 billion according to the International Coffee Organization (ICO). Produced in over 70 countries, coffee cultivation has played a key role in the economic development of countries as diverse as Colombia, Ethiopia, Brazil, Indonesia, Vietnam, Kenya and Costa Rica, with the first 5 of these countries dominating production. Coffee-producing nations like Indonesia and Mexico are also becoming important consumers, following the example of Brazil, which is now the second largest coffee consumer in the world after the United States.

The coffee sector has been targeted by a range of certification schemes seeking to improve the overall sustainability of coffee production and farmer incomes. These range from general schemes including commodities such as coffee such as those of the Rainforest Alliance (https://www.rainforest-alliance

http://dx.doi.org/10.19103/AS.2021.0096.01

.org/insights/rainforest-alliance-certified-coffee/) and Fairtrade International (https://www.fairtrade.net/product/coffee) to sector-specific schemes such as 4C (https://www.4c-services.org/). These have been seen to have some benefits in improving incomes, though evidence suggests they have been most beneficial to mid-size to larger farms with better resources to access and implement these schemes than individual smallholders (Raynolds et al., 2007; Crabs et al., 2016; Bray and Neilson, 2017; Winter et al., 2020; Jena and Grote, 2022). Life cycle analysis (LCA) and other techniques have also been used to assess the environmental impact of coffee production and to identify areas of improvement (Bockel and Schiettecatte, 2018). However, results are mixed whilst, at the same time, coffee has been seen to be vulnerable to the likely effects of climate change (Bunn et al, 2015).

However, despite initiatives to support coffee farmers, the value of the trade in coffee has not always been translated into a sufficient and sustainable income for farmers growing coffee, especially smallholders. It has also not fed through to investing resources to make coffee production more sustainable.

2 Coffee farmer incomes and sustainability

As Fig. 1 shows, around 98% of the estimated 12.5 million farmers growing coffee are smallholders who are farming in less than 5 hectares. This group is responsible for 60% of overall production. However, despite their importance in coffee production, evidence suggests this group does not receive sufficient

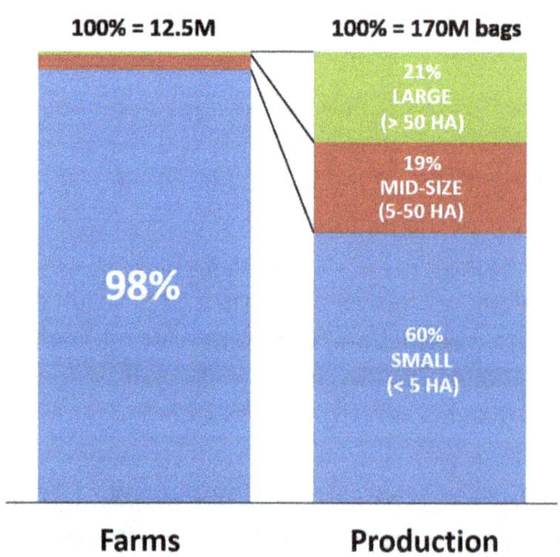

Figure 1 Coffee farm size and global production. Source: Enveritas.

income from the coffee they grow. According to the Coffee Barometer 2020, the coffee retail market generates annual revenues of US$200–250 billion (https://coffeebarometer.org/). However, the total annual export value of green (unprocessed) coffee has been estimated at an average US$20 billion over the period of 2015–2020, accounting for less than 10% of overall revenue from coffee. In general, allowing for intermediaries in the supply chain, coffee growers receive less (much less in some countries) than the export value of coffee (see Discussion later in this chapter).

Even using the figure of US$20 billion and dividing it by 12.5 million farms growing coffee, we arrive at an average income of US$4.38/day per farm. This is below the World Bank's international poverty line of US$1.90 a day per person, assuming that an average farm has to sustain a family of three or four people. In addition, it is reasonable to assume that larger farms and countries with higher average levels of productivity will account for a higher percentage of this income (see the subsequent paragraphs), leaving even less for smallholders in less well-resourced countries. This evidence suggests coffee cultivation is not providing an adequate living income for many smallholder farmers, let alone one that allows them to invest in being environmentally more sustainable and resilient in the face of climate change.

Figure 2 shows the 14 countries producing more than 1 million bags of coffee in 2019–2020 (a bag represents 60 kg of green coffee). Other than Brazil and India, which produce both Arabica (A) and Robusta (R) coffees in significant quantities, all other countries mainly produce only one of these types (shown by the first letter next to the country). Brazil and Vietnam account for a little over 50% of the world's production with roughly 25% of the global area planted with coffee. The two countries have an average production of about 30 bags (1800 kg) per hectare which leaves other coffee producers with average productivity of about 10 bags (600 kg) per hectare. While a few of these countries produce substantial volumes of coffee and have average productivities between 15 and 20 bags (900 kg and 1200 kg) per hectare, there are many countries that produce only an average of 5 bags (300 kg) per hectare. Since mid-size and larger coffee farms are typically more productive than small farms (with better resources and economies of scale), many smallholders will have an even lower productivity than five bags per hectare. Such low levels of productivity both reduce incomes and make sustainable production more difficult.

As has already been noted, the income and profitability of smallholder coffee farmers in many countries suffers from the difference between farm gate and free on board (FOB) prices for export (a standard measure of export prices). Figure 3 shows the average difference between the export price and the amount of money that coffee farmers actually receive.

Crop year		2010/11	2011/12	2012/13	2013/14	2014/15	2015/16	2016/17	2017/18	2018/19	2019/20
Brazil	(A/R)	55.428	48.592	55.418	54.689	53.305	52.871	56.788	52.740	65.131	58.211
Viet Nam	(R/A)	20.000	26.500	23.402	27.610	27.241	31.090	27.819	33.432	30.283	30.487
Total A (1-2)		**75.428**	**75.091**	**78.820**	**82.299**	**80.545**	**83.961**	**84.607**	**86.172**	**95.414**	**88.698**
Colombia	(A)	8.523	7.653	9.927	12.124	13.333	14.009	14.634	13.824	13.866	14.100
Indonesia	(R/A)	9.129	6.889	13.070	12.901	10.946	12.585	11.541	10.852	9.618	11.433
Total B (3-4)		**17.651**	**14.542**	**22.997**	**25.025**	**24.280**	**26.594**	**26.175**	**24.676**	**23.484**	**25.533**
Ethiopia	(A)	5.560	5.650	6.132	6.242	6.383	6.515	7.143	7.347	7.511	7.343
Honduras	(A)	4.331	5.882	4.686	4.583	5.268	5.786	7.457	7.560	7.153	5.931
India	(R/A)	5.600	5.334	5.403	5.075	5.450	5.830	6.161	5.813	5.325	4.988
Uganda	(R/A)	3.267	3.115	3.914	3.633	3.744	3.650	4.962	4.597	4.704	5.509
Total C (5-8)		**18.758**	**19.981**	**20.135**	**19.534**	**20.846**	**21.781**	**25.723**	**25.317**	**24.693**	**23.771**
Mexico	(A/R)	4.001	4.561	4.327	3.916	3.636	2.772	3.635	4.485	4.351	3.985
Peru	(A)	4.069	5.373	4.450	4.106	2.883	3.304	4.223	4.279	4.263	3.836
Guatemala	(A/R)	3.950	3.850	3.763	3.189	3.310	3.410	3.684	3.734	4.007	3.606
Total D (9-11)		**12.020**	**13.784**	**12.540**	**11.212**	**9.830**	**9.485**	**11.542**	**12.497**	**12.621**	**11.427**
Nicaragua	(A)	1.638	2.193	1.873	2.060	1.898	2.130	2.555	2.642	2.879	2.882
Côte d'Ivoire	(R)	837	2.216	1.994	2.598	1.728	1.289	1.117	1.624	2.175	1.929
Costa Rica	(A)	1.614	1.831	1.823	1.552	1.475	1.440	1.372	1.561	1.427	1.472
Total E (12-14)		**4.088**	**6.241**	**5.690**	**6.210**	**5.101**	**4.860**	**5.045**	**5.827**	**6.481**	**6.283**
Total (1-14)		**127.946**	**129.639**	**140.182**	**144.279**	**140.601**	**146.681**	**153.093**	**154.489**	**162.693**	**155.712**
Total World		**140.078**	**141.327**	**151.184**	**153.910**	**150.302**	**156.126**	**162.320**	**163.693**	**172.461**	**165.053**

A = Arábica R = Robusta

Figure 2 Total production in the largest coffee-producing countries (thousand 60 kg bags). Source: International Coffee Organization. - ICO.

Figure 3 Coffee farm gate prices vs. FOB prices. Source: Enveritas.

As Fig. 3 shows, the prices paid to coffee farmers range widely, being:

- above 85% of the FOB price in Brazil, Costa Rica and Vietnam;
- close to 80% of the FOB price in Colombia and Guatemala;
- between 60% and 70% of the FOB price in Uganda and Honduras;
- around 55% in Ethiopia.

This difference between countries is a result of a number of factors including:

- differences in farmers' bargaining power in negotiating with wholesalers and exporters;
- differences in costs of out-of-the-farm processing;
- differences in efficiency of the coffee supply chain;
- differences between countries in transportation and shipping costs;
- differences in taxes imposed by governments.

All of these issues need to be addressed in supporting smallholders.

3 Concentration and diversity in global coffee production

Figure 4 assesses production in the 14 leading coffee-producing countries. The choice of 5-year periods to compare shares of world production accounts

Countries		Average 1 2010/11-2014/15	World share (%)	Average 2 2015/16-2019/20	World share (%)	Change 1 - 2	Growth 2 vs. 1 (%)
Brazil	(A/R)	53.486	36%	57.148	35%	3.662	7%
Viet Nam	(R/A)	24.950	17%	30.622	19%	5.672	23%
Total A (1-2)		**78.437**	**53%**	**87.770**	**54%**	**9.334**	**12%**
Colombia	(A)	10.312	7%	14.087	9%	3.775	37%
Indonesia	(R/A)	10.587	7%	11.206	7%	619	6%
Total B (3-4)		**20.899**	**14%**	**25.293**	**15%**	**4.394**	**21%**
Ethiopia	(A)	5.994	4%	7.172	4%	1.178	20%
Honduras	(A)	4.950	3%	6.777	4%	1.827	37%
India	(R/A)	5.372	4%	5.624	3%	251	5%
Uganda	(R/A)	3.535	2%	4.684	3%	1.150	33%
Total C (5-8)		**19.851**	**13%**	**24.257**	**15%**	**4.406**	**22%**
Mexico	(A/R)	4.088	3%	3.845	2%	-243	-6%
Peru	(A)	4.176	3%	3.981	2%	-195	-5%
Guatemala	(A/R)	3.613	2%	3.688	2%	76	2%
Total D (9-11)		**11.877**	**8%**	**11.515**	**7%**	**-363**	**-3%**
Nicaragua	(A)	1.932	1%	2.618	2%	685	35%
Côte d'Ivoire	(R)	1.875	1%	1.627	1%	-248	-13%
Costa Rica	(A)	1.659	1%	1.454	1%	-205	-12%
Total E (12-14)		**5.466**	**4%**	**5.699**	**3%**	**233**	**4%**
Total (1-14)		**136.530**	**93%**	**154.534**	**94%**	**18.004**	**13%**
Total World		**147.360**	**100%**	**163.931**	**100%**	**16.571**	**11%**

A = Arábica R = Robusta

Figure 4 Total production in largest coffee-producing countries (thousand 60 kg bags). Source: International Coffee Organization.

for variations in production from year to year. As the figure shows, 25% of producing countries (out of a total of 55 countries tracked by the ICO) account for almost 95% of world coffee production. However, the numbers in Fig. 4 challenge the common perception that Brazil and Vietnam have significantly increased their share of world production in recent years and that the largest producing countries have also increased their overall share. From 2010 to 2020, the share of the 14 largest producers did increase but only marginally, from 93% to 94%, and the share of production by Brazil and Vietnam remained largely unchanged.

Figure 4 shows countries in various groups according to market share. These groups show varying rates of growth:

- A: 12%;
- B: 21%;

- C: 22%;
- D: -3%;
- E: 4%.

As this shows, growth was higher in Group B and C countries than in the two leading countries (Brazil and Vietnam) in Group A. Group E countries showed modest growth whilst Group D countries lost production. The countries with the largest percentage increase in production were, in descending order, Honduras, Colombia, Nicaragua and Uganda, suggesting a more diverse and dynamic market. Recent trends indicate that the large volumes of commercial and differentiated coffees required by the market are going to be supplied by these largest producing countries that also supply large volumes of specialty coffee. The rest of the producing countries will tend to focus on the specialty and differentiated coffee markets and domestic consumption in the producing countries themselves. This will not necessarily decrease diversity and may even increase it.

4 Specialty and differentiated coffees

Adding value to green coffee is a potential source of increased income for growers. One way of adding value is to develop specialty coffees able to command a higher price. Specialty coffees have been defined as high-grade coffees scoring 80 points or more out of 100 for sensory quality by a certified coffee taster or a licensed coffee grader (https://sca.coffee/). They are associated with very high-quality beans typically grown in unique micro-climates with particular care by growers, often linking them to particular regions and groups of farmers (see Chapter 4 on specialty coffees in this book). An example is the Geisha variety of Arabica bean from Panamá, currently one of the world's most expensive coffees. Specialty coffees now play a key role in production from small producing countries, e.g. Panamá, Jamaica and Rwanda, as well as an important and growing role in mid-size and larger producing countries like Guatemala, India, Colombia and even Brazil.

Specialty coffees are known for their unique sensory qualities, small volumes and reliability of supply. It is a niche market where consumers often want to know the name of the grower that produces the coffee they drink. As a result, there may be a close relationship between the producer and players at the end of the coffee supply chain: processors, coffee shop chains and even retailers. Specialty coffees were a factor in the creation of coffee shops (both global chains such as Starbucks (founded in 1971) and independent coffee shops) and the development of the so-called third and fourth coffee waves as consumers started to become more aware of and demand higher-quality coffees. They are associated with developments such as the supply of micro-lots of very specific coffees that can often be traced back to a specific grower or farm. One potential area of development to help producers of specialty and other coffees is the development of regenerative agricultural techniques to reduce reliance on synthetic fertilizers and pesticides and reduce negative impacts on the environment. With appropriate training, these low-cost techniques may be of particular benefit to smallholders in ensuring more sustainable production.

The demand for much larger volumes of higher-quality coffee led to the development of so-called differentiated coffees consisting of large volumes of consistent quality coffees. The growth of coffee-shop chains and single-serve capsules has been a key factor in the development of differentiated coffees. These are usually supplied by large growers, cooperatives and traders who have formed close relationships with coffee roasters and national and international coffee shop chains in order to set standards and ensure quality is maintained in differentiating these coffees from cheaper coffees of more standard quality (see Fig. 5).

As has been noted, the potential to exploit the development of value-added products such as specialty and differentiated coffees depends critically on developing relationships with players at the end of the supply chain, as well as implementing good quality and traceability systems. The production and supply of nano- and micro-lots of specialty coffees e.g. provides a way to identify and add value to produce from individual growers by securing a higher price per bag. The supply of differentiated coffees is another source of value addition, with the smaller premium per bag offset by higher volumes going e.g. to larger coffee chains or the specific, higher-quality blends going into capsules which may well be sold globally.

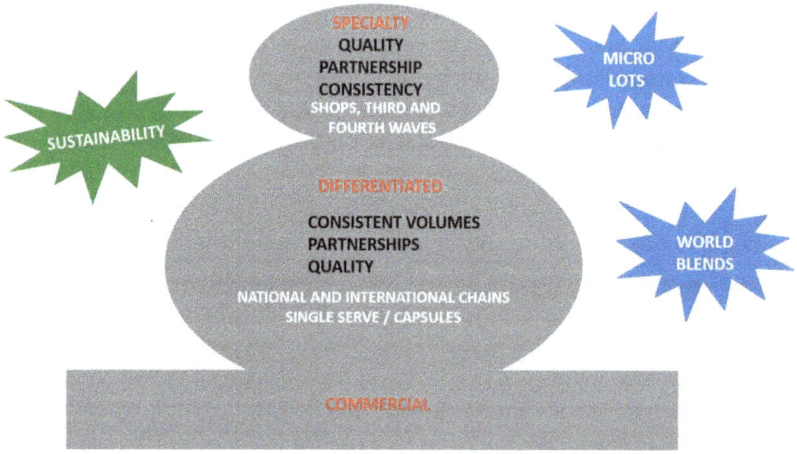

Figure 5 Coffee market segmentation: commercial, differentiated and specialty coffees.

Linked to these changes is continued growth and diversification in consumer markets. This includes the growth of coffee consumption in producing countries, often linked to urbanization and increasing affluence. A domestic market has advantages for growers that it is more accessible, may create a new outlet for coffees that are less exportable and can help to provide a safety net to offset low international coffee prices. Whilst more exclusive specialty coffees can help to make drinking coffee more fashionable, the bulk of growth in consumption will have to occur in middle- and low-income social groups who account for the majority of consumption in any country. These groups require moderate- and low-price coffee products, e.g. standard roast-and-ground and soluble coffee products. Consumption growth can result from either pre-competitive institutional programs to promote coffee consumption or promotion by coffee brands or both, as shown by the ICO Step-by-step Guide to Promote Coffee Consumption in Producing Countries (https://www.ico.org/promotion_guide.asp). Each country is a different case, but promotion of domestic consumption opens up good opportunities to increase growers' incomes.

5 Trends in Arabica and Robusta coffees

In spite of the growing role of specialty and differentiated coffees, which are mostly Arabica varieties, the proportion of Robusta varieties in global production has been rising steadily: from about 25% in 1980 to around 35% in 2005, over 42% in 2020 and may currently be moving toward 45%. This is mainly related to the introduction of coffee in traditional tea-drinking countries like Indonesia, China, India and Vietnam where the initial coffee beverage of choice is cheaper and lower-quality instant/soluble coffee, mainly produced with Robustas. In addition, improvements in the quality of Robusta, including the development of wet processing, are increasing its inclusion in roast-and-ground and single-serve blends offered by coffee roasters. Finally, there is a new focus on differentiating the quality of different Robusta varieties, with cupping competitions and single-origin and 100% Robusta blends.

The Coffee Quality Institute (CQI - https://www.coffeeinstitute.org/) now has schemes for both certified Q Arabica and Q Robusta Graders to grade and score coffees according to standards developed by the Specialty Coffee Association (see Fig. 6). CQI has also recently introduced Q Processing Robusta courses to certify stages of competence in Robusta coffee processing. These courses were previously only available for Arabica. There are now also moves to improve quality standards in soluble coffee. The Brazilian Soluble Coffee Roasters Association (ABICS, from its initials in Portuguese) has e.g. recently released a white paper with a methodology to evaluate the quality of soluble coffees and to describe their sensory quality attributes.

Figure 6 Q Arabica and Q Robusta Graders.

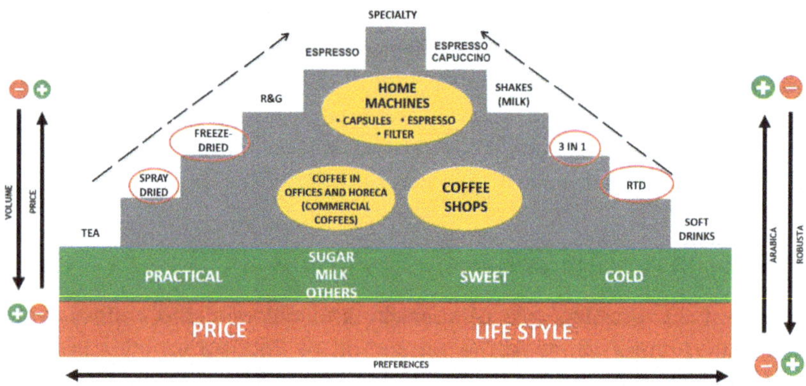

Figure 7 P&A Coffee Ladder.

The P&A Coffee Ladder shown in Fig. 7 describes how consumers first enter the coffee market and evolve in their preferences and purchasing patterns as a result of factors such as income growth and greater knowledge of coffee quality. This is an oversimplification of a more complex picture, and it provides an overview of the market and the roles of Arabica and Robusta coffees in it.

Tea drinkers tend to enter the coffee market via soluble coffee which is similar in preparation to tea as well as cheaper and more convenient to use. From spray-dried soluble coffee, they tend to move to higher-quality freeze-dried soluble coffee. Depending on income and perceptions of coffee quality, consumers may move to roast-and-ground and espresso coffee and eventually specialty products.

Young soft-drink consumers may start by preferring ready-to-drink or '3-in-1' coffee beverages: small single-serve envelopes combining soluble coffee, non-dairy milk and sugar. The next step may be milk-based coffee beverages like shakes or cappuccinos and eventually specialty coffee, again depending on income and perceptions of coffee quality.

The factors that determine a consumer's journey up include factors such as price, practicality and convenience, products mixed with coffee, lifestyle, etc.

The way coffee is supplied also plays a role in the choice of coffee product to consume. Office coffee services, coffee shops and machines for home use can help to pave the way for the consumption of different coffee beverages of differing quality. Robusta-based soluble products open up new markets and pave the way for the consumption of more Arabica as consumers climb up the coffee ladder. This dynamic and evolving market, with its wide range of coffee qualities, market segments and quality attributes, provides significant new opportunities to coffee growers, including smallholders.

6 Creating an enabling environment to make coffee farming more sustainable

The continuing low levels of productivity achieved by many smallholders (who account for such a significant share of overall coffee production) is a major, ongoing challenge for the coffee sector as a whole. Low productivity means both an insufficient income and an inability to invest in making coffee production both more efficient and sustainable over the long term. Improving productivity requires developing and implementing good agricultural practices (GAP) capable of delivering more resource-efficient and climate-smart production. As an example, a number of chapters in Part 2 of this book explore more integrated, resource-efficient and environmentally friendly techniques for managing insect pests and diseases of coffee effectively. Reducing losses from these biotic threats using such techniques has the potential to significantly improve productivity.

Achieving such improvements in productivity requires a number of steps. Since the costs of research are high, it should be carried out primarily by the larger and more successful coffee-producing countries. Smaller or less-developed producing countries should be able to rely on improved practices and technologies developed elsewhere and suitably adapted to their own conditions. However, a critical further step is how to ensure that smallholders in less-productive countries are informed of these improvements, incentivized and given the resources to adopt them in practice in a way that meets their needs and local conditions.

Smallholders face a number of obstacles to implementing improved cultivation practices. They lack the returns to re-invest in innovations such as improved varieties, equipment, fertilizers and other agrochemicals to boost productivity. This is exacerbated by poor access to finance and credit. They also lack access to training and extension services to help them understand and implement new techniques effectively. Low and fluctuating incomes make smallholders understandably risk averse and short term in outlook, reducing the willingness to invest time and resources in new methods. More broadly, individual smallholders typically lack bargaining power in selling their product to intermediaries such as wholesalers, locking them into a cycle of low returns, investment and productivity. Addressing these challenges requires support from government and the private sector as well as action by smallholders themselves e.g. in forming growers' associations and cooperatives to pool resources and improve bargaining power.

Creating an enabling environment for coffee growers to improve production is in the hands of both government and the private sector. Governments may provide some of these services. They can fund research, provide training and extension services as well as access to finance and inputs. They can improve infrastructure to facilitate the transport of coffee and reduce its costs, and they can sometimes intervene to regulate markets or alter the tax system to favor (smallholder) producers. On the other hand, a properly regulated and competitive private sector can also play a key role in providing training services, financing, processing facilities, trading and logistics support, etc. The public-private sector mix will be different from country to country, but their joint actions to improve the enabling environment are critical to increase both farm productivity and the share of the export price received by coffee growers, both of which have a direct impact on farmers' income and their ability to become more sustainable.

The creation or development of grower's associations may depend on both government regulation, e.g. legislation to provide a framework for cooperatives, and support from the private sector or non-governmental organizations advising smallholders. Growers' associations and cooperatives can pool resources to buy inputs, set up their own processing facilities to add value to the harvested coffee or improve their bargaining power in the market. Farmers' associations and cooperatives may also play an advocacy role at a regional and national level, especially on behalf of smallholder growers who suffer the most from market inefficiencies, access to services, etc.

A critical look at the countries that have successfully improved production shows that they have benefited from improvements in the enabling environment that paved the way for both higher productivity and greater grower participation in the share of the FOB export price that reaches them. A third factor is political and government support of the coffee sector that is difficult to measure but has been and/or is behind the growth of coffee production in several countries.

7 The case of Brazil in improving coffee grower incomes

The United Nations Food and Agriculture Organization has described the coffee market as 'characterized by recurrent supply-demand imbalances and asymmetric income distribution among actors in the supply chain' (https://www.fao.org/markets-and-trade/commodities/coffee/en/). For smallholders, this has meant a sometimes disastrous combination of facing widely fluctuating green coffee market prices and receiving a very small share of an already low price, resulting in an inadequate and unstable income from year to year. However, many countries have been able to increase their coffee production despite these very volatile coffee prices. An efficient enabling environment in countries like Brazil and Vietnam has facilitated their gain in world market share at times of low coffee prices. High coffee prices can stimulate increased coffee production in a country, but this will be sustainable only if structural changes in the enabling environment take place at the same time.

What has happened in Brazil over the last 30 years demonstrates how structural changes can make a great difference to a country's coffee sector in spite of low coffee prices. Figure 8 starts in 1990, after the end of quota system the ICO had previously used to regulate supply, and stops in 2019 before coffee prices started to go up. It clearly shows that both production and productivity in Brazil increased significantly, despite volatile and often low coffee prices.

The dismantling of the Brazilian Coffee Institute (IBC), set up by the government to regulate the coffee business and to conduct research as well as provide extension and marketing services, occurred at the same time as the end of the ICO quota systems. This meant the country's coffee sector started the 1990s facing falling coffee prices (only briefly alleviated by widespread frosts in 1994 which cut back production) and loss of the coffee research, extension, marketing and other activities previously carried out by IBC. Paradoxically, this may have created the challenges that the Brazilian coffee business needed to become much more efficient.

A combination of initiatives by the private sector and government actions succeeded in improving the efficiency of the coffee supply chain both before and beyond the farm gate. Productivity nearly doubled in the first 10 years after 1990 and again in the first 20 years of this century: an almost four-fold increase in 30 years. These changes are summarized in the following paragraphs.

One change was that individual universities, research institutes and foundations became much more active in coffee research, with the initial problems of lack of focus, strategic planning and duplication of effort. This was eventually corrected by the creation of the Coffee Research Consortium that brought all these institutions together to develop a coordinated coffee research agenda. At the same time, extension and training services performed by the IBC were progressively transferred to state governments and farmer cooperatives, actually improving overall outreach.

In addition, growers themselves 'woke up' from a period of controlled prices before 1990 that allowed inefficient production to prevail. New varieties, greater planting densities, mechanization and GAP were introduced to significantly improve productivity. The pulped natural system (also called the CD or honey system) was developed and on-farm processing increased. Gains in on-farm efficiency were coupled with actions beyond the farm gate. These

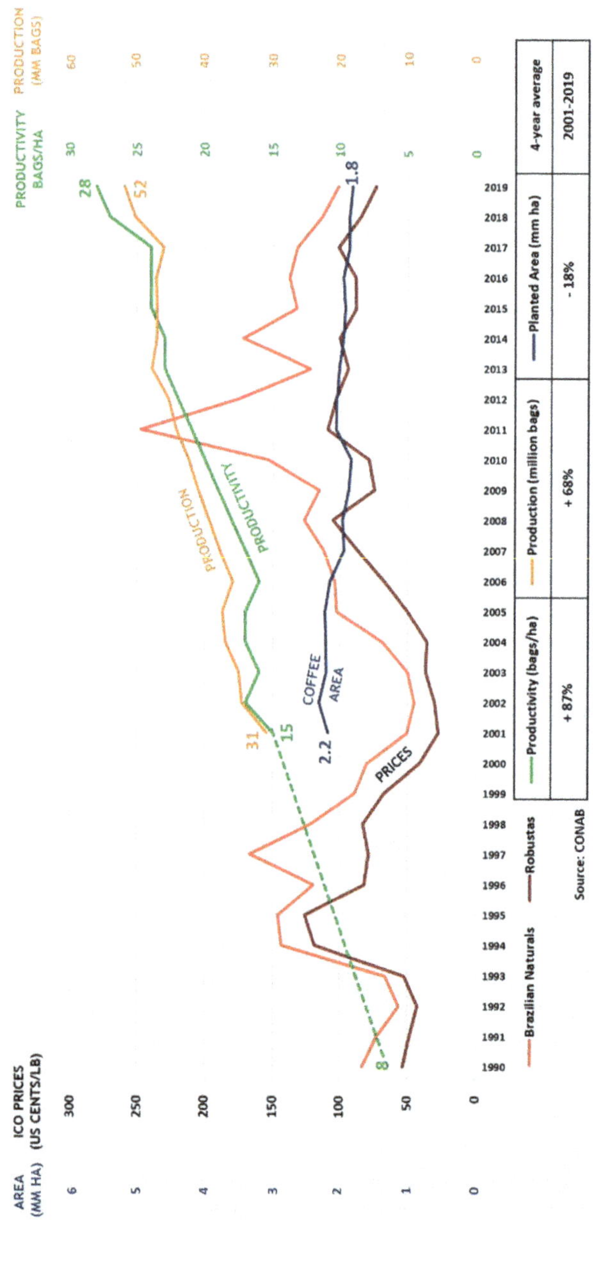

Figure 8 Coffee prices, production, productivity and area under production in Brazil. Source CONAB.

included streamlining the supply chain from farm gate to harbor, as well as rebranding Brazilian coffees from a standard fair commercial quality to a range of quality standards, including specialty coffees. The Brazilian Specialty Coffee Association was created. The brand Café do Brasil became Cafés do Brasil, 'one country many flavors,' with the country divided into producing regions that offered a range of coffee qualities. At the same time, the private sector started to promote domestic coffee consumption. In 30 years, Brazilian consumption increased more than three times, from 6.5 to about 21 million bags per year, creating a huge market for locally grown coffees.

The remaining coffee inventory at the end of the quotas period was progressively sold to create the Brazilian Coffee Fund (Funcafé) that is primarily used to finance coffee growers and, to a lesser extent, the coffee supply chain. Funcafé financing has however now been surpassed by that provided by private banks that are, by law, required to channel a given percentage of deposits to low interest financing of agricultural activities. Government also supports agribusiness through exemptions from sales taxes for agriproducts that are exported.

The example of Brazil shows how producing countries can bring about durable structural change to increase efficiency and competitiveness. Analyses similar to Brazil can be made for Vietnam, Colombia, Uganda and Honduras to show how government and private sector initiatives can create an enabling environment to increase production, productivity and to gain world market share.

8 Farm size, income and sustainability

The average coffee farm size in Brazil is between 7 and 8 hectares, compared with less than 1.5 hectares for the rest of the world. This may be an important factor supporting Brazilian competitiveness. However, Vietnam has so far been competitive in Robusta production with an average farm size of about 1.5 hectares. The disadvantages of being a small coffee farmer are significant. Fixed costs are typically too high for the size of the operation. Smallholders have limited access to inputs, technology or finance. Because they have poor bargaining power, there is 'unfair' trading of inputs and tools, equipment and coffee itself. Small farmers usually pay more for inputs and equipment and sell their coffee for less than their larger competitors. Economists call this 'diseconomies of scale'. An additional problem is that some farms become even smaller as they change hands from generation to generation.

The focus should be on bringing growers together to buy and sell as a group, thus benefiting from economies of scale. This has benefits in areas such as discounted bulk purchase of inputs, equipment sharing, joint processing facilities and pooling of coffee to sell at a better price. There are examples of successful national organizations that support small coffee growers, like the Colombian Coffee Growers Federation, but they are the exception rather than the rule. There are also some examples of successful cooperatives in Brazil and a few other countries that show that there are ways to support the work of small coffee farmers. These initiatives to bring small growers together to become more competitive and also to prevent further division of existing coffee farms must be disseminated and new initiatives must be developed. Cooperatives have existed for well over 100 years, and growers' associations also have a long history, but their use is not as widespread or effective as it could be. The main barriers may well be more behavioral and sociological than technical and should be dealt with accordingly. This is a challenge for governments, agencies and companies concerned with the plight of the small coffee growers and the sustainability of the coffee business.

One important aspect of the problem is retaining younger generations in coffee farming. They will only do so if they see a viable future and are given the opportunity to bring in new ideas that both give them new business opportunities and a stake in the sector. There also need to be ways of giving those younger members of the family who cannot work on a farm, because it can only support a certain number of people, other opportunities in the rural economy. This shows that supporting small farmers requires looking beyond the coffee sector to look more broadly at regional development, e.g. to provide employment for coffee growers' children and promote rural–urban diversification with members of coffee growers' families having diverse jobs or small businesses in their coffee-producing regions.

9 Conclusion

9.1 Environmental, social and governance criteria and sustainability

ESG is an acronym for environmental, social and governance. ESG is a holistic concept covering an organization's ability to create and sustain long-term value in a rapidly changing world and manage the risks and opportunities associated with these changes. ESG criteria cover the performance of an organization in areas such as the environmental impact of its operations; the way an organization manages relationships with its employees, suppliers, customers and the communities where it operates; and, finally, the robustness and integrity of an organization's decision-making structures and processes in ensuring it acts responsibly for all relevant stakeholders (whether shareholders, employees, customers, suppliers or the wider community).

In theory, ESG should include both the organization itself and its wider role in the supply chain within which it operates. In reality, ESG is often applied narrowly to a company's direct operations but not to the broader supply chain of which it is a part. As an example, it is not enough for a coffee retail chain to look after its own carbon footprint, the energy it uses, the effluents it generates and the diversity of its board. The company should also address its role as a purchaser of coffee and its role in promoting environmental and social sustainability at earlier stages in the coffee supply chain back to the farmers involved in cultivation. Having been a pioneer, e.g. in the implementation of sustainability certification, coffee should pave the way for the integration of ESG and sustainability along its supply chain from cup to seed. This integration should also minimize the risk of the association of ESG with 'greenwashing' that has been popular in the media. Achieving sustainable and climate-smart production of coffee requires shared responsibility from cup back to seed.

9.2 Shared responsibility in achieving sustainability

It is known that the majority of coffee consumers would like to consume coffees produced in a sustainable way but only a minority is prepared to pay more for sustainable coffees. This brings up the issue of shared responsibility across the whole coffee supply chain including both producing and consuming countries. Governments in producing countries and foreign aid and donors from importing/consuming countries both have a role to play in helping not only to create an enabling environment for coffee smallholders but also to promote wider regional development in coffee-growing areas to increase efficiency and sustainability in coffee cultivation.

One example of this shared responsibility is sustainability certification schemes developed in importing countries to improve both environmental and social sustainability. There is scope to expand the volume of coffee certified, especially that produced by small farmers, and to expand the scope of certification to take in landscape approaches. Certification organizations are an essential part of these discussions since they best know the strengths and challenges of current models of certification. There are also issues about defining sustainability. An important recent move has been the release of the Coffee Sustainability Reference Code by the Global Coffee Platform (GCP) with the support of the largest roasters in the world, some of the largest coffee traders and the GCP National Platforms in producing countries (https://www.globalcoffeeplatform.org/latest/2021/announcing-the-coffee-sustainability-reference-code/).

Pre-competitive multistakeholder initiatives like the GCP, which include players from all stages of the coffee supply chain, are a positive step in the direction of shared responsibility from seed to cup. The price premiums for certified coffees are a positive way to help growers bear the costs of being sustainable, but their impact has been eroded by the falling premiums resulting from overproduction of certified coffees in relation to consumer demand for them. Current actions to promote a greater demand for sustainable coffees are Sustainable Coffee Challenge members' commitments to use sustainable coffee, GCP's annual reports of sustainable coffee purchases by coffee roasters and the Coffee Barometer's regular reports that analyze a range of sustainability parameters. Another potential route is paying farmers for the environmental services they supply rather than just the produce they grow. This is still in its infancy but has great potential for the future.

Living, prosperous and sustainable income for coffee farmers, climate-smart regenerative agriculture and shared responsibility from seed to cup are key challenges to achieve long-term sustainability of coffee production. These challenges should be addressed together under the umbrella of regional, economic and sustainable development in a holistic landscape approach that goes beyond the farm and incorporates neighboring villages, towns and cities.

Enjoy the book and apply its suggestions!

10 Where to look for further information

- Enveritas;
- Global Coffee Platform (GCP);
- International Coffee Organization (ICO);
- United Nations Food and Agricultural Organization (FAO).

11 References

Bockel, L. and Schiettecatte, L.-S. 2018. Life cycle analysis and carbon footprint of coffee value chains. In: Lashermes, P. (Ed.), *Achieving Sustainable Cultivation of Coffee: Breeding and Quality Traits*, pp. 359-382. Burleigh Dodds Science Publishing, Cambridge, UK.

Bray, J. and Neilson, J. 2017. Reviewing the impacts of coffee certification programs on smallholder livelihoods. *Int. J. Biodiv. Sci. Ecosyst. Serv. Man.* 13(1): 216-232.

Bunn, C., Läderach, P., Ovalle Rivera, O. and Kirschke, D. 2015. A bitter cup: climate change profile of global production of Arabica and robusta coffees. *Clim. Change* 129(1-2): 89-101.

Crabs, J., Kilian, B., Hernández, D. C. and Dietz, T. 2016. Understanding coffee certification dynamics. *Int. Food & Agribus. Man. Rev.* 19(3): 31-55.

Jena, P. R. and Grote, U. 2022. Do certification schemes enhance yields and coffee household income? *Front. Sustain. Food Syst.* 5. DOI: 10.3389/fsufs.2021.716904.

Raynolds, L. T., Murray, D. and Heller, A. 2007. Regulating sustainability in the coffee sector - a comparative analysis of third-party environmental and social certification schemes. *Agric. Hum. Values* 24(2): 147-163.

Winter, E., Marton, S. M. R. R., Baumgart, L., Curran, M., Stolze, M. and Schader, C. 2020. Evaluating the sustainability of certified coffee production systems in Brazil and Ethiopia. *Front. Sustain. Food Syst.* 4. DOI: 10.3389/fsufs.2020.00049.

Chapter 2

The coffee sector and smallholder farmers

Gabriele Regio, Oxfam, Italy

1 Introduction

Coffee is key to the livelihood of millions of rural families in many underdeveloped countries that produce and export this major commodity. Small producers in these countries are often not well organized, making it difficult for them to design or influence national or regional strategies that take their interests into account. These producers generally do not have access to basic services and infrastructure supporting production and marketing. In addition, education and employment opportunities are more limited, contributing to increased poverty. Consequently, building a sustainable global coffee economy is a long-term commitment that faces multiple challenges.

Small- and medium-scale coffee farmers share the following characteristics:

- Producing coffee in areas smaller than 5 hectares (ha) and producing less than five bags (300 kg or 660 pounds) of coffee per season. Considering a medium price of US$1.5/pound, a family's annual gross income has been estimated at around US$990 (US$2.71/day).
- Farms are typically more than 15–20 years old with farmers, on average, 55–60 years of age.

http://dx.doi.org/10.19103/AS.2021.0096.02

Figure 1 Farm sizes for Arabica coffee production.

- Family members (average four or five members) provide most of the labor.
- Farmers typically use very simple tools such as hoes and machetes in cultivation.
- Farmers use inputs such as organic manure to improve fertility (since it is cheap and immediately available) but rarely have the resources to pay for organic certification which might provide a premium for such practices.
- Farms are often mixed: the farmer plants different types of crops on the farm, including grains like maize and vegetables, to feed the family.
- Yields per unit area are typically low with high relative production costs and low profitability.

These characteristics clearly show that, in most cases, small and medium coffee producers are subsistence farmers with a net income from coffee production that often does not meet the basic needs of coffee farmer's families.

According to the International Coffee Organization,[1] in 2018/2019 coffee farmers produced a record crop of almost 170 million 60 kg bags of which 60% was Arabica and 40% was Robusta. A total of around 10.5 million hectares are used for coffee production. Seventy-five per cent of total coffee production was exported, representing a market of more than US$20 billion. The majority of coffee (60%) is produced by smallholder farmers in more remote tropical regions, representing the main family revenue for millions of people.

According to the report by Enveritas *Bringing Smallholder Coffee Farmers out of Poverty* (2019), 21% of the world's coffee is produced on estates and large farms over 50 hectares, while19% comes from estates and farms between 5 and 50 hectares (Fig. 1). The remaining 60% comes from smallholder coffee farmers with less than 5 hectares. These smallholder farmers face unique challenges in

1 ICO Annual Coffee Market report. The International Coffee Organization (ICO) is the main intergovernmental organization for coffee, bringing together exporting and importing Governments to tackle the challenges facing the world coffee sector through international cooperation. Its Member Governments represent 98% of world coffee production and 67% of world consumption.

improving production. This is apparent in differences in yields in the 2018/2019 coffee harvest. While average yields on Brazil farms for Arabica coffee were over 1500 kg of green coffee per hectare, their counterparts in Ethiopia, Haiti, Dominican Republic achieved an average of just over 250 kg.

Based on analyzing over 20000 interviews in 20 countries, Enveritas estimate that there are 12.5 million smallholder coffee farmers[2] (in contrast to the previous estimate of 25 million). These 12.5 m farmers live predominantly in 20 countries where the climate and soil are suitable for growing coffee. Almost half of these smallholder farmers can be found in Ethiopia (2.2 m), Uganda (1.8 m) and Indonesia (1.3 m). Vietnam, Burundi, Kenya and Colombia each have over 500000 smallholder farmers. Of the 12.5 million smallholder coffee farmers, Enveritas estimate that at least 5.5 million live below the international poverty line of $3.20 a day. The highest levels of poverty are observed in Africa and Oceania.

The majority of coffee farmers living in poverty are concentrated in six East African countries, accounting for approximately 63% of the world's coffee farmers living in poverty and 71% of coffee farmers living in extreme poverty. Major East African producers are: Ethiopia, Kenya, Tanzania, Rwanda, Burundi and Uganda. An example can be seen in a 2002 Oxfam interview with Peter and Salome Kafuluzi, who live on a farm in Kituntu in Uganda with 13 of their children and grandchildren. Peter told Oxfam that he could not support his family anymore, because coffee prices were the lowest he had ever seen. Salome said: '*We are failing in everything. We cannot have any meat, fish, rice, just sweet potatoes, beans and matoke [cooking bananas] ... We can't send the children to school.*' East Africa is characterized by particularly low coffee yields that result in low incomes for these farmers.

Many experts argue that low coffee yields are generally solvable through simple and organic techniques (such as pruning and stumping of trees), but these methods require appropriate training programs which many smallholders cannot access. International organizations such as World Coffee Research and national coffee associations such as Anacafe in Guatemala suggest solutions could include new (e.g. rust-resistant) varieties, education on farming economics and improved techniques from planting through to coffee mill wastewater management. If delivered properly, these solutions provide the potential for an economical, effective and sustainable approach to improve yield, revitalize family coffee farms and the livelihoods of the communities to which they belong. It is also important to note that the health of coffee-growing sectors also affects national economies. However, the importance of coffee growing is not limited to the jobs and income it generates. It also has a fundamental ecological role in the protection of soils and watersheds in the highlands of coffee-growing countries.

2 Map of the Month: Bringing Smallholder Coffee Farmers out of Poverty, Dic 2019

2 Challenges facing small-scale coffee farmers: an overview

Whether a farmer can make a profit depends on whether their total income exceeds their costs. There are different models for farming coffee. Some models, like high-intensity monoculture farming, produce more coffee but also cost more because they require more inputs. Some cost very little but also produce very little. Many factors influence quality which also affects the end price the farmer earns. Whether coffee farming is profitable is a question of what resources farmers have available and how they allocate these resources. In order to develop strategies to make coffee farming more attractive and profitable, it is first necessary to identify how and why farmers make the choices they do. What constraints do they face? Can a small- or medium-scale farmer live through just producing coffee?

Smaller-scale coffee farmers face a range of well-known challenges, including increasing input costs (e.g. fuel and fertilizers) and low yields. Despite growing global consumer demand, poor market access and bargaining power often mean low prices for smaller producers, resulting in poor profitability. In terms of markets, smallholders also face fluctuating coffee prices and the depreciation of national currencies in poorer countries more vulnerable to global trade fluctuations. Global coffee production varies from year to year according to weather conditions, disease and other factors, resulting in a coffee market that is inherently unstable and characterized by wide fluctuations in price. As suggested by Raphael Kaplinsky (University of Sussex), the coffee market is characterized by low barriers to entry which risks oversupply and therefore lower prices. This price volatility has significant consequences for those who depend on coffee for their livelihood, making it difficult for growers to predict their income for the coming season and budget for their household and farming needs.

Over 90% of coffee production takes place in developing countries, while coffee is consumed mainly in industrialized economies. The coffee supply chain is therefore complex as beans pass through the hands of growers, traders, processors, exporters, roasters, retailers until finally reaching the consumer. Most farmers have little idea of where their coffee goes or at what price it is finally sold.

Raphael Kaplinsky argues that the increasing concentration of ownership and power at the downstream end of the coffee value chain has improved the bargaining power of large companies processing and selling coffee at the expense of upstream actors such as farmers, helping to explain the growing asymmetry of incomes in the coffee chain. As analyzed by John Talbot in his book *Grounds for Agreement: The Political Economy of the Coffee Commodity Chain* (2004), transformations in the commodity chain during the 1990s allowed transnational corporations based in coffee-consuming countries to capture an increasing proportion of the profits from coffee. It has been estimated that in the last two decades, the producers' share of the total value of the coffee market has declined from approximately 30% to less than 10%.

Farmers often have to accept the price offered by traders and have very little, if any, power to negotiate. Although traders may squeeze extra margins from farmers to improve their returns, the highest margins in the market are made, after export, by the roaster companies. Even in the worst years of the coffee crisis in the 1990s, when prices collapsed, roaster companies in the United States and Europe made significant profits on their retail coffee business. Analysts estimated that operating profit margins for roaster companies at this time varied from 17% to 30%, while farmers and exporters in coffee-exporting countries made losses or (at best) tiny margins.

Roasters gain from advantages such as the strength of their brands and products, their ability to mix and match blends and their ability to buy in bulk from around the world at the most competitive prices. Small- and medium-scale farmers, even when organized in groups or cooperatives, also lack the bargaining power to bargain with large exporters with consolidated processing facilities and established relationships with international buyers. This situation has not been improved by the more recent trend toward direct coffee buying by multinational companies in developed countries. Rather than bringing increased benefits to producers, roasting companies cut costs by sidestepping intermediaries and paying the same price as local buyers. As Oxfam has pointed out, this sort of direct market access is not beneficial to smallholders since it replicates existing low prices for producers. The unequal distribution of value across the coffee supply chain has sometimes had a disastrous effect on small- and medium-scale farmers, millers and even exporters in developing countries.

A further challenge facing the sector is that coffee-producing communities are found in the rural areas. These regions are often isolated and relatively neglected by government authorities, have limited services (e.g. in health and education) and experience widespread poverty. Isolation and lower education levels mean they have less access to information about new techniques that can be applied to boost productivity. Insecurity due to volatile prices and exchange rates, inconsistency in crop quality and lack of finance can deter investments to improve productivity. Increasingly there is a shortage of labor due to migration out of rural areas with low wages and opportunities. A lack of safety nets such as pension or insurance arrangements makes the coffee sector a less attractive proposition for the next generation of coffee farmers. These factors also lead parents to encourage their children to find opportunities elsewhere.

Another challenge coffee growers face is climate change which some analysts predict could halve the area suitable for agricultural production by

2050. Warmer temperatures are also increasing pests and diseases. Indeed, in recent years the coffee sector has suffered significantly from diseases such as coffee rust and a mix of prolonged dry seasons and hurricanes, which have dramatically reduced coffee production and increased price volatility.

Table 1 Concerns of small- and medium-scale coffee farmers

Social problems	Institutional problems	Technical and environmental problems
Women are represented in the associations and participate in the coffee value chain, but are often not directly involved in the decision-making process or in the marketing of coffee. They do not have access to resources and therefore do not enjoy economic autonomy.	Low organizational, commercial and financial capacity in local cooperatives, and limited power to influence national governments, demanding the development of stronger and more efficient public policy to support the sector.	Agricultural production systems and post-harvest and processing equipment are obsolete; transport systems are highly deficient; lack of accessible technological packages, quality extension mechanisms and available inputs cause low productivity and profitability of the sector.
Socio-cultural barriers and certain weaknesses within women's and mixed producer organizations hinder women's economic empowerment.	Weaknesses in public sector technical assistance to small-scale producers.	Lack of an adaptation and mitigation strategy to climate change, which causes the development of conditions that put coffee cultivation at risk (diseases that are more resistant, prolonged periods of drought, stronger hurricanes, etc.).
Social and economic inequality continues to increase, driving rural (especially youth) and mountain populations into urban areas due to lack of opportunities.	Lack of national policies supporting local coffee consumption and promoting national production on local markets.	Lack of an outreach system that considers the specific needs of women and the effect of climate change on women coffee farmers.
Absence of generational relay due to the continuous exodus of young qualified farmers toward the big cities in search of work opportunities.		
Additional factors such as the effects of climate change on agriculture, increased violence in communities (specifically gender violence), lack of connectivity or services and inability to access productive assets and inputs were mentioned as amplifiers for migration.		

Small-scale producers are particularly vulnerable to challenges such as climate change. The low profitability of small- and medium-sized producers makes it difficult to invest to improve production e.g. in affording improved varieties with better yields, quality characteristics or resistance to abiotic and biotic threats (e.g. coffee rust). The lack of ability to invest and improve production potentially makes farming progressively less sustainable both economically and environmentally in the long term.

These conditions also result in a lack of interest in coffee farming among the younger generation. Coffee is mostly grown on smallholdings by farmers whose average age is increasing - for example the average age of a Latin American farmer is 55. This creates a serious problem of generational replacement that could result in many coffee farms being abandoned in the next 10 years. Organizations of coffee growers are still led in many cases by men over 60 who have often been in post for decades. This also creates a problem of succession as well as potentially limiting the flow of new talent and ideas into the sector. Recent research summarizing common concerns among small- and medium-scale coffee farming families is shown in Table 1.[3]

3 Challenges facing smallholder farmers: market trends

Analyzing the ICO data on coffee production for 2018/2019 (based on information from 56 coffee-producing countries), we can see a production increase from 129 million 60 kg bags in 2009/2010 to 170 million 60 kg bags in 2018/19, indicating an increase of over 40 million 60 kg bags. It is interesting to see that just six countries were responsible for most of this increased production by 2018/019:

1 Brazil with 18 948 000 60 kg bags;
2 Colombia with 5 760 000 60 kg bags;
3 Honduras with 3 725 000 60 kg bags;
4 Vietnam with 13 349 000 60 kg bags;
5 Uganda with 1 810 000 60 kg bags;
6 Peru with 977 000 60 kg bags.

Some countries have seen a production decrease, for example:

- Venezuela - 688 000 60 kg bags;
- Indonesia - 1 962 000 60 kg bags;
- The Philippines - 524 000 60 kg bags;

3 INVESTING IN YOUTH IN COFFEE GROWING COMMUNITIES, Hanns R. Neumann Stiftung, SAFE & Sustainable Food Lab (2018).

- Cameroon - 632 000 60 kg bags;
- El Salvador - 314 000 60 kg bags;
- Thailand - 313 000 60 kg bags;
- Guinea - 251 000 60 kg bags;
- Ecuador - 212 000 60 kg bags;
- Togo - 128 000 60 kg bags.

In some countries, the balance between national production and domestic consumption is negative, which means these countries are obliged to import coffee:

- The Philippines - 3 094 000 60 kg bags;
- Cuba - 102 000 60 kg bags;
- Thailand - 918 000 60 kg bags;
- Venezuela - 1 025 000 60 kg bags.

In other countries, we see minimal fluctuations in production with around 50–70% of local production used for domestic consumption. These figures suggest some countries have more successful and efficient coffee-growing sectors than others, potentially putting additional pressure on less competitive counties. However, these more successful countries also show what can be done to boost the coffee sector with the right strategy and support.

However, there is little incentive to increase production or productivity if prices are low. While coffee prices have always been volatile, the collapse in coffee prices in the 1980s and 1990s represented one of the worst market shocks producers had ever experienced. Between 1989 and 1996, the approximate profit per cup of coffee paid to producers fell from over ten cents to around two cents.[4] Ripple effects from the crisis continue to shape markets and policies in coffee-producing regions. Analysts predicted (correctly) that coffee prices might start to recover as global production fell to match demand in the wake of the crisis.[5] However, prices have remained volatile. This can be seen in the price reduction in the last 10 years (from US$3/lb in 2010 vs. US$1.2 in 2019), partly as a result of the increase in production in that period, representing an over 50% reduction in prices which disproportionately affects smaller producers.

4 John M. Talbot (1997), Coffee crisis: the case for a regulated market, *Georgetown Journal of International Affairs*, 3(1): 45-52.

5 Bryan Lewis, Daniele Giovannucci and Panos Varangis (2004), Coffee Markets: New Paradigms in Global Supply and Demand, *Agriculture and Rural Development Discussion Paper 3*, The World Bank, Washington (DC), USA).

4 The role of women and young people in the coffee-growing sector

In 2019 Oxfam conducted a study in the Dominican Republic (in the Caribbean): *Diagnóstico sobre la situación de jóvenes y mujeres en el sector rural de zona fronteriza de la República Dominicana*. This study suggested the main problems specifically affecting women in coffee growing relate to broader gender inequalities in a particular society e.g. the culture of 'machismo' associated with Hispanic countries such as the Dominican Republic.

Women play multiple roles in coffee growing. Few are farm owners, but they are involved in the entire production process, from planting to harvesting crops. Between 20% and 30% of coffee farms are female-operated and up to 70%[6] of labor in coffee production is provided by women, depending on the region. In regions such as Africa women can do up to 90%[7] of field and harvest work, particularly as most small-scale coffee production is not mechanized. Women must participate in production activities without abandoning traditional roles such as managing the household budget, caring for children, elderly relatives and the sick, preparing food and fetching water or firewood.

However, despite their importance to coffee growing, women often fare significantly worse than men economically. When the farm is owned by others, they often receive lower remuneration than men while, when they or other family members do own farms, the money is usually managed by men in the family. Family coffee income is channeled mainly through men who may then decide what money to give their wives for household expenses. The evidence suggests women have less access to resources such as land, credit, inputs, education and training than men. This often results in a measurable gender gap in economic outcomes, including in yields, productivity and farm income for female-owned coffee farms.

Female empowerment would not just benefit female-owned farms in improving productivity and resilience against economic shocks from volatile coffee markets as well as the impact of climate change but would result in wider social and economic benefits for rural families and communities. An increased female say in household decisions is associated with better nutritional and health status of children, a higher share of income spent on education and better overall financial decision-making.

Turning to young people in coffee-growing areas, the multiple obstacles in accessing good-quality education is a recurring problem identified in all the focus groups of young people organized by non-governmental organizations

6 Gender equality in the coffee sector, International Coffee Organization (2017) (https://www.ico.org/documents/cy2017-18/icc-122-11e-gender-equality.pdf).

7 https://www.thecoffeeguide.org/coffee-guide/niche-markets-environment-and-social-aspects/Sustainability-and-gender/.

(NGOs) such as Oxfam. This is reflected in poor provision of schools in rural areas, especially high schools, the long distances between communities and existing educational centers, overcrowded classrooms and variable teaching quality. Those who are able to access and complete a high school education are rarely able to proceed to the next level and enroll in universities, which are typically long distances from coffee-growing areas. University fees, transport and housing costs are often beyond many household budgets.

Lack of educational opportunities is compounded by lack of employment opportunities and high levels of unemployment in rural areas, problems which disproportionately affect young people trying to enter the local labor market. Even young people who have received technical and/or professional training suffer from unemployment while, in some cases, deploring the fact that the few jobs in the local public administration are reserved for members of the government party and their families. There is often little investment by government in rural areas, whether in local infrastructure; services for local people; or technical, financial and marketing support for agriculture.

According to the FAO publication: *Promoting rural youth employment in Uganda's coffee sector,*[8] rural youth unemployment and underemployment are one of the biggest development challenges the country is currently facing. As coffee is cultivated all over the country, the coffee value chain has considerable potential to provide income and employment to rural youth even in remote, mountainous areas. In order to create opportunities for rural youth, policy options to promote youth employment and entrepreneurship in the coffee value chain include the need to:

- facilitate sustainable access to markets for youth-led businesses and equip young people with technical and financial skills in coffee growing, selling and processing;
- support the establishment of well-organized youth groups;
- foster youth investments in coffee businesses by facilitating their access to capital;
- increase young women's involvement in the coffee industry and promote fair compensation for female labor;
- promote inter-generational transfer of land and develop by-laws which make it easier for young people, especially female youth, to access land for coffee production either through leasing or share cropping; and
- invest in the upgrading of physical and digital infrastructures to support the coffee sector.

8 http://www.fao.org/3/cb0918en/CB0918EN.pdf.

Possible solutions to support the coffee sector as a whole and, more specifically, to increase the participation of women and young people in the value chain are discussed later in the chapter.

5 The impact of the COVID-19 pandemic on small- and medium-scale coffee farmers

The global outbreak of the COVID-19 virus in 2019 has had a major impact on international trade. This section provides a short overview of what is happening in the coffee supply chain. It offers practical steps for small- and medium-sized enterprises (SMEs), like farmer groups and exporters in coffee-producing countries, to deal with the immediate effects of COVID-19 on the sector.[9]

An estimated 125 million people depend on coffee for their livelihood, from growing, transporting and roasting the beans to selling the final product. Baristas and coffee shop owners were some of the first to close the doors to their businesses as lockdowns began. The coronavirus crisis caused sales volumes in 2020 to decrease by an estimated 50% in Europe and an estimated 25% in the United States. This led to roasters ordering less coffee from importers in 2020 and 2021 with warehouses remaining full with existing stock.

In providing an overview, José Sette, the Executive Director of the International Coffee Organization (ICO), commented in 2020:

> The impact of Covid on coffee production is relatively minor, Governments moved to safeguard workers by declaring coffee production an essential economic activity. Despite the difficulties finding sufficient migrant labour, even much smaller coffee-producing countries like Ethiopia, which exports about four million bags of coffee per year, have largely been able to continue production as planned; but with consumption, the uncertainty is greater. We do not know how the global economy will settle after this shock. When government support is no longer possible, we may see mass unemployment.

An analysis of coffee demand by the ICO in April 2020 found out-of-home consumption[10] decreased significantly as offices, coffee shops and restaurants were closed in order to reduce the spread of the virus. However, panic buying and stockpiling by consumers from supermarkets and other retail outlets actually increased demand in some countries in the short term.[5] This effect has been reinforced by some countries also stockpiling coffee.

However, a more negative, longer-term effect on global coffee demand can be expected as the result of a global recession triggered by the direct and indirect effects of the COVID-19 pandemic. Reduced household incomes

9 International Coffee Organization, Coffee Break Series No. 1 APRIL 2020.

10 In 2018, coffee consumed out-of-home represented approximately 26.1% of the total quantity of coffee consumed worldwide according to an estimate by CoffeeBi, a consultancy organization.

could translate into lower demand for coffee in the longer term. In addition, price-sensitive consumers may substitute higher-value coffee with lower-value blends or brands. Indeed, while the market price for coffee has increased in response to demand, prices for fair trade coffee have stagnated, creating a risk of coffee producers abandoning fair trade schemes that promote better income and working conditions for those cultivating coffee. In addition, coffee sold at cafés is generally high-end or premium specialty coffee, suggesting the premium and specialty coffee segments of the coffee market may suffer.

ICO has analyzed the relationship between GDP growth and coffee consumption (in volume terms) in order to estimate the demand-side shock resulting from the COVID-19 pandemic. Their analysis suggests that even a modest decline in GDP growth as a result of the COVID-19 pandemic could have a significant impact on global coffee demand. The model predicts that a larger drop in GDP growth or a global recession would have a proportionally greater effect. As a result, coffee consumption levels could stagnate (or even decline) compared to pre-crisis years that were characterized by steadily increasing demand for coffee (at a rate of 2–3% per annum).

In July 2020, International Coffee Partners (ICP), a group of mainly European family-owned coffee businesses seeking to support smallholders supplying them with coffee to process, published a report: *How COVID-19 impacts smallholder coffee farmer families*. This found that, while 74% of farmer families expected to harvest their coffee crop as normal, most expected reduced prices (77%) with some expecting to delay selling coffee given an unfavorable market (33%). 71% of families interviewed already reported a negative impact on household income and with 47% of farmer families having already had to change cultivation practices because of COVID-19. The biggest management challenges were the availability of farm workers (64%), access to inputs (62%) and access to financial services in a sector still reliant on cash payments (46%). This had affected practices such as planting (26%), fertilization (51%) and pruning of trees (37%). The impact of the pandemic has tended to increase overall production costs. An additional issue is the extra costs in precautions to prevent the spread of COVID-19 such as personal protective equipment (PPE) and sanitation materials.

Coffee farmers in communities where COVID-19 has spread are being affected in three ways. First, farmers who hire outside labor (medium-scale farmers that hire seasonal workers for harvesting and whose economic resilience is lower than the large farmers) are reporting labor shortages as farm laborers are either afraid or unable to travel. For example, farmers in Guatemala are reporting that labor costs have increased by 30%. In Peru, exporters are projecting that up to 30% of the coffee crop may not be harvested. Harvests in Brazil are also threatened by lack of labor. Labor shortages affect both volume and quality of supply since harvesting relies on skilled and experienced pickers.

In addition, reduced transportation in rural areas due to travel restrictions is making it more difficult for farmers to get their produce to market, which is resulting in income losses for coffee farmers in Guatemala who also sell perishable crops such as avocados. Farmers around the world are paying higher transportation costs to get their coffee to market. Finally, agrochemical inputs such as fertilizers are not readily available in some rural communities because of supplier closures. In addition, in countries like Mexico, fertilizers are imported, creating both shortages due to disrupted supply chains and increased prices, an issue that can be exacerbated by currency fluctuations, particularly devaluation of currencies in smaller and poorer countries (though this can also make coffee exports from those countries cheaper).

These issues are highlighted by a June 2020 report published by the NGO Verité which promotes better labor conditions: *COVID-19 in the Coffee Sector: Challenges for Workers and Farmers*. Verité identifies issues such as:

- Forced and child labor: Verité has found that the pandemic is increasing the risk of child labor and forced labor, especially in sectors such as agriculture which have been declared essential services, meaning that workers are forced to continue working in these sectors despite the risk of exposure to COVID-19.
- Travel restrictions: As states and municipalities have sought to limit travel to contain the spread of the virus, farmers are being prevented from hiring migrant workers. Coffee harvesting is highly time-sensitive, so the lack of access to migrant labor means that many coffee producers will have to turn to other sources of harvesting labor, such as inexperienced workers who have been laid off from other jobs, workers recruited by unvetted labor brokers and family labor. This will increase forced labor risks like deceptive recruitment and recruitment fees, as well as risk of child labor. As coffee harvesters are typically paid based on their production, inexperienced workers will likely earn far less than experienced harvesters, further increasing their vulnerability to exploitation.
- Quarantine: The mandatory quarantine for migrant workers for a period of up to two weeks, posing restrictions on their freedom of movement and resulting in an increased risk of debt bondage if they are expected to cover their food and housing costs during quarantine.
- Hours of work: Social distancing may require a reduced workforce, increasing production demands on hired workers and increasing their working hours.
- Health and safety: Many workers employed in the coffee sector will be forced to continue working whether or not there are factors that increase their risk of COVID-19, such as advanced age (with workers employed in the coffee sector in some countries averaging over 50 years of age),

pre-existing health conditions, working for employers that lack PPE, inadequate sanitary and social distancing measures or a lack of COVID-19 diagnosis and treatment options.

- Personal protective equipment (PPE) and sanitary measures: Given that even prior to the pandemic workers were rarely provided with the PPE needed to safely harvest and process coffee - such as hats, overalls/coats, boots, facemasks, goggles and gloves - it is unlikely that farms will provide workers with sufficient PPE and sanitary measures to effectively protect workers during the pandemic. A lack of infrastructure for proper handwashing and disposal of gloves or masks is an additional risk.
- Social distancing: While it is possible to ensure adequate social distancing in coffee harvesting by carefully communicating guidance and monitoring compliance, the crowded nature of worker housing and dining facilities on many coffee farms makes social distancing especially challenging, particularly in bathrooms, dining rooms and bedrooms. Social distancing is also difficult to implement when workers who are paid by the piece have to bring their coffee to weighing stations, which are centralized and few in number, meaning that many workers must congregate around them at the end of each day.
- Identification and treatment of workers with COVID-19: Most coffee farms are located in rural areas where workers lack access to advanced medical facilities, creating an obstacle for the testing and treatment of those infected with the coronavirus.
- Loosening of regulations and enforcement: There are indications that labor shortages will be severe in the coffee sector during the pandemic due to restrictions on the movement of migrant workers and worker illnesses. In response, some coffee-producing countries are already beginning to loosen protective regulations (around social distancing, worker recruitment, working hours and even child labor). In addition, governments will have fewer resources to carry out labor inspections, and inspectors already facing challenges carrying out inspections in rural areas will face additional barriers and risks due to the pandemic.

Some solutions to these problems are discussed in the following section.

6 Solving the challenges facing small- and medium-scale coffee growers

The complex dynamics of the coffee sector do not currently allow for the economic, social and environmental sustainability of coffee cultivation in many parts of the world. In these areas, coffee cultivation is in a vicious circle of low prices, productivity and returns in which coffee growers, especially young people, no longer want to participate. Creating a fairer and more sustainable system to reverse this downward cycle requires change in some key areas:

- dedicated policies by national and state/local governments to support the coffee sector;
- enhancing the efficiency and sustainability of coffee cultivation together with the quality of the harvested product;
- where regions allow (e.g. conditions for cultivation have the greatest potential), developing high-quality, added-value products (such as specialty coffees) and effectively promoting them; and
- where higher quality may not be feasible, promoting diversification into other agricultural or non-agricultural products and services that spread economic risk and enhance resilience to climate change.

To begin with, national governments need to have longer-term (e.g. 10 years) plans to support their coffee-growing regions. These need to:

- develop a broad research program that helps to increase productivity, reduce the effects of pests and diseases and adapt to the effects of climate change;
- create dedicated funds to finance improved materials and technologies in planting, production (including management of pests and diseases), processing and commercialization of coffee;
- expand and renew the network of agricultural professionals supporting the coffee sector through training and advice;
- create programs to promote coffee processing for national and international markets, including coffee roasting and grinding. This will help to add value to coffee products;
- develop campaigns to promote value-added products such as specialty coffees;
- support diversification into cultivating other crops in coffee-growing areas such as sapote, macadamia, citrus, bananas and plantains; and
- include a cross-cutting strategy to promote the role of women and young people in the coffee sector (in part to stem the loss of talent from the sector).

A good example of what governments can achieve both individually and collectively to support coffee growers is the Early Warning System for Coffee Rust (EWS)[11] which operates across Central America to alert farmers to disease risk so they can take preventative action.

11 https://www.fontagro.org/new/proyectos/sistema-de-alerta-temprana-roya-cafe/en

In particular, there needs to be full recognition of how much the coffee sector depends on the contribution of women and young people who are rarely taken into account in the design of public policies. This is despite the fact that they are often victims of discrimination, exploitation and even violence. A place to start is open discussion of such questions as:

- Do women have the same opportunities and receive the same treatment as men?
- What are the reasons why there are so few people under 30 in an organization?

Existing institutional structures and decision-making processes may well need scrutiny and revision to make them more representative, inclusive, democratic and accountable. One specific recommendation is to explore the adoption of a positive quota system by gender and age in management positions. A starting point, for example, would be to establish a rule that at least 40% of positions should be held by women and 35% by young people under 35. Beyond quotas, governments should have specific policies to encourage the participation of young people and women, listening to their views and defending their demands for equity. Governments should also support women and young people in forming their own support groups since these encourage participation and leadership through mutual support. An important aspect of such groups is effective communication so that their voice can be heard both within the coffee-growing sector and more broadly in society. Another measure would be to provide funding for young people to undertake technical training or attend university on the understanding these skills would be deployed back in their local, coffee-growing areas.

Recognizing the contribution of women and young people, confronting the ingrained attitudes, power relationships and institutions which prevent them having an equal voice and opportunities, and providing the support to give them a more equal position are not just the right course on the grounds of equity and social justice. It is critical to the health and survival of the communities involved in coffee growing as well as unlocking new talent to help address the many challenges the sector faces.

7 Diversification in the coffee sector

Given the volatility in the coffee market, the potential for global overproduction leading to lower prices, and the range of biotic and abiotic threats to coffee cultivation (exacerbated by climate change), a sensible strategy for some coffee growers is diversification away from overreliance on a single crop. This can be achieved e.g. by intercropping: growing other crops alongside coffee.

Diversification offers the potential of spreading risk and ensuring a more stable and continuous income, particularly if different crops are harvested and sold at differing times. It can also contribute to a more diverse and nutritious diet for a farming family.

Diversification has other benefits such as allowing a more agroecological approach to farming in which a variety of crops can contribute to improving soil physical, chemical and biological health, potentially reducing the need for fertilizers as well as reducing the risk of soil erosion. Agricultural diversification makes it possible to achieve an ecological balance within farms and potentially improve ecosystem services such as pollination and pest control as well as improving soil fertility.

However, a shift from a monocultural to more diverse cropping system requires planning, appropriate knowledge and resources in growing new crops, and carries with it risks during the transition period. All these pose significant barriers to change. Governments can play a key role in designing crop diversification programs focused on the coffee sector to allow farmers to make a transition over 3-4 years. They may need also to provide appropriate support in terms of agronomic advice, planting materials and equipment together with financial support such as extended credit to cover extra costs and a potential dip in income during the initial years of transition. More broadly, governments also need to consider the potential markets for new crops and ensure farmers have the means to access and secure equitable returns from those markets, particularly export markets for new produce.

There are a range of options that can form part of a diversification strategy:

- forest trees;
- fruits and vegetables;
- musaceous species, perennial and annual fruit tree species; and
- livestock.

An obvious area of diversification is forest plants. Traditionally coffee is shade grown under a canopy of trees. The canopy protects coffee plants from extremes of sunshine, rain and temperature (e.g. frost). Trees can also provide nutrients to plants and soil and protect the latter from erosion. They also provide a habitat for beneficial insects such as predators preying on insect pests. More broadly forest shade provides a range of ecosystem services such as supporting biodiversity (e.g. shelter for migratory and native birds), carbon capture, water retention and soil protection. However, shade species are also potential sources of high-quality wood and fuel. Among forest species that are commonly used in shade coffee cultivation, coffee are native and species (e.g. laurel, walnut and pine) as well as species that can be used, e.g. for wood chip or pulp.

Traditionally many smallholder coffee farms were polycultures in which coffee was grown alongside a range of fruits, vegetables, grains and other crops such as mushrooms. These provided food sources for smallholder families and the possibility of additional income from selling any surplus in local markets. This diversity of crops grown also had wider benefits in areas such as improving soil health. However, there is potential to build on these traditional practices to grow a range of commercial crops alongside coffee.

There is significant potential in planting musaceous crops (e.g. bananas and plantains), perennial fruit trees (e.g. avocados and cacao) and annual fruit trees (e.g. strawberry and pineapple). Solanaceous and other vegetable species also have great potential, given the temperature, relative humidity, soil and altitude conditions in which coffee is grown. These species include potatoes, tomatoes, onions, celery, carrots, beets, broccoli, cauliflower and peas. Ornamental crops provide also provide considerable market potential. Ferns and flowers are currently grown mainly on several medium-size and large farms as they require a significant capital investment per hectare. Orchids also have potential as income generators in coffee areas.

In addition to crops, livestock farming also offers opportunities for diversification. This includes dairy cattle, sheep, goats, pigs, chickens and rabbits. Livestock also provide a source of organic fertilizer through manure. Another option is beekeeping which can not only provide honey but, through pollination, promote coffee plant growth and even improve cup quality.

Successful diversification requires planning which needs to take into account considerations such as the following:

- deciding on the appropriate balance between crops for family consumption and those for sale;
- clearly identifying the requirements for a new crop (e.g. appropriate soil and climatic conditions, risks from pests and diseases, planting materials and equipment, likely yields and market potential);
- choosing crops that require labor at different times of the year, matching the available resources (e.g. within the family);
- choosing crops that are ready to harvest at differing times of the year to ensure a steady income stream; and
- looking for synergies where possible e.g. using surplus grains or tubers (where market demand might be low) as livestock feed e.g. for chickens or pigs.

On balance, it is better for smallholders to focus on a few crops that meet commercial diversification requirements over the medium and long term and which they can grow successfully rather than overextend themselves with too many.

However, to make a significant difference to smallholders in a coffee-growing area, diversification requires government support. This may include technical guidance and support, access to subsidized materials and credit to reduce the costs and risk in making a transition. It also requires potential support in marketing, particularly if produce has potential beyond local markets, e.g., in supplying more distant urban or export markets. This support might include help in estimating likely sales and production costs, how to meet the quality criteria required by a particular market, any primary processing and packaging required as well as linking to retail or manufacturing partners further on in the supply chain.

8 Environmental services provided by coffee smallholders

There is a widespread recognition that the intensive, monocultural production of crops such as coffee is contributing to climate change, and that alternative agroecological approaches are needed to reverse this trend and ensure a more sustainable agricultural sector for the future. Small- and medium-scale coffee farmers potentially have a significant role to play in helping to achieve this transition in their areas.

One possible mechanism that could contribute to reversing climate change, while supporting small- and medium-scale coffee farmers through an alternative source of income, is a system of payments for environmental services (PES). Such systems are designed to give incentives to land users to provide an environmental (ecological) service that benefits society as a whole. An example is planting trees to promote carbon sequestration. If this is seen to have a broader societal benefit (e.g. in combating climate change), society (e.g. through government agencies) can place a value on this public good and pay for its delivery. There are four categories of environmental services to which small- and medium-scale coffee farmers can potentially contribute (and for which they can potentially be rewarded):

- **Carbon sequestration and storage**: Trees provide a wide range of ecological services such as producing oxygen, promoting soil health and biodiversity. They are also able to store significant amounts of carbon. There is now a wide range of both government and private-sector schemes to incentivize coffee and other farmers to plant more trees to promote carbon sequestration.
- **Protection of biodiversity**: Biodiversity involves genetic diversity both within and among species which is seen to depend on promoting landscape diversity. Landowners can enter into legal agreements (sometimes called conservation easements or covenants) in which they

undertake to manage land for conservation purposes e.g. by measures such as areas where wild plant species are encouraged or trees are planted to create a more diverse landscape to support greater local species biodiversity.
- **Watershed protection**: Watersheds are landscape basins in which water collects and is funneled into streams and rivers and eventually into a larger water body such as a lake or major river flowing to the sea. Upstream land in a watershed can play a critical role in collecting, purifying and channeling water as it moves downstream. If the land is farmed, farmers are an integral part of watershed management in such areas as drainage management, improving soil water retention and modifying fertilization practices to avoid runoff into and pollution of watercourses. They can be incentivized to improve these practices either by governments or private bodies (e.g. companies supplying water to domestic and commercial customers).
- **Cultural/recreational benefits**: Agricultural landscapes provide areas for recreation and tourism; they may also have a broader aesthetic and cultural value (e.g. some types of landscapes are seen integral to the identity of a region). Farming activities over generations have shaped these landscapes and helped give them this broader cultural value. Farmers can benefit directly from tourism or from government schemes to maintain key features in a landscape.

There are, for example, well-established voluntary certification schemes established by NGOs (e.g. Rainforest Alliance) which pay coffee growers a premium for coffee grown according to agroecological principles, funded by a premium paid by consumers willing to pay extra for coffee with this certification and labeling.

PES schemes have also been established by governments. An example is Law 64-00 from the Dominican Republic in the Caribbean which sets out a general framework for compensation and/or remuneration of environmental services in order to contribute to the conservation, recovery and sustainable use of the country's biological diversity and natural resources. The law sets out the main environmental services as:

- water regulation for urban, rural or hydroelectric power;
- conservation of biodiversity;
- soil and water quality management;
- fixation, reduction, storage and absorption of greenhouse gas (GHG) emissions;
- contribution to the diversity and beauty of the landscape, for tourism, ecological and scientific purposes; and
- conservation of local flora and fauna.

An example is the Yaque del Norte River Basin Project. This initiative is the result of an agreement between a number of key stakeholders in this watershed: the Corporación de Empresas Eléctricas Estatales (CDEEE), the Corporación de Acueducto y Alcantarillado de Santiago (CORAASAN) and the Ministry of Environment and Natural Resources. This project has benefited owners and users of forest land in the Upper Yaque del Norte River Basin who receive a bi-annual payment in return for cultivation practices such as implanting and maintaining agroforestry systems for coffee. In total, the project has made payments that have covered an area of 3300 hectares, with an investment of US$ 27000.

Another example is the municipality of Tacuba in the Ahuachapán Department in the Republic of El Salvador. The population of Tacuba faced serious water supply problems in 1997 and consequently started initiatives to protect the local water basin. In 2001 a project was started by EMSAGUAT, a body responsible for providing water to many residents of the area. In 2002 a mechanism for payment for environmental services was created through a participatory consultation process and began to function in 2003. More than 22000 people benefit from water services and 32% participate in payment mechanisms. Water users contributed 12 cents per month over a period of five years (2003–2008), raising $12500. The principal beneficiaries were the Las Colinas Cooperative of farmers as well as a number of individual agricultural producers. In return for the subsidy, the Cooperative committed to abandoning environmental damaging agricultural practices, shifting to organic coffee production and maintaining shade trees. These practices have reduced contamination from agrochemical use and improved water flow and quality. Coffee producers have benefited from the higher price of organic coffee on the market and also from the guarantee of permanent access to resources such as firewood and other wood products for use in the home. Potential future improvements include establishing alliances between the different water administration entities in the territory to extend the scope of the scheme. Research has shown that, although Tacuba is a relatively poor municipality, it is possible to achieve both successful collection of funds and their use to improve coffee-growing practices to benefit watershed function.

A final example of what can be achieved can be seen in Costa Rica in Central America. The government has been praised for its pioneering work from the 1990s e.g. through Forest Law 7575 which established a framework for compensation systems for environmental services. The program seeks to encourage forest protection and management by paying forest owners for environmental services. Environmental services covered include:

- watershed protection;
- carbon sequestration;

- biodiversity conservation; and
- aesthetic features of the natural landscape.

The main source of financing is a national tax on fossil fuel consumption. 5% of the program is funded by the World Bank, grants from the Global Environmental Facility (GEF) and other sources. The National Forest Finance Fund (Fondo Nacional de Financiamiento Forestal–FONAFIFO) collects funds and issues certificates for PES. Owners of forest properties of 2-300 ha, and owners that intend to establish or manage plantations of at least 1 ha, can apply for the program through regional offices of FONAFIFO if they can demonstrate legal land tenure and hire a certified forester to attest to a land use plan to implement agroecological practices. Contracts last 5–10 years, and payment levels and schedules vary according to management activity. From 1997 to 2007, the scheme covered a total of approximately 500 000 ha (FUNDECOR, n.d.). Farmers receive US$ 1.30 per tree, with 350–3500 trees allowed in each contract.

According to the 2003 Report *The social impacts of payments for environmental services in Costa Rica* by the Centro Internacional de Política Económica para el Desarrollo Sustenible (CINPE) of the National University of Costa Rica, the program has delivered a number of financial benefits for farmers:

- an average increase of approximately 15% in household disposable income (equivalent to an average of approximately $4200/year per property, with a range of $880–$11 200);
- a higher level of investment especially in forested areas of farms; and
- a variable level of job creation, mostly in relation to hiring of occasional workers (average pay $13.5 per person per day).

There have also been broader social benefits, including a reduction in centralization and bureaucracy with more local decision-making as well as more voluntary agreements and better coordination between different government and other agencies. There has also been a substantial improvement in environmental education and understanding of waste management, involving schools, parents and civil society. Landowners benefit directly from advice from FUNDECOR and CNFL on planting, fertilization and crop management through to harvesting, as well as the design and maintenance of paths (e.g. to promote eco-tourism) and measures to minimize the risk of illegal hunting the properties. There have been significant improvements in agro-ecological practices such as agroforestry, use of organic compost and biofertilizers, as well as better grass species for feeding livestock.

At a national level, the PES program has contributed to:

- the protection of approximately 16 500 ha of primary forest;
- the sustainable management of 2000 ha of forest;
- the reforestation of 1 300 000 ha.

Over one million nursery trees provided by CNFL have been used in environmental education programs with children and youth in public areas and gardens. These activities, together, more sustainable agriculture and livestock practices, help guarantee the protection of existing forest, add new trees as well as other benefits such as protecting biodiversity and prevention of soil erosion.

As shown by the examples described here, PES can be a tool to finance reforestation, restoration, conservation and changes in land use that enhance rural development. In addition, PES programs can induce attitude changes in farmers, which is a major goal of rural development programs. The application of PES requires evaluation of the ecosystem services provided, determination of payments and monitoring of success. The evaluation of environmental services provided by each land use type requires multidisciplinary teams, including ecologists, economists and sociologists. Payment level can be determined based on indices including data on evaluations of environmental services and opportunity costs.

9 Summary

This chapter has shown the complexity and volatility of the coffee market. This has disproportionately disadvantaged smallholder coffee farmers and, in many cases, made coffee growing economically, environmentally and socially unsustainable, with young people unwilling to participate in a sector which seems to offer few opportunities. The scale of the challenges requires a coordinated response from all those involved, starting with governments who need to develop, resource and implement comprehensive, long-term strategies for their coffee sectors.

Two elements are fundamental to these strategies. The first is that such strategies must be based on agro-ecological principles that ensure that coffee cultivation plays a role in mitigating rather than contributing to climate change. This approach (including diversification into producing other crops commercially) also provides the basis for sustainable improvements in yields, productivity and profitability. Where appropriate, coffee farmers should be remunerated for delivery of ecosystem services which benefit society as a whole.

The second element is to involve and promote the role of women and young people in the coffee sector. Women play a huge role in coffee cultivation but are often marginalized in decision-making processes. Young people who represent the future of coffee growing face a sector in which there seem few opportunities and in which they too may struggle to be heard. Listening to and empowering them will bring new talent and ideas into the sector and help secure its longer-term future.

Chapter 3

Assessing and managing the environmental and social impact of coffee production

H. A. Jürgen Pohlan and Marc J. J. Janssens, University of Bonn, Germany; and Dennis José Salazar Centeno, Universidad Nacional Agraria Managua, Nicaragua

1 Introduction

Increasing focus by governments, nongovernmental organizations (NGOs) and consumers has highlighted the sustainability of coffee production. This chapter reviews ways of measuring and improving the sustainability of coffee production systems, including good agricultural practices (GAP), life cycle assessment (LCA), ways of calculating the carbon footprint of coffee as well as methods for assessing social impacts. It also summarizes current certification schemes and initiatives. The chapter includes a summary of the range of environmental and social impacts of coffee production.

Building on this framework, the chapter also discusses ways of improving sustainability in the coffee value chain, particularly by understanding and optimizing coffee production as an agroecosystem. The chapter reviews techniques to support agroecosystem function and resilience such as shade, soil and biodiversity management. All these approaches need to take account of the complexity of the coffee value chain which is discussed in the next section.

http://dx.doi.org/10.19103/AS.2021.0096.09

2 The complexity of coffee value chains

Coffee production value chains are characterized first by the production of Arabica and Robusta coffees, with Arabica grown most widely. Both coffee species are ideally managed using GAP and good post-harvest practices (GPP). The life cycle of Arabica coffee plantations has four main phases (Pohlan and Janssens, 2010):

1. Seedbed and nursery establishment (5–12 months);
2. planting and development of a young immature plantation (12–36 months);
3. the first production cycle (6–10 years or more depending on the pruning system used); and
4. rejuvenation and subsequent production cycles (4–10 years each).

This cycle results in an average economic life cycle of a plantation of 20–25 years for traditional varieties as Bourbon, and 10–15 years for highly productive varieties cultivated at high density without shade.

It is important to take account of the many different coffee farming systems. The first detailed description of coffee farming systems was provided by Moguel and Toledo (1999) for Mexico and there are now similar descriptions for many other regions and different socioeconomic conditions (Anhar et al., 2021; Rahn et al., 2018; Läderach et al., 2011; Muschler et al., 2006; Soto-Pinto et al., 2000).

Five main cropping systems can be differentiated in the production of Arabica and Robusta coffee though, in reality, there is often a continuum from forest coffee production in Ethiopia to intensive, large-scale and mechanized production in Brazil. The five main systems are as follows:

1. Smallholder family farms, typically using existing forest resources, including low coffee planting densities (under 1200 coffee plants per hectare), 70–100% of shade cover and very extensive cropping;
2. Traditional polycultures including a mixture of native trees and planted shade trees with 60–90% shade cover and typically 1200–1500 coffee plants per ha;
3. Commercial polycultures using 30–60% shade, including other temporary or permanent cash crops, together with an increased density of coffee plants. Typical planting densities are 1600–2400 plants per ha for Arabica and 900–1200 for Robusta;
4. Shaded monocultures include Arabica coffee with densities between 2400 and 4000 plants per ha and only 10–30% of shade cover; and
5. Sun-grown coffee trees planted without any shade. These involve high-yielding strategies based on high coffee planting densities, high inputs of chemical fertilizers and pesticides.

Each one of these five different cropping systems has innumerable regional and local variants that have developed according to local environmental and social conditions. Particular influences on coffee farmer production strategies are their economic situation and the sociocultural traditions. Successful production needs adaptable coffee farmers ready to use resilient approaches which are in harmony with their particular site conditions, taking into account GAP and GPP.

Coffee growing is, of course, part of a broader value chain. This includes four main phases:

1 cultivation (farm level with growing and harvesting);
2 processing (post-harvest management);
3 roasting; and
4 consumption.

The structure and components of this value chain are described by Quiñones-Ruiz (2020), Murphy and Dowding (2019), Grabs (2017), Rahn et al. (2014) and Killian et al. (2013).

3 Ways of assessing the sustainability of coffee production

As discussed previously, the coffee value chain is a very broad and complex system, which includes relationships between many partners including coffee farmers and their associations, processors, roasters and consumers. There are many different agroecosystems, with very different production areas and types of agronomy, and with an enormous diversity of socioeconomic and cultural backgrounds. Successful coffee growing may vary with a number of factors including environmental conditions, phenology, genotype-species and varieties, as well as plant and management practices, and their social cultural traditions. This complex situation has prompted the development of many regional and international initiatives to develop new thinking and practice about improving sustainability. During the last three decades, LCA, environmental and social sustainability standards, and the development of methods to measure the carbon footprint of coffee and other crop value chains have become popular. Growing pressure from consumers in coffee-drinking countries in particular has accelerated interest in these standards and methods. This is fueling the transition of coffee from a colonial staple good to a more varied and differentiated specialty product (Bager and Lambin, 2020; Raynolds et al., 2007; Ponte, 2004). This section looks at a range of approaches, including GAP, LCA, broader frameworks such as the 'coffee flower system', current certification schemes and ways of calculating the carbon footprint of coffee.

GAP, managed using a traffic light system, have been developed with and for smallholder coffee growers (Pohlan and Salazar Centeno, 2012). This GAP system is based on ensuring effective control of 13 key issues/cultivation stages (from site selection and planting through nutrition, weed and pest management to harvesting) using 50 control points of which 11 are classified as red (very high risk of noncompliance if not correctly implemented), 15 as yellow (high risk of noncompliance) and 24 as green (low risk of noncompliance but may result in economic loss if not correctly implemented). GAP and GPP provide a foundation of good practice for improving efficiency (and thus sustainability) along the coffee value chain.

The concept of environmental LCA originated in the 1960s and 1970s[1] (Guinée et al., 2011; Klöpffer, 1997). The International Organization for Standardization (ISO 1440:2006) describes the principles and framework for LCA.[2] The four main components are as follows: inventory analysis, impact analysis, improvement analysis and interpretation. The goal of LCA is a 'cradle-to-grave' approach to measure and evaluate products, materials, processes, services and industrial systems with respect to their environmental impacts, but not only in providing data but also in facilitating decisions to reduce impacts.[3] The product lifecycle generally consists of five phases: raw material extraction, manufacturing and processing, transportation, usage and retail, as well as waste disposal.[4] LCA has subsequently been developed into life cycle sustainability assessment, which is a transdisciplinary integration framework of models rather than a model in itself. A particular challenge has been the integration of different life cycle approaches to cover all the stages in the value chain from coffee farmers to coffee consumers (Murphy and Dowding, 2019; Pohlan et al., 2019a; Bockel and Schiettecatte, 2018; Ponte, 2004; Salomone, 2003). It is particularly important to define the boundaries of coffee LCA systems and all the inputs and outputs from the various coffee life cycle stages and steps.

The so-called 'coffee flower system' describes a framework which combines agronomic (e.g. GAP), environmental (e.g. LCA) as well as social-cultural and governance issues that are all part of the complete coffee value chain (Pohlan et al., 2006). Coffee growing can be represented as a flower with the petals showing the various ecosystem benefits it delivers. The coffee flower system combines key elements of growing coffee and its ecological benefits and impacts (Fig. 1), creating a bridge between producers and consumers that aims to deepen and reinterpret the experience of drinking coffee.

Part of this system focuses on the social as well as economic and environmental aspects of coffee production. These include: low levels of

1 https://www.semtrio.com/en/history-of-life-cycle-assessment.
2 https://www.iso.org/standard/37456.html.
3 https://www.e-education.psu.edu/eme807/node/690.
4 https://ecochain.com/knowledge/life-cycle-assessment-lca-guide/.

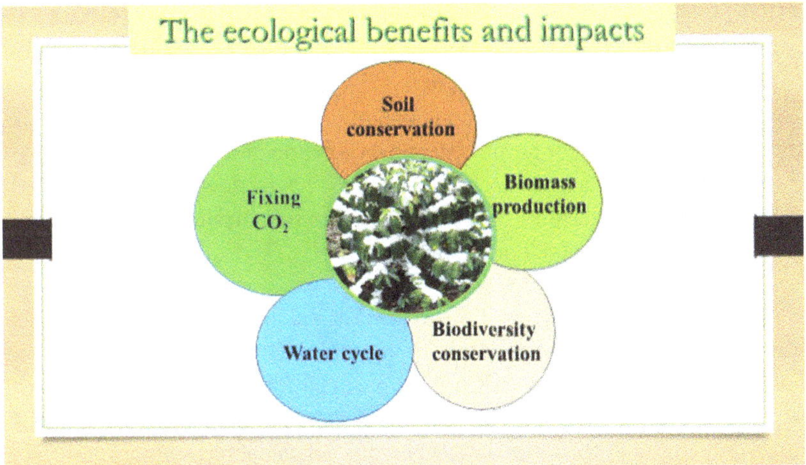

Figure 1 The coffee flower system.

educational and training among many producers, the emigration of the most talented youth out of coffee growing and the lack of recognition of the role of women in coffee production. These social issues make it difficult to improve coffee cultivation sustainably to optimize its economic, environmental and social benefits.

The range of voluntary certification systems in the coffee industry addressing environmental and social sustainability can be categorized as follows (Murphy and Dowding, 2019; Giovannucci and Ponte, 2005; Muradian and Pelupessy, 2005):

- First-party certification with a voluntary regulatory system developed by the leading coffee retail chain Starbucks (2001) based on sustainable production, ethical sourcing, environmental stewardship and community involvement[5];
- Second-party certification initiated by multinational food companies involved in coffee processing (Nestle´, Kraft Foods, Sara Lee and Tchibo) who created the Sustainable Agriculture Initiative (SAI) platform in October 2002 under the guide of Nestle, Unilever and Danone and which involves different stakeholders along the food chain;
- Third-party certification based on voluntary schemes developed by NGOs that respect World Trade Organization nontariff barriers and non-discrimination principles. The most important NGO-based coffee certification schemes are Fair Trade, Organic coffee, Rainforest Alliance (which now incorporates Utz Kapeh) and Shade/Bird Friendly;

5 https://www.greenbiz.com/article/starbucks-announces-coffee-sourcing-guidelines

- Fourth-party certification launched in 2003 as a multi-stakeholder voluntary scheme through the Common Code for the Coffee Community (4C) covering areas such as: avoiding deforestation/biodiversity loss; promoting GAPs to protect soil, air and water resources; respecting labor and land rights; and adequate training for farmers to improve productivity and profitability; and
- Fifth-party certification schemes which have developed from 2018, involving new business models directed to technically engaged younger audiences such as Millennials and Gen Z.[6,7]

It also important to recognize the broader global initiatives that have created a framework for understanding environmental and social governance issues. These include the concept of Corporate Social Responsibility developed in the 1970s; the development of Voluntary Sustainable Standards by NGOs particularly in the 2000s; the United Nations Sustainable Development Goals set out in 2016; the incorporation of the 'three Ps' (People, Profit and Planet) into ways companies (and investors) measure their performance in social and environmental as well as financial terms.

Some of these issues have been brought together in the Global Coffee Platform (GCP) set up in 2016. The GCP has been supported by the Alliance for Coffee Data Standards established in 2016 which uses common indicators based on the Sustainability Progress Framework developed jointly by the Sustainable Coffee Challenge and the GCP (Meems, 2019). The GCP Baseline Coffee Code currently contains 27 principles (economic, social and environmental), based on current good agricultural and management practices and international conventions/guidelines and targeting ten unacceptable production practices (e.g. improper use of pesticides).[8]

With more than 11 million hectares under cultivation area and average production of 10 million tons of green coffee per year, coffee cultivation has a significant global carbon footprint. This has resulted, particularly in the last decade, in many studies of ways to measure the carbon footprint of coffee supply chains. Important publications include those by Usva et al. (2020), Bockel and Schiettecatte (2018), Rahn et al. (2014) and Killian et al. (2013). Calculating a carbon footprint must take into account many elements in the coffee value chain and the diversity of coffee production systems.

It has been estimated that the production of 1 kg of green coffee, including subsequent processing from roasting to consumption and disposal, emits between 1.2 kg CO_2e and 4.98 kg CO_2e into the atmosphere. It can be hard for coffee consumers to relate this statistic to the cup of coffee they drink and thus evaluate their own footprint (see Box 1). Figures 2 and 3 provide a

6 https://www.allegraceoforum.com.
7 https://verocoffee.ie.
8 https://www.globalcoffeeplatform.org/latest/2020/work-begins-to-strengthen-gcp-baseline-coffee-code/.

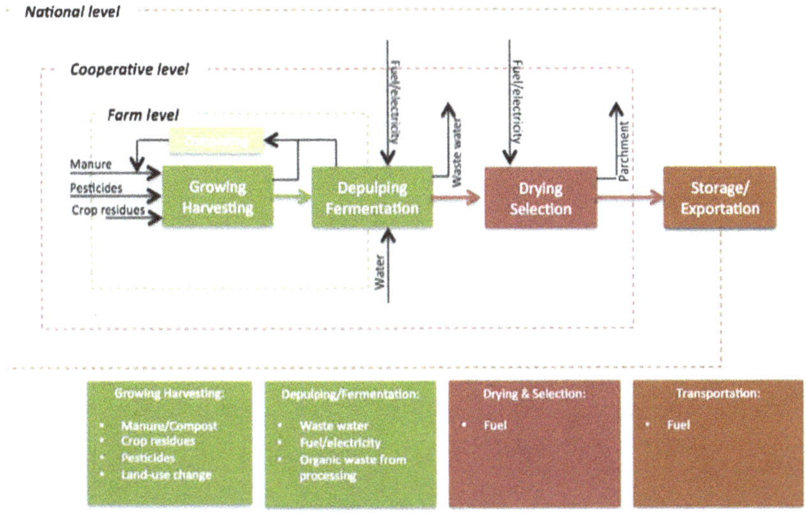

Figure 2 System boundaries with emission sources for the carbon footprint assessment (Rahn et al., 2014).

simplified overview of this problem. They show the distribution of the carbon footprint across the supply chain with 20% accounted for by cultivation, 13% accounted for by processing and 67% for the rest of the coffee value up to coffee consumption. A key idea behind measuring the carbon footprint in the coffee supply chain is, indeed, to determine the role of the different actors in producing environmentally damaging emissions. This provides a starting point in developing measures to minimize the carbon footprint of coffee production and moving towards carbon-neutral coffee production.

Box 1: The footprint of one cup of coffee

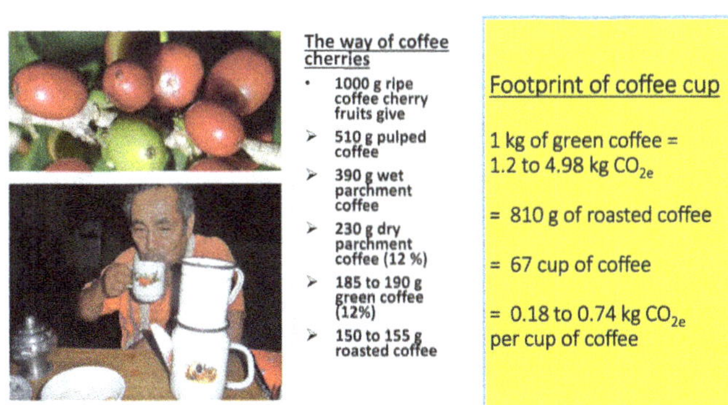

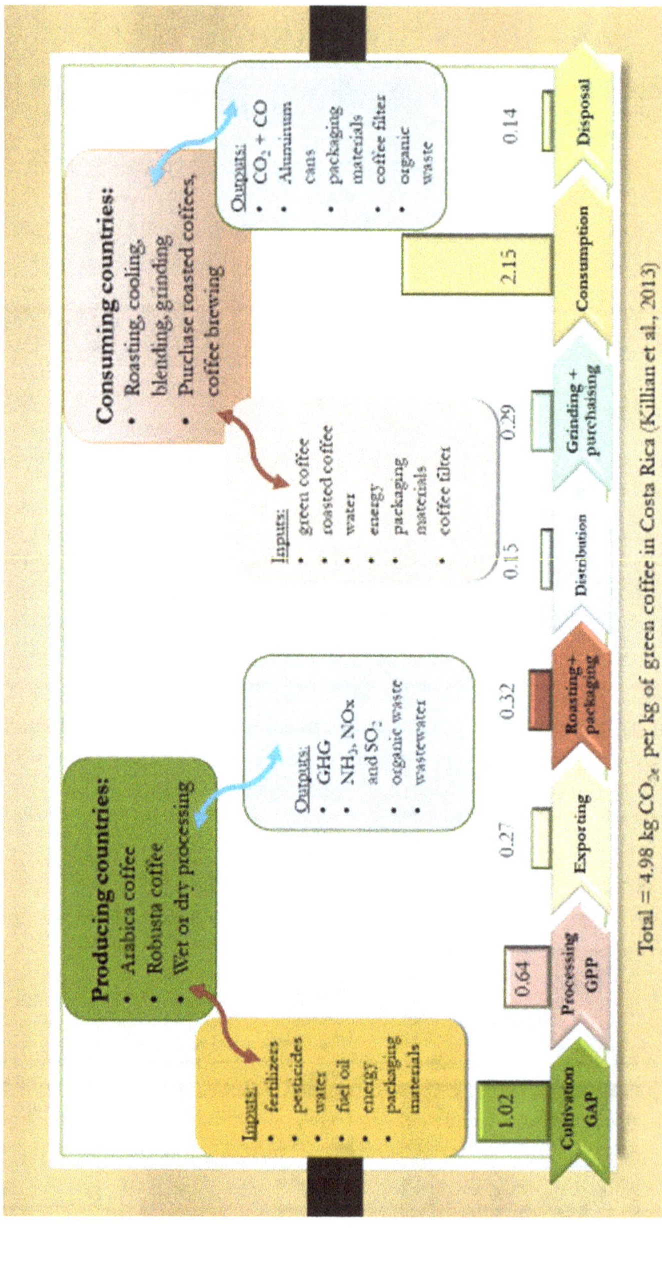

Figure 3 Coffee supply chain structure, core elements, inputs and outputs, indicators and footprint.

4 Environmental and social impacts of coffee production

Approaches such as LCA seek to identify environmental impacts of coffee agroecosystems. Coffee cultivation based on GAP, together with processing and post-harvest management using GPP, must be monitored for possible harmful environmental impacts. The main environmental impacts are as follows:

- GHG emissions from coffee farms, especially intensive, large-scale, shade-free growing systems, including from use of fertilizers and crop protection products as well as energy for irrigation; GHGs include carbon dioxide (CO_2);
- Emissions from other harmful chemicals such as ammonia (NH_3), nitrous oxide (NOx) and sulfur dioxide (SO_2) in coffee production;
- Water consumption and water contamination;
- Coffee waste and ways it is reused and detoxified, e.g. through microbial degradation;
- Toxins resulting from coffee growing, processing, roasting, packaging and end of life disposal, including ochratoxin, acrylamide and formaldehyde; and
- Deforestation caused in coffee areas including negative effects such as: increased GHG emissions, soil degradation, flora and fauna losses, effects on water storage and regulation.

4.1 Greenhouse gas emissions on coffee farms

GHG emissions are analyzed in very different ways using different methods. This means very different assessments of the carbon footprint (in CO_2e) of fresh or green cherry (Birkenberg et al., 2021; Moore, 2020; Giraldi-Díaz et al., 2018; Ratchawat et al., 2017; Hassard et al., 2014; Rahn et al., 2014, 2018; Killian et al., 2013; Noponen et al., 2013). Differences also reflect different growing systems as well as the specific conditions for growing Arabica or Robusta coffee. Measurements typically include the following:

- the amount of CO_2 and other (even more damaging) greenhouse gases such as methane and nitrous oxide;
- direct and indirect soil N_2O emissions, resulting from applications of organic manure, mineral fertilizer (N, P_2O_5) and synthetic crop protection products (pesticides); and
- energy consumption from fuels (e.g. gasoline, diesel) and electricity.

A factor that is sometimes forgotten is the production of organic by-products or waste such as coffee pulp, coffee husk, coffee dry skin, coffee silver skin, parchment and spent coffee grounds (SPG).

The calculations of carbon footprints vary as follows:

- for 1 kg of fresh coffee cherries, they vary between 0.10 kg CO_2e and 0.70 kgCO_2e; and
- for 1 kg of green coffee, they vary between 1.02 kg CO_2e and 5.10 kg CO_2e.

Calculations also vary across different coffee growing systems (Chatterjee et al., 2020; Rikxoort et al., 2014; Evizal et al., 2009; Jende et al., 2007):

- The CO_2 balance for all types of shaded cropping systems result in net positive carbon sequestration, taking into account factors such as litter fall, living soil cover and pruning residues in improving soil organic matter which, together with retention of trees, promotes carbon sequestration; and
- Non-shaded coffee cultivation areas, in contrast, have a negative CO_2 balance.

N_2O emissions represent between 25% and 92% of the total footprint and are caused by over use of mineral fertilizers and organic manure (Noponen et al., 2013). Emissions of methane (which can be caused by phosphorous fertilization or mixed coffee-livestock systems (e.g. involving sheep) are not reported.

4.2 Coffee waste

As noted in the previous section, the annual production 10 million tons of green coffee is accompanied by approximately 50 million tons of coffee pulp and coffee husk, 0.9 million tons of coffee silver skin and 0.7 million tons of SCG. This organic waste is challenging given the high volumes generated in growing areas, presenting a potential contamination issue.

Many farmers lack the technologies and infrastructure (e.g. effective composting techniques) to convert these waste materials into useful by-products such as biofertilizer. However, it is possible to reuse these materials to create value-added products like substrates for biogas, bioethanol, biodiesel or fruit spirit production, composting material, adsorbents for the removal of heavy metals and dyes, substrates for mushroom production, as source of natural phenolic antioxidants or as novel food products. Details about the challenges and opportunities presented by coffee wastes are discussed by Blinová et al. (2017), Cerino-Córdova et al. (2020), Einfalt et al. (2020), Klingel et al. (2020), Blinová and Sirotiak (2019), Torres-Valenzuela et al. (2019), Janissen and Huynh (2018), van Dam and Harmsen (2010).

4.3 Water consumption and wastewater

Coffee production uses a significant amount of water and thus generates a lot of contaminated wastewater. Most Arabica and Robusta coffee is cultivated in areas with sufficient rainfall though irrigation is used, e.g. to secure higher yields. Estimates of the water footprint vary between 3 000 and 20 000 m^3 per ton of coffee cherry (Giraldi-Díaz et al., 2018; Moberg, 2016; Hoekstra et al., 2011; Chapagain and Hoekstra, 2007). It has also been estimated, in the case of semi-wet and wet processing, that each ton of coffee cherry needs between 1 and 20 m^3 of water. After processing, this water is contaminated and needs effective treatment if it is not to create a pollution risk. Characteristics of contaminated water include low pH (2.8-4.4) and the presence of toxic chemicals such as tannins and alkaloids. Wastewater recycling and reuse are of the highest priority in any wet mill (Endris et al., 2020; Ijanu et al., 2020; Brando, 2013; Siu et al., 2007). A standard cup of coffee is estimated to require 140 liters of water in its production.

4.4 Emissions of NH_3, NOx and SO_2 in coffee production

Air pollution from the coffee chain is concentrated in roasting and consumption phases. The roasting process as well burning of coffee silver skin to dry wet parchment coffee can produce airborne pollutants such as particulate matter, volatile organic compounds, organic acids as well as gases such as NOx, CO_2. Other toxic compounds emitted during the roasting process are acrolein, acetaldehyde and formaldehyde. Burning of biomass is the main source of NH_3 emissions, particularly as a result of deforestation for new coffee plantation. SO_2 is emitted by use of biogas and liquefied petroleum gas for coffee drying and roasting.

4.5 Toxins from coffee growing, processing, roasting and consumption

Toxins can occur in all main phases of the coffee supply chain especially in the form of pesticides, Ochratoxin A, packaging materials, chlorogenic acid, caffeine, tannins, acrylamide, formaldehyde or aluminum (from coffee capsules). One of the key roles of GAP and GPP covering stages such as cultivation and roasting is safe use and disposal of hazardous chemicals (GCP, 2021; Leitão, 2019; Janissen and Huynh, 2018).

4.6 Deforestation in coffee areas

Arabica and Robusta coffees are often produced in regions with tropical rainforests. In the last decade, major coffee producers such as Vietnam, Honduras, Ethiopia, Uganda and India have increased their coffee production areas by between 20% and 45 %, often as a result of illegal deforestation (Treanor and Saunders, 2021; Baker, 2014). By destroying trees, deforestation exacerbates global warming processes, destroys ecosystems and biodiversity, exacerbates soil and water erosion, whilst coffee cropping adds to these problems (e.g. burning of firewood for coffee drying) (Gomes et al., 2020; Pohlan et al., 2006). However, coffee growing also provides a potential solution through shade cultivation and as a component in agroforestry, practices which offer a way to grow crops commercially whilst conserving forest resources. Other elements include promoting GAP and GPP, particularly amongst smallholder coffee farmers as well as regulation and certification schemes in coffee consuming countries to support deforestation-free cultivation.

4.7 Assessing social impacts

Social impact assessment is part of numerous studies and certification schemes related particularly to the incomes and well-being of smallholder coffee growers and their families (Pohlan, 2019b; Ibanez and Blackman, 2016; de Mello Bliska et al., 2013; Trimarchi, 2014). To be effective, SAI needs to take account of factors such as the following (Vanclay et al., 2015; Vanclay, 2003):

- characteristics of local populations;
- local community resources and institutional structures; and
- family structures and status, including their culture, health and well-being.

The history of the last 50 years has shown very clearly that projects for social improvement only work with sustained leadership over time and when project managers work in direct and permanent contact with the target group, for example in the case of successful projects in Mexico, Angola, Nicaragua, Bolivia, Guatemala and Vietnam (Bello Baltazar et al., 2019; Folch and Planas, 2019; Pohlan et al., 2019c; Gaitan-Cremaschi et al., 2018; Giovannucci and Ponte, 2005).

Success also depends on participatory approaches which allow local communities to express their views and codesign schemes to help them. This depends on understanding not just the economic aspirations and constraints of smallholder coffee farmers but their broader sociocultural beliefs and traditions. A key objective is to reinforce communal management of land and natural resources both to support local community structures and improve chances of success through mutual support. There are many examples of positive interventions by certification bodies (such as Rainforest Alliance), environmental groups and other NGOs as well national and regional agencies in individual countries. However, there are unresolved issues such as costs of compliance and certification for smallholders well as possible conflicts of interest with a particular certification body (Benítez Kánter et al., 2019; Bray and Jeffrey Neilson, 2017; Markoulakis, 2012).

5 Improving the sustainability of coffee production

Developing coffee production so that it can both grow to meet demand whilst meeting environmental and social needs, particularly for smallholders farming more marginal land, is a huge challenge. This is particularly so given the diversity and complexity of the coffee value chain which means no single standard or approach will fit all contexts.

However, this very diversity means there are also innumerable opportunities to develop carbon neutral coffee production that is economically, environmentally and socially sustainable (Vogt, 2020; Pohlan et al., 2019a; Sebatta et al., 2019; Gaitan-Cremaschi et al., 2018; Nguyen and Sarker, 2018; Rahn et al., 2018). These approaches need to meet a number of objectives, including:

- adapting coffee production so it is more resilient to climate change;
- redesigning coffee production according to agroecological principles so it supports and benefits from ecosystem services (reducing costs and ensuring more sustainable, long-term improvements in yields);
- reducing deforestation and promoting restoration of degraded landscapes, e.g. through agroforestry; and
- creating more circular production systems that minimize energy use, nutrient and other losses as well as recycling and repurposing waste materials.

The benefits of agroecological approaches to coffee cultivation have been widely discussed (Gomes et al., 2020; Piato et al., 2020; Pohlan and Salazar Centeno, 2012; Reijntjes et al., 1992).

The success of these approaches depends first on identifying suitable land which then provides the foundation for environmentally and socially sustainable cultivation (Fig. 4). A key problem for the coffee sector has been smallholders growing coffee on poor-quality, marginal land when coffee prices are higher (with high input costs relative to poor productivity and yields), only to find coffee growing become rapidly unsustainable when prices fall. Since many coffee consumers do not want to pay a sufficient premium for such products, shifting to organic coffee production and certification does not resolve this problem.

As an example, Arabica coffee and Robusta require different agroecological conditions according to elevation and latitude (Fig. 5). Higher market prices for Arabica coffee has meant extending cultivation to lower elevations which reduces quality, yields and increases the risk of diseases such as coffee rust. Some characteristics of coffee friendly landscapes are discussed next in terms of managing carbon, water and nutrients (Pohlan et al., 2006).

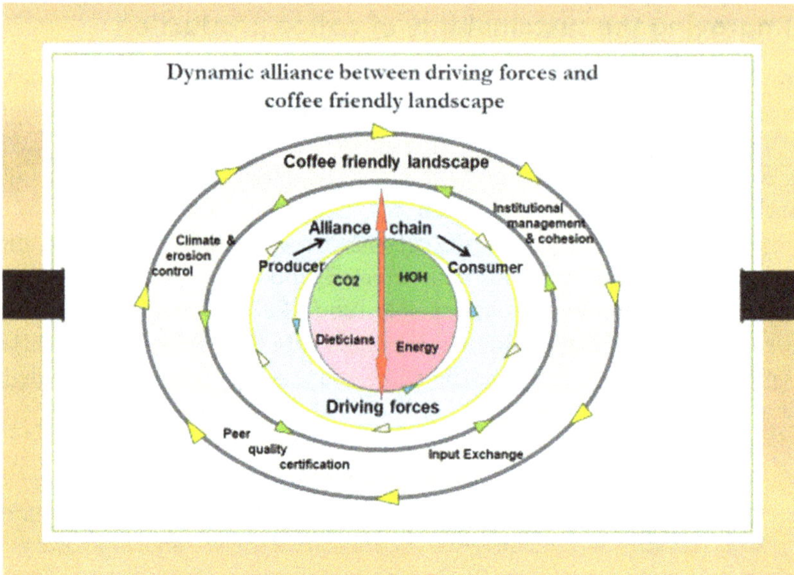

Figure 4 Green Deal Turnover in land use systems.

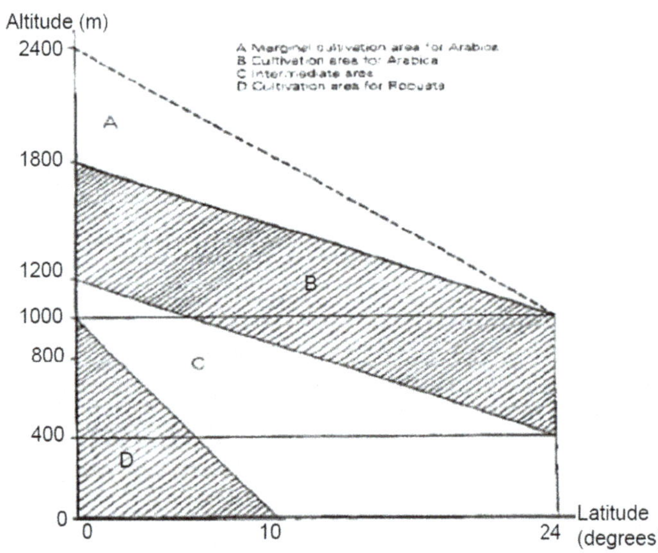

Figure 5 Coffee adaptation according to species, altitude and latitude: A = marginal cultivation area for Arabica; B = cultivation area for Arabica; C = intermediate area; D = (Pochet, P. and Flémal, J. in Raemaekers, 2001).

5.1 Managing carbon

One of the most valuable resources for the farmer is carbon. It is, however, a major challenge to manage this resource efficiently to achieve carbon-neutral cultivation. The size and number of shade trees are the intrinsic part of carbon management. Carbon storage is related to factors such as total tree height multiplied by basal area of the canopy. This can be seen in estimations of carbon stored in biomass, necromass and soil in coffee plantations in Cundinamarca (Colombia), with three different shade levels (low, medium and high) in three municipalities using IPCC recommended measures (Andrade and Zapata, 2019). Soil stored 75% of the total C (93.9–137.7 Mg C ha^{-1} in the upper 30 cm), followed by trees (19%). Carbon increases with a rise in shade (55.8 vs 42.0 vs 23.0 Mg C ha^{-1} for high, medium and low shade, respectively). These coffee plantations fixed a mean of 2.3 Mg C ha^{-1} $year^{-1}$, with a maximum value of 7.1 Mg CO_2 ha^{-1} $year^{-1}$ under a shade of 30–40% considering 3.67 Mg CO_2 equiv./Mg C (Andrade and Zapata, 2019).

Shading effects can best be appreciated over a long period. In a long-term experiment spanning from 1960 until 1982 with Robusta coffee at INERA-Bongabo in the Democratic Republic of Congo, the positive effect of shading (in this case *Croton mubango* at a density of ca. 20 trees/ha) only became apparent from the second decade onwards (Table 1). Owing to the turnover effect of the shading trees, coffee yields were not only more sustainable in the long run but also 8% higher averaged over a 22-year period. Shading can also improve quality. In the Yunnan Province of China, the shading system has been proven to improve cup quality and particularly the aroma and the total dry weight of parchment beans/fresh coffee fruit weight (Chen et al., 2007).

Another way of managing carbon is by combining different planting densities and using different pruning systems over a long period (e.g. 11 years). Multiple-stem pruning can improve yields over the first 2–3 years although the first year is lost in training the multiple stems. Thereafter, single-stem pruning should be employed multiple-stem pruning is invasive, may exhaust the coffee tree and can promote diseases like coffee wilt (Janssens, 1987).

Table 1 Twenty-two yearlong shading experiment with Robusta coffee in Congo-Kinshasa (average yearly yield in kg green beans/ha) (Janssens, 1987)

	Unshaded		Shaded		Mean
Period	kg/ha	%	kg/ha	%	kg/ha
1960–1970	1411	100	1347	95.5	1379
1970–1980	734	100	965	131.5	849.5
1980–1982	305	100	368	120.6	336.5
1960–1982	1003	100	1084	108.1	1043.5

5.2 Managing transpiration

As well as carbon storage, shade cultivation can also create conditions promoting coffee plant growth. These include the following:

- litter fall which promotes soil organic matter;
- providing cooler temperatures which slows ripening and can improve coffee berry quality;
- retaining water;
- providing a wind break effect;
- promoting pollinators and beneficial insects (e.g. pest predators); and
- extending the length of the growing season by providing a stable microclimate.

Shading reduces excessive coffee fruit load and hence improves physical coffee quality (Pohlan et al., 2019a; Kassaye Tolessa et al., 2017; Vaast et al., 2006). Rain in excess of 1500 mm/year will diminish cup quality. Yearly transpiration of shaded coffee is 10% more than coffee in full sun while total transpiration of coffee is 85% higher than that of unshaded coffee, also promoting improved coffee quality (Van Kanten and Vaast, 2005) (Table 2). However, when comparing transpiration on a leaf area base, shaded coffee leaves will transpire less than those in full sun, particularly during the dry season (Vaast et al., 2006).

5.3 Managing coffee plant nutrition

Fertilizers amount to around 50% of the global energy input of a crop. Nitrogen fertilizers alone, account for about half the total fertilizer budget, suggesting the value of organic alternatives both as a source of nutrition and in reducing the overall energy budget. In shaded cultivation, litter fall, e.g. is a major source of plant nutrition. Leaf litter can also help to smother weeds and, in some cases, has a herbicidal effect (e.g. *Grevillea* sp.). Returning processing wastes to the field is part of GAP. Processing wastes can also be enriched with manure from

Table 2 Estimated annual transpiration (mm year^{1}) of coffee in full sun and of coffee plus associated trees in the three agroforestry systems (Van Kanten and Vaast, 2005)

		Agroforestry system with		
Tree variation	Coffee in full sun	*Eucalyptus deglupta*	*Terminalia ivorensis*	*Erythrina poeppigiana*
Coffee	483	689	538	702
Shade tree	-	210	519	195
Total	483	899	1057	897

livestock. The enriched composted material can be turned into liquid humus by mixing with potassium hydroxide (KOH) and water. This liquid humus can then be sprayed in the orchard. Using fertilizer products with insecticidal effects (e.g. KNO_3, NH_4NO_3, $(NH_4)_2SO_4$, wood ashes) can also be beneficial.

Leguminous cover crops promote improved soil health and carbon storage as well as coffee plant nutrition through nitrogen fixation; aromatic cover crops also improve coffee cup quality (Pacheco Bustos et al., 2008). Leguminous cover crops include: *Mucuna, Canavalia, Crotalaria, Desmodium, Stylosanthes, Centrosema, Brachiaria, Calopogonium, Pueraria (invasive), Arachis pintoi* and *Flemingia congesta*. Leguminous shade trees include *Inga, Erythrina, Gliricidia* and *Leucaena leucocephala*. Leguminous fodder trees include *Gliricidia* and *Leucaena leucocephala*.

Cover crops also have a range of other benefits as follows:

- They reduce soil temperatures, maintain soil humidity and improve rainwater infiltration rates;
- Coffee soil temperatures of 35°C+ have been recorded at 50 mm depth in bare soils; under cover crops, temperatures can be 10°C cooler or more; high temperatures cause high mortality of feeder roots and cause organic matter to break down very quickly, reducing its availability; and
- Cover crops with deep roots may be particularly effective in improving soil organic matter.

As dry seasons become longer and/or more intense, it is likely that competition for soil moisture (between cover crop and coffee trees) will become more significant. Covering the cover crop with cut residues (e.g. from pruning) will slow cover crop growth, which is advantageous because this will reduce soil moisture evaporation and also cause root breakdown to improve soil organic matter.

6 Shade management in coffee production

Shade management depends on the specific site conditions, coffee planting density and the vigor of the coffee variety. The use of temporary shading (e.g. with Musa spp. or *Ricinus communis*) and/or permanent shade trees before transplanting young coffee is part of GAP. Shade trees most commonly used in Arabica and Robusta systems are *Acacia abyssinica*, *Albizia* sp., *Afzelia* sp., *Brachystegia* sp., *Cedrela* sp., *Cordia africana*, *Cordia alliodora*, *Croton* sp., *Erythrina* sp., *Ficus* sp., *Gliricidia sepium*, *Grevillea robusta*, *Inga* sp., *Pinus* sp., *Tabebuia* sp., *Terminalia amazonia* and *Solanum multiflorum*. These trees provide different canopy levels in coffee agroecosystems. As noted, they protect understory coffee plants from intense rainfall and sunshine, help maintain soil fertility, reduce the need for weeding and aid in pest control. Organic matter from shade trees also provides a natural mulch of litter, which reduces the amount of inorganic fertilizers required, reduces erosion, contributes important nutrients to the soil and increases the agrobiodiversity (Haggar et al., 2011; Muschler et al., 2006). Valuable timber species are often preferred as shade trees as they provide a further source of income.

Many coffee farmers prefer nitrogen fixing leguminous trees (*Inga, Leucaena leucocephala, Gliricidia, Albizia, Erythrina* sp., *Acacia mangium*, etc.). A minority of farmers will look for valuable wood trees (Cordia, Teak, *Melia azedarach, Grevillea robusta*) or deciduous trees with good litter fall (*Erythrina poeppigiana*).

A particularly important shade tree species is *Eucalyptus deglupta*. Advantages include a small stem basal area (2.79 m^2/ha) which avoids too much shade. *Eucalyptus* also sheds litter including a large portion of bark with slow carbon release and hence with a longer lasting mulch cover as compared to a species like *Terminalia*. Table 3 compares the effects of *Eucalyptus, Terminalia, Erythrina* and full sun cultivation on coffee yields and quality characteristics.

The Columbian coffee board recommends 33% of shade cover, which is a good yardstick in avoiding shade trees becoming dominant. At canopy closure, a Robusta plantation at an average height of 2.5 m would reach a total eco-volume of 10 000m^2 × 2.5 = 25 000 m^3/ha (eco-volume measures species

Table 3 Coffee crop characteristics compared with beverage scores under Eucalyptus, Terminalia, Erythrina or full sun under lowland conditions below 700 masl (after Vaast et al., 2006) *(additional parameters in italic)*

2002			Coffee under		
			Eucalyptus	*Terminalia*	*Erythrina*
Parameters		Full sun	6 × 6 m	6 × 6 m	8 × 8 m
Yield	kg ha^{-1}	700c	1890a	1160b	814c
Height shade tree	M	0	9.7b	9.5b	5.2a
Crown basal area	*m^2/ha*	*0*	*9174b*	*12 232a*	*4212c*
Stem basal area shade	*m^2/ha*	*0*	*2.79c*	*6.16a*	*4.1b*
Stem basal area coffee	m^2ha^{-1}	6.91b	8.16a	9.02a	8.04a
Stem basal area total	*m^2ha^{-1}*	*6.91c*	*10.95b*	*15.18a*	*12.14b*
Large beans	%	56b	67a	67a	62a
Fat	g kg^{-1}	125b	130a	133a	122b
Acidity		2.21b	2.45a	2.41a	2.27b
Bitterness		1.88a	1.65b	1.73b	1.75b
Preference		2.28b	2.80a	2.78a	2.62b

richness in a given area). Covering 33% of the field with 10 m high shade trees will augment the eco-volume by 10 m × 3300m^2 = 33 000m^3/ha, adding up to a total of 58 000m^3/ha. Eventually, it will increase the evaporative cooling leaf area exponentially.

Fine-tuning requires annual monitoring both of the total basal stem area as well as that of canopy basal area. It is good practice to mix shade trees with different growth cycles (e.g. starting with bananas). In principle, leguminous species are good, but nonleguminous trees with abundant litter fall can supply more nitrogen than leguminous trees. Some species like, e.g. *Albizia* sp., *Croton* sp., etc. are sparsely covered with leaves, allowing dimmed light under the shade canopy.

Shade trees require yearly monitoring and maintenance. Pollarding (i.e. severe pruning of shade trees) is a great tool to control shading by vigorous tree species such as *Erythrina poeppigiana* (leguminous species) from once to up to three times a year. Results showed that, by pollarding once a year, 18 470 kg of dry matter per hectare are produced; pollarding twice per year produces an additional 11 800 kg/ha while pollarding three times per year produces a further 7850 kg/ha. The amount of nitrogen removed was approximately 230 kg/ha/year in the case of pollarding once or twice/year and 170 kg/ha/year if pollarding three times/year (Russo and Budowski, 1986).

7 Soil health monitoring and management

Soil is fundamental both for production of food, feed, fuels and fibers and for the delivery of many essential ecosystem services (FAO, 2015a,b). Soil management requires identifying and measuring key physical, chemical and biological properties which collectively determine overall soil health (Astier-Calderón et al., 2002; SQI, 2000). These properties must be easy to measure accurately in field conditions as well as sensitive to variations in climate and soil management (Estrada-Herrera et al., 2017; Navarrete Segueda et al., 2011). It is also essential that coffee farmers develop the skills to analyze the quality of their soils plot by plot.

García Centeno (2017) has validated field methodologies for farmers to use to determine four physical parameters (depth, texture, porosity and infiltration of water in the soil) and two chemical parameter (pH and organic matter). Subsequently, two further chemical parameters were added: cation exchange capacity (CEC) and percentage of base saturation (BS), together with a biological parameter (earthworms per square meter). To evaluate chemical indicators of the soil, it is necessary to make a soil sample in each lot and send them to a laboratory for analysis (e.g. Ca/K, Mg/K and Ca/Mg), as illustrated in Tables 4 and 5.

Table 4 Physical and chemical characteristics of the soils of the Los Mangos agroecosystem, San Carlos, Río San Juan, Nicaragua (Castro Guzmán, 2021)

	Physical indicator	Chemical indicators								Intercationic relationships			Soil liming technique
			pH	P	K	Ca	Mg						
Plot or lot	Depth (cm)	OM (%)	H_2O	ppm	meq g^{-1} de suelo			CEC	BS%	Ca/Mg	Ca/K	Mg/K	t ha^{-1}
Basic grains	27	3.2	6.1	0.0	0.2	10.26	2.63	24.66	53.1	3.9	51.3	13.2	Not necessary
Cocoa	46	3.04	7.14	3.89	0.25	21.61	7.14	28.89	100.4	3.0	86.4	28.6	Not necessary
Improved pasture	39	2.93	6.51	2.69	0.08	21.9	7.57	33.49	88.2	2.9	273.8	94.6	Not necessary
Natural grass	99	3.31	6.16	0.48	0.03	20.42	8.69	37.54	77.6	2.3	680.7	289.7	Not necessary
		Ideal value range of interaction ratios								**2 a 5**	**5 a 25**	**3.5 a 15**	

OM: Organic matter; BS: base saturation.

Table 5 Category of four physical, four chemical and one biological parameter or indicators of soil quality (Díaz Torres, 2019; García Centeno, 2017)

	Soil quality parameters or indicators								
	Physicals				Chemicals				Biological
Categories	Depth (cm)	Texture	Total porosity (%)	Infiltration (cmh^{-1})	pH	CEC	BS%	OM	Earthworms m^{-2}
1	<25	Clayey	>70	<1.95	<5.2	<10	<20	Null	≤16
2	25–50	Sandy	<39	>25	>7.5	10–20	21–35	Come down	16 a 32
3	50–100	Sandy clay loam	51–55	12.1–25	5.3–5.9	21–35	36–45	Half	33 a 64
4	100–150	Silty clay loam	56–69	2–6	6.6–7.4	36–45	46–85	High	65 a 99
5	>150	Loamy	40–50	6.1–12	6.0–6.5	>45	>86	Very high	>99

CEC: Cation exchange capacity; BS: base saturation; OM: organic matter.

The four physical soil indicators are related to how the soil absorbs, retains and provides water to the plants, conditions that limit the growth of the roots, the emergence of the seedlings, the infiltration and the movement of the water within the plant profile and conditions promoting optimal gas exchange. The chemical parameters affect soil-plant relationships (water quality, the buffering capacity of the soil and the availability of water and nutrients for plants and soil microorganisms) (Navarrete Segueda et al., 2011). The only biological indicator is the number of earthworms per square meter since they improve soil structure and function, help with incorporation of organic matter and influence porosity, aeration, infiltration as well as fertility (Ortiz Gamino and Ortiz Ceballos, 2018; García et al., 2012).

Each of the nine parameters has a scale from 1 to 5 where the maximum value (5) represents the ideal value for optimum growth of the crop in each plot. An example of scores for the nine parameters is shown in Fig. 6. This scoring provides a very useful tool that allows the coffee farmer to identify very easily the most limiting parameters or indicators, e.g. availability of phosphorus (P) or potassium (K) (Table 6). This helps in designing and implementing appropriate soil management strategies. This approach ensures balanced crop nutrition, whether using organic or synthetic fertilizers (or a combination of both), taking into account the 4Rs (right source/rate/time/place) (IPNI, 2012).

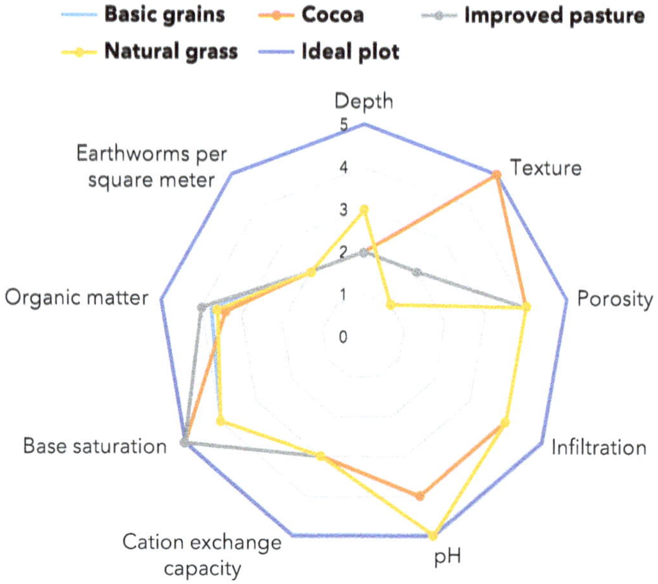

Figure 6 Nine parameters or indicators of soil quality in the Los Mangos agroecosystem, owner Ángel Martínez, Los Chiles, San Carlos, Rio San Juan, Nicaragua, 2017-2018 (Castro Guzmán, 2021).

Table 6 Level of complexity of the designs and management of biodiversity that the farmer or coffee grower has achieved in his agroecosystem (Vázquez Moreno, 2013)

BMC	Degrees of complexity of the agroecosystem
0.1-1.0	Simplified (s)
1.1-2.0	Little complex (lc)
2.1-3.0	Medium complex (mc)
3.1-3.5	Complex (c)
3.6-4.0	Highly complex (hc)

As noted, one useful biological indicator farmers can use to assess soil health are edaphic macroinvertebrate fauna. This group is made up of organisms more than two centimeters long or more than two millimeters in diameter which spend all or part of their life cycle within the soil, on the soil surface or in the organic material covering the soil (Bignell et al., 2012). These invertebrates perform a number of functions as follows:

- Detritivores, involved in the decomposition of organic matter and the recycling of nutrients;
- Herbivores, which feed on living parts of plants;
- Predators that consume diverse invertebrates; and
- Soil engineers influencing the physical, chemical and biological properties of soils.

Among this group are the earthworms (Clitellata), slugs and snails (Gastrópoda), mealybugs (Isopoda), millipedes (Diplopoda), centipedes (Chilopoda), spiders, paton spiders and false scorpions (Arachnida), cockroaches, beetles, earwigs, flies and mosquitoes, bed bugs, leafhoppers, ants, termites, butterflies, caterpillars, crickets and grasshoppers (Insecta). Manuals have been developed so that farmers, with appropriate support, can develop the skills to identify and quantify edaphic macroinvertebrates as a measure of biological soil health (Cabrera-Davila, 2014). Information about these organisms contributes to management strategies to improve soil quality and fertility.

To promote ecosystem function, the flora in coffee plantations should be rich (diverse) and abundant but without prejudice to the productivity of the coffee tree. A rich phytomass, e.g. provides habitat for edaphic invertebrate which, in turn, contribute to ecosystem function. A high amount of biomass produced in the agroecosystem (e.g. leaf litter, crop residues or manure) can be used to increase soil organic matter. Ways to increase organic matter include litter or residues from shade trees, coffee trees and cover crops as well as application of biofertilizers (Potočnik, 2010; Pohlan et al., 2006).

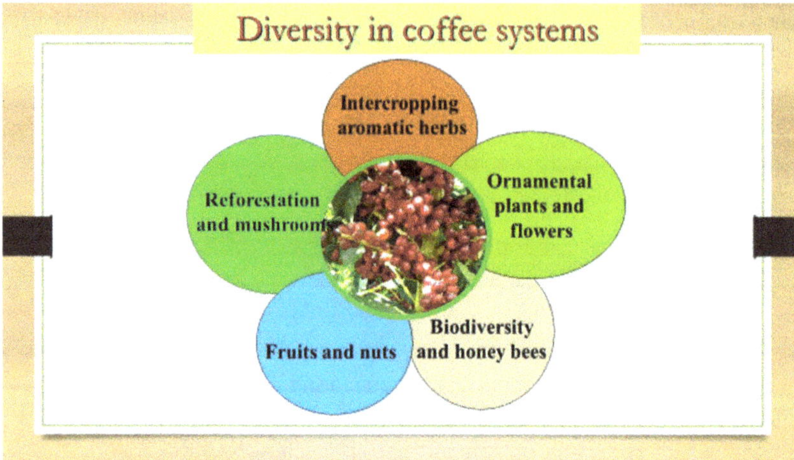

Figure 7 Important elements of the biodiversity flower in coffee growing.

A difficulty for coffee farmers is the long (two to three year) period between planting coffee and obtaining a steady income stream. With intercropping, cover crops can provide income to farmers before coffee harvesting and improve coffee cup quality in the case of intercropping with aromatic plants like oregano (Pacheco Bustos et al., 2008). Smallholders, in particular, intercrop coffee with locally important annual food crops and/or perennial cash crops (e.g. banana, papaya, Rambutan, oil palm, pineapple) to overcome the lack of income in the first years after coffee planting (Pohlan et al., 2006). Intercropping with short period annual crops, such as maize, rice, Phaseolus beans or groundnuts and sweet potato, is also possible after rejuvenation pruning (Fig. 7).

Further diversification options may include production of ornamentals, production of wood trees, production of Macadamia nuts or honey production. In areas such as the Democratic Republic of Congo and Mexico, shade trees can be combined with vanilla winding plants. All these types of diversification will contribute significantly to the species richness and biodiversity of coffee farms. Finally, biodiverse coffee landscapes provide other opportunities such as ecotourism.

8 Managing biodiversity

An essential element in successful agroecosystems is biodiversity. Coffee growers can plan and manage conservation programs to promote biodiversity (Vázquez et al., 2014). There are practical methodologies for rapid characterization of biodiversity such as that proposed by Vázquez Moreno (2013). This includes the following components:

- Design and management of productive biodiversity (DMPrB), i.e. the biota that are introduced or cultivated to promote crop growth;
- Soil management and conservation (SMC), given the importance of soil to the biota that inhabit it or which it supports;
- Water management and conservation, given the importance of this resource to all biological activity in an agroecosystem;
- Management of sanitary interventions in productive areas, which refers to products or other techniques to reduce the incidence of organisms harmful to cultivated plants and farm animals;
- Design and management of auxiliary biodiversity, which refers to creating appropriate habitat to support biodiversity (e.g. shade canopy trees, windbreaks, living fences and ecological corridors to allow movement of fauna); and
- The state of the elements of the associated biodiversity, which is made up of organisms (be they animals, plants and microorganisms) which are associated with cultivated plants and livestock, with both positive or negative interactions, including pollinators and beneficial species as well as pests and disease-causing organisms with different functions in the agroecosystem.

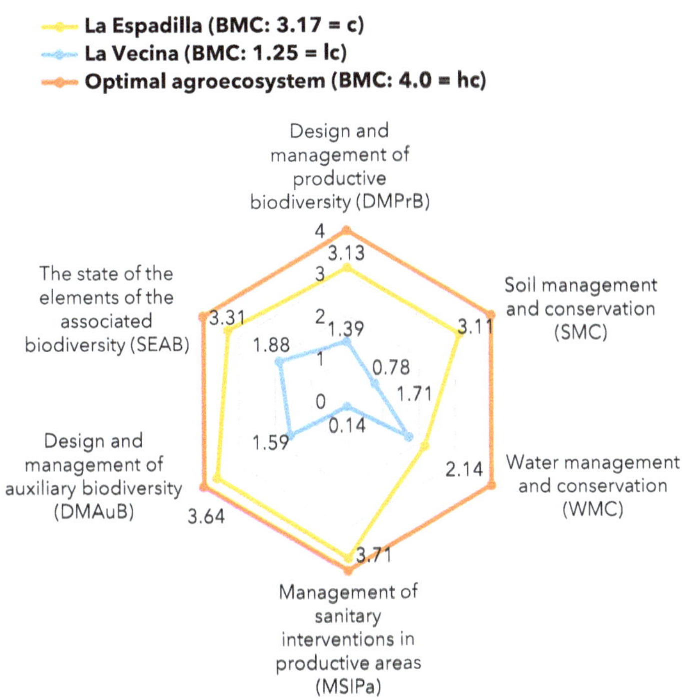

Figure 8 Representation of the application of the methodology of Vázquez Moreno (2013) in two agroforestry agroecosystems with coffee, San Ramón, Matagalpa, Nicaragua, 2017 (Rodríguez González, et al., 2017).

Each component has its indicators and these are categorized on a scale from 1 to 4, where 4 is optimal. The average of these components, which ranges between 0.1 and 4, is known as the biodiversity management coefficient (BMC). This allows the agroecosystem to be classified in terms of its complexity, strengths and weaknesses. This provides the foundation to develop and implement an improvement plan, allowing coffee agroecosystems to evolve over time from less (c) up to highly complex (hc) biotopes. Table 6 illustrates levels of complexity in designs and management of biodiversity while Figure 8 shows the application of this methodology in two agroforestry-based coffee agroecosystems. Figure 9 illustrates examples of the biodiversity possible in shade-grown coffee cultivation. In summary, higher levels of biodiversity in coffee agroecosystem promote more stable, resilient systems which can deliver a range of regulating, provisioning and cultural ecosystem services which are vital for society, the local community and for the producer families.

9 Conclusion

Pressure from governments, NGOs and, in particular, consumers has focused attention on achieving carbon neutral coffee production. The major players in the coffee value chain are now also competing with each other in this space. Ways to achieve this goal include the following:

- improve nutrient management, especially reducing over application of nitrogen-based fertilizers;
- use more natural nutrient sources such as intercropping, shade trees, cover crops, litter fall, rhizobium and mycorrhiza;
- reduce excessive use of pesticides and promote integrated pest management (IPM) programs to make full use of more sustainable cultural, physical and biological means of control;
- improve water use, e.g. using drip irrigation systems;
- develop biodiversity strategies to improve agroecosystem function and resilience; and
- use green energy in post-harvest and processing activities;
- recycle wastes back into the system, e.g. as biofertilizers or process into value-added coproducts.

This will require multidisciplinary knowledge, skills and abilities to create a more resilient, sustainable, climate-smart coffee sector benefitting coffee growers, local communities and society as a whole.

Figure 9 Examples of flora and fauna in coffee plantations. (a) Centennial trees in coffee area of Selva Negra Coffee Estate, Nicaragua 2011 (Courtesy H. A. J. Pohlan). (b) Larva of flannel moth (Lepidoptera: Megalopygidae) in coffee area, Pichanaki, Peru 2012 (Courtesy H. A. J. Pohlan). (c) Beetle of *Oncideres* (Coleoptera: Cerambycidae: Lamiinae) in coffee area, Pichanaki, Peru 2012 (Courtesy H. A. J. Pohlan). (d) Honey bee in coffee flowers, Finca Las Promesas de San Blas, Dipilto, Nicaragua 2021 (Courtesy G. Castillo). (e) Flowering *Tabebuia donnel-smithii* in coffee area, Tapachula 2005 (Courtesy H. A. J. Pohlan). (f) Howling monkey in agroforestry coffee system, El Palacio Estate Jinotega, Nicaragua 2010 (Courtesy J. Solorzano). (g) *Idiarthron subquadratum* (Orthoptera: Tettigoniidae) in coffee area, Chiapas 2005 (Courtesy J. F. Barrera). (h) Swarm of wasps (Hymenoptera: Vespidae) in coffee area, Finca Linda Vista, Chiapas 2008 (Courtesy H. A. J. Pohlan). (i) Litter and parasol mushroom in Argovia coffee plantation, Chiapas 2006 (Courtesy H. A. J. Pohlan).

10 Where to look for further information

The aim of this chapter is to combine both the practical experiences of the coffee farmers as well as the real cropping conditions of the different agro ecological coffee systems with the environmental models of coffee production in so far related to their social impact. The multipurpose coffee growing systems and their development plans pursue more efficient consumption of limited resources. This is based on the four major driving forces of coffee growth (CO_2, HOH, nutrients, energy). Additionally, a very high diversity in the flora and fauna of coffee ecosystems is waiting for much more coordinated implementation by all coffee production units.

Achieving the strategic goals for a successful evaluation and validation of both environmental and social impacts in the coffee production is a challenging task. Indeed, the coffee value chain is extremely diverse and presents innumerable opportunities for managing a carbon neutral coffee production devoid of harmful impacts, be it either energy dissipation or inequity between marketing and coffee producers.

This implies (i) the characterization of climate-smart and socially equitable production; (ii) the presentation of several concepts and ideas for the continuous monitoring of coffee production value chains; (iii) the demonstration of both the life cycle and the carbon footprint of the coffee value chains and last but not least (iv) the important tools enabling the improvement of the turnover for all driving forces behind any Green Deal.

The most important in-time actualized information are given by:

- https://ecochain.com.
- https://verocoffee.ie.
- https://www.allegraceoforum.com.
- https://www.conservation.org.
- https://www.e-education.psu.edu.
- http://www.fao.org.
- https://www.globalcoffeeplatform.org.
- https://www.greenbiz.com.

Relevant guidelines to all these issues are offered by:

- The Global Coffee Platform (GCP): https://www.globalcoffeeplatform.org/.
- The Sustainable Coffee Challenge: https://www.sustaincoffee.org/.

11 References

Andrade, H. J. and Zapata, P. C. (2019). Mitigation of climate change of coffee production systems in Cundinamarca, Colombia, Floresta e Ambiente 26(3), e20180126. doi:10.1590/2179-8087.012618.

Anhar, A., Abubakar, Y., Widayat, H. P., Muslih, A. M., Baihaqi, R. A. and Baihaqi, A. (2021). Altitude, shading, and management intensity effect on Arabica coffee yields in Aceh, Indonesia, Open Agriculture 6(1), 254-262. doi:10.1515/opag-2021-0220.

Astier-Calderón, M., Maass-Moreno, M. and Etchevers-Barra, J. (2002). Derivación de indicadores de calidad de suelos en el contexto de la agricultura sustentable, Agrociencia 36(5), 605-620. Obtenido de. Available at: https://www.redalyc.org/pdf/302/30236511.pdf.

Bager, S. L. and Lambin, E. F. (2020). Sustainability strategies by companies in the global coffee sector, Business Strategy and the Environment 29(8), 3555-3570. doi:10.1002/bse.2596.

Baker, P. S. (2014). Global coffee production and land use change, *Conference Paper*, 4 pp, September 2014. Available at: https://www.researchgate.net/publication/273445289.

Bello Baltazar, E., Soto Pinto, L., Huerta Palacios, G. and Gomez Ruiz, J. (Eds.) (2019). *Caminar el cafetal. Perspectivas socioambientales del café y su gente*. ECOSUR, Juan Pablos Editor, México. 469 pp.

Benítez Kánter, M., Soto Pinto, L., Erin, I. J., Estrada Lugo, E. I. J. and Pat Fernández, L. (2019). Jóvenes y cultivo de café orgánico (C. arabica) en la Sierra Madre de Chiapas: panorama general y retos. In Bello Baltazar, E., Soto Pinto, L., Huerta Palacios, G. and Gomez Ruiz, J. (Eds), *Caminar el cafetal. Perspectivas socioambientales del café y su gente*. ECOSUR, Juan Pablos Editor, México, Capítulo III 17, pp. 253-264.

Bignell, D. E., Constantino, R., Csuzdi, C., Karyanto, A., Konaté, S., Louzada, J. and Zanetti, R. (2012). Macrofauna. In Moreira, F. M., Huising, E. J. and Bignell, D. E. (Eds), *Manual de biología de suelos tropicales: Muestreo y caracterización de la biodiversidad bajo suelo*, Instituto Nacional de Ecología, México, DF, pp. 91-148.

Birkenberg, A., Narjes, M. E., Weinmann, B. and Birner, R. (2021). The potential of carbon neutral labeling to engage coffee consumers in climate change mitigation, Journal of Cleaner Production 278, 123621. doi:10.1016/j.jclepro.2020.123621.

Blinová, L. and Sirotiak, M. (2019). Utilization of spent coffee grounds for removal of hazardous substances from water: a review, Research Papers (Faculty of Materials Science and Technology in Trnava, Slovak University of Technology in Bratislava) 27(44), 145-152, doi:10.2478/rput-2019-0015.

Blinová, L., Sirotiak, M., Bartošová, A. and Soldán, M. (2017). Review: utilization of waste from coffee production, Research Papers (Faculty of Materials Science and Technology in Trnava, Slovak University of Technology in Bratislava) 25(40), 91-101, doi:10.1515/rput-2017-0011.

Bockel, L. and Schiettecatte, L.-S. (2018). Chapter 18: Life cycle analysis and the carbon footprint of coffee value chains. In Lashermes, P. (Ed.), *Achieving* Sustainable Cultivation *of* Coffee, pp. 359-382. doi:10.19103/AS.2017.0022.20.

Brando, C. H. J. (2013). The use of water in processing. Available at: https://scanews.coffee/2013/07/08/the-use-of-water-in-processing-treatment-conservation-and-impacts-on-quality/.

Bray, J. G. and Jeffrey Neilson, J. (2017). Reviewing the impacts of coffee certification programmes on smallholder livelihoods, International Journal of Biodiversity Science, Ecosystem Services and Management 13(1), 216-232. doi:10.1080/21513732.2017.1316520.

Cabrera-Davila, G. (2014). Manual práctico sobre la macrofauna edáfica como indicador biológico de la calidad del suelo, según resultados en Cuba, Fundación Rufford (RSGF, para la Conservación de la Naturaleza), La Habana, 34 pp.

Castro Guzmán, J. N. (2021). Calidad de suelo y macrofauna invertebrada en dos diseños y manejos de la biodiversidad en agroecosistemas con cacao (Theobroma cacao L.), Universidad Nacional, Managua, Nicaragua Agraria, tesis de ingeniería.

Cerino-Córdova, F. J., Dávila-Guzmán, N. E., García León, A. M., Salazar-Rabago, J. J. and Soto-Regalado, E. (2020). Revalorization of Coffee Waste. doi:10.5772/intechopen.92303.

Chapagain, A. K. and Hoekstra, A. Y. (2007). The water footprint of coffee and tea consumption in the Netherlands, Ecological Economics 64(1), 109–118.

Chatterjee, N., Ramachandran Nair, P. K., Nair, V. D., Bhattacharjee, A., de Melo Virginio Filho, E., Muschler, R. G. and Noponen, M. R. A. (2020). Do coffee agroforestry systems always improve soil carbon stocks deeper in the soil?–a case study from Turrialba, Costa Rica, *Forests* 11, 49. doi:10.3390/f11010049.

Chen, L., Pohlan, J., Jihong, L. and Janssens, M. J. J. (2007). Comparison in cup quality of Arabica coffee from Yunnan, China, Deutscher Tropentag 2007, October 9–11, Witzenhausen, International Research on Food Security, Natural Resource Management and Rural Development. Available at: http://www.tropentag.de/2007/abstracts/links/Chen_O9yQOu7l.php.

de Mello Bliska, F. M., Nogueira Turco, P. H., Bliska Júnior, A. and Daniel Carvalho Nepomuceno, D. (2013), Impacts of coffee production in agroforestry system for sustainable development, *Journal of Agricultural Science and Technology B* 3, 535–544. doi:10.1016/j.fcr.2010.10.006.

Díaz Torres, K. R. (2019). Evaluación agroecológica de dos agroecosistemas con cacao (Theobroma cacao L.) en Siuna, Nicaragua, Universidad Nacional Agraria, tesis de maestría, 115. Managua, Nicaragua. Obtenido de. Available at: https://repositorio.una.edu.ni/3870/1/tnf08d542e.pdf.

Einfalt, D., Meissner, K., Kurz, L., Intani, K. and Müller, J. (2020). Fruit spirit production from coffee cherries–process analysis and sensory evaluation, Beverages 6(3), 57, doi:10.3390/beverages6030057.

Endris, S., Kebede, T., Senbet, B. S., Legesse, W., Kassahun, Y., Yakob, T. and Sualeh, A. (2020). Environmental contamination from by-products of wet coffee processing, coffee diversity & knowledge. Proceeding of a National Workshop Four Decades of Coffee Research and Development in Ethiopia, pp. 335–343.

Estrada-Herrera, I. R., Hidalgo-Moreno, C., Guzmán-Plazola, R., Almaraz Suárez, J. J., Navarro-Garza, H. and Etchevers-Barra, J. (2017). Indicadores de calidad de suelo para evaluar su fertilidad, Agrociencia 51, 813–831. Obtenido de. Available at: http://www.scielo.org.mx/pdf/agro/v51n8/1405-3195-agro-51-08-813.pdf.

Evizal, R., Prijambada, I. D., Widada, J. and Widianto, D. (2009). Biomass production of shade-grown coffee agroecosystems, *Proceeding International Seminar on Sustainable Biomass Production and Utilization Challenges and Opportunities (ISOMASS)*, pp. 3–4.

FAO, Organización de las Naciones Unidas para la Alimentación y la Agricultura (2015a). El suelo un recurso no renovable: su conservación es esencial para la seguridad alimentaria y nuestro futuro sostenible. Obtenido de. Available at: http://www.fao.org/3/a-i4373s.pdf.

FAO, Organización de las Naciones Unidas para la Alimentación y la Agricultura (2015b). Suelos y biodiversidad: los suelos albergan una cuarta parte de la biodiversidad de nuestro planeta. Obtenido de. Available at: http://www.fao.org/3/a-av127s.pdf.

Folch, A. and Planas, J. (2019). Cooperation, fair trade, and the development of organic coffee growing in Chiapas (1980–2015), Sustainability 11(2), 357. doi:10.3390/su11020357.

Gaitan-Cremaschi, D., van Evert, F. K., Jansen, D. N., Meuwissen, M. P. M. and Oude Lansink, A. G. J. M. (2018). Assessing the sustainability performance of coffee farms in Vietnam: A social profit inefficiency approach, Sustainability 10(11), 4227. doi:10.3390/su10114227.

García Centeno, L. J. (2017). Manual: metodologías de campo para determinar profundidad, densidad aparente, materia orgánica, infiltración del agua, textura y pH en el suelo. Grupo SEVEN Nicaragua, Managua. Obtenido de. Available at: https://cenida.una.edu.ni/documentos/NP33G216m.pdfe.

García, Y., Ramírez, W. and Sánchez, S. (2012). Indicadores de la calidad de los suelos: una nueva manera de evaluar este recurso, Pastos y Forrajes 35(2), 125–138. Obtenido de. Available at: http://scielo.sld.cu/pdf/pyf/v35n2/pyf01212.pdf.

GCP, Global Coffee Platform (2021). Baseline Coffee Code Revision. Pesticides Lists, 12 pp.

Giovannucci, D. and Ponte, S. (2005). Standards as a new form of social contract? Sustainability initiatives in the coffee industry, Food Policy 30(3), 284–301. doi:10.1016/j.foodpol.2005.05.007.

Giraldi-Díaz, M. R., De Medina-Salas, L., Castillo-González, E. and León-Lira, R. (2018). Environmental impact associated with the supply chain and production of grounding and roasting coffee through life cycle analysis, Sustainability 10(12), 4598. doi:10.3390/su10124598.

Gomes, L. C., Bianchia, F. J. J. A., Cardosoc, I. M., Fernandesc, R. B. A., Fernandes Filhoc, E. I. F. and Schulte, R. P. O. (2020). Agroforestry systems can mitigate the impacts of climate change on coffee production: a spatially explicit assessment in Brazil, Agriculture, Ecosystems and Environment 294, 106858. doi:10.1016/j.agee.2020.106858.

Grabs, J. (2017). The rise of buyer-driven sustainability governance: emerging trends in the global coffee sector. ZenTra Working Papers in Transnational Studies. No. 73 / 2017. Available at: https://ssrn.com/abstract=3015166.

Guinée, J. B., Heijungs, R., Huppes, G., Zamagni, A., Masoni, P., Buonamici, R., Ekvall, T. and Rydberg, T. (2011). Life cycle assessment: past, present, and future, Environmental Science and Technology 45(1), 90–96.

Haggar, J., Barrios, M., Bolaños, M., Merlo, M., Moraga, P., Munguia, R., Ponce, A., Romero, S., Soto, G., Staver, C. and Virginio, E. de M. F. (2011). Coffee agroecosystem performance under full sun, shade, conventional and organic management regimes in Central America, Agroforestry Systems 82(3), 285–301. doi:10.1007/s10457-011-9392-5.

Hassard, H. A., Couch, M. H., Techa-Erawan, T. and McLellan, B. C. (2014). Product carbon footprint and energy analysis of alternative coffee products in Japan, Journal of Cleaner Production 73, 310–321 doi:10.1016/j.jclepro.2014.02.006.

Hoekstra, A. Y., Chapagain, A. K., Aldaya, M. M. and Mekonnen, M. M. (2011). *The Water Footprint Assessment Manual Setting the Global Standard*. First published in 2011 by Earthscan, London and Washington, DC, ISBN: 978-1-84971-279-8 hardback, pp. 228.

Ibanez, M. and Blackman, A. (2016). Is eco-certification a win-win for developing country agriculture? Organic coffee certification in Colombia, World Development 82, 14–27.

IPNI, International Plant Nutrition Institute (2012). 4R Plant Nutrition Manual: A Manual for Improving the Management of Plant Nutrition. Bruulsema, T. W., Fixen, P. E., Sulewski, G. D. (Eds.), International Plant Nutrition Institute, Norcross, GA.

Ijanu, E. M., Kamaruddin, M. A. and Norashiddin, F. A. (2020). Coffee processing wastewater treatment: a critical review on current treatment technologies with a proposed alternative, Applied Water Science 10(1), Article 11.

Janissen, B. and Huynh, T. (2018). Chemical composition and value-adding applications of coffee industry by-products: a review, Resources, Conservation and Recycling 128, 110-117, doi:10.1016/j.resconrec.2017.10.001.

Janssens, M. J. J. (1987). Rapport de la mission effectuée dans les stations de Bongabo et de Boketa du 27.12.86 au 3.1.87. INERA, Kinshasa, DR Congo-Kinshasa.

Jende, O., Pohlan, H. A. J. and Janssens, M. J. J. (2007). Processo de conversão das plantações de café de Chiapas (México) através de sistemas agroflorestais em plantações florestais, e suas implicações ecológicas, Ecossistema 30(2), 3-5.

Killian, B., Rivera, L., Soto, M. and Navichoc, D. (2013). Carbon footprint across the coffee supply chain: the case of Costa Rican coffee, Journal of Agricultural Science and Technology B 3(3), (Serial Number 23), 151-170.

Klingel, T., Kremer, J. I., Gottstein, V., Rajcic de Rezende, T., Schwarz, S. and Lachenmeier, D. W. (2020). A review of coffee by-products including leaf, flower, cherry, husk, silver skin, and spent grounds as novel foods within the European Union, Foods 9(5), 665, published online. doi:10.3390/foods9050665.

Klöpffer, W. (1997). Life cycle assessment: from the beginning to the current state, Environmental Science and Pollution Research International 4(4), 223-228. doi:10.1007/BF02986351.

Läderach, P., Oberthür, T., Cook, S., Estrada Iza, M., Pohlan, J. A., Fisher, M. and Lechuga, R. (2011). Systematic agronomic farm management for improved coffee quality, Field Crops Research 120(3), 321-329.

Leitão, A. L. (2019). Occurrence of ochratoxin A in coffee: threads and solutions–a mini-review, Beverages 5(36), 14 pp. doi:10.3390/beverages5020036.

Markoulakis, A. R. (2012). Corporate Social Responsibility in the Coffee Industry, MSc. Thesis, Modul Vienna University, 107 pp.

Meems, P. (2019). *Global Coffee Data Standard Documentation*. Release 1.0.0, Chapter 1 Partners of this project, 36 pp.

Moberg, E. (2016). The Water Footprint of Coffee Production in Miraflor, Nicaragua. Examensarbete, Department of Earth Sciences, Program for Air, Water and Landscape Sciences, Uppsala, University, Uppsala, 78 pp.

Moguel, P. and Toledo, V. M. (1999). Biodiversity conservation in traditional coffee systems of Mexico, *Conservation Biology*, 13(1), 1-11.

Moore (2020). Available at: https://www.sustainablebusinesstoolkit.com/environmental-impact-coffee-trade/.

Muradian, R. and Pelupessy, W. (2005). Governing the coffee chain: the role of voluntary regulatory systems, World Development 33(12), 2029-2044, doi:10.1016/j.worlddev.2005.06.007.

Murphy, M. and Dowding, T. J. (2019). The Coffee Bean: A Value Chain and Sustainability Initiatives Analysis, University of Connecticut, Stamford CT, 19 pp.

Muschler, R. G., Yépes, C., Ródriguez Camacho, A., Peters Grether, W. and Pohlan, H. A. J. (2006). Capítulo IV 1: Manejo y valoración de la biodiversidad de flora y fauna en cafetales. In Pohlan, J., Soto, L., and Barrera, J. (Eds), *El cafetal del futuro: Realidades y Visiones*, Shaker Verlag Aachen, Germany, pp. 333–360.

Navarrete Segueda, A., Vela Correa, G., López Blanco, J. and Rodríguez Gamiño, M. (2011). Naturaleza y utilidad de los indicadores de calidad del suelo, Contactos 80, 29–37. Obtenido de. Available at: http://www2.izt.uam.mx/newpage/contactos/anterior/n80ne/suelo.pdf.

Nguyen, G. N. T. and Sarker, T. (2018). Sustainable coffee supply chain management: a case study in Buon MeThuot City, Daklak, Vietnam, International Journal of Corporate Social Responsibility 3(1), 1. doi:10.1186/s40991-017-0024-x.

Noponen, M. R. A., Haggar, J. P., Edwards-Jones, G. and Healey, J. R. (2013). Intensification of coffee systems can increase the effectiveness of REDD Mechanisms, Agricultural Systems 119, 1–9.

Ortiz Gamino, D. and Ortiz Ceballos, Á. I. (2018). Belleza extravagante y funcionalidad: lombrices de tierra, Biodiversitas 138, 12–16. Obtenido de. Available at: https://www.researchgate.net/profile/Angel_I_Ortiz-Ceballos/publication/325320585_Belleza_extravagante_y_funcionalidad_Lombrices_de_tierra/links/5b058d54a6fdcc8c25228f24/Belleza-extravagante-y-funcionalidad-Lombrices-de-tierra.pdf.

Pacheco Bustos, A., Pohlan, J. and Schulz, M. (2008). Allelopathic effects of aromatic species intercropped with coffee: investigation of their growth stimulation capacity and potential of caffeine uptake in Puebla, Mexico, Allelopathy Journal 21(1), xx–xx.

Piato, K., Lefort, F., Subía, C., Caicedo, C., Calderón, D., Pico, J. and Norgrove, L. (2020). Effects of shade trees on Robusta coffee growth, yield and quality: a meta-analysis, Agronomy for Sustainable Development 40(6), 38. doi:10.1007/s13593-020-00642-3.

Pochet, P. and Flémal, J. (2001). Coffee. In Raemaekers, R. (Ed.), *Crop Production in Tropical Africa*. DGIC, Directorate General for International Co-operation, Brussels, Belgium, pp. 921–960.

Pohlan, J., Salazar Centeno, D., Janssens, M. J. J. and Torrico Albino, J. C. (2019a). Chapter 2.1: implementation and monitoring of good agricultural practices. In Oberthür, Th., Läderach, P., Pohlan, H. A. J., Lee Tan, P. and Cock, J. H. (Eds), *Specialty Coffee - Managing* Quality, Fully revised 2nd ed., Cropster GmbH, pp. 68–85.

Pohlan, H. A. J. (Ed.) (2019b). Manual de Bom Cafeicultor das terras altas. Diagnóstico, monitoria e auditoria de Boas Práticas Agrícolas (BPA) através da Metodologia do Sistema de Semáforo (SdS) em Cafezais de Angola, Shaker Verlag Düren, Germany, 110 pp.

Pohlan, H. A. J., Salazar Centeno, D. and Torrico Albino, J. C. (2019c). La comunicación y sus medios: experiencias en la formación de recursos humanos en el sector cafetalero. Reflexiones sobre Nicaragua, Bolivia y Angola, Capítulo 29, In Bello Baltazar, E., Soto Pinto, L., Huerta Palacios, G. and Gomez Ruiz, J. (Eds), *Caminar el cafetal. Perspectivas socioambientales del café y su gente*. ECOSUR, Juan Pablos Editor, México, pp. 397–408.

Pohlan, H. A. J. and Salazar Centeno, D. (2012). Diagnóstico, monitoreo y auditoría de buenas prácticas agrícolas a través del sistema de semáforo, Managua, Nicaragua, 63pp. Available at: http://repositorio.una.edu.ni/id/eprint/2423.

Pohlan, H. A. J. and Janssens, M. J. J. (2010). Growth and production of coffee. In Verheye, W. H. (Ed.), *Soils, Plant Growth and Crop Production*, Encyclopedia of Life Support

Systems (EOLSS), Developed under the Auspices of the UNESCO, Eolss Publishers, Oxford, pp. 102–134. Available at: http://www.eolss.net

Pohlan, J., Soto, L. and Barrera, J. (2006). El cafetal del futuro: Realidades y Visiones, Shaker Verlag Aachen, Germany, 462 pp.

Ponte, S. (2004). Standards and Sustainability in the Coffee Sector: A Global Value Chain Approach, International Institute for Sustainable Development, Winnipeg, Manitoba, Canada, 49 pp.

Potočnik, J. (2010). La biodiversidad del suelo: la fábrica de la vida. Unión Europea. Obtenido de. Available at: https://ec.europa.eu/environment/archives/soil/pdf/soil_biodiversity_leaflet_es.pdf.

Quiñones-Ruiz, X. F. (2020). The diverging understandings of quality by coffee chain actors–insights from Colombian producers and Austrian roasters, Sustainability 12(15), 6137. doi:10.3390/su12156137.

Rahn, E., Läderach, P., Baca, M., Cressy, C., Schroth, G., Malin, D., van Rikxoort, H. and Shriver, J. (2014). Climate change adaptation, mitigation and livelihood benefits in coffee production: where are the synergies?, Mitigation and Adaptation Strategies for Global Change 19(8), 1119–1137. doi:10.1007/s11027-013-9467-x.

Rahn, E., Liebig, T., Ghazoul, J., van Asten, P., Läderach, P., Vaast, P., Sarmiento, A., Garcia, C. and Jassonge, L. (2018). Opportunities for sustainable intensification of coffee agro-ecosystems along an altitudinal gradient on Mt. Elgon, Uganda, Agriculture, Ecosystems and Environment 263, 31–40. doi:10.1016/j.agee.2018.04.019.

Ratchawat, T., Chiarakorn, S. and Chidthaisong, A. (2017). Greenhouse gas emissions from roasted and ground coffee productions in Thailand, Journal of Sustainable Energy & Environment 8, 1–3.

Raynolds, L. T., Murray, D. and Heller, A. (2007). Regulating sustainability in the coffee sector: A comparative analysis of third-party environmental and social certification initiatives, Agriculture and Human Values, Springer 24(2), 147–163. doi:10.1007/s10460-006-9047-8.

Reijntjes, C., Haverkort, B. and Waters-Bayer, A. (1992). Farming for the Future: An Introduction to Low-External-Input and Sustainable Agriculture. McMillan Press Ltd., ILEIA; Leusden, Netherlands, 273 pp.

Rikxoort, H., Schroth, G., Läderach, P. and Rodríguez-Sánchez, B. (2014). Carbon footprints and carbon stocks reveal climate-friendly coffee production, Agronomy for Sustainable Development 34(4), 887–897.

Rodríguez González, R. H., González Merlo, H. L., Herrera Moncada, J. H., Vargas Urbina, J. E., Laguna Ramírez, M. J., López Montenegro , G. and Medina Acuña, R. I. (2017). Diagnóstico de los diseños y manejos de la biodiversidad de los agroecosistemas. In Salazar Centeno, D. J., García Centeno, L. J., Rodríguez González, H. R., Calero, C. A., Morales Navarro, A. M. and Valverde Luna, L. O. (Eds), *Evaluación agroecológica de dos agroecosistemas con café (Coffea arabica L.) en San Ramón y dos en Condega, Nicaragua*, Grupo SEVEN Nicaragua, Managua, Nicaragua, pp. 20–32. Obtenido de. Available at: https://cenida.una.edu.ni/documentos/NF08U58ea.pdf.

Russo, R. O. and Budowski, G. (1986). Effect of pollarding frequency on biomass of *Erythrina poeppigiana* as a coffee shade tree, Agroforestry Systems 4(2), 145–162.

Salomone, R. (2003). Life cycle assessment applied to coffee production: investigating environmental impacts to aid decision making for improvements at company level, Journal of Food, Agriculture and Environment 1(2), 295–300.

Sebatta, Ch., Mugisha, J., Bagamba, F., Nuppenau, E. A., Domptail, S. E., Kowalski, B., Hoeher, M., Ijala, A. R. and Karungi, J. (2019). Pathways to sustainable intensification of the coffee-banana agroecosystems in the MT.Elgon region, Cogent Food and Agriculture 5(1), 1611051. doi:10.1080/23311932.2019.1611051.

Siu, Y., Mejia, G., Mejia-Saavedra, J., Pohlan, J. and Sokolov, M. (2007). Heavy metals in wet method coffee processing wastewater in soconusco, Chiapas, Mexico, Bulletin of Environmental Contamination and Toxicology 78(5), 400-404.

Soto-Pinto, L., Perfecto, I., Castillo-Hernandez, J. and Caballero-Nieto, J. (2000). Shade effect on coffee production at the northern Tzeltal zone of the state of Chiapas, Mexico, Agriculture, Ecosystems and Environment 80(1-2), 61-69.

SQI, instituto de calidad de suelo (2000). Guía para la evaluación de la calidad y salud del suelo. Argentina. Obtenido de. Available at: https://www.nrcs.usda.gov/Internet/FSE_DOCUMENTS/stelprdb1044786.pdf.

Tolessa, K., D'heer, J., Duchateau, L. and Boeckx, P. (2017). Influence of growing altitude, shade and harvest period on quality and biochemical composition of Ethiopian specialty coffee, Journal of the Science of Food and Agriculture 97(9), 2849-2857. doi:10.1002/jsfa.8114.

Torres-Valenzuela, L. S., Serna-Jiménez, J. A. and Martínez, K. (2019). Coffee By-Products: Nowadays and Perspectives, doi:10.5772/intechopen.89508.

Treanor, N. B. and Saunders, J. (2021). *Tackling (Illegal) Deforestation in Coffee Supply Chains: What Impact Can Demand-Side Regulations Have?* Forest policy trade and finance initiative, Report, 12 pp. Available at: www.forest-trends.org.

Trimarchi, E. (2014). The Environmental, Social and Economic Impact of Different Certification Programmes of Coffee in Santander, Colombia, MSc. Thesis, Universiteit Utrecht, 92 pp.

Usva, K., Sinkko, T., Silvenius, F., Riipi, I. and Heusala, H. (2020). Carbon and water footprint of coffee consumed in Finland–life cycle assessment, The International Journal of Life Cycle Assessment 25(10), 1976-1990. doi:10.1007/s11367-020-01799-5.

Vaast, P., Bertrand, B., Perriot, J. J., Guyot, B. and Génard, M. (2006). Fruit thinning and shade improve bean characteristics and beverage quality of coffee (Coffea arabica L.) under optimal conditions, Journal of the Science of Food and Agriculture 86(2), 197-204.

Van Dam, J. E. G. and Harmsen, P. (2010). *Coffee* Residues Utilization - *Literature* Survey, Report 1146, Wageningen UR Food & Biobased Research, 75 pp.

Van Kanten, R. and Vaast, P. (2005). Transpiration of Arabica coffee and associated shade tree species in sub-optimal, low-altitude conditions of Costa Rica, Agroforestry Systems 00, 1-16.

Vanclay, F. (2003). International principles for social impact assessment, Impact Assessment and Project Appraisal 21(1), 5-12.

Vanclay, F., Esteves, A. M., Aucamp, I. and Franks, D. (2015). Social Impact Assessment: Guidance for Assessing and Managing the Social Impacts of Projects, International Association for Impact, Fargo ND, 98 pp.

Vázquez Moreno, L. L. (2013). 'Diagnóstico de la complejidad de los diseños y manejos de la biodiversidad en sistemas de producción agropecuaria en transición hacia la sostenibilidad y la resiliencia, Agroecología 8(1), 33-42. Obtenido de. Available at: file:///C:/Users/addis/Downloads/182951-Texto%20del%20art%C3%ADculo-665021-1-10-20130923%20(1).pdf.

Vázquez, L. L., Matienzo, Y. and Griffon, D. (2014). Diagnóstico participativo de la biodiversidad en fincas en transición agroecológica, Fitosanidad 18(3), 151–162. Obtenido de. Available at: https://www.redalyc.org/pdf/2091/209143451003.pdf.

Vogt, M. A. B. (2020). Developing stronger association between market value of coffee and functional biodiversity, Journal of Environmental Management 269, 110777.

Chapter 4

Specialty coffees as drivers of change

H. A. Jürgen Pohlan, University of Bonn, Germany; Dennis José Salazar Centeno, UNA Managua, Nicaragua; Juan Carlos Torrico-Albino, UMSA La Paz, Bolivia; and Marc J. J. Janssens, University of Bonn, Germany

1 Introduction

Between 2019 and 2020, coffee production increased from 10.2 million (m) hectares to 11.1 m and from 7.76 m to 10.04 m tons of green coffee. Around 12.5 million coffee farms and 25 million people depend for their income on growing coffee (Panhuysen and Pierrot, 2020; Sachs et al., 2019; FAO, 2019; Pohlan et al., 2006). In 2019, coffee market was worth US$27.0 billion with an estimated compound annual growth rate of 6.7% from 2019 to 2025.[1] The global coffee market is predicted to reach US$134.25 billion in 2024.[2] The market can be segmented in various ways, e.g. by product type (whole-bean, ground coffee, instant coffee, coffee pods or capsules), distribution channel (on-trade and off-trade) or by region (North America, South America, Europe, Middle East, Africa and Asia-Pacific).[3]

One segment in the coffee market is specialty coffees. This category covers the highest grades of coffee typically with distinctive sensory properties and often from a single estate (https://sca.coffee/research/what-is-specialty-coffee). One

1 https://www.grandviewresearch.com/industry-analysis/coffee-beans-market.
2 https://www.businesswire.com/news/home/20200513005323/en/Global-Coffee-Market-2020-to-2024---Insights-Forecast-with-Potential-Impact-of-COVID-19---ResearchAndMarkets.com.
3 Sample - Global Coffee Market (2020-2025). Mordor Intelligence.pdf.

http://dx.doi.org/10.19103/AS.2021.0096.10

definition is 'a coffee or coffee experience recognized for its distinctive attributes and, because of these attributes, has significant extra value in the marketplace' (Giuliano et al., 2021). The global specialty coffee market is expected to reach US$80 billion by 2024.[4] Specialty coffee has been estimated at 12% of total coffee production although, given the price differential from conventional coffee, the sector is estimated at over 50% of the coffee industry by value.

An accompanying chapter in this book explores the range of issues in both measuring and improving the environmental and social sustainability of coffee production. This chapter discusses the particular role of specialty coffees in improving the sustainability of coffee production (Oberthür et al., 2019).

2 The emergence of specialty coffees

It has been suggested that Erna Knutsen was one of the first to coin the term specialty coffee in a speech in the year 1978 to the delegates of an international coffee conference in Montreuil, France.[5] The concept of specialty coffees can, perhaps, be best understood in terms of different phases or waves in the development of the coffee market (Fischer, 2017; Oberthür et al., 2012; Skeie, 2003). These waves have focused on growing sophistication in consumer attitudes to coffee quality and the increasing differentiation in the market that has resulted. The waves been characterized as follows:

- **1st Wave**: *Growing coffee consumption through mass marketing to increase consumption exponentially (1800 to 1960).*
- **2nd Wave**: *The emergence of specialty coffee to satisfy an emerging group of more discerning coffee consumers (1962 to 2002).*
- **3rd Wave**: *The shift by a growing number of consumers to purchasing coffee based on attributes such as origin and artisan methods of production linked to perceptions of superior quality (starting 2002).*
- **4rd Wave**: *Increasing segmentation based on more refined graduations in defining and measuring quality/taste, together with a greater focus on environmental and social responsibility (beginning 2010).*[6,7]
- **5th Wave**: *New business models for new customer segments, e.g. technically engaged millennials and Gen Z (as from 2018).*[8,9]

Whilst these trends have helped to grow, diversify and add value to the global coffee market, change has not always happened in a coordinated

4 https://www.prnewswire.com/news-releases/.
5 https://thosecoffeepeople.com/what-is-specialty-coffee/.
6 https://www.baristainstitute.com.
7 https://mtpak.coffee/understanding-the-different-wave-of-coffee/.
8 https://www.allegraceoforum.com.
9 https://verocoffee.ie.

way, threatening to erode the concept, distinctiveness and value of specialty coffee and its potential benefits in supporting smallholder coffee growers in particular. This has meant different definitions of specialty coffee around the world (Panhuysen and Pierrot, 2020; Pohlan, 2019; Oberthür et al., 2012[10]). There have been more recent attempts to produce standardized guidelines defining the intrinsic characteristics of 'specialty coffee' (Giuliano et al., 2021).

An important step in clearly defining specialty coffee has been the establishment of specialty coffee associations. One of the most influential has been the Specialty Coffee Association of America (SCAA) founded in 1982. The SCAA has played a leading role in establishing standardized definitions and measurement of quality attributes. Another important step has been consolidation in what had been a fragmented specialty coffee sector. This can be seen in the merger of the SCAA and the Specialty Coffee Association of Europe into a single Specialty Coffee Association (SCA) in January 2017. The SCA now has more than 10 000 active members from around the globe and has established multidisciplinary strategies around the slogan: 'Make Coffee Better', bringing together representatives across the coffee value chain from farming, processing, exporting, importing and roasting to brewing.[11] This new platform for the global coffee industry connects buyers and sellers and is organized using the coffee systems map (which maps the coffee value chain, its processes and partners and the ways they interconnect).[12] Important members include the Brazilian Specialty Coffee Association (BSCA) and the Asociación Cafés Finos of Costa Rica.[13,14]

The SCAA, which has played a leading role in setting quality standards, defines specialty coffee in its green stage as coffee that is:

- free of primary defects;
- includes no quakers (defective beans);
- is properly sized and dried;
- presents in the cup free of faults and taints; and
- has distinctive taste attributes.

In practical terms, this means that the coffee must be able to pass aspect grading and cupping tests which assess different aspects of quality.

In terms of sensory quality, a specialty coffee typically offers a combination of complex flavors.[15] The SCAA has accordingly created a rating system in order to evaluate the quality of flavors in each type of coffee and gives each coffee a quality rating on a 100-point scale. Coffees with cupping scores (measuring

10 https://www.cbi.eu/market-information/coffee/specialty-coffee/market-potential.
11 https://sca.coffee/.
12 https://sca.coffee/coffee-systems-map.
13 http://bsca.com.br/en/cafes-especiais.php, May 21, 2017.
14 https://www.sca.cr/quienes-somos.
15 https://www.maxicoffee.com/blog/en-gb/specialty-coffee/.

sensory quality) of 80 and above are considered specialty coffees. A specific tool for defining and measuring different coffee flavors and aromas has been published by the World Coffee Research Sensory Lexicon (2017). This includes global reference standards for 24 separate flavor attributes: Sour, Bitter, Salty, Apple, Grape, Coconut, Pineapple, Acetic acid, Butyric acid, Isovaleric acid, Fermented, Peapod, Fresh, Papery, Musty/Earthy, Musty/Dusty, Moldy/Damp, Phenolic, Petroleum, Brown Spice, Almond, Vanillin, Floral and Jasmine. This lexicon provides descriptive, quantifiable and replicable definitions:

- allowing the identification of all 110 types of flavors, aromas and texture attributes present in coffee; and
- setting out the corresponding techniques for identifying these attributes objectively and measuring their intensity.

The Coffee Taster's Flavor Wheel[16] shows these sensory attributes in visual form and has become a standard tool both in defining sensory quality and in understanding the effects of coffee genetics/breeding, brewing techniques and roasting methods on coffee flavor (Ahmed et al., 2021; Maspul, 2021; Songer, 2019; Traore et al., 2018).

By creating a clear definition, supported by a quality rating system, SCAA has helped to prevent the term specialty coffee potentially losing its meaning. These developments have also made it possible to expand the range of specialty coffees to include those with less well-known origins, reputations or traditions but which can be carefully selected and processed to obtain a clean cup with distinctive and exceptional attributes.[17] This provides potential opportunities for small-scale coffee growers and processors to develop value-added specialty coffees reflecting unique local conditions and techniques to help build more sustainable livelihoods (Rieser, 2019; Rhinehart, 2009). It also helps those involved in roasting and retailing specialty coffees to develop new niches in growing the market.[18]

In addition, a specialty coffee must have a traceable origin: it must be sourced from a particular area or region with its own identity, meaning that each cup of coffee has a story to tell. Those origins are reflected in the taste. Since these play a key role in the final taste, a specialty coffee must also be traceable in terms of how it is processed and roasted to produce the final cup for a customer to drink. This interest in how a coffee is processed has also become linked to increasing consumer concerns about the environmental and social sustainability of coffee production. The sustainability of production has therefore become part of the story that can help define the distinctive quality of

16 https://sca.coffee/research/coffee-tasters-flavor-wheel.
17 https://elpurica.com/what-is-so-special-about-specialty-coffee-2/.
18 https://www.cbi.eu/market-information/coffee/specialty-coffee/market-potential.

a specialty coffee. The definition of specialty coffees can now also encompass being certified within different sustainability standards such as organic, Fair Trade, Rainforest Alliance and others (Rieser, 2019).

Types of origin associated with specialty coffees include the following:

- *Single origin*: associated with high quality and uniqueness from a certain region or country. Recognized examples of single-origin coffees are Jamaican Blue Mountain, Hawaii Kona, Kenya AA, Antigua Guatemala, Marcala Honduras, Sidamo Ethiopia, Kopi Luwak Java (Indonesia) and Toraja Indonesia.
- *Single farm or estate*: these coffees from a particular farm provide detailed information about the farm history and how and by whom the coffee was grown, harvested and processed. Famous examples are Kifaru Coffee from Tanzania, Hacienda La Esmeralda from Panama, Argote of Colón Génova from Colombia.
- *Micro lots and nano lots:* these products from small coffee areas which produce extremely high quality or exclusive coffee beans in very small volume between 5 bags (nano lot) and 10–75 bags (micro lot), often selling at very high prices.

Figure 1 illustrates how these various sensory quality attributes can be combined into a sensory profile, including qualitative standards covering, e.g. aroma, taste and flavor as well as quantitative standards that place the intensity of the qualitative assessment on a rating scale (Songer, 2019).

3 Markets and pricing for specialty coffees

There are two main global commodity markets for coffee beans:

- the market for Arabica coffee trade in New York; and
- the market for Robusta in London.

The prices are set by in three main ways (ICO, 2020):

- The *International spot market* with which provides indicator prices for representative types of coffee: Colombian Milds, Other Milds, Brazilian Naturals and Robustas;
- *National spot market prices* referring to farm-gate prices paid to growers, reported in local currency by International Coffee Organization (ICO) Member countries; and
- *Futures market prices* with quotations from the New York (Arabica) and London (Robusta) futures exchanges.

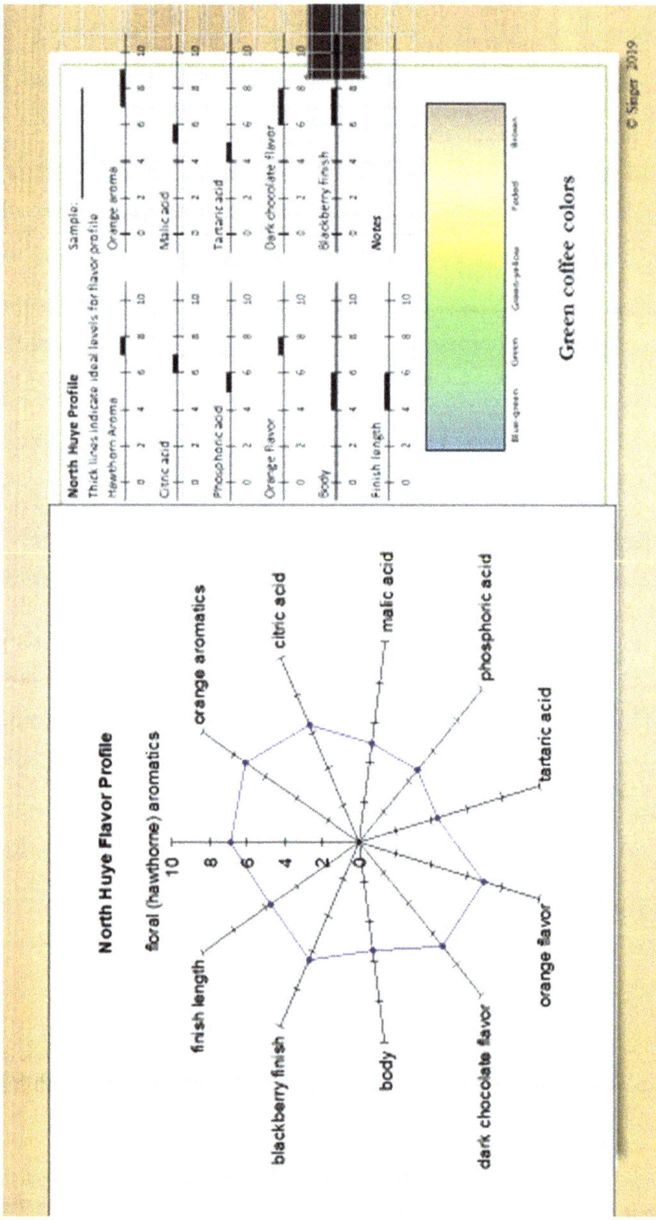

Figure 1 Flavor profile for a regional Rwandan coffee (North Huye district).

Besides these markets for conventional coffee, specialty coffee markets have developed their own market mechanisms. Dedicated retailers, including specialty coffee shops and baristas, often work directly with specialty coffee farmers. Pricing of specialty coffee depends on a mix of intrinsic qualities (cupping score, physical appearances, size/grade, roast color, descriptive profile) and extrinsic attributes (origin, certification, name of farm, brand) (Giuliano et al., 2021; Lundy et al., 2019). Pricing of specialty coffees takes into account grading and classification attributes such as[19]:

- Altitude and/or region;
- Botanical variety;
- Bean size (screen size), sometimes also bean shape and color;
- Density of beans;
- Number of defects (imperfections);
- Preparation (wet or dry process = washed or natural); and
- Roast appearance and cup quality (flavor, characteristics, cleanliness).

This can result in prices many times higher than for conventional coffees.

Traditionally, most specialty coffees are Arabica, ranging from traditional varieties such as Tipica and Bourbon as well as new cultivars with high quality potential such as Geisha, Java or SL. However, there is increasing interest in high-quality Robusta specialty coffees with the development of standards to define high-quality Robustas (e.g. Q grading schemes and training). An example of a specialty Robusta is 'Petit Kwilu' from the Mayumbe forest of the Atlantic coast of Guinea in Africa, which has been promoted as a single origin specialty coffee or blended with Arabica.

The coffee market uses a classification systems based on detailed quality criteria which affects the price paid for each coffee. There are five classes of coffee[20]:

- Class 1 Specialty Coffee - 0 to 5 defects;
- Class 2 Premium Grade - 6 to 8 defects;
- Class 3 Exchange Grade - 9 to 23 defects;
- Class 4 Below Standard Grade to 24 to 86 defects; and
- Class 5 Off Grade - More than 86 defects.

Classes 1 and 2 can command price premiums whereas Class 4 and 5 coffees are discounted.

19 https://www.intracen.org/coffee-guide/world-coffee-trade/grading-and-classification/
20 http://www.coffeeresearch.org/market/coffeemarket.htm

The most important green coffee traders globally are: The Neumann Kaffee Gruppe (NKG), Ecom Agroindustrial, Olam, Volcafe Ltd./ED&F Man, Louis Dreyfus Company (LDC), Sucafina and Mercon Coffee. These worldwide players are also involved in the market for specialty coffees.[21] There is also a group of independent specialty traders and importers focused only on the high-quality coffees (e.g. cupping scores of 84+) whether classified as high commercial, specialty grade or micro lot coffee.[22] The most famous are Belco, Falcon Specialty, Red Fox Coffee Merchants, Nordic Approach, Tropiq, Mare Terra, CCS and Algrano.

The major global roasters and retailers representing major brands in the coffee market are (Panhuysen and Pierrot, 2020): Nestlé SA (907 000 t), JDE Peet's (730 000 t), J.M. Smucker Company (360 000 t), Starbucks (310 000 t), Strauss (282 000 t), Lavazza (267 000 t), Melitta (195 000 t), UCC (190 000 t) and Tchibo (180 000 t). Many of these companies support voluntary sustainable standards (VSS) certification. The most successful and recognized brands worldwide are Nescafe, Keurig, Nespresso, Jacobs, Folgers, Lavazza, Tchibo, Carte Noire, Starbucks and Maxwell House.

The specialty market is continuing to grow and diversity, influenced by the inventiveness of specialty coffee farmers and producers, the creativity of roasters and retailers in educating consumers, and the growing interest of consumers understanding the story behind their favorite coffee (CAFEMALIST, 2021; Madaro, 2018;).[23,24] Recent trends in the coffee market include the following (McCauley, 2020):

- Interest in regional coffee flavors (e.g. Turkish coffee, Vietnamese coffee)
- A growing range of techniques to make coffee of differing quality at home (instant coffee, canned coffee (RTD) and recyclable coffee pods from table-top coffee makers);
- Interest in 'healthy' coffee chain and substitutes (matcha, golden milk latte, brewed cacao) and healthy coffee additives (mushroom, turmeric, collagen creamer); and
- The market for coffee subscription boxes for roasted or unroasted coffee beans and cold brew coffee.

Together with broader innovations (e.g. in traceability), these developments are making it easier for coffee consumers to access and enjoy specialty coffees. There is evidence of an increasing interest by consumers in moving from established

21 https://commoditytrading.guru/commodities/who-moves-the-coffee-markets-meet-the-worlds-largest-green-coffee-traders/.
22 https://commoditytrading.guru/commodities/agricultural-products/coffee/best-specialty-coffee-traders-importers/.
23 https://www.cbi.eu/market-information/coffee/trends.
24 https://verocoffee.ie.

coffee types and brands to experiment with new types and flavors, linked to broader issues such as health, environmental awareness and social responsibility.

The overall market for coffee has continued to expand. Established markets (e.g. the United States, Japan and the European Union) are being complemented by emerging markets (Africa, Central and Eastern Europe, the Middle East and East and South Asia) with the development of a more affluent urban middle class and a 'coffee culture' which may previously have not existed (ICO, 2020; Canet Brenes and Soto Víquez, 2016).

A particularly important driver of growth and diversification is young coffee drinkers (IICA, 2019). This is particularly the case with the Millennial generation, those born between 1980 and 2000 (Samoggia et al., 2020; Marca and Ríos, 2018; Silva Monteiro Boaventura et al., 2018). This generation is hyper-connected digital natives (Ferrer, 2018). In general, millennials are health conscious, interested in other cultures and in issues such sustainability, as well in more naturally produced, authentic and high-quality products and services (Ferris, 2020; Grant, 2020).

There are three main categories of specialty coffee:

- Specialty coffees defined by cupping standards for aroma/flavor;
- Specialty coffees from particular regions certified by accredited certifying agencies in the relevant country of origin; and
- Specialty coffees defined by other attributes defined by the international Trade-Related Aspects of Intellectual Property Rights (TRIPS) Agreement which defines and protects different types of intellectual property.

Many producers or producer groups certify their products under more than one type of standard with the Rainforest Alliance (now including Utz) one of the most popular (Rainforest Alliance, 2020). Another important area is certification of organic coffee. Global organic standards are set by the International Federation of Organic Agriculture Movements (IFOAM), taking into account relevant regulations from the main importing markets (USA, Japan and the European Union) and the respective legislation in each exporting country. Key national organic regulations are USDA Organic, JAS (Japanese Agricultural Standard) and EU Organic (European Union). Internationally recognized organic certification bodies include California Certified Organic Farmers (CCOF), OCIA International, Naturland, Eco-LOGICA, Kiwa BCS Öko Garantie, Demeter or BIO LATINA SCA (see Table 1).

The VSS system is very important as a key element of corporate social responsibility (CSR) strategies across the coffee industry (Panhuysen and Pierrot, 2020). VSS differ widely in certification criteria and audit methodologies. Generally, the aim is to offer guidelines for producing and selling coffee variously identified as 'sustainable', 'responsible', 'ethical', etc. (ICO, 2020). Table 1 summarizes some of the key standards used in the coffee sector.

Table 1 Certification organizations of main attributes, intrinsic organoleptic attributes and positive externalities for the environment producers and consumers (ICO report 2020; Panhuysen and Pierrot, 2020; special organizations and recent trends)

Labels/organization and the private sector standards	Program/labels	Objectives/effects/description	Benefits/criteria benefits
4C Certification System climate-friendly coffee	Common Code for the Coffee Community (CCCC or 4C). Founded in 2006 https://www.4c-services.org/	Promote the sustainability of the main bulk of the green chain of coffee or mainstream coffee and increase the quantities of coffee provide independent, credible and innovative solutions for sustainable supply chains	Coffee plus environmental, social and economic benefits, internationally recognized sustainability standard for the entire coffee sector
Fairtrade	Fairtrade Labeling Organizations International (FLO) https://www.fairtrade-deutschland.de/	Ethical certification system with a social focus	Strategy for poverty reduction and sustainable development
Organic coffee	International certification agencies (CCOF, OCIA, Naturland, Eco-LOGICA, Kiwa BCS Öko Garantie, Demeter, etc.) and regional certification agencies (BIO LATINA SCA, etc.)	Involving the verification and certification of coffee production according to a certain set of environmental and social standards	Allow producers to gain premium prices by providing a certain guarantee to concerned consumers regarding the sustainability of the production
Rainforest Alliance	Rainforest Alliance certification program is managed by the Rainforest Alliance and the Sustainable Agriculture Network (ITC, 2011) https://www.rainforest-alliance.org/	The program focus on sustainable agriculture, crop productivity and profitability, and natural resources and ecosystem services	Agroecosystems and community groups mitigate and adapt to climate change and increase their resilience by implementing sustainable practices and diversifying
UTZ Certified (now part of Rainforest Alliance)	This program fully incorporates the GLOBAL G.A.P. standard (ITC, 2011) https://utz.org/	UTZ Certified has developed two standards: the Code of Conduct and the Chain of Custody. Producers, producer groups and Supply Chain Actors (SCA)	These outcomes include climate-smart agriculture and food security goals
Starbucks' C.A.F.E. Practices	The Practices C.A.F.E. and Nespresso ecolaboration by Starbucks is an initiative that seeks to lead coffee producers, processors and suppliers toward more sustainable production methods https://www.starbucks.com/	The program focuses on product quality, financial responsibility, social responsibility and environmental leadership	

Nespresso's AAA Guidelines	In 2003, Nespresso launched the 'Nespresso AAA Sustainable Quality Program' https://www.nespresso.com/	Ensure coffee of the highest quality and at the same time promote environmental, social and economic sustainability throughout the entire value chain, from the farmer to the consumer	
Bird-friendly coffee	SMBC "Bird Friendly" certification program belongs to the Smithsonian Migratory Bird Center started in the early 1990s https://global.si.edu/	Promotes the cultivation of coffee under shade that is economically, environmentally and socioculturally viable	Protect structural biodiversity, integral part of agroforestry systems, contribute to the conservation and sustainable use of natural resources
New trends and their niches			
Café Femenino and gender friendly coffee	https://www.globalcoffeeplatform.org/latest/2018/is-your-coffee-value-chain-gender-friendly-why-does-it-matter/	Empower women in the international coffee community to achieve meaningful and sustainable lives	Promote women's rights through women's associations and guarantee gender equity
Low-carbon coffee	NAMA Support Project: Low-Carbon Coffee Costa Rica; initiated 2015 https://www.nama-facility.org/	Introduce and use climate-friendly cultivation and processing methods	Priority is reducing emissions in the coffee industry, from cultivation through to processing
Zero Carbon Coffee	https://www.zerocarboncoffee.com	Do Good. Drink the Best	
Whale-friendly coffee	https://whitewhalecoffee.com.au	White *Whale Coffee* is a specialty *coffee* roaster, espresso bar and barista school in Cairns, Queensland	Supply cafes, businesses and homes with delicious effects
Bee-friendly coffee Eco-friendly coffee	*Bee Friendly Coffee* (@bfriendlycoffee) IsaBee *Coffees*, directly from the Fazenda Paiol, Serra das Abelhas, Minas Gerais, Brasil https://ecofriendlycoffee.org	Improve coffee quality, restoring bee habitat and improving livelihoods of coffee farmers; understand the significant role played by honey bees and other insect pollinators	Enhancing crop production and conservation of biological diversity
Coffees transformed by animal species	Kopi Luwak, Java by civics https://www.dejavacoffee.com Black Ivory, Thailand by elephants Monkey Coffee, India by rhesus monkeys Bat Crop Coffee, Caribbean by bats https://kaffeeklingler.de	Ripe coffee cherries are eaten, digested and defecated for these animals and then by farmers collected and processed	Exotic and scurrile coffees with a very high price level, quick additional income for coffee farmers

4 Geographical indicators for specialty coffees

Indicators of geographical origin for specialty coffees (e.g. indicating it is from a particular geographical area) can be a key way of differentiating a product and giving it a premium price (Canet Brenes and García González, 2010). They provide a unique opportunity for coffee farmers and coffee cooperatives in a particular region to improve their livelihoods and can be a significant boost to regional development (Ram, 2017; Hindorf et al., 2015; Pohlan and Salazar Centeno, 2012).

Such indicators are governed by TRIPS Agreement related to the Design and Geographical Indications Law Section of the World Intellectual Property Organization (WIPO). The use of geographical indicators is challenging, partly because it needs robust traceability of a product back to a particular location, and partly because all those growing and processing coffee in the defined area must act collectively in relation to production methods, quality and control standards (e.g. GAP and GPP), final product quality as well as in the distribution and commercialization of products (OMPI, 2017). Geographical origins are often associated with particular types and quality of flavor and/or methods of cultivation or processing associated with that region. All those involved in the coffee in that region must work together to maintain this distinctive regional identity and attributes.

Famous coffees from a particular location include Café Veracruz from Mexico, Indian Monsooned Malabar Arabica and Monsooned Malabar Robusta, Buon Ma Thout Coffee from Vietnam, Tenom Coffee from Malaysia, the Sulawesi Coffees (from Indonesia) Toraja, Kalosi, Mamasa and Gowa, as well as Mandheling, Lintong and Gayo also from Indonesia.

The National Federation of Coffee Growers of Colombia has successfully developed the denominations of origin of Café de Colombia, Café Santander, Café Cuaca, Café Nariño, Café Huila, Café del Norte de Colombia and Café Tolima (Trujillo et al., 2021; Oberthür et al., 2011). These coffees are marketed under a protected designation of origin or protected geographical indication in national and international markets (National Federation of Coffee Growers of Colombia, 2014).

Other examples in Latin America and the Caribbean of specific regional or national designations of origin can be found in Mexico (Café de Veracruz, Café de Chiapas, Café Pluma Oaxaca); Guatemala (Café Antigua and Café Acetenango); Honduras (Café Marcalá and Cafe del Occidente Hondureño); El Salvador (Café Apeneca-Ilamatepec, Café Cacahuatique, Café Alotepec, Café Balsamo-Quezaltepec, Café Tecapa-Chinameca and Café Chintontepec); Nicaragua (Café Dipilteño); Costa Rica (Café Tarrazú and Café de Costa Rica); Panama (Café Renacimiento and Café de Palmira). The Dominican Republic has obtained the geographical designation of origin for Café de Valdesia. Peru has registered the geographical indication of Café Villa Rica and Café Machu Picchu-Huadquiña (Ayala Durán and Radomsky, 2020; IICA, 2019; Escobedo Paredes, 2013; Campos Azofeifa, 2011; Pérez Akaki, 2011).

5 Good agricultural practices and good processing practices in specialty coffee production

Factors that influence coffee cup quality include genotype, environmental conditions, agronomic, harvest and post-harvest practices as well method of coffee preparation (Pohlan et al., 2019; Läderach, et al., 2011). The coffee bean is the result of a complex sequence of biochemical, physiological and morphological events that determine the growth, flowering and fruiting of the coffee plant. These events are influenced by eco-physiological factors such as temperature, radiation, soil depth and composition, availability of water, carbon-to-nitrogen ratio, crop load and the multiple interactions between these factors.

An understanding of these interactions and how to use them to optimize yield and quality underlies the development and use of good agricultural practices (GAP). GAP, managed using a traffic light system (TLS), have been developed with and for smallholder coffee growers (Pohlan and Salazar, 2017). This GAP system is based on ensuring effective control of 13 key issues/ cultivation stages (from site selection and planting through nutrition, weed and pest management to harvesting) using 50 control points of which 11 are classified as red (very high risk of noncompliance with recommended practice if not correctly implemented, and requiring immediate improvement where not properly implemented), 15 as yellow (high risk of noncompliance) and 24 as green (low risk of noncompliance but may result in economic loss if not correctly implemented) (Fig. 2).

The TLS not only steers the management control points, but also allows information to be collected on whether growers execute the recommended actions and on the quality of the coffee produced. The TLS is thus an excellent system for disseminating recommended practices and validating their effectiveness in a continuous improvement process. The TLS is a virtuous circle in which feedback on its use and recognition by stakeholders of its effectiveness refines the system in identifying those practices which are optimal for producing high-quality coffee.

Based on reviewing the literature on environmental and agronomic factors that affect coffee quality, we have subjectively ranked the effects of GAP within the TLS in relation to coffee quality rather than productivity/ yield (Table 2). For each key practice, the impact on coffee quality is rated between four (very high) and 1 (low). In the case of control points for export requirements, red (3) means that the coffee is not exportable, yellow (2)

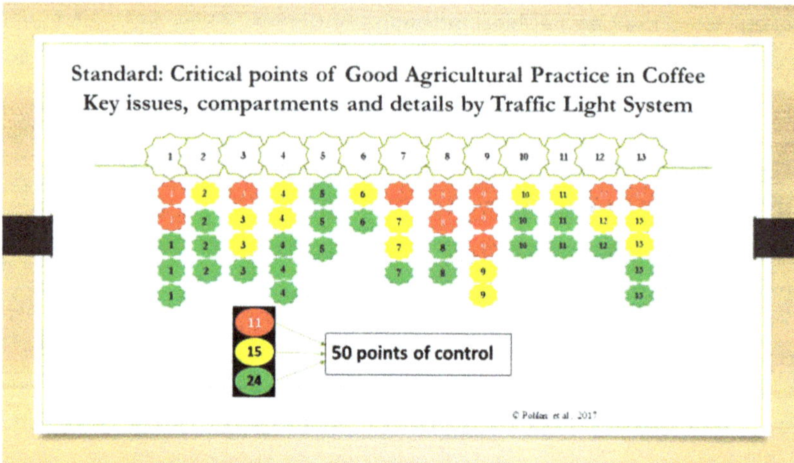

Figure 2 Key issues and components and their GAP risk levels for coffee. From initial site selection (key issue 1) to harvest and post-harvest activities (key issue 12) and correct management of CSR (key issue 13).

means that changes are required and green (1) that the coffee meets export requirements. The relationship between management practices and yield and quality under different conditions (based on controlled experiments and the observations from farmers' fields) can be used to develop a list of recommended actions that can be shared with farmers. Using systems like Cropster and CropCheck (Wiesinger and Oberthür, 2019; Pohlan et al., 2017), farmers submit information about their production activities to a centralized database where they can be analyzed and evaluated and hence provide information on the effectiveness of different management decisions in specific environments.

Two major challenges for GAP and the TLS are as follows:

- to adapt practices and management strategies to obtain maintain yields and quality in areas where coffee is currently grown but where future climatic conditions may become less favorable; and
- to develop practices for new coffee growing areas that will become similar to current coffee growing areas as a result of climate change, but which do not have a traditional coffee culture to build on.

The GAP and TLS approaches consider the spatial and temporal diversity of coffee-growing areas and are thus well placed to meet these two major challenges.

As important as GAP in cultivating coffee are good processing practices (GPP) in ensuring post-harvest operations can turn high-quality raw product

Table 2 Good agricultural practices indicators and its impact on coffee quality, weighting system and most important control points for export.

GAP indicator/key issue	Impact on coffee quality[a]	Weighting system by TLS[b]			Control point for export standards[c]
		Red	Yellow	Green	
1. Site selection, plot history, mapping, data recording by plot and environmental impacts	4	1	5	10	3
2. Seed origin, propagation and nursery practices	4	1	5	10	2
3. Selection of highest quality plants from nurseries	2	1	5	10	3
4. Land clearing and soil conservation	2	1	5	10	2
5. Shade trees, planting methods and planting period	1	1	5	10	1
6. Shade management and plant and fruit density control	3	1	5	10	2
7. Weed management	1	1	5	10	3
8. Nutrition management	2	1	5	10	3
9. Integrated pest and disease management	2	1	5	10	3
10. Irrigation, intercropping and diversification	1	1	5	10	2
11. Coffee pruning	1	1	5	10	2
12. Harvest management	4	1	5	10	3
13. Social and human rights, labor conditions, CSR and compliance, education/training	3	1	5	10	3
Maximal GAP points				**130**	
Need strong/urgent improvement				*13-55*	
Need medium improvement				*56-80*	
Need minimum improvement				*81-100*	
Optimum				*101-130*	

[a] 4 very high, 3 high, 2 medium, 1 low.
[b] TLS for field control.
[c] (3) Determinant for export, if red is not possible to export. (2) Yellow need urgent change to green. (1) Normal control point.

into an equally high-quality cup of coffee. There exists an extensive literature on post-harvest management of coffee (Haile and Kang, 2019; Subedi, 2010; Wintgens, 2009; Rothfos, 1980; Coste, 1969; Haarer, 1969[25]). There are three main traditional methods of coffee processing:

- Dry or natural processing;
- Semi-dry or honey processing; and
- Wet or washed processing.

A key quality requirement is only harvesting fully ripe coffee cherries. With mass production of conventional coffee, little attention was paid to the maturity stages of the coffee cherries at harvest, resulting in poorer quality. Specialty coffee growers have returned to this traditional, skilled but labor-intensive of selective harvesting as a key element of GAP and driver of quality.

The last three decades has seen significant research into the different biotic and abiotic factors contributing to the intrinsic and extrinsic attributes of coffee quality (Giuliano et al., 2021; Montagnon et al., 2019; da Matta and Martins, 2019; Boot, 2019). This has led to the development of new techniques such as anaerobic coffee processing[26,27] as well as new roasting processes to improve the coffee value chain (Ahmed et al., 2021; Bazame et al., 2021; Craparo et al., 2021; Assefa and Kufo-Obso, 2019; Selvamurugan et al., 2010). Very important in these developments in the case of specialty coffee have been direct relationships between coffee farmers and processors with those at the end of the coffee value chain such as retailers and baristas preparing and serving coffee to customers. This has made it possible to better understand and meet customer needs for quality and variety of their coffee-drinking experience.

6 Conclusion

Specialty coffees have evolved from a niche product cultivated by a few, poor smallholders to a globally recognized brand category. They offer a potential avenue for coffee farmers to bridge the often unstainable gap between farm gate and retail prices. To be sustainable, specialty coffee needs to be produced by farms with high economic efficiency, good yields and excellent cup quality. This requires close attention to GAP which need to continue to develop to reflect the latest research linking quality and yield to agronomic practice. GAP also need to evolve to meet the challenge of climate change, reducing

25 https://perfectdailygrind.com/2016/07/washed-natural-honey-coffee-processing-101/.
26 https://www.backyardbeans.com/blog/2020/8/21/anaerobicprocessing.
27 https://www.cafeimports.com/europe/blog/2019/09/10/whats-anaerobic-fermentation-and-why-is-it-so-popular-all-of-a-sudden/.

environmentally damaging practices whilst developing more sustainable, ecologically based alternatives.

As the scope of specialty coffee has expanded to include aspects of environmental and social sustainability, there is an urgent need to consolidate the innumerable standards and types of certification. A key aspect is to allow more independent local certification rather than relying on expensive and cumbersome international schemes. An additional requirement is to develop robust but practical traceability systems that can track the journey of a specialty coffee from a particular plot to the cup a consumer drinks. This requires digital technologies to manage all the steps of the supply chain in a transparent and traceable way.

7 Where to look for further information

As a standard and up-to-date information compendium for 'specialty coffee' production, we recommend the *Specialty Coffee* book by Oberthür et al. (2019), which gives an insight into the manifold and complex changes, both in conventional and specialty coffee cultivation, processing and trading since early on. The reader will appreciate learning more about the progress and ongoing alternate approaches facing the two major challenges and risks i.e. climate change and the volatile and underpaid farm gate prices.

Today, coffee consumers are highly interested in helping small farmers improve their social life conditions. They also want to be informed about the farmers' health conditions and their environment work. Hence, a detailed understanding of GAP, GPP and ESG guidelines by each coffee grower and their neighbors is necessary so they can eventually transform a whole landscape into more biodiverse and eco-friendly production areas.

Information can be found at the following websites:

- https://bsca.com.br/index/home.
- https://perfectdailygrind.com.
- https://sca.coffee.
- https://thosecoffeepeople.com.
- https://verocoffee.ie.
- https://worldcoffee.info.
- https://www.allegraceoforum.com.
- https://www.baristainstitute.com.
- https://www.cbi.eu/market-information/coffee.
- https://www.grandviewresearch.com.
- https://www.prnewswire.com/news-releases.

There exist a number of coffee organizations and institutions with current research projects designed to support coffee farmers in the world and to identify current problems and ways of tackling them, which are listed as follows:

- Coffee World Organization (ICO): https://www.ico.org/.
- Specialty Coffee Association (SCA): https://sca.coffee/.
- World Coffee Research (WCR): https://worldcoffeeresearch.org/.

8 References

Ahmed, S., Brinkley, S., Smith, E., Sela, A., Theisen, M., Thibodeau, C., Warne, T., Anderson, E., Van Dusen, N., Giuliano, P., Ionescu, K. E. and Cash, S. B. (2021). Climate change and coffee quality: systematic review on the effects of environmental and management variation on secondary metabolites and sensory attributes of Coffea Arabica and Coffea Canephora, Frontiers in Plant Science 12, 708013. DOI: 10.3389/fpls.2021.708013.

Assefa, A. and Kufo-Obso, T. (2019). Effect of methods of processing on raw and intrinsic quality of Sidama and Yirga Cheffee coffee types, The Second General Conference of Ethiopian Coffee Science Society, pp. 1-12, 24-25 May, Bonga. Available at: https://hdl.handle.net/10568/105996.

Ayala Durán, C. and Radomsky, G. (2020). Indicaciones Geográficas en Centroamérica: un crecimiento poco diversificado, RIVAR 7(20), 1-21. Available at: https://scielo.conicyt.cl/pdf/rivar/v7n20/0719-4994-rivar-7-20-1.pdf.

Bazame, H. C., Molin, J. P., Althoff, D. and Martello, M. (2021). Detection, classification, and mapping of coffee fruits during harvest with computer vision, Computers and Electronics in Agriculture 183, 106066.

Boot, W. (2019). Chapter 2.6. Post harvest processing of coffee beans. In: Oberthür, T., Läderach, P., Pohlan, H. A. J., Lee Tan, P. and Cock, J. H. (Eds), Specialty Coffee - Managing Quality (fully rev. 2nd edn.), Cropster GmbH, Innsbruck, pp. 206-221.

CAFEMALIST (2021), Las Tres Olas Del Café: ¿Qué Son? Available at: https://cafemalist.com/tres-olas-del-cafe/.

Campos Azofeifa, A. (2011). En el contexto de los acuerdos con la UE: avances en América Latina en la implementación de las indicaciones geográficas y denominaciones de origen, Inter-American Institute for Cooperation in Agriculture. Available at: http://repositorio.iica.int/bitstream/handle/11324/4137/BVE17089198e.pdf;jsessionid=B185F0F676E332CF9CDFA57B4DFD56CB?sequence=1

Canet Brenes, G. and García González, A. (2010). PROMECAFE: 30 años de experiencia. In: IICA (Ed.), Guatemala. Available at: http://repiica.iica.int/docs/B2177e/B2177e.pdf.

Canet Brenes, G. and Soto Víquez, C. (2016). Comercialización de café: alternativas de innovación. In: Canet Brenes, G., Víquez, C. S., Ocampo Thomason, P., Rivera Ramírez, J., Navarro Hurtado, A. and G. M. (Eds), Guatemala Morales and S. Villanueva Rodríguez, La Situación y tendencias de la producción de café en América Latina y el Caribe. IICA/CATIE, confederación unitaria de trabajadores, San José, pp. 71-96. Available at: https://www.google.com/search?q=La+Situaci%C3%B3n+y+tendencias+de+la+producci%C3%B3n+de+caf%C3%A9+en+Am%C3%A9rica+Latina+y+el+Caribe

+2016+IICA%2FCATIE&rlz=1C1SQJL_esNI936NI936&oq=La+Situaci%C3%B3n+y+tendencias+de+la+producci%C3%B3n+de+caf%C3%A9+en+Am%C3%A.

Coste, R. (1969). El café. Editorial Blume, Barcelona, 285 pp.

Craparo, A. C. W., Van Asten, P. J. A., Läderach, P., Jassogne, L. T. P. and Grab, S. W. (2021). Warm nights drive Coffea arabica ripening in Tanzania, International Journal of Biometeorology 65(2), 181–192, DOI: 10.1007/s00484-020-02016-6.

da Matta, F. M. and Martins, S. C. V. (2019). Chapter 2.2: Crop physiology. In: Oberthür, Th., Läderach, P., Pohlan, H. A. J., Lee Tan, P. and Cock, J. H. (Eds), Specialty Coffee - Managing Quality (fully rev. 2nd edn.), Cropster GmbH, Innsbruck, pp. 86–108.

Escobedo Paredes, C. (2013). La Propiedad Intelectual y los beneficios para productores y consumidores. Available at: https://www.redinnovagro.in/documentosinnov/escobedo.pdf.

FAO (2019). Available at: http://www.fao.org/faostat/en/#data/QC. Accessed March 16th, 2021.

Ferrer, R. (2018). DOSSIER: la generación del Milen. Caixa Bank Research 422, 32–33. Available at: http://www.gadeso.org/files/2018/07/milenials.pdf.

Ferris, N. (2020). El café y los jóvenes, presente y futuro de una nueva generación de consumidores, Fórumcafé 82, 32–33. Available at: https://www.revistaforumcafe.com/el-cafe-y-los-jovenes.

Fischer, E. F. (2017). Quality and inequality: Taste, value, and power in the third wave coffee market, *MPIfG Discussion Paper*, No. 17/4, Max Planck Institute for the Study of Societies, Cologne, pdf. 36 p. Available at: http://hdl.handle.net/11858/00-001M-0000-002C-AA63-2.

FNC (2014). Comportamiento de la Industria Cafetera 2014, Federación Nacional de Cafeteros de Colombia. Available at: https://federaciondecafeteros.org/static/files/Informe_Industrial_2014_Web.pdf.

Giuliano, P., von der Lieth, K. J., Fernández-Alduenda, M. R., Apostolopoulos, Y., Ionescu, K. E. and Rugolo, J. (2021). Towards a definition of specialty coffee: building an understanding based on attributes towards a definition of specialty coffee, *An SCA White Paper*, 12 pp . Available at: https://sca.coffee/sca-news/just-released-new-sca-white-paper-towards-a-definition-of-specialty-coffee.

Grant, T. (2020). Exploring millennial trends in specialty coffee, Perfect Daily Grind. Available at: https://perfectdailygrind.com/2020/02/exploring-millennial-trends-in-specialty-coffee-2/.

Haarer, A. E. (1969). Capítulo 12. Cosecha y preparación del café. In: Revolucionaria, E. (Ed.), Producción moderna de café, La Habana, pp. 343–406.

Haile, M. and Kang, W. H. (2019). The Harvest and Post-harvest Management Practices´ Impact on Coffee Quality. DOI: 10.5772/intechopen.89224.

Hindorf, H., Tedesse, M., Pohlan, J., Weedon, O. and Finckh, R. (2015), chapter 5.9. Organic coffee disease management. Specific crops. Part 5. In: Finckh, M. R., van Bruggen, A. H. C. and Tamm, L. (Eds) Plant Diseases and Their Management in Organic Agriculture, pp. 367–382.

ICO (2020). *The Value of Coffee – Sustainability, Inclusiveness and Resilience of the Coffee Global Value Chain*, International Coffee Organization. Coffee Development Report (2020). ICO FR/01/19E, ISBN: 978-1-5272-8440-1, 108 p.

IICA (2019). Manual de producción sostenible de café en la República Dominicana, Instituto Interamericano de Cooperación para la Agricultura, Santo Domingo, República Dominicana. Available at: https://repositorio.iica.int/handle/11324/8726.

ITC (2011). Tendencias del comercio de café certificado, Centro de Comercio Internacional. Available at: http://infocafes.com/portal/wp-content/uploads/2016/03/Rainforest_Alliance_Tendencias_de_cafes_certificados.pdf.

Läderach, P., Oberthür, T., Cook, S., Estrada Iza, M., Pohlan, J. A., Fisher, M. and Lechuga, R. (2011). Systematic agronomic farm management for improved coffee quality, Field Crops Research 120(3), 321-329.

Lundy, M., Bexc, G., Rodriguez, F. and Oberthür, Th. (2019). Chapter 3.2. Business models for quality coffee. In: Oberthür, Th., Läderach, P., Pohlan, H. A. J., Lee Tan, P. and Cock, J. H. (Eds), Specialty Coffee - Managing Quality (fully rev. 2nd edn.), Cropster GmbH, Innsbruck, pp. 254-280.

Madaro, M. (2018). Exploring the 3 waves of coffee in Costa Rica, Perfect Daily Grind. Available at: http://coffeegeekio.blogspot.com/2018/08/exploring-3-waves-of-coffee-in-costa.html.

Marca, Y. and Ríos, C. (2018). Importancia de la adaptación del marketing y los recursos humanos en tiendas de café a las nuevas tendencias del mercado, Fundación Universitaria Argentina de la Empresa (Trabajo final de licenciatura). Available at: https://repositorio.uade.edu.ar/xmlui/bitstream/handle/123456789/7714/TIF%20FINAL%20INDUSTRIA%20DEL%20CAF%C3%89.pdf?sequence=1&isAllowed=y.

Maspul, K. A. (2021). Applying the sustainable knowledge and behaviour to specialty Barista professional in the UAE, Academia Letters preprint. Open Access - Distributed under CC by 4.0.

McCauley, C. (2020). Top coffee trends of 2021. Available at: https://www.webstaurantstore.com/blog/2356/top-coffee-trends.html, accessed March 29th, 2021.

Montagnon, Ch., Marraccini, P. and Bertrand, B. (2019). Chapter 2.3. Breeding for coffee quality. In: Oberthür, T., Läderach, P., Pohlan, H. A. J., Lee Tan, P. and Cock, J. H. (Eds), Specialty Coffee - Managing Quality (fully rev. 2nd edn.), Cropster GmbH, pp. 109-143.

Oberthür, T., Läderach, P., Pohlan, H. A. J. and Cock, J. (2012). Specialty Coffee - Managing Quality, International Plant Nutrition Institute, Southeast Asia Program (IPNI SEAP), ISBN 978-983-44503-1-1, 347 pp .

Oberthür, T., Läderach, P., Pohlan, H. A. J., Lee Tan, P. and Cock, J. H. (2019). Specialty Coffee - Managing Quality (fully rev. 2nd edn.), Cropster GmbH, ISBN 978-983-44503-1-1, 370 pp.

Oberthür, T., Läderach, P., Posada, H., Fisher, M. J., Samper, L. F., Illera, J., Collet, L., Moreno, E., Alarcón, R., Villegas, A., Usma, H., Perez, C. and Jarvis, A. (2011). Regional relationships between inherent coffee quality and growing environment for denomination of origin labels in Nariño and Cauca, Colombia, Food Policy 36(6), 783-794

OMPI (2017). Las indicaciones geográficas, Organización Mundial de la Propiedad Intelectual, Suiza, Ginebra. Available at: https://www.wipo.int/edocs/pubdocs/es/geographical/952/wipo_pub_952.pdf.

Panhuysen, S. and Pierrot, J. (2020). Coffee Barometer 2020, Published by the Coffee Collective 2020, Conservation International, Hivos, Oxfam Wereldwinkels, Solidaridad, Pdf, 56 p.

Pérez Akaki, P. (2011). Denominaciones de origen (DO) y marcas colectivas (MC) en el café Méxicano, ¡Estrategia para el desarrollo regional? Revista Geográfica de América

Central, pp. 1-22. Available at: https://www.redalyc.org/pdf/4517/451744820506.pdf.

Pohlan, H. A. J. (Ed.) (2019). Manual de Bom Cafeicultor das terras altas. Diagnóstico, monitoria e auditoria de Boas Práticas Agrícolas (BPA) através da Metodologia do Sistema de Semáforo (SdS) em Cafezais de Angola, Shaker Verlag Düren, Germany, ISBN 978-3-8440-6515-2, 110 pp.

Pohlan, H. A. J. and Salazar Centeno, D. (2012). Diagnóstico, monitoreo y auditoría de buenas prácticas agrícolas a través del sistema de semáforo, Managua, Nicaragua, 63p, ISBN 978- 99924-1-0118-9. Available at: http://repositorio.una.edu.ni/id/eprint/2423.

Pohlan, H. A. J. and Salazar Centeno, D. J. (2017). Chapter 5. Gerencia de fincas cafetaleras a través de la metodología del sistema de semáforo y buenas prácticas agrícolas. In Memorias de los Talleres de Agroecología y Roya del Café en Mesoamérica y República Dominicana, Food and Agriculture Organization, Panamá, pp. 19-27. Available at: http://www.fao.org/3/a-i7697s.pdf.

Pohlan, H. A. J., Salazar Centeno, D. and Torrico Albino, J. C. (2017). Experiences in the exchange and sharing information through Best Agricultural Practices (BAP) in Angola, Bolivia and Nicaragua: an outlook for the cooperation between Asian coffee enterprises. Proceedings of the 2017 Asian Coffee Annual Conference (ACAC), Mangshi, Yunnan, PR China, p. 44.

Pohlan, H. A. J., Salazar Centeno, D., Janssens, M. J. J. and Torrico Albino, J. C. (2019), Chapter 2.1. Implementation and monitoring of good agricultural practices. In: Oberthür, Th., Läderach, P., Pohlan, H. A. J., Lee Tan, P. and Cock, J. H. (Eds), Specialty Coffee - Managing Quality (fully rev. 2nd edn.), Cropster GmbH, pp. 68-85.

Pohlan, H. A. J., Soto, L. and Barrera, J. (2006). El cafetal del futuro: Realidades y Visiones, Shaker Verlag Aachen, Germany, 462 pp.

Rainforest Alliance (2020). Informe de datos de la certificación del café 2019: programas de Rainforest Alliance y UTZ. Available at: https://www.rainforest-alliance.org/business/wp-content/uploads/2020/05/Coffee-Certification-Data-Report-2019-ESPA%C3%91OL.pdf.

Ram, A. S. (2017). Geographical Indications on Coffee, Proc. 2017 First Dali Binchuan Zhukula International Coffee Forum, Specialty Coffee Society of Yunnan, Kunming, China, pp. 257-280.

Rhinehart (2009). What is Specialty Coffee? Available at: https://scanews.coffee/2017/03/17/what-is-specialty-coffee/.

Rieser, K. (2019). Chapter 1.1. The evolution of specialty coffees with specific quality characteristics. In: Oberthür, T., Läderach, P., Pohlan, H. A. J., Lee Tan, P. and Cock, J. H. (Eds), Specialty Coffee - Managing Quality (fully rev. 2nd edn.), Cropster GmbH, Innsbruck, pp. 2-18.

Rothfos, B. (1980). Coffee Production, Gordian Max Rieck GmbH, Hamburg, Germany, 366 pp.

Sachs, J., Cordes, K. Y., Rising, J., Toledano, P. and Maennling, N. (2019). Ensuring economic viability and sustainability of coffee production, Columbia Center on Sustainable Investment. Available at: http://ccsi.columbia.edu/work/projects/coffee/.

Samoggia, A., Del Prete, M. and Chiara Argenti, Ch. (2020). Functional needs, emotions, and perceptions of coffee consumers and non-consumers, Sustainability 12(14), 5694; DOI: 10.3390/su12145694.

Sample - Global Coffee Market (2020-2025). *Mordor Intelligence*. 37 pp, pdf.

Selvamurugan, M., Doraisamy, P., Maheswari, M. and Nandakumar, N. B. (2010). High rate anaerobic treatment of coffee processing wastewater using upflow anaerobic hybrid reactor, Iranian Journal of Environmental Health Science and Engineering 7(2), 129–136.

Silva Monteiro Boaventura, P., Caires Abdalla, C., Lobo Araújo, C. and Sarkis Arakelian, J. (2018). 'Value co-creation in the specialty coffee value chain: the third-wave coffee movement, RAE-Revista de Administração de Empresas (Journal of Business Management), RAE Eletrônica DOI: 10.1590/S0034-759020180306.

Skeie, T. R. (2003). Norway and coffee, The Flamekeeper, Spring, 2–5. Available at: https://web.archive.org/web/20031011091223/http://roastersguild.org/052003_norway.shtml.

Songer, P. (2019). Chapter 3.4. Laboratory principles, practices and setup. In: Oberthür, T., Läderach, P., Pohlan, H. A. J., Lee Tan, P. and Cock, J. H. (Eds), Specialty Coffee - Managing Quality (fully rev. 2nd edn.), Cropster GmbH, Innsbruck, pp. 298–312.

Subedi, R. N. (2010). Comparative Analysis of Dry and Wet Processing of Coffee with Respect to Quality in Kavre District, Nepal, Larenstein University, Wageningen, Netherlands, 77 pp.

Traore, T. M., Wilson, N. L. W. and Fields III, D. (2018). What explains specialty coffee quality scores and prices: A case study from the cup of excellence program, Journal of Agricultural and Applied Economics 50(3), 349–368. DOI: 10.1017/aae.2018.5.

Trujillo, H. A., Mitidieri, F. J. and Hashimoto, M. E. (2021). 'Denomination of origin 'Café del Huila' and dynamics of coffee growing in Colombia, Coffee Science 2021, e161827, DOI: 10.25186/.v16i.1827.

Wiesinger, M. and Oberthür, T. (2019). Chapter 3.5. Opportunities for digital information to support the specialty coffee value chain. In: Oberthür, T., Läderach, P., Pohlan, H. A. J., Lee Tan, P. and Cock, J. H. (Eds), Specialty Coffee - Managing Quality (fully rev. 2nd edn.), Cropster GmbH, pp. 315-331.

Wintgens, J. N. (2009). Coffee: Growing, Processing, Sustainable Production: A Guidebook for Growers, Processors, Traders, and Researchers, Wiley-VCH Press, Weinheim, 984 pp.

Chapter 5

Fair-trade coffee: how fair is fair?

Ruud Bronkhorst, InfoBridge Foundation, The Netherlands

1 Introduction

An important aspect of sustainability that needs discussion is a fair income for producers and labourers involved in agriculture. In this chapter, we will discuss what 'fair' means and the approach used by some major Voluntary Sustainability Standards (VSS) to calculate 'fair' prices for producers. This is followed by two alternative approaches to calculate prices: the Living Income Reference Price Formula and the Living Income/Fair Price Approach, a bottom-up approach for defining what is a 'fair' price from the farmers' perspective.

Fair trade is often thought to include a 'fair price', but that is not always correct. Fair trade covers more than a fair price alone. It is also about conditions of work, terms of trade, etc. In this chapter, the payment of a fair producer price will be highlighted. The notion of a fair price has until recently not played an important role in fair trade organisations with no agreement about the question of what a fair price is. The issue is now receiving more attention in the fair trade movement.

We will first discuss the meaning of the word 'fair' and some underlying concepts concerning 'fair' prices. Basic to these concepts is the Universal Declaration of Human Rights, adopted in 1948 by the United Nations (UN).

http://dx.doi.org/10.19103/AS.2021.0096.11

We will see how these fundamental human rights are put into practice by some VSS, notably Fairtrade International (FT) and Rainforest Alliance (RA).

Two other approaches are also discussed:

- the Living Income Reference Price Formula; and
- the Living Income/Fair Price Approach.

The Living Income/Fair Price method is further explained by means of a model showing how to calculate fair prices for a coffee producer using the Living Income/Fair Price methodology. Finally, some conclusions are drawn.

2 Defining what is 'fair'

There are a lot of possible ways to interpret 'fair'.[1] Should 'fair' prices and wages be paid because it is an economic or social necessity, a moral right or as an act of charity? Does the worker have a moral or economic 'right' to fair remuneration or is it something we grant him/her because of our 'goodness'? As we will see below, the answer to this question is important for 'fair' price calculations.

Debates on pricing usually revolve around market prices, which are determined by supply and demand. In conventional economic theory, the market price is the price at which the supply of and demand for a product reach equilibrium. In practice, however, market prices are determined by a wide range of other factors that have nothing to do with supply and demand. Indeed, market prices often diverge from equilibrium prices for several reasons:

- Some markets are imbalanced because they have only a single buyer or a few buyers.
- Markets may be dominated by a few large-scale producers.
- Prices can be distorted by subsidies, import/export restrictions and other government interventions.
- Demand from poorer consumers, who lack purchasing power, does not figure in the equilibrium-price calculation.

Considering these distortions in the market, more attention should be paid to an approach grounded in ethics and fairness.

The importance of the influence of prices on coffee producers can be illustrated by the number of people involved. Coffee is produced on approximately 12.5 million coffee farms. The structure and size of the farms varies by producer country. Almost 95% of coffee farms are smaller than 5 ha and 84% of all coffee farms are smaller than 2 ha. It is estimated that smallholder

1 For more details on how economists have looked at the concept 'fair', see Bronkhorst 2020, Annex I.

farms produce up to 73% of all coffee, with the remaining 27% produced by large coffee estates (Coffee Barometer, 2020). Assuming an average household consists of four people would suggest that some 50 million people are affected by coffee prices. This figure does not take into account all other people who are dependent on these households such as shopkeepers, suppliers, etc.

When talking about fair producer prices, we must distinguish between fair producer prices and fair consumer prices. A fair price to the producer basically means that he/she is not paid too little, while a fair price to the consumer means that he/she is not paying too much for an acceptable product. A fair price to the consumer is based on the consumer's willingness to pay which can be constrained by lack of monetary means. A fair consumer price implies the absence of excess profits in the chain, due to monopolies, oligopolies and other factors. A fair price to the consumer implies not that the lowest amount possible is paid, but that not too much or too little is paid. This means that a fair price to the consumer implies a fair price to the producer.

The following definition of a 'fair price' is the definition I use in my book *The Economics of Human Rights* (Bronkhorst, 2020):

> A **'fair' price** is the minimum price the producer should receive in order to cover all production costs (fixed and variable costs) and be able to pay Fair Wages to all his/her workers on this product and him/herself. It should also allow the producer to make necessary investments. In the case of coffee and other agricultural activities it should provide at least a 'Living Income' for the farmer.

A 'fair price' must include the payment of 'fair wages': wages sufficient to enable a worker on a coffee farm to have a decent life for himself and for his family. This includes food, shelter, education, health care, etc. Payment is on condition that a complete working week is spent on the product and that the product is considered useful. A common way to determine the usefulness of a product is through the demand for it.[2] Besides coverage of production costs, a reasonable profit must be part of the fair price. A reasonable profit is a profit that permits the firm to continue its operations and make the necessary investments to be able to continue in the future as well. This includes investment, e.g. needed to prepare for climate change impacts.

When fair prices are not paid, and income from coffee falls below a certain minimum, producers suffer and may be forced to switch to other sources of income. Figure 1 shows highly fluctuating international coffee prices during the period 1995–2020. Figure 1 shows that there are long periods during which coffee prices are very low, significantly reducing producers' incomes and pushing them into poverty.

2 Usefulness cannot be measured by demand alone though. There are many products that are in high demand (such as certain drugs) but that are harmful both to the user as well as to the general health system. Besides demand implies only demand by people with sufficient purchasing power. The demand of the poorest is not taken into account.

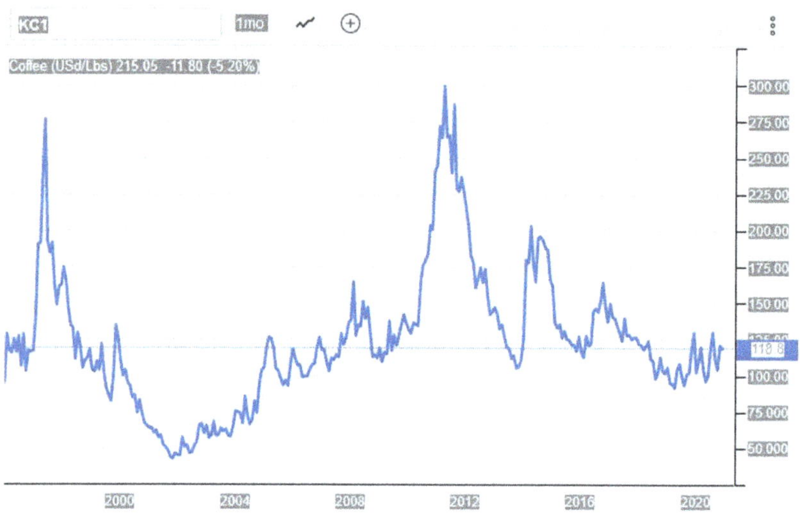

Figure 1 International coffee prices 1995–2020. Source: https://tradingeconomics.com/commodity/coffee.

When producer prices are too low, the impact is not just economical but social and environmental as well. Xavier (2020) has identified a number of sub-strategies that producers used to reduce costs when income is insufficient. Among these strategies are reducing operating expenses, reducing expenditure on land conservation (e.g. reducing use of fertilisers and maintenance work on plantations), reducing wages or numbers of workers employed, intensifying the use of household labour, including children, and reducing household expenses.

Producers are often small-scale farmers working on marginal lands, including steep slopes and degraded soils. If they cannot afford to maintain or improve soil health, the result is increased erosion, lower soil fertility and increased incidence of pests and diseases that make plantations less and less productive. If productivity drops too much, plantations may be given over to cattle grazing or abandoned altogether (Xavier 2020). In the case of Costa Rica, e.g. the drop in prices for coffee beans in the period 2000–2008 led to an abandonment of coffee production and a change to keeping livestock. Trees were cut down, there was overgrazing by livestock and the effect was deforestation and significant land degradation with serious consequences for the environment. We may conclude that a sufficient income for the farmer is necessary not only for his livelihood but also to enable him/her to make investments to protect land and the ecosystem services it delivers, change to more sustainable production methods and prepare for, and adapt to, climate change.

When talking about 'fair' prices to be paid to producers, it is important to recognise the work of the Max Havelaar Foundation and FT. The Max Havelaar Foundation was founded in 1988 in the Netherlands in response to the plea from Mexican coffee farmers that, rather than being given assistance, they would be better served by being paid a 'fair' coffee price. Max Havelaar is now part of FT which covers many regions and products beyond coffee. They have done an important job in making people think about fair prices and have put the subject in the spotlight. Other organisations such as UTZ and Rainforest Alliance (RA) (now merged into one organisation) also work for the benefit of the producers. However, only FT has, in coffee, developed a system focused specifically on providing the producer with a fair price. Defining and achieving 'fair' prices by organisations such as FT have continued to prove challenging through to the present.

It is important to keep in mind that, for developing countries, the concept of 'fair trade' can be seen as derived from the global North and imposed top-down on the global South instead of being developed bottom-up. The interpretation a coffee farmer in the South may have of a 'fair' price may be quite different from the idea of what is 'fair' developed in the North. It is often assumed that a deal between buyer and seller is fair when both parties agree, without considering the relative situation in which both parties operate. For the producer, it often is a case of 'take it or leave it', where the livelihood of his/her whole family is at stake, while the buyer can easily go to another producer to get the cheapest deal. Differences in access to information also play an important role in seller–buyer relationships. The relative bargaining power of buyer and seller can lead to very different so-called 'fair' agreements.

3 Defining 'fair' wages and incomes

There are lots of difficulties when trying to calculate a fair price. These include differences in price levels and living standards between different countries, differences in farm sizes and households, costs and profitability, etc. We will now discuss the approach to fair or reference prices developed by FT, as well as the 'Living Income Reference Price Formula' and the 'Living Income/Fair Price' methodology. We will start by discussing some underlying concepts for 'fair' prices.

Basic to our definition of 'fair' is the Universal Declaration of Human Rights adopted by the UN in 1948. Article 23 of the Universal Declaration of Human Rights states that:

> Everyone who works has the right to just and favourable remuneration ensuring for himself and his family an existence worthy of human dignity, and supplemented, if necessary, by other means of social protection.

Article 25 states that:

> Everyone has the right to a standard of living adequate for the health and well-being of himself and of his family, including food, clothing, housing and medical care and necessary social services, and the right to security in the event of unemployment, sickness, disability, widowhood, old age or other lack of livelihood in circumstances beyond his control.

These articles are on the basis of the sustainable development goals (SDGs) adopted in 2015 by the UN. Goals 1 'No Poverty' and 2 'Zero Hunger' are about ending poverty and hunger, as well as achieving food security. A further application to these principles is the development of a number of related concepts:

- poverty line,
- minimum wage,
- living wage (LW), and
- living income (LI).

We will first discuss these concepts and then their application to fair prices.

3.1 Poverty lines

There are several different definitions of poverty lines. The most common are the poverty lines calculated by the World Bank. These are based on the concept of 'purchasing power parity' (PPP). PPPs are based on the cost of purchasing a representative basket of goods, calculated in US$ to allow comparisons between countries. Examples are:

- International Extreme Poverty Line: US$1.90 PPP,
- Lower Middle Income Class Poverty Line: US$3.20 PPP, and
- Upper Middle Income Class Poverty Line: US$5.50 PPP.

PPPs have been criticised in various ways, e.g. by not taking into account the conditions facing groups such as the poorest or basing PPPs on unrepresentative selections of goods based on middle-class incomes. Another frequently used poverty line is the national poverty line. This is calculated by individual national governments so is not affected by problems associated with PPPs. A danger here is political interference; it may be in the interest of the government to define a high or a low poverty line to suit a particular political agenda.

3.2 Minimum wage

Minimum wages can be implemented in a single firm, a specific industry or at a national level by governments in the form of a legal minimum wage which firms must pay all employees. A legal minimum wage may be different for specific industries or regions. Like national poverty lines, setting minimum wages can be subject to political influence, e.g. the relative influence of interested parties such as employers, trade unions, etc., in setting a particular figure.

3.3 Living wage

The development of the concept of 'living wage' is quite recent. Although there is no international labour convention relating specifically to an LW (Vaughan-Whitehead), the International Labour Organisation (ILO) made a clear reference to it from 1919 in its constitution and has included an LW as a human right in several major ILO declarations (Anker, 2011). There is no universal agreement about ways of defining and calculating an LW. The most widely used definition is that agreed upon in 2013 by the Global Living Wage Coalition, consisting of FT, Forest Stewardship Council, GoodWeave, Sustainable Agriculture Network/RA, Social Accountability International and UTZ Certified. Their definition of an LW is:

> The remuneration received for a standard work week by a worker in a particular place sufficient to afford a decent standard of living for the worker and her or his family. Elements of a decent standard of living include food, water, housing, education, health care, transport, clothing, and other essential needs including provision for unexpected events.
>
> *(Global Living Wage Coalition)*

This definition provides a foundation for the rest of this chapter.

The standard work on LWs by Richard and Martha Anker 'Living Wages Around the World: Manual for Measurement' (2017) says:

> Estimating a living wage begins by estimating the cost of a basic but decent lifestyle for a worker and his/her family. This involves adding up the cost of three expenditure groups: food (for a low-cost nutritious diet), housing (for basic healthy housing), and other essential expenses for a family, and then adding a small margin for sustainability and emergencies
>
> *(Anker and Anker, 2017).*

When this cost of a basic but decent life for a family is divided by the number of full-time workers per family, the outcome is the net living wage. Anker and Anker use labour force participation rates for persons aged 25–59 to estimate the number of full-time equivalent workers per couple (Anker and Anker, 2017).

In their approach, living costs are divided into three categories:

- food,
- housing, and
- other essential needs.

Food costs are estimated based on:

- a low-cost nutritious diet that meets World Health Organization recommendations for calories, macronutrients and micronutrients and is consistent with local food preferences and a country's development level, and
- local food prices for the types, qualities and quantities of foods that workers typically buy.

Housing costs are estimated using international (UN-HABITAT) and national standards for decency. The cost of other essential needs is estimated using an extrapolation method based on secondary household expenditure data. This is then 'post checked' to make sure that sufficient funds are included for needs such as health care, education and transportation. The Ankers approach is important because it has been accepted by major organisations organised in the global living wage coalition. It has therefore evolved into a de-facto standard.

3.4 Living income

The concept of an LW applies to paid workers only. For agriculturalists who work on their own account, earning a living by selling their produce on the open market or to customers such as large companies, another concept is needed. This new concept is 'Living Income (LI)'. A definition of LI by the Living Income Community of Practice (an alliance of organisations including FT and RA) is:

> A Living Income is the net annual income required for a household in a particular place to afford a decent standard of living for all members of that household [3]

whereby:

- Net income = total income minus all costs, and
- Household is a group of people (often family) who form an economic unit (pooling income and other livelihood assets), often (but not always) living under the same roof or within the same compound.

3 https://www.living-income.com/.

Further important insights on decent work and income are provided by ILO and the Food and Agriculture Organisation of the United Nations (FAO). ILO defines decent work as follows:

> Decent work sums up the aspirations of people in their working lives. It involves opportunities for work that is productive and delivers a fair income, security in the workplace and social protection for families, better prospects for personal development and social integration, freedom for people to express their concerns, organize and participate in the decisions that affect their lives and equality of opportunity and treatment for all women and men.

The FAO defines decent employment as:

> work that provides a living income and reasonable working conditions. Work should be remunerative and dignified. It should enable people-whether through self-employment or wage labour-to provide for themselves and their families. Workers should be able to perform their work under safe and healthy conditions and have a voice in the workplace.

The definition of LI used later in this chapter in discussing the Living Income/Fair Price methodology is an extension of this definition by Bronkhorst (2020). The extension implies adding to the calculation a percentage to enable additional investments to improve future yields. This percentage might be between 5% and 10%, depending on local circumstances. The reason for adding this percentage is that farmers in particular need more income to prepare for and adapt to challenges such as climate change and soil exhaustion. Extra funds may be essential to help them to prepare for the future and ensure sustainable production over the long term.

4 Voluntary Sustainability Standards relating to fair prices

After this explanation of different definitions, we will now discuss some approaches on how to calculate the fair prices needed to achieve LWs and LIs. The two major players in the field are FT and the RA (including UTZ which merged with the RA in 2018). The chapter will then discuss the 'Living Income Reference Price Formula' developed by Krain et al. and the 'Living Income/Fair Price' approach developed by Bronkhorst. Both RA and FT certification programs include requirements relating to the three pillars of sustainability: social, environmental and economic. However, each has a different approach to implementation.

4.1 Rainforest Alliance

Before its merge with RA, UTZ required a premium to be paid to producers. For every purchase of a UTZ product, the first buyer in the supply chain, normally a trader, had to pay a cash premium on top of the market price to the producer or producer group they bought the product from. This was recorded in the UTZ traceability system.

The new 2020 Certification Program of The Rainforest Alliance came into force in July 2021. The new program includes 'critical criteria' in three areas: social, economic and environmental improvement. These requirements must be met by farms and supply chain actors in order to earn and maintain certification. Improving farmer livelihoods is seen as an important goal in a holistic approach to help farmers grow their businesses and become more profitable and resilient, e.g. through training in farm management, financial literacy and market access.[4]

Their new program requires buyers to pay a Sustainability Differential, a mandatory additional cash payment to certified farms over and above the market price. This payment is designed to be completely free of restrictions or requirements on how it is used. The amount of the Sustainability Differential is not fixed. The new program has also introduced a Sustainability Investment requirement for buyers, who must make cash or in-kind investments to farmers based on the needs identified in their certificate holders' own investment plans. Sustainability Differentials and Sustainability Investments must be reported in the RA system. It is important to note that the amount of the Sustainability Differential is not fixed for coffee. A minimum amount is set for cocoa but has not yet been set for coffee. There is also no fixed amount for Sustainability Investments.

4.2 Fairtrade International

FT's approach to minimum and premium pricing has changed over time. In 1988, a minimum price was based on the reference price used by the International Coffee Organisation (ICO) to stabilise coffee prices. The premium above the minimum price varied between 0% and 10%, depending on the market price. As an example, when the minimum price was US$115 cents per pound, there was a premium of 10% which increased the price to US$126 cents per pound (5% above the minimum ICO price). When the price rose to US$165 cents per pound, there was no additional premium.

In 1995, a fixed premium of US$5 cents per pound was established, calculated from the same minimum price of US$126 cents per pound. The reference point for the price was changed to free on board (FOB) which takes account of shipping costs. The Fairtrade Minimum Price (FMP or FTMP) serves as a safety net to protect producers in times of very low market prices. At that time, FLO (Fairtade Labelling Organisation International) had not yet

4 https://www.rainforest-alliance.org/faqs/difference-between-rainforest-alliance-certified-fair-trade.

determined how a 'living' wage should be calculated or implemented but was working on doing so.

Prices in 2011 were:

	Prices in USD cents/lb
FTMP Arabica washed	140
Fairtrade Premium	20 out of which at least 5 cents are to be invested in productivity and/or quality
Total Fairtrade price	160
Organic differential	30
Total Fairtrade organic price	190
FTMP Arabica natural	135

In 2018, a new document was published to replace previous guidance documents: 'Guidance Document for Fairtrade Coffee Pricing' (FLO, 2018).[5] This remains based on a price floor combined with a premium.[6] FT also supports fair pricing in other ways, e.g. encouraging long-term contracts for greater stability.

Setting an FMP is the outcome of two processes:

- First, information is collected from producers about their cost of production which provides the base for preparing a proposal for an FMP.
- Second, the outcome of research is shared with all stakeholders and a process of consultation is organised.

The FMP is the outcome of this process.

A point of departure in deciding an FMP is that hired workers have the right to an LW and self-employed farmers should be able to make an LI from their farming activities. Following this principle, a farm should be big enough to fully absorb the average household labour force to generate a LI. This would be considered a viable farm size or a 'full-employment farm size'.

The costs associated with producing sustainable yields are likely to be higher than current production costs, since many farms have low levels of investment and productivity. Costs of production are projected at the level of a sustainable yield. This calculation is based on fully employed non-remunerated household labour. Additional hired labour needs are factored in on the basis of an LW. Calculated in this way, the Fairtrade living income reference price (FLIRP) covers an LI for the household members (provided the available household labour is effectively employed on their farm) from the farm profits and an LW for hired workers.

5 https://files.fairtrade.net/standards/Guidance_Document_for_Fairtrade_Coffee_Pricing_2_0_EN_April_2018.pdf.
6 Source: www.fairtrade.net.

The Fairtrade price for green coffee is defined at FOB level. The Fairtrade price is the reference market price or the FMP, whichever is higher, plus the Fairtrade Premium of 20 cents/pound and the Fairtrade organic differential of 30 cents/pound (in the case of organic coffee). The reference market price is the sum of the terminal market price (ICE New York C contract for Arabica, ICE London RC contract for Robusta) in the relevant month, plus or minus the prevailing differential. A negative differential can never be applied to the FMP. The Fairtrade Premium and the Fairtrade organic differential are not subject to negotiation and are meant for the producer organisation only.

The **FMP** is the minimum that producers are paid when selling their products through FT. It aims to cover the average costs of sustainably producing a crop and acts as a safety net when market prices drop. Producers (cooperatives, not individual farmers) get the market price when this is higher and can also negotiate for more.

The **Fairtrade Premium** is an extra sum of money, paid on top of the selling price, that farmers or workers invest in projects of their choice.[7] The use of the Fairtrade Premium is restricted to investment in the producers' business, livelihood and community (for a small producer organisation or contract production set-up) or investment in improving the lives of workers and their community (for a hired labour situation). Its specific use is democratically decided by the producers.

Organic differential: for some certified organic products, a minimum organic differential is set. It is the lowest possible differential that producers must receive in addition to the FMP or market price, whichever is higher. If a premium is set for the conventional product variety, this premium also applies to the organic product, unless stated differently.

In the case of FT coffee, e.g. buyers pay the FMP or the market price, whichever is higher, and a premium for conventional coffee. For organic coffee, an organic differential needs to be paid on top of price and premium.

In addition to the FMP, a FLIRP has also been developed. This FLIRP indicates the price needed for an average farmer household with viable farm size and an adequate productivity level to make a LI from the sales of their crop. It is based on the following key parameters:

- cost of a decent standard of living (Living Income benchmark),
- sustainable yields (productivity benchmark),
- viable farm size (to fully employ the available household labour), and
- cost of sustainable production (in order to achieve above-mentioned yields).[8]

7 Source: https://www.fairtrade.net/standard.
8 Source: https://files.fairtrade.net/2019_FairtradeLivingIncomeReferencePrice_Model.pdf.

While the FMP is mandatory and part of FT certification standards, the FLIRP is a price indicator for partners who voluntarily want to go a step further in enabling LIs in their supply chains. The FMP is a safety net to ensure farmers do not go deeper into poverty, but it will not necessarily lift farmers out of poverty. The FLIRP concept has been developed to help achieve this more ambitious goal. Payment of a FLIRP is currently voluntary. The aim is to implement the FLIRP with willing partners and generate evidence of its effectiveness to move towards LIs.

As the previous discussion suggests, whilst beneficial, the FMP may only have a modest impact on already low incomes for coffee smallholders. Producers still face a range of costs such as sending children to school, paying for health services, etc. This means producers will still not earn the minimum income defined under the Universal Declaration of Human Rights.

In addition, the Fairtrade Premium is paid to the farmers' organisation, so the individual farmer cannot spend it the way he/she likes. Individual members jointly decide on the way the premium is spent in the general assembly of their organisation. They may decide to distribute (part of) the premium in cash to members directly, but it is often agreed to invest the premium in common goods or services or to address priority needs for the community as a whole. The individual farmer also has to contribute something to the group in order to receive these and other benefits of being in a cooperative (such as gains of scale, better access to markets and therefore higher prices).

The cooperative a farmer belongs to has to apply for FT certification which costs €565 per customer. On top of the application fee, a customer pays an annual fee. In the first year of certification, this fee starts at €1545 for the smallest possible small producer organisation (SPO, below 50 members) and €2940 for traders. This sum covers all costs, including the audit, for the first 12 months. The annual fee is lower in subsequent years, if the customer can be classified into a lower fee category, something that will be determined after the first audit. Examples of typical annual certification fees are small SPO (< 50 members): €1260 average SPO (~260 members): €2120 large SPO (> 500 members): €2595 .[9] A tool that calculates the cost of initial/annual Fairtrade certification can be found on: https://www.flocert.net/solutions/fairtrade/cost-calculator/. A detailed description of Fairtrade's Living Income Reference Price model for coffee applied to Colombia can be found at https://www.fairtrade.net/news/fairtrade-launches-its-first-living-income-reference-prices-for-colombian-coffee.

9 Source: mail correspondence with Flocert, the global certification body for FT

5 Other approaches: The Living Income (and Other Benchmarks) Reference Price Formula and the Living Income/Fair Price approach

Two different approaches that are not part of the 'Voluntary Sustainability Standards' should be mentioned as well. These are:

- the Living Income (and other benchmarks) Reference Price Formula (the 'Linnert Steffens' formula) developed by Krain et al. (2021), and
- the Living Income/Fair Price approach developed by Bronkhorst (2020).

These are discussed in the following sections.

5.1 The living income (and other benchmarks) reference price formula (the 'Linnert Steffens' formula)

This is a general method that allows the setting of prices against any benchmark including, e.g. poverty lines. The following formula is used (Krain et al. 2021):

$$BM_{rp} = \frac{\left(\frac{BM_w}{WD} * NoWDS\right) + CP}{Y}$$

where:

Variable	represents
BM_{rp}	a LI (or other benchmark) reference price in monetary value (e.g. a currency unit) for a particular crop per usual weight or volume unit (e.g. per kg, ton or litre) per unit area of production (e.g. per ha)
$\frac{BM_w}{WD}$	a benchmark 'wage' per workday
$NoWDS$	number of workdays needed to produce a certain yield per unit area.
CP	cost of production (input and service cost, but no labour cost) per unit area
Y	yield of the crop per unit area for which the benchmark reference price is calculated e.g. in kg.

Source: Krain et al. (2021)

In this approach, a reference price is calculated for two scenarios:

- 'current production',
- 'improved production'.

The current production scenario reflects the costs and incomes of the producer's existing situation and reference prices are calculated against this scenario. The 'improved production' scenario allows recalculation of reference prices based on raising productivity and income. For more details on this approach, see Krain et al. (2021) where a detailed description and calculation are given of its application to cocoa and cashew production in Ghana.

5.2 Living Income/Fair Price approach

Another approach that takes into account the actual conditions that a coffee grower faces is the Living Income/Fair Price approach. The definition used for living income is an extension of the definition by the Living Income Community of Practice:

> Living Income is the net annual income required for a household in a particular place to afford a decent standard of living for all members of that household. [10]

Although this is a good definition, an extension of the definition is needed for a farm household to improve its situation. This means that a percentage to enable additional investments to improve future yields must be added. This percentage might be between 5% and 10%, depending on local circumstances. The reason for adding this percentage is that farmers in particular need additional income to develop improved production methods to meet such challenges as climate change and soil exhaustion. Extra funds for investment help them to prepare for a more sustainable and secure future.

The LI of the farmer must include 'Fair or Living Wages' for all workers involved, both hired labour and family labour, as well as funds for additional investments. The payment should be based on full employment by the household in coffee production, and the product should be one that is useful (i.e. for which there is a reasonable demand). Because of the assumption of full employment in coffee production, it is not necessary to separately calculate the number of hours spent on each crop. Equally important is that the concept of a LI considers the local context in defining a decent standard of living.

To calculate fair prices, a number of steps have to be taken.

- First, living wages must be calculated:[11]
    - This means that the average family size must be known: adults, children and people above the age of 60.

10 https://www.living-income.com/.

11 For a detailed methodology on how to calculate LWs according to the Anker methodology, see Anker, Richard and Anker, Martha, Living Wages Around the World, Edward Elgar 2017.

- The total expenses of the household during the year on food, housing and other essential needs to ensure a basic but decent life must be calculated.
- To this amount of total expenses, 5–10% (depending on region, cultural values, etc.) must be added for additional, and often unforeseen, expenses.
- The average number of adult earners per household must be known.
- Number of working days per year must be calculated = calendar days - Sundays - festivals - social obligations - emergencies - sick days.
- In the case of agriculture, attention must be paid to the fact that working hours are not equally divided over the year; there are peak and slack periods.

From this, the LW per adult earning family member per working day can be calculated:

$$\text{LW per person(pp) per year} = \text{family income needed to be divided by number of adult income earners.}$$

$$\text{LW pp per day} = \text{LW per year divided by number of working days.}$$

This is shown in Fig. 2.

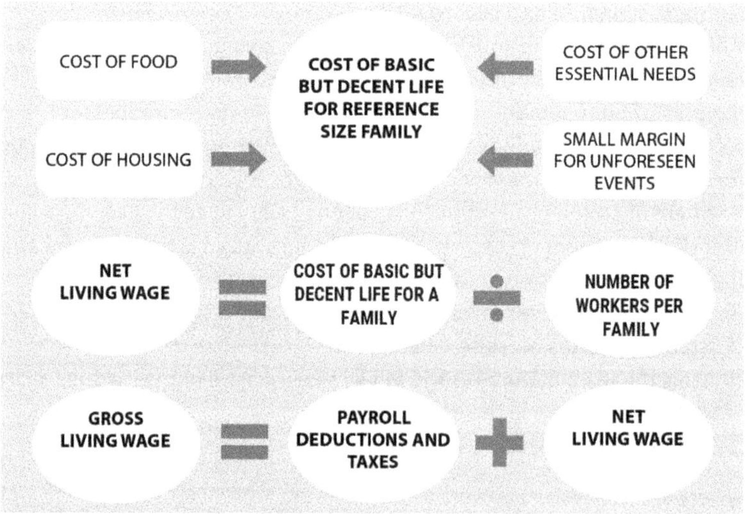

Figure 2 Methodology for estimation of living wage. Source: https://www.globallivingwage.org/about/anker-methodology/.

Based on those calculations, the level of fair wages for labourers can be calculated, in consultation with representatives of the workers themselves, as recommended by Vaughan-Whitehead. These benchmarks for LWs, as well as any legal minimum wage levels established by governments, provide an absolute minimum below which fair wages cannot fall. It is important to be aware that, for the producer to be able to pay fair wages, the LW must be known, there must be an effective demand for the product and the consumer must have the financial resources needed to purchase the product at a price that can fund fair wages. LWs have now been calculated for many countries and regions and can be used. Calculated LWs can be found at https://www.globallivingwage.org/, https://wageindicator.org/ and https://align-tool.com/.

When the LW is known, the LI can be calculated. LI/year = (number of adult earners × LW) × 1.1, whereby 1.1 is an additional 10% for investment costs. This percentage can be adapted to a specific situation. A higher percentage can be used to enable the farmer to invest in items like better seeds and better equipment. Additional investments needed to adopt more sustainable methods of production may necessitate a higher percentage.

When LWs and LI are known, several other data have to be collected as well:

- A typical smallholder farm size must be determined. Which farm size is chosen depends on the chosen target group: marginal farmers or small farmers, fertility of the soil, mode of production, distance to the market, etc.
- Production costs must be known for all products, including investment costs and depreciation. These costs may differ even for one crop, depending on method of cultivation, e.g. use of manual labour or tractors and other machinery, with or without fertiliser and/or improved seeds, etc. Also included in production costs are rent payments to a landowner or allowance for sharecropping where part of the produce is taken by the landowner *in lieu* of rent. These costs can be considerable.
- For fixed investment costs that must be made once in a couple of years, the level of depreciation can be used in the calculations. This also applies to livestock that provides returns (e.g. milk) over a certain period, after which they are slaughtered or sold. In this case, the whole productive life of the animal must be considered with costs and revenues spread over this period.
- Production per hectare must be known for each type of cultivation method, e.g. with/without fertiliser and/or improved seeds. The use of machinery can make a large difference to productivity.
- Additional income must be known, both from agricultural and from non-agricultural sources.

These data are necessary to calculate fair prices.

5.2 Fair price

The minimum price the producer should receive in order to attain a LI is called the 'fair' price. In this context, 'fair' does not imply that a higher price cannot be more beneficial, but that any price below this fair price is not sufficient, making this the absolute minimum price the producer must receive. As mentioned before, 'fair' can be interpreted in different ways. 'Fair' here is derived from the 'Universal Human Rights', adopted by the UN in 1948, and in particular, Articles 23 and 25 (see earlier). We can see the concepts of LW and LI as a way of fulfilling these human rights. On this basis, we use them here as a minimum basis for the calculation of fair prices.

This price is based on the assumption of full employment working on the specific crop. When we know the average or mean size of the farm plots needed for full employment, all production costs and the LI, we can calculate the fair price for a specific crop. A fair price for coffee can be calculated using the following formula:

$$\text{Fair Price} = \frac{(\text{Total cost involved in coffee production} + \text{Living income} - \text{Additional income})}{\text{Coffee production}}$$

Additional income means all income from other sources, such as other crops, work as a labourer, etc. Because production costs are different in each situation, there is also a different 'fair' price in each case. This means that there is a different fair price for each type of production. For policy reasons, it is advisable to agree on a typical smallholder cost structure on which to base calculations.

When there is a gap between income needed and actual income, the underlying reasons should be analysed. If a full-time working farmer does not receive the income for his/her crops that he/she needs for a 'decent life', this is a reason to further analyse the pricing system in the production chain that the farmer depends upon.

5.3 Living Income/Fair Price Calculation model

Since 95% of the coffee farms are smaller than 5 ha, and 84% of all coffee farms are smaller than 2 ha, we will look at the example of a coffee farm of 2 ha. What can be considered a 'fair' price to the producer?

First, we will have to find out the average size of the family and how many full-time workers there are in the family. It can be assumed e.g. that a working adult works 250 days per year (calendar days – Sundays or any other weekly holiday, such as Friday or Saturday – festivals - social obligations - emergencies – sick days). The total number of working days for a household per year can then be calculated (number of adult workers × number of working days).[12]

Where LWs for the area have already been calculated, the LI can be calculated '(LW/month × number of adult workers × 12) + 10%)'. In the case of LWs, we must differentiate between the net LW that the worker receives, and the gross LW the employer has to pay. The amount of the net LW is used to assess workers' income. The gross LW is used for the costs of hired labour paid by the employer which are included in the production costs.

Yield and production cost per hectare may be known or must be investigated. We must multiply by two to determine the yield and costs for an area of 2 ha. The average farm for which we want to calculate the fair price for coffee must be carefully chosen. There may be great differences in production costs per hectare for small and large farms. Whilst a 'fair' price for coffee will be calculated, the farmer often also grows secondary crops. When no fair prices are known for these secondary crops, they may be valued at opportunity costs.[13] Home-grown food that is consumed by the family may also be valued as an opportunity cost.

We assume that there is full employment on the farm. In terms of the price calculation, this implies that the revenues should equal the total production costs + LI. The calculation of a fair price is shown below:

Living Income/Fair Price Calculation model of coffee

Average number of adult earners per household	to be determined
LW pp/year (per earning adult)	amount
Total working days/year (calendar days – weekly holidays – festivals - social obligations – emergencies– sick days)	number of days
LW per adult earning family member per working day	amount
(LW pp × number of adult earners) / working days	
Size of the farm in ha	to be determined
Living income per year	to be determined
Production in kg	to be determined
Production costs	to be determined

12 It is not possible to give a uniform number of days since much depends on cultural values, climatic circumstances, etc.

13 Opportunity cost: the value of that which must be given up to acquire or achieve something (*Penguin Dictionary of Economics*).

Revenue other products	***product A***	***product B***	***other***	total
	to be determined	idem	idem	amount
Production costs other products	***product A***	***product B***	***other***	total
	to be determined	idem	idem	amount
Additional income (revenue - costs)	***product A***	***product B***	***other***	total
	amount	idem	idem	idem

$$\text{Fair price per kg} = \frac{(\text{Total cost involved in coffee production} + \text{Living income} - \text{Additional income})}{\text{Coffee production in kg}}$$

We can also calculate what the area of coffee plantation should be to obtain a LI with actual prices. This calculation is made assuming that the family can work full-time on coffee, without having to employ other labour. When additional labour is required, this must be calculated at gross LWs. The base of the calculations is:

$$\text{price}(p) \times \text{quantity}(q) \text{must equal living income} + \text{total costs.}$$

In making these calculations, we face the problem that costs can change by area, even if the production method does not change. In addition, when prices change, the quantity to produce (q) changes, together with the costs to produce the required quantity, so careful calculation is necessary.

The required size of the landholding does not only depend on price; production costs are very important as well. Economies of scale in the case of larger or smaller landholdings will change the proportion of production costs. When we know production costs, LI and price, the size of the landholding can be calculated. On the other hand, if the size of the landholding is known and the associated production costs are also known, fair prices for this landholding can be calculated, assuming that the landholding has full employment.

After calculation of a 'fair' price, the next step is to compare the fair price with the actual price the producer receives. When there is a gap between these two, the underlying reasons for the gap should be analysed. Are there blockages in the supply chain such that monopolistic or oligopolistic players are able to secure higher profits than other actors in the chain (such as roasters, importers, traders and retailers),[14] a lack of cooperation among producers to improve their bargaining power etc.? Where this is the case, who should take responsibility for tackling these issues (governments, companies, international organisations,

14 An example of how different ways of calculation have different results on final consumer prices is given in Bronkhorst (2020) Table 8.1.

NGO's, cooperatives of coffee producers, labour unions)? A fair price does not appear as by magic but requires a lot of effort to change existing structures.

6 Conclusion

An important aspect of fair trade should be the price that is paid to the producer. To allow the farmer to lead a decent life and to pay for additional investments to cope with factors like changing to more sustainable production methods and adjusting to climate change, the price the farmer receives for his/her products should be a 'fair' price. To define 'fair', we have taken the Universal Declaration of Human Rights as a base. Application of these basic human rights leads to the concepts of an LW and LI that are used to define fair prices. A 'fair' price for the producer can be defined as the price for a product that includes all production costs and at least a 'Living Income' for the farmer and 'fair wages' for all workers involved, plus a small percentage for necessary investments.

Payment of fair prices is such an important aspect of fair trade that we have taken a closer look at its application in this chapter. This has been done by discussing the approaches taken by FT, RA, the 'Living Income Reference Price Formula' and the 'Living Income/Fair Price' approach.

RA and FT have their own reasons to base their policies on more than fair prices. RA sees their approach as holistic and focuses on helping farmers grow their businesses and become more profitable and resilient through training in farm management, financial literacy and market access. The mission of FT is to connect disadvantaged producers and consumers, promote fairer trading conditions and empower producers to combat poverty, strengthen their market position and take more control over their lives.

Both FT and RA are working hard to achieve LIs for farmers. They must be credited for the fact that they function as guides for other organisations and have put the position of the farmer on the agenda. The two organisations have a leading role within the group of major VSS organisations.

The approach of FT to fair prices is very promising and their approach to calculate fair prices could serve as an example for other organisations. However, it is possible to criticise FT approach:

- The producer is faced with too low an income now. Supporting a better income in the future is good, but producers face immediate costs such as sending their children to school, paying for healthcare services and all kinds of other bills. FT approach is not yet sufficient for the producer to earn the income that he/she is entitled to on the basis of the Universal Declaration of Human Rights.
- FT should take into consideration that the cost of sustainable production is not the same as current costs faced by the producer. First, the farmer

needs the resources to achieve truly sustainable production. That means that calculations with more realistic costs are required, as in the living income/fair price approach.

- The Fairtrade premium is paid to the farmers' organisation, so the individual farmer cannot spend it the way he/she likes.
- The application fee for the Fairtrade certification and the annual fee make production costs even higher. It would seem more reasonable for these costs to be paid by the customer instead of by the producer.

In the case of the RA, the amount of the Sustainability Differential is not fixed. This makes it very difficult to assess the final income of the producer and therefore impossible to see whether the amount is sufficient. An advantage of the RA approach is the fact that the farmers can use the Sustainability Differential as they wish.

The price the farmer receives for his/her products should be a 'fair' price. Trade conditions, support to farmers, subsidies on inputs, better access to markets and other support measures help to lower production costs, but when the prices paid for their products are not sufficient, this does not solve the basic problem for farmers and workers of too low an income. Whatever other support coffee farmers may receive, the bottom line is what they gain with their work.

Important work on LI reference prices has been done by GIZ (Krain et al. (discussed earlier). This methodology has the advantage that it is a general method that allows the setting of prices against any benchmark including, e.g. poverty lines. Both current and improved farming practices can also be taken into account.

The chapter has also discussed the Living Income/Fair Price method in detail. It presents an approach to calculate fair prices based on the real conditions the farmer finds him/herself in. Rather than theoretical or optimal yields, this method accounts for actual yields on the ground.

From these calculations, conclusions can be drawn about whether prices should be raised or not. If this appears to be necessary, these calculations can provide a guide on what amount prices should be raised to. When fair prices are paid, this will have effects on a whole society. Income of both farmers and workers will rise, leading to more purchasing power in the rural areas. Because of multiplier effects, this will lead to more economic activity in the coffee-producing regions.

Fair prices should not be limited to VSS only. Governments and companies also have a role to play. Much can be learnt from the ongoing introduction of the Living Income Differential (LID) in cocoa. This LID was introduced in 2019 by the governments of Ghana and Cote d'Ivoire, to apply a premium on the export price of cocoa from the 2020/21 crop. This additional revenue was intended to increase the incomes of farming families to help them achieve a LI. A negative effect is that more cocoa is sold on the future market so that companies can avoid direct purchase. However, it will be interesting to follow the development of this project and see if, and if so how, this model can be applied to coffee as well. This means that responsibility for fair prices will be down to not VSS alone but also governments and companies.

It should be kept in mind, however, that payment of fair prices to producers will only achieve its goal of economic and social development when workers are paid fair wages. If not, the money will remain with the (large) producers, both within and outside the country of production, and not help the poorer segments of the population.

7 Abbreviations

FAO	Food and Agriculture Organization of the United Nations
FLIRP	Fairtrade Living Income Reference Price
FLO	Fairtade Labelling Organisation International
FMP	Fairtrade Minimum Price
FT	Fairtrade International
FOB	Free on Board
GIZ	German Agency for International Cooperation
ILO	International Labour Organization
LI	Living Income
LW	Living Wage
NGO	Nongovernmental Organisation
PPP	Purchasing Power Parity
RA	Rainforest Alliance
SDG's	Sustainable Development Goals
SPO	Small Producer Organisation
UN	United Nations
VSS	Voluntary Sustainability Standards
WB	World Bank
WHO	World Health Organization

8 Where to look for further information

A book on Fair Prices is Bronkhorst, Ruud: *The Economics of Human Rights: Using the Living Income/Fair Price Approach to Combat Poverty*, Palgrave Macmillan, 2020. In this book, the Living Income/Fair Price Approach is analysed, and Living Income Reference Prices and the effects of the payment of living wages, income and fair prices on producers, labour, economy and environment are discussed. The book starts with history of the concept 'fair'.

On the subject of Living Wages, the standard work of Richard and Martha Anker is recommended: *Living Wages Around the World: Manual for Measurement*, Edward Elgar, 2017.

The *Global Living Wage Coalition* is engaged in both research and action on living wage, with the ambition of ensuring achieving a living wage for all.

Calculations of Living Wages can be found on their website https://www.globallivingwage.org/.

Calculations of Living Wages and minimum wages can also be found at the website of *WageIndicator.org*: https://wageindicator.org/.

The goal of the *Living Income Community of Practice* is to support activities focused on improving smallholder incomes towards living incomes, aiming to enable smallholder farmers to achieve a decent standard of living. Living Income calculations for selected areas can be found at their website https://www.living-income.com/.

InfoBridge Foundation (https://infobridge.org/) is developing a pilot training project on Fair Prices for agricultural researchers, practitioners and policy makers in developing countries.

9 References

Anker, R. (2011). *Estimating a Living Wage*. International Labour Organization.

Anker, R. and Anker, M. (2017). *Living Wages Around the World: Manual for Measurement*. Edward Elgar Publishing.

Bronkhorst, R. (2020). *The Economics of Human Rights: Using the Living Income/Fair Price Approach to Combat Poverty*. Palgrave Macmillan.

Coffee Barometer. 2020 https://coffeebarometer.org/.

FLO. Available at: https://www.fairtrade.net/.

FLO (2018). Available at: https://files.fairtrade.net/standards/Guidance_Document_for_Fairtrade_Coffee_Pricing_2_0_EN_April_2018.pdf.

FLO. Available at: https://files.fairtrade.net/Fairtrade_Minimum_Price_and_Premium_Table_EN_PUBLIC.pdf.

FLO. Available at: https://files.fairtrade.net/2019_FairtradeLivingIncomeReferencePrice_Model.pdf.

FLO. Available at: https://www.fairtrade.net/news/fairtrade-launches-its-first-living-income-reference-prices-for-colombian-coffee.

Global Living Wage Coalition. Available at: https://www.globallivingwage.org/.

Krain, E., Gyimah, J. O., Pumpuni, I., Kwapong-Akuffo, N. Y. and Kuntze-Fechner, M. (2021). *Analysis and Report of a Baseline Study for a Living Income (and Other Benchmarks) in Cashew- and Cocoa-Growing Regions of Ghana*. GIZ.

Panhuysen, S. and Pierrot, J. *Coffee Barometer 2020*.

Rainforest Alliance. Available at: https://www.rainforest-alliance.org/faqs/difference-between-rainforest-alliance-certified-fair-trade.

UN. 'Universal Declaration of Human Rights'. Available at: http://www.un.org/en/universal-declaration-human-rights/.

UN Sustainable Development Goals. Available at: http://www.un.org/sustainabledevelopment/hunger/.

Vaughan-Whitehead, D. (2010). 'Fair Wages'. *Edward Elgar World Bank*. Available at: http://povertydata.worldbank.org/poverty/home/World Bank.

Xavier, V. A. (2020). *True Price as Condition of Sustainability: The Global Coffee Crisis (1999–2003) and the Brazilian Amazon as Case Studies*. Campinas.

Chapter 6

Advances in Arabica coffee breeding: developing and selecting the right varieties

Álvaro Gaitán-Bustamante, Juan Carlos Arias-Suarez and Claudia Patricia Flórez-Ramos, Colombian National Coffee Research Center (CENICAFE), Colombia

1 Introduction

Coffee plantations are found in 78 countries and cover 10.6 million hectares worldwide, with 85% used for the *Coffea arabica* species, and are mostly cultivated by smallholder farmers that produced 171 million bags of green coffee in 2019 (ICO, 2020). These plantations face many challenges, ranging from global warming to devastating diseases such as coffee leaf rust (CLR) or stricter commercial requirements and new consumer trends. While more features are demanded from a historically narrow genetic pool of commercial coffee genotypes, in order to become more accessible and economically viable for farmers, the effort to widen the genetic background available for breeding purposes is becoming increasingly scarce. The majority of main crops have active breeding and seed-producing systems, supported by the private sector; however, coffee can be considered an 'orphan' crop, where breeding is not an economically attractive activity and investment involves a very long

http://dx.doi.org/10.19103/AS.2021.0096.13

period prior to any return. For decades, the difficulties in sustaining long-term breeding programmes, and delivering affordable and large quantities of improved planting material to coffee producers, have resulted in unchanging genetic composition in the varieties planted. It also casts a shadow on coffee sustainability for producers worldwide, the related business and personal enjoyment for millions of consumers every day.

Since the collapse of the International Coffee Agreement in 1989, funding for many research institutions devoted to coffee agronomy was severely cut or even discontinued, which greatly affected longstanding efforts such as coffee-breeding programmes, the prospection of germplasm banks and production of certified seeds. Meanwhile, more complex races of the rust pathogen *Hemileia vastatrix* have been evolving, diseases, such as coffee berry disease (CBD) periodically spread into new areas, climate variability continuously threatens extreme growing conditions, and the cost of land, agricultural input and labour place a permanent demand for more efficient and economically successful farming practices. Other factors include health and environmental concerns, in addition to strict legislation progressively calling for cleaner and more sustainable production, while more sophisticated customers continuously look to the market for richer cup profiles.

Although the solutions to ensure coffee supply for years to come must be tailor-made for every country or even for every coffee-producing area, in general, they have to be comprehensive, integrating agronomy with other fields of science, funding research for longer terms, increasing the availability of agricultural extension, providing special bank lines of credit for farmers and implementing fairer trading schemes. Nonetheless, in this complex landscape, genetically improved Arabica coffee varieties remain a key component of the toolbox necessary for the coffee chain to be innovative, environmentally friendly and profitable.

This review will present the main components and achievements of the global effort focused on breeding Arabica coffee, and the perspectives for present and future initiatives.

2 Reproductive behaviour of Arabica coffee

From their original sites in what is now known as Ethiopia, growing between 6° and 9° north, at altitudes ranging from 900 m to 1950 m above sea level, at an average temperature of 18–20°C, with a marked dry season lasting 4-5 months and a median annual rainfall of 1500–1800 mm, selected coffee seeds were transported to Yemen and cultivated under controlled conditions for many years until the Europeans dispersed those genotypes to their colonies in southern Asia (Jaramillo, 2018; Davis et al., 2018).

The historical genetic bottleneck was accentuated when some coffee plants were moved, around 1706, to botanical gardens in the Netherlands and France. A few plants reached the New World and spread from the Caribbean islands to the rest of the Americas (Smith, 1985; Wellman, 1961), in a genetic line later known as Typica, a tall plant (2.5 m in height after 24 months in the field), with young bronze leaves at the tip of the branches and a relatively large bean size. In 1860, a second genetic line, originating from coffee plantations on Reunion Island, arrived in Brazil, and later moved across the Americas. Known as Bourbon, this variety was also a tall plant, with young apical green leaves, smaller bean size and distinct cup quality. Typica and Bourbon have been the basic varieties of multiple breeding initiatives, and for breeding purposes, the following characteristics that apply to Arabica coffee genetics must be considered.

2.1 Allotetraploidy

The *Coffea* genus has a basic chromosome number of $n = 11$, which is also shared by other genera of the Rubiaceae family (Chevalier, 1947), and all species in the *Coffea* section are diploid with $2n = 22$, except *Coffea arabica,* which displays a tetraploid cytogenetic map with $2n = 4x = 44$ chromosomes (Hendre and Aggarwal, 2007). Based on the location of its diversity centre and because of its simple meiotic behaviour, Carvalho (1952) suggested this species was an allotetraploid, i.e. derived from the chromosomal combination of distinct genomes. Narasimhaswamy (1962) proposed that the species resulted from an interspecific cross between *C. eugenioides* and *C. canephora*, which was later confirmed using Amplified Fragment Length Polymorphism (AFLP) markers (Lashermes et al., 1995), Random Fragment Length Polymorphism/Genetic in situ Hybridization (RFLP/GISH) markers (Lashermes et al., 1999), and more recently through a combined Genotyping-by-sequencing (GBS)/multi-labelled phylogenetic tree (Bawin et al., 2021) and the complete genome sequencing of the Caturra variety (National Center for Biotechnology Information (NCBI) Assembly GCA_003713225.1). *C. eugenioides* was determined to be the maternal progenitor from chloroplast analyses (Lashermes et al., 1999). Scalabrin et al. (2020), using genome-wide single nucleotide polymorphism (SNP) genotyping on 736 *C. arabica* accessions, found extremely low variability levels, which is compatible with the occurrence of a single polyploidisation event for the whole species. The speciation process in *C. arabica* is estimated to have occurred between 1.08 million and 543 thousand years ago, matching the late quaternary period (Bawin et al., 2021), explaining why the two sub-genomes behave as a genetic diploid for breeding purposes.

2.2 Autogamy

For many years, the arabica species has been considered to be mostly autogamous, making *C. arabica* exceptional among the *Coffea* genus, where only *C. heterocalix* and *C. anthonyi* are also self-compatible species (Davis et al., 2006). Based on the field studies carried out using phenotypic markers such as the wax mutant (yellow colour endosperm), Carvalho and Krug (1949) determined cross-pollination frequencies between 7.3% and 9% under Brazilian conditions, and less than 10% in Colombian experiments as reported by Castillo (1976). Jiménez and Castillo (1976) observed a natural self-pollination frequency of 92%. More recently, Philpott et al. (2006) and Jaramillo (2010) measured the pollination rates in coffee plantations under both shade and full sun exposure and found no difference between natural and self-pollination. On the other hand, autogamy rates ranging from 47.9% to 59% have been reported, with an important role of flower visitor insects and wind in cross-pollination (Raw and Free, 1977; Hutchinson, 2012); finally, using molecular markers, Berecha et al. (2014) reported a cross-pollination rate of up to 76% in two agroforestry coffee systems in Ethiopia. Self-compatibility is important for breeding as it promotes the easy distribution of genetically homogeneous material, and facilitates planting conditions of single genotypes in commercial plots, obtaining numerous viable seeds from single plants in a clonal-like propagation.

2.3 Homozygosity

Due to the low genetic diversity among cultivars, a high degree of homozygosity for each sub-genome is present in cultivated varieties (Lashermes et al., 1999); therefore, minimal hybrid vigour or phenotypic novelty can be expected after controlled pollination across genotypes derived from the Typica and Bourbon families (Sybenga, 1960). At the same time, the genetic composition of the arabica cultivars has profound implications on the fixation of traits introgressed by the breeder in a tetraploid genome requiring few generations to obtain offspring with very low segregation rates.

3 Arabica genetic resources

By definition, genetic resources refer to the total set of plants from which gene interchange is possible based on the degree of sexual compatibility (Harlan and De Wet, 1971), although current advances in genomics and plant biotechnology can overcome gene transfer issues among distant organisms. Brockhaus and Oetmann (1996) described genetic resources as plant material with current or potential value for food, agriculture or forestry. The conservation of these reserves depends on the sustainability of the germplasm banks that ensure the identity of the material, their survival, availability and the preservation of a genetic diversity that may hold the key to solving current and future challenges (FAO, 2010). Genetic resources in *C. arabica* may be classified into the following groups:

1. Varieties derived from Typica and Bourbon types and their mutants.
2. Varieties derived from crosses with the Timor Hybrid.
3. Introgressed lines derived from interspecific hybrids.
4. Wild and semi-domesticated genotypes.
5. Other plant species or organisms.

3.1 Varieties derived from Typica and Bourbon types and their mutants

As mentioned above, Typica and Bourbon are the main ancestors of the majority of the coffee cultivars existing today (Fig. 1), displaying good adaptation to many different environmental conditions worldwide, ranging from latitudes close to the Tropic of Cancer (23.5° north) to near the Tropic of Capricorn (23.5° south), in what is known as the 'Coffee Belt' (Jaramillo, 2018). Descendants of Typica and Bourbon come in two forms: single point mutants (Table 1), with mild to severe phenotypic changes, resulting in dwarfism (Pacas, Caturra, Villa Sarchí and San Bernardo), fruit colours (varying from deep red to yellow), bean size (Maragogipe, Mokka) and caffeine content (Bourbon Pointu or Laurina); or crosses among them such as the widely used Mundo Novo (Leon, 1962). Although unstable and hardly used, additional variations can be found in aneuploids, or plants with an abnormal number of chromosomes. These individuals show phenotypic features such as thinner and slimmer leaves, narrower branches, smaller flowers and fruits with a single seed (Sybenga, 1960), or in the case of Monosperma, 1 seed per fruit and only 22 chromosomes (Mendes and Bacchi, 1940), and Bullata, with thick leaves and an octoploid kariotype (Chinnappa, 1969).

The onset of severe CLR epidemics, first in India and Ceylon (1869–1885) and much later (1970) in the Americas, became the main reason for developing coffee-breeding programmes around the world. Finding resistant arabica plants, such as the Kent genotypes, selected in India by L.P. Kent in 1911, enabled intraspecific crosses that took advantage of the genes later identified as SH1 (Susceptibility to Hemileia 1), SH2 and SH4, which together with SH5 (present in Bourbon and Typica) then completed the available pool for rust resistance in the species (Rodrigues Jr. et al., 1975).

3.2 Varieties derived from crosses with the Timor Hybrid

Unfortunately, resistance from Arabica genotypes was not sufficient to generate durable resistance to CLR. Other sources of resistance were needed and

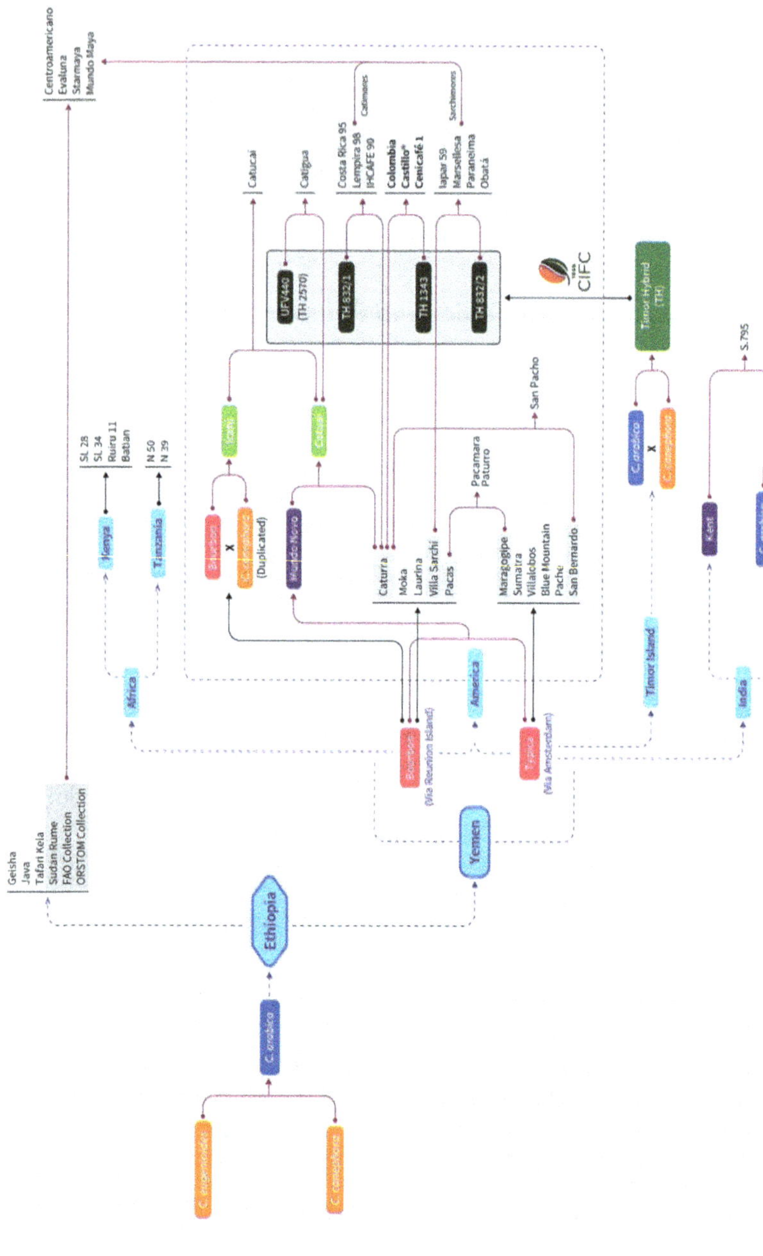

Figure 1 Main cultivars of *C. arabica* and their relationship with Typica and Bourbon types.

Table 1 Major single gene variations that define the *C. arabica* phenotype

Trait	Condition	Name	Variants					Dominance
Leaf shape and colour	Shape	Angustifolia	***AgAg***		***Agag***		***agag***	***Recessive***
			Normal leaf				Thin leaf	
	Apical leaf colour	Bronze	***BrBr***	***Brbr***			***brbr***	
			Bronze	Light bronze			Green	***Partial dominance***
	Leaf colour	Purpuracens	***PrPr***	***Prpr***			***prpr***	***Recessive***
			Green				Purple	
	Defective chloroplasts	Variegated	*Cytoptasma*					***Cytoplasmic inheritance***
Plant growth	Plant height	Caturra	***CtCt***	***Ctct***			***ctct***	
		San Bernardo	***SbSb***	***Sbsb***			***sbsb***	***Dominant***
		San Ramon	*SrSr*	*Srsr*			*srsr*	
			Dwarf				Tall	
	Angle of insertion of branches		***ErEr***		***Erer***		***erer***	***Dominant***
		Erects	Acute				Normal	
	Organ size	Laurina	***LrLr***			***Lrlr***	***lrlr***	***Recessive***
			Normal				Small	
		Maragogipe	***MgMg***			***Mgmg***	***mgmg***	***Dominant***
			Very large				Normal	
		Mokka	***LrLrMoMo***			***LrlrMomo***	***lrlrmomo***	***Recessive***
			Normal				Moka	
	Fertility and bean size		***NaNa***	***Nana***[a]			***nana***	***Recessive***
		Nana	Normal	Normal (Murt)			Sterile	

(*Continued*)

Table 1 (*Continued*)

Trait	Condition	Name	Variants				Dominance
Fruit and seed	Fruit colour	Xantocarpa	***XcXc***	***Xcxc***[b]		***Xcxc***	***Recessive***
			Red			Yellow	
	Endosperm colour	Wax[c]	***CrCr***		***Crcr***	***crcr***	***Recessive***
			Green			Amber	
	Calix in fruit	Developed sepals	***SdSd***	***Sdsd***		***sdsd***	***Partial dominance***
			No calix	Intermediate		Developed	
Fertility	Absence of pollen	Androsterility	***MSMS***	***Msms***		***msms***	***Recessive***
			Androfertile			Androsterile	
Disease resistance	CBD resistance	Genes T and R	***TT***	***Tt***		***tt***	***Dominant***
			RR	***Rr***		***rr***	
			Resistant			Susceptible	
		Gene k	***KK***	***Kk***		***kk***	***Recessive***
			Susceptible			Resistant	
	CLR resistance	SH genes	***SH5H***		***SHsh***	***shsh***	***Dominant***
			Resistant			Susceptible	
	M. exigua resistance	Mex-1	***Mex-1 Mex-1***		***Mex-1 mex-1***	***mex-1 mex-1***	***Partial dominance***
			Resistant			Susceptible	

[a]When originated from crosses with the Bourbon variety.
[b]Sometimes result in reddish colours.

[c]The Xenia effect occurs. Sources: Krug and Carvalho, 1951; Georget et al., 2019; Rodrigues et al., 1975; van der Vossen and Walyaro, 1980; Noir et al., 2003.

providentially found in the Timor Hybrid and its parental species *C. canephora*, with the genes SH6 through to SH10, and probably more (Barka, 2017). This hybrid originated in a Typica plantation established in 1927, at an altitude of 800 m above sea level on Timor Island, using seeds taken from a single vigorous tree found with evident resistance to CLR, leaf morphology comparable to *C. canephora* var. robusta, apical leaves with bronze colouring and a secondary branching similar to *C. arabica* (Bettencourt, 1973; Goncalves et al., 1977). The Timor Hybrid's accessions have been the most widely used material to generate new coffee varieties via introgression in controlled crosses through several generations.

According to the Portuguese Missão de Estudos Agronómicos do Ultramar (Commission for Agronomic Studies Overseas), segregant populations were identified among the descendants. Their combinatorial ability with arabica trees led to the deduction that a natural interspecific cross had occurred, with a non-reduced gamete of *C. canephora* combined with a normal *C. arabica* gamete, to generate a tetraploid plant (Bettencourt, 1973). The Centro de Investigação das Ferrugens do Cafeeiro - CIFC (Coffee Rusts Research Center), at the time located at Oeiras, Portugal, played a crucial role in identifying and using this population of the Timor Hybrid to generate parental material to be used worldwide. The first set of seeds arrived in 1957 and gave rise to the clones CIFC 832/1 and 832/2, resistant to all the rust races that were known at the time; a second set of seeds was received in 1960, denominated by CIF 1343, and later in 1966 with the introductions CIFC 2252, 2255, 2256 and 2257 (Rodríguez et al., 2004).

In addition to leaf rust resistance genes, resistance to CBD (*Colletotrichum kahawae*) and to nematodes (*Meloidogyne exigua*) have also been identified, which makes this genotype particularly interesting (Bettencourt, 1973; Goncalves and Pereira, 1998). Commercially, the varieties used have been derived from very specific lines generated by CIFC, i.e. the clones 832/1 (hybrid HW26) and 832/2 (hybrid H361), the parentals of the Catimores (such as Costa Rica 95, Lempira 98) and Sarchimores (such as IAPAR 59 and Marsellesa), and 1343/I.574, the parent of the Colombian varieties Colombia and Castillo.

The full potential of the Timor Hybrid awaits further exploration. Observations made by Silva et al. (2018) on 152 genotypes at the germplasm bank at the Universidade Federal de Viçosa (Brazil) suggest additional rust resistance genes in the Timor Hybrid, a much-needed resource as more virulent races of the fungus continue to develop.

3.3 Introgressed lines derived from interspecific hybrids

Interspecific coffee plants began to appear after the coffee rust epidemics in Ceylon and Java, from crosses between local *C. arabica* material and introduced commercial diploid species (Charrier and Eskes, 2004). In India, hybridisations among *C. arabica* and *C. liberica* date back to 1850, which later gave rise to tall varieties such as S.288 and S.795 with the rust-resistance factor SH3, and the addition of SH2 with a cross back to arabica with the Kent variety (Narasimhaswamy, 1960). Similarly, interspecific material has been obtained from crosses with *C. canephora* (Alvarado and Cortina, 1997). Since crosses between diploid and tetraploid parents result in chromosome instability if a triploid descendant is obtained, alternative methods have been used to restore the tetraploid condition, including duplication of chromosomes in the diploid parent or haploidisation of the tetraploid parent before pollination, and the backcrossing to arabica (Gimase, 2015).

3.4 Wild and semi-domesticated genotypes

Considering Ethiopia as the centre of origin and diversity of *C. arabica*, and Yemen as the secondary centre of diversification, several official prospecting missions were undertaken in order to collect wild and cultivated genotypes, including Lejeune, Bechtel & Meyer 1956, Meyer, Monaco, Narasimhaswamy, Fernie & Greathead 1964–1965, Charrier 1965–1975, Guillaumet & Hallé in 1966 in Ethiopia, Berthaud, Guillaumet & Lourd in 1977 and Eskes in Yemen in 1989. The main goal was to find new sources for resistance to CLR (Vega et al., 2008). Among these genotypes, three main groups could be defined according to Guillaumet and Hallé (1978): natural populations preserved *in situ*, plots established from local plants by farmers and selected populations brought from other locations.

Individual collecting trips were also performed by Archer (1951), Sylvain (1952–1954), Lejeune (1954–1956), Krug (1957), Anderson (1959), Bechtel (1957–1960), Meyer (1961–1962), in addition to the trip made by Wellman and Cowgill in the early 1950s (Eskes, 1989a). Exploring other species of the *Coffea* genus was motivated by the rapid destruction of tropical forests in continental Africa, Madagascar, and the Comoros and Mascarene Islands (Anthony et al., 1999). They were highly successful in collecting more than 20 000 accessions, representing approximately 70 *Coffea* species.

Ethiopian institutions have made outstanding efforts to preserve coffee genetic resources in *in-situ* and *ex-situ* collections for breeding purposes (Labouisse and Adolphe, 2012). Several studies indicate that a large portion of the diversity present in Ethiopian wild accessions remain unexplored. Recently, a group of 24 Yemeni accessions was identified with significant differences from the wild Ethiopians and from cultivated varieties derived from Typica and Bourbon (Montagnon et al., 2021). These findings constitute an unexplored reserve of germplasm with large potential for Arabica coffee breeding (Al-Murish et al., 2013; Scalabrin et al., 2020; Montagnon et al., 2021).

4 Coffee germplasm banks

The preservation of genetic diversity relies on germplasm banks, which also allow characterisation of the accessions in order to identify traits of interest (Anthony et al., 2007). In many species, this preservation is carried out through orthodox seeds that can be stored for years. Coffee seeds are neither orthodox nor recalcitrant but are the biological model of an intermediate classification where germination is not affected when the water content during storage remains either over 10% at low temperatures or at 20% with temperatures close to 18°C (Ellis et al., 1990, 1991; Walters, 2015). This biological condition makes coffee seeds viable for only a few months, therefore *ex-situ* plantations are required to maintain the collections that also preserve used and potential parental plants, as well as pre-breeding materials and released varieties.

Although *ex-situ* field collections have advantages for characterisation and distribution purposes (Reed et al., 2004), over the last 50 years they have also demonstrated a high susceptibility to rapidly losing genetic diversity (Dulloo et al., 2009). Pest and diseases, climate variability, lack of funding and changing agricultural policies constantly challenge *ex-situ* coffee germplasm collections. In this respect, seed storage is the most efficient method to preserve plant material. Efforts towards the cryopreservation of *C. arabica* seeds by Dussert et al. (1997, 1998, 2000) indicated that they could tolerate liquid nitrogen when dehydrated in a saturated solution of ammonium sulphate (81% relative humidity), obtaining 89% germination after a rehydration procedure (Dussert and Engelmann, 2006). These findings introduce a more efficient and secure preservation of arabica genetic resources. In fact, a cryogenic collection of *Coffea* germplasm has been set up at the Center of Genetic Resources in Brasilia (Dulloo et al., 2009).

According to the United Nations Food and Agriculture Organization -FAO- (2010), a total of 30 307 accessions of the *Coffea* genus are preserved worldwide, with the Ivory Coast holding the largest representation of 8000 accessions. Among the most outstanding arabica field collections worldwide are the Ethiopian collections at the Jimma Agriculture Research Center and the Ethiopian Biodiversity Institute (EBI), with over 5000 accessions in each institute, the collection at the Coffee Research Foundation (CRF) in Kenya with 592 accessions, the Central Coffee Research Institute (CCRI) germplasm bank in India with 329 accessions, the germplasm collection of the Tropical Agricultural Research and Higher Education Center (Centro Agronómico Tropical de Investigación y Enseñanza - CATIE) in Costa Rica with 2117 accessions, the Colombian Coffee Collection at the National Coffee Research Center (Centro Nacional de Investigaciones de Café - CENICAFE) with 881 accessions, and the germplasm banks in Brazil that include the Campinas

Agronomy Institute (Instituto Agronômico de Campinas - IAC) with 4000 accessions, the Parana Agronomy Institute (Instituto Agronômico do Paraná - IAPAR) with 1000 accessions of *C. arabica*, and a new collection established in 2003 at the Patrocinio Experiment Station by the Minas Gerais Agricultural Research Enterprise (Empresa de Pesquisa Agropecuaria do Minas Gerais - EPAMIG), with over 1000 accessions of *C. arabica*, including 200 of the Timor Hybrid (Bettencourt and Konopa, 1988; Anthony, 1992; Ebert, 2008; Labouisse et al., 2008; Dulloo et al., 2009; FAO, 2010).

Given the low genetic diversity available in *C. arabica* (Scalabrin et al., 2020), the diploid species constitute the largest source for coffee breeding. Out of the 124 known *Coffea* species (Davis et al., 2011), only 55% are represented in *ex-situ* collections (Davis et al., 2019). Among those, *C. canephora* is the best represented with more than 4000 accessions mainly preserved in Brazil (IAC), India (CCRI), Indonesia (Indonesian Coffee and Cocoa Research Institute - ICCRI) and French Guyana (L'organisme français de recherche agronomique et de coopération internationale pour le développement - CIRAD). For the rest of the genus, the current collections include continental African species in the Ivory Coast (Centre National de Recherche Agronomique - CNRA) and native species of Madagascar (National Centre for Applied Research and Rural Development - FOFIFA), accounting for around 6000 accessions belonging to 70 species (Dulloo et al., 2009).

An important activity in coffee banks, besides conservation, is accession characterisation in order to expand the genetic basis for improving commercial varieties, as first indicated by Nikolai Vavilov. Phenotypic characterisations of traits with high heritability and interest, such as yield and resistance to biotic or abiotic factors, will depend greatly on environmental conditions and the presence or absence of the disease-causing agents (Engels and Visser, 2007; Anthony, 1992). The current use of genetic markers and genomics allows the detection of misidentifications, the measurement of genetic diversity and the early identification of traits of interest (Engels and Visser, 2007; Bretting and Wirdlechner, 1995).

Berthaud (1997) proposed pre-breeding activities at germplasm banks to transfer genes of interest and generate new combinations of plant genotypes that can be easily used as parents for the improvement of particular commercial varieties that are currently planted. These activities could save time and resources in breeding perennial species such as coffee and facilitate local solutions for countries, regions or markets.

5 Breeding methods in Arabica coffee

C. arabica has been subjected to several breeding methods as a prevalent autogamous species, among which selection and hybridisation stand out (Krug and Carvalho, 1951). Selection works on highly heritable traits and does not generate variability in the offspring. On the other hand, hybridisation consists of controlled crosses of genetically different parents, to bring together the desirable characteristics of both parents into a single genotype. This can be done through simple crossings, when two parents are involved, or by complex crossings, when more than two parents are used (Ferh, 1987).

5.1 Conventional crossing

The genealogic or pedigree method has been the most widely used in *C. arabica* to obtain new varieties, requiring long periods for evaluation and selection that could require up to five filial generations and 25 years in order to fix the desired characters. The use of either genealogical or pedigree methods in coffee has been highly successful in the breeding of new varieties. This method starts with crossing of the selected parents (Fig. 2), followed by identifying the best plants in the progeny and successive self-pollination for 4-5 generations to reduce the segregation frequency, resulting in fixed traits in lines that can be released as seeds (Castillo and Moreno, 1988). The introgression of desired traits can be achieved in two ways. One, and the most common, is by perfecting the background of elite advanced materials with additional parent traits. The second consists of the large-scale development of populations adapted to local conditions from germplasm that has not been improved throughout long and repetitive processes of self-pollination, progeny evaluation and selection.

Since the fixation of a character using the pedigree method requires 15-25 years, while testing the plants in the field for complete production cycles, the generation of intraspecific F1 hybrids can take much less time, and the concentration of desired agronomical traits, heterosis or the enhanced expression in the offspring of certain characteristics when compared to its parents, is also possible.

5.2 Intraspecific hybrids

Intraspecific hybrids usually require the identification of genetically distant parents, defined by several parameters that include agronomical performance and phenotypic variation, and differential expression of chemical compounds (Villegas et al., 2014), diversity in geographic origin or polymorphism in molecular markers (Lashermes et al., 1996; Anthony et al., 2001; Chaparro et al., 2004). The evaluation of F1 populations resulting from crosses of accessions collected by ORSTOM in 1966, in terms of growth, yield and caffeine content, demonstrated effects of dominance and super dominance, with gains in hybrid vigour (Charrier, 1978).

Figure 2 Cross-pollination stages in the development of *C. arabica*-bred varieties. Female parent: (a) branches within 2–3 days of pre-anthesis flower buds are selected, and the stamens are removed (emasculation). (b) Before covering the branch with a waxed paper bag, any other floral structure is eliminated. (c) Just before artificial pollination, the top of the bag is cut out. Male parent: (e) flowers in pre-anthesis are removed and kept in a humidity chamber. (f) Pollen-donor flowers are taken to the field once they open. (g) The stamens are rubbed on the stigmas of the receptor. (h) Branches are covered again with a paper bag for 5 days and the number of thickened ovaries are counted.

In the evaluation of F1 hybrids obtained from wild Sudan-Ethiopian accessions and cultivated coffee varieties, created to improve adaptability and yield, as well as increase resistance to pests and diseases, and preserve cup quality, Bertrand et al. (1997) found that the hybrids were more productive than some commercial varieties, with a 30% increase in productivity compared with the best parent. Similarly, the comparison of 21 F1 hybrids with American commercial varieties at different altitudes and under two illumination systems, full sun exposure and shade, resulted in 58–34% higher yields, respectively, by the hybrids (Bertrand et al., 2011). More recently, when Marie et al. (2020) compared the agronomic performance and the physical and cup quality of nine *C. arabica* hybrids against the Caturra and Marsellesa varieties, the best hybrid outperformed their yields by 50%.

However, it must be noted that the F1 hybrid route requires additional effort to propagate and distribute the selected materials since, at the F2 stage of breeding, seed propagation results in the maximum frequency of progeny segregation. Hybrid propagation is currently achieved using three methods: manual pollination, as used in Africa for Ruiru 11 (Opile and Agwanda, 1993), tissue culture propagation through somatic embryogenesis, followed by microcutting multiplication (Georget et al., 2017; Etienne et al., 2018) and the use of androsterile lines for seed production (Georget et al., 2019; Berry et al., 2021).

5.3 Coffee tissue culture

Tissue culture allows the replication of identical plant material under laboratory conditions, a useful resource when a single plant (an accession or a mutant of interest) needs to be preserved and multiplied.

The most developed technique for coffee propagation using tissue culture is somatic embryogenesis. This method allows the multiplication of plants using leaf explants, which generate embryogenic tissue that later differentiates into embryos that, in turn, become whole plants for the nursery, with the attribute of being genetically identical to the parental plant. In short, calluses are induced from leaf explants using a solid medium with 2,4-dichlorophenoxyacetic acid -2,4 D- and isopentenyl adenine -2iP- hormones. After 1 month, explants are moved into embryogenic-inducing medium (Murashigue salts, 2,4 D and 6-benzylaminopurine -BAP-), incubated in the dark and later transferred into proliferation medium (MS, 2,4 D and Kinetine) until globular embryos are obtained. Finally, the torpedo embryos are incubated in regeneration medium (Murashigue-Skoog salts, Kinetine and 1-naphthaleneacetic acid) until leaves are formed, and they are exposed to a photoperiod of 12 h for complete differentiation (Murashige and Skoog, 1962; Murashige, 1974; Samson et al., 2006; van Boxtel and Berthouly, 1996), in a process that takes at least 8 months.

Propagation of F1 hybrids requires tissue culture and micrografting for commercial purposes, which makes the method expensive and limits its adoption. The most used method in *Coffea* sp. for massive propagation of hybrids is the Automated Container for Temporal Immersion, known as RITA® system for its acronym in Spanish, or temporary immersion bioreactor (Teisson and Alvard, 1995; Georget et al., 2017; Etienne et al., 2018). Depending on the genotypes, the performance varies from 15 000 to 50 000 somatic embryos per gram of embryogenic suspension cells (Ducos et al., 2007). A more efficient system was reported by Etienne et al. (2013) with the MATIS® system, where an increase in the incubation surface favours embryo dispersion while the light is easily transmitted, resulting in a better conversion of embryos to plants (91%). Finally, Ducos et al. (2007) developed another system called Box-A-Bag, where up to 20 000 embryos can be obtained in 10-litre bags, compared with 1000 embryos per container with RITA®. Large-scale somatic embryogenesis and micropropagation have resulted in remarkable increases in efficiency and cost reduction, allowing the delivery of more than 20 million plants to coffee producers in the decade before 2018 (Etienne et al., 2018); however, the volumes and low costs of seed production remain more economical.

5.4 Use of androsterile parents

Seeds originating from the pollination of androsterile ovules will ensure natural cross-pollination, and will originate large quantities of F1 hybrid plants without the use of tissue culture or micropropagation. Androsterility is the incapacity to produce functional male gametes and includes all the irregularities that affect the male reproductive system (Mazzafera et al., 1989). Adverse growing conditions may cause androsterility, as well as diseases or mutations (Budar and Palletier, 2001), with three forms known: structural, with the absence or malformation of male organs; sporogenic, when anthers are present, but pollen is absent; and functional, when pollen is incapable of germinating (Kaul, 1988).

Although androsterility has been observed in the *Coffea* genus, dating as far back as 1961 in Brazil and India, it was only after the evaluation of more than 5500 plants, in the coffee germplasm collection in the Instituto Agronômico de Campinas (Brazil), that Mazzafera et al. (1989) reported six androsterile plants in *C. arabica*. These plants came from the FAO Ethiopian collection and from a cross between Blue Mountain and the Timor Hybrid originated at CIFC; they have normal anthers but no pollen. Genetically, the androsterility gene (ms) displays recessive behaviour. More recently, Dufour et al. (1997) reported five more androsterile plants among 7000 evaluated at CATIE in Costa Rica, which were also of Ethiopian origin and had sporogenic androsterility. Georget et al. (2019) reported the success on seed multiplying an F1 hybrid by using the androsterile line CIR-SM01 with the commercial variety Marsellesa (from the

Sarchimor line), to obtain the hybrid population Starmaya. In addition, Berry et al. (2021) verified a trueness to type close to 94% of the seed obtained in multiplication plots of two commercial F1 varieties using androsterile lines. This finding will significantly impact volume and cost per plant of hybrid varieties in the future.

5.5 Interspecific hybrids

Interspecific breeding incorporates traits from other species of the same genus into *C. arabica*. The closest species of interest is *C. canephora*, where this intercrossing occurs naturally, resulting in hybrids with some fertility. Since the ploidy difference poses a barrier for the use of this genetic source, three methods have been developed to handle the F1 generation: (a) duplication of the diploid donor with colchicine before crossing with *C. arabica*, and later a backcross with arabica was performed in Brazil to generate the Icatu variety; (b) duplication of the F1 after the interspecific cross, before backcrossing to Arabica; (c) direct backcrossing of the F1 triploid to arabica, which is highly sterile, due to extensive meiotic abnormalities, but with a recovery of fertility after a few generations (Gimase, 2015). Interspecific protoplast fusion in *C. arabica* has been reported for *C. arabica* without further developments (Cordeiro et al., 2000).

5.6 Genome modifications

For a perennial crop, with late reproductive stages, the coffee-breeding methods described above can be slow, costly and highly inefficient when rapid changes are required. Deep exploration of *C. arabica* genomes is paving the way to identify key genes and chromosomal regions that will facilitate the early selection for traits of interest, reducing the number of plants to be tested in the field and saving economic resources. Currently, three methods are helping in gene function studies to unravel the potential of the genetic resources for coffee and eventually translate that information into new varieties: transgenesis, with its ability to introduce new traits from distant species, in a very controlled manner; gene silencing, turning off selected features that result in advantages for the plants and gene editing, introducing single point changes that could result in major gains in the field. Genetically modified *C. arabica* plants have been obtained by introducing the *Cry1Ac* gene from *Bacillus thuringiensis* that confers resistance to *Perileucoptera coffeella* (Leroy et al., 2000), and with the genes α-AI1 from the common bean (*Phaseolus vulgaris*) and Cry10Aa that generate resistance against the Coffee Berry Borer (CBB; Barbosa et al., 2010; Valencia-Lozano et al., 2019). However, low transformation efficiencies and poor public perception towards GMOs (genetically modified organisms) are delaying the release of these developments as commercial varieties. Breitler et al. (2018) reported the first proof of concept for gene editing in *C. canephora* using CRISPR/Cas9, and similar applications in *C. arabica* are expected soon.

6 Selection criteria for breeding

Although many traits have been genetically characterised, as indicated in Table 1, only genes T and R for CBD resistance and the presence of the rust resistance factor SH3 have associated molecular markers to be used in assisted selection (Gichuru et al., 2008; Mahé et al., 2008; Gimase et al., 2020a,b). For all the other traits of interest, field trials are necessary to confirm their presence and fixation in the progenies. The main traits selected from field measurements are discussed in the following sections.

6.1 Yield

Yield potential is a key target in coffee breeding. The price of land and the pressure on using environmental resources have driven the development of both new varieties and the agronomic practices that accompany them to promote higher yields in less area. Precocity, or a plant's ability to rapidly enter into its reproductive physiology, is an important, although elusive, trait. Other aspects, such as quality and disease or pest resistance, become less interesting if the yield does not meet expectations. Studies indicate that heritability of this trait is high, ranging from 0.35 to 0.88 (Walyaro, 1983; Castillo and Moreno, 1988; Fanelli Carvalho et al., 2020), with starting selection for this trait being from F3, since plants are still highly heterozygous before that generation (Castillo and Moreno, 1988).

Once coffee plants enter their productive stage (18–24 months after planting, depending on the average temperature and the number of thermic hours), *accumulated yield* must be determined, accounting for the total weight of cherries per plant in the observation plot for a 4-5-year production cycle. This requirement makes yield evaluation very expensive, therefore, alternative tools to increase selection efficiency are becoming imperative for coffee breeding. In this respect, Castillo and Moreno (1988) and Walyaro and Van der Vossen (1979) indicated that the data from the harvest of a few years, in early generations, is enough to determine the yield potential of a genetic line, conclusions corroborated more recently by Andrade et al. (2016) and Filho et al. (2018). Other methods based on molecular genetics include characterisation of quantitative trait loci (QTL; Moncada et al., 2016) or genomic selection (Sousa et al., 2019), which can increase the efficiency of breeding programmes by 331%, mostly by reducing the time in the cycle for field evaluation (Fanelli Carvalho et al., 2020).

Along with yield, it is important to estimate the *conversion factor*, which is the mass of cherries required to obtain 1 kg of parchment coffee, and that must correspond to 5 kg or less. Similarly, the *performance factor* must be calculated, which corresponds to the mass of parchment coffee needed to obtain 70 kg of green coffee, and that should be approximately 94 kg or less. Since yield is a relative response that depends on several factors beyond genetics, including agronomic management and environmental conditions, comparative trials are performed against the best local variety grown under the same conditions.

6.2 Bean physical and sensory qualities

A high content of abnormal beans in green coffee results in a significant reduction in yield and quality. According to Moreno (1977), the most important economic defects are empty seeds and peaberries. Other defects include monsters and triangle beans, which are infrequent and, therefore, irrelevant (Moreno, 1977). Castillo and Moreno (1981) indicated that the Timor Hybrid, the most frequent parent used for new varieties, has a large percentage of empty (16%) and peaberry beans (19.5%) that remain in the F1 stage after crossing it with commercial varieties, but that can be removed in further generations via selection. Another important feature is the percentage of Supremo beans, which are those retained by mesh measuring 17/64 of an inch. Although the percentage of large beans comes from heterotic effects (Georget et al., 2019) and may diminish over further generations, there is enough heritability to select materials with >70% of Supremo beans.

For selection purposes, the percentage of empty beans during the middle of the harvest season must be determined from a sample of 100 cherries (expecting less than 6%), and the percentage of peaberries (or "Caracolillo" beans) from a sample of 400 beans of dried parchment coffee (accepting less than 6%), and optionally, the percentage of Supremo beans from a sample of 100 green beans (values depend on the breeders' target).

In progenies originating from interspecific crosses, bean defects caused by both pre-zygotic and post-zygotic barriers (peaberries and empty fruits, respectively) are perhaps the most frequent (Narasimhaswamy, 1960); hence, selection against these traits or additional backcrossing for normalisation are necessary. Although occurring less frequently, the percentages of these defects are often higher in F1 hybrids when compared with their parents, leading to a possible fertility reduction (Bertrand et al., 2005).

Chemical differences have been clearly established between arabica and robusta beans, particularly in their major storage compounds including polysaccharides and sucrose, lipids, proteins and chlorogenic acids (De Castro and Marraccini, 2006), but the presence and relative quantity of many other compounds (including precursors for the roasting Maillard reactions) need to be studied to characterise the sensory attributes that define speciality Arabica coffees. Tran et al. (2018) reported SNPs related to caffeine and trygonelin content, 2 of the 5 most important non-volatiles in coffee. Similarly, Sant'Ana et al. (2018) identified candidate genes related to lipid and diterpene (Cafestol and Kahweol) composition, with potential use for improving cup quality. Employing traditional breeding methods, Silvarolla et al. (2021) crossed caffeine-free or low-caffeine *C. arabica* plants with elite varieties, to obtain F3 progenies with high yield potential and caffeine content of 0.1% or less.

For cup quality evaluation, 3-6 kg samples are taken from the plot during the middle of the second and third harvesting seasons. Ripe cherries are selected and processed until the humidity reaches 10-12%. A 500-g sample of dried parchment coffee is prepared for cupping. Nowadays, it is very common to use the cupping grades defined by the Specialty Coffee Association (SCA), but wide range determinations of chemical composition are also assessed using techniques such as near infrared reflective spectrometry (NIRS) and metabolomics (Montagnon et al., 1998; Bertrand et al., 2006; Kathurima et al., 2010; Pérez-Míguez et al., 2020).

The coffee chain, in general, must realise that even though genetics plays a role in the potential cup quality of a variety, final taste and fragrance/aroma result from many other factors, including environmental conditions, agronomical management, harvesting quality, cherry processing, milling, storage conditions, roasting curves and cup preparation method.

6.3 Growth variables and wide environmental adaptability

The development of dwarf varieties has been significant for the green revolution that coffee plantations experienced during the 1970s. This motivated the change in plot density, from 2000 plants per hectare up to 10 000, in addition to the use of edaphic fertilisers and regulation of shading conditions, resulting in a production jump that also allowed a more rational use of land in coffee-growing areas. According to Carvalho (1958), three independent genetic factors have been identified in coffee that reduce plant height by affecting the internode distance (Table 1): Caturra (Ct), San Bernardo (Sb) and San Ramon (Sr). Measurements for height selection are taken at month 24 of field planting, from the base of the trunk to the apex of the plant. The diameter of the canopy can also be determined if a particular architecture is desired. Caixeta et al. (2021), applying genome-wide association, found 110 SNPs significantly associated with five morphological traits, including plant height, for which they suggest a major QTL is located in chromosome 6 of *C. arabica*.

Bred-variety success depends not only on the gathering of appealing attributes for producers and consumers but also on their consistency through time (stability) and under different environments (adaptability). A variety has wide adaptation when it displays limited variation in its performance at several locations (Roy and Kharkwal, 2004). For this reason, the last step in the development of a commercial variety is the evaluation of the performance of candidate genotypes in a representative range of environmental conditions where it is expected to be cultivated (Allard and Bradshaw, 1964; Ferreira et al., 2006). Wide adaptation has been characteristic of coffee, considering the narrow genetic diversity that has been used in commercial plantations. In the case of Brazil, Mundo Novo and Catuaí varieties, developed by the Instituto Agronômico de Campinas (IAC), are planted in almost 1.6 million hectares at latitudes ranging from 4° to 25° south, and between 800 m and 1300 m in altitude. In Colombia, the varieties Colombia and Castillo General, released by CENICAFE, represent more than 80% of the 850 000 hectares cultivated in the country, located between 800 m and 2000 m in altitude (Garcia et al., 2014). Field evaluations require observations for at least 1 complete production cycle, considering the variations from year to year in the weather patterns, that in the tropical regions are heavily influenced by the El Niño Southern Oscillation (ENSO), which imposes changes in average relative humidity, rain distribution and quantity, thermal hours, and in turn, the incidence of plagues and diseases. Wide adaptation varieties have the advantages of potentially reaching more farmers and reducing the costs and logistics of seed production.

6.4 Resistance to CLR

Historically, resistance to CLR has been the main goal of arabica breeding. The economic and environmental impact of the disease, which for susceptible varieties can only be controlled with the use of fungicides or planting at high altitudes, has been calculated worldwide to be 2 billion dollars annually (Talhinhas et al., 2017), justifying the efforts to incorporate genetic resistance into commercial varieties.

Hemileia vastatrix Berk. & Broome (Basidiomycota, Pucciniales) is a fungus described in 1869 in West Africa as the causal agent of the orange leaf rust of coffee. Since then, it has spread to every corner of the coffee world. Large epidemics in Asia and Africa made this the main disease of *C. arabica* (Talhinhas et al., 2017), hence its name 'vastatrix', which means 'destructive'. America was free of the disease until 1970 when it arrived in Brazil and spread all over the continent in the next 16 years (Kushalappa and Eskes, 1989).

Around 1925, the Mysore Coffee Experiment Station was created in Balenhonnur, India, in order to develop varieties resistant to CLR through a systemic process of selection, obtaining S.795 and S.288, displaying high productivity and cup quality, in addition to rust resistance (Narasimhaswamy, 1960). In 1955, the CIFC (Centro de Investigação de Ferrugens do Cafeeiro or Coffee Rust Research Center), in Oeiras, Portugal, was created in order to facilitate international cooperation towards research on CLR and produce pre-breeding resistant materials (Rodrigues Jr. et al., 1975). A key contribution from CIFC was the characterisation of the Timor Hybrid (HdT), and the development of its derivate populations, mainly CIFC 832-1, CIFC 832-2 and CIF 1343, which were widely used in Latin America to generate bred lines known as Catimores (Caturra x HdT_CIFC 832-1), Sarchimores (Villa Sarchi x HdT_CIFC 832-2) and H3001- H3005 (Caturra x HdT_CIFC 1343), which gave rise to most of the rust-resistant varieties in the Americas and some in Africa.

The work of Mayne at Balehonnur allowed the identification of four physiological races of the pathogen, explaining why resistant varieties later became susceptible (Rodrigues Jr. et al., 1975), as the disease exhibits a gene-by-gene interaction. Testing 779 rust samples at the CIFC, brought from different regions worldwide, on a large collection of coffee germplasm, resulted in the characterisation of 30 rust races, with 15 further described afterward (Pinto and Marques, 2005). Due to the genetic bottleneck in arabica cultivars, which hold only the SH5 resistance gene, CLR race II is the widest distributed rust race globally, as it carries the corresponding virulence gene V5 (Rodrigues Jr. et al., 1975). Besides SH5, nine other resistant genes were later determined by CIFC; three originated from *C. arabica* (SH1, SH2 and SH4), one from *C. liberica* (SH3) and five more from *C. canephora*, via the Timor Hybrid, corresponding to SH6, SH7, SH8, SH9 and SH10 (Hisoshi Sera et al., 2007). Breeding programmes attempt to introgress those genes into commercial varieties. SH3 has been widely used in India since the 1920s, with arabised materials from interspecific crosses with *C. liberica* (Lashermes et al., 2010), which resulted in famous selected lines such as S.26, S.31, S.288, S.353, and the BA (Balehonnur Arabica) series BA8, BA10 and BA16 (Eskes, 1989).

Even with the availability of multiple major SH resistance factors, the high capacity of the fungus to mutate in a short time, compared with the time required to develop a new variety, makes it difficult to sustain resistance in the field. For selection under controlled conditions, complex rust races are required to artificially inoculate nursery plants in order to identify the presence of multiple resistance genes and probable durable resistance. Since rust race identification and amplification is itself a laborious process (*H. vastatrix* cannot be cultivated *in vitro*), selection is performed using natural inoculum in the field, evaluating rust incidence once or twice per year, depending on the dynamics of the CLR epidemics in the area, starting when plants are 24 months old and using the field scale of Eskes and Thoma-Braghini (1981). To date, only SH3 can be detected by molecular markers, one codominant (Sat244) and one dominant (BA-124-12k-f) as described by Mahé et al. (2008) and Ruas et al. (2017), and further work is needed.

Once major genes have been defeated by new races of CLR, the presence of residual incomplete resistance was evident, which retarded the infection process and reduced the number of spores per pustule. Observations indicate that this horizontal resistance is polygenic in nature, is largely influenced by environmental and agronomic conditions (Alvarado, 2011) and could also be considered in breeding strategies.

6.5 Resistance to CBB

Hypothenemus hampei is the most severe plague for coffee cherries around the world. Originating in Africa, it arrived in Brazil in 1930 and slowly spread to the rest of the continent, reaching the remainder of South America in the 1980s and Central America in the 1990s.

Generally, genetic resistance to insects is ineffective, and resistant plant varieties rarely offer a complete solution (Painter, 1951). However, even low levels of resistance can be useful in an integrated pest management programme as reducing the number of generations or the fitness of the individuals during the season will facilitate and enhance the efficiency of other control measures, maintaining the level of plague infestation below the economic threshold (Adkinsson and Dyck, 1984).

Three mechanisms have been described for plant resistance to insects, namely antixenosis (deterrence of the insect or lack of attraction to the plant), antibiosis (affecting the development and number of individuals) and tolerance, which is not significant in CBB as the organ attacked is of economic importance (Painter, 1951). Vuillet (1925) found that fruits from several species of the *Coffea* genus were evenly perforated by the insect, although certain genotypes of *C. liberica* were less susceptible. Sabino (1971) in Brazil found no statistical differences among the varieties Mundo Novo, red Bourbon, yellow and red Typica, red Caturra and red Maragogipe, all tested under laboratory conditions. Koch (1973), evaluating the Robusta Ineac, Robusta Ebobo and Conilon (Kouillou) varieties, found that the latter variety did not favour insect development. Villagran (1991) observed similar results with the species *C. kapakata*, and Duarte (1992), using olphatometric methods, determined that *C. arabica* was more attractive to CBB than *C. liberica* (Dewevrei) or *C. kapakata*. Finally, Idarraga et al. (2012) identified differentially expressed genes between *C. arabica* var. Caturra and *C. liberica* during the early response to CBB.

In field and lab trials performed at IAPAR, Sera et al. (2010) also found less severe attacks on *C. kapakata*, *C. eugenioides* and *Psilanthus bengalensis*. The use of artificial field infestations with entomological nets on the branches (Arcila et al., 1997), the complete coverage of the plant (Montoya, 1992), feeding the CBB larvae on meridic diets (Portilla, 1996), or the infestation of individual fruits or beans under controlled conditions enabled the identification of genotypes of *C. arabica* with 20–30% less infestation when compared with Caturra. The heritability of these traits must be evaluated by crossing the potential

CBB-resistant genotypes of interest with current elite varieties (Romero and Cortina, 2004).

6.6 Resistance to CBD

The fungus *Colletotrichum kahawe* var. kahawae is a coffee pathogen that attacks flowers and leaves, but particularly on green fruits, causing premature fall and mummification (Silva et al., 2006), and is responsible for major economic damage. While CLR is present all over the world, CBD is currently confined to Africa, where it was first reported in 1922 in Kenyan coffee plantations above 1700 m in altitude, with an increasing dispersion pattern inside the continent, and yield losses between 20% and 50% (Bieysse et al., 2002).

Besides the heavy use of fungicides, genetic resistance is an alternative. *C. arabica* cv. Ruiru 11, cultivated in Kenya, is resistant to CBD and CLR (Omondi et al., 2001), as well as the Batian variety (Gichimu and Omondi, 2010). However, testing for resistance in countries where the disease is not present is difficult. By performing tests in the field, as well as via hypocotyl inoculation, Van der Vossen and Walyaro (1980) identified three resistance genes in the varieties Rume Sudán (genes R and K) and K7 (gene K), and in the Timor Hybrid (gene T). While K7 is a commercial variety of the Kent type, Rume Sudan comes from the Boma plateau in Sudan, with small beans and poor yield, and therefore, requires a longer breeding procedure (Agwanda et al., 1997).

Since the disease is absent in many coffee-cultivating countries, it has been crucial to perform hypocotyl inoculation tests at CIFC in Portugal, using multiple isolates of the fungus and qualifying resistance using the 12-degree scale proposed by Van der Vossen and Walyaro (1980) or the 5-degree scale from Van der Graaff (1981, 1982). Currently, it is also possible to perform assisted marker selection, as indicated by Gichuru et al. (2008) and Gimase et al. (2020), using microsatellites and Single Nucleotide Polymorphisms (SNPs) linked to genes T and R, respectively, which may facilitate the selection procedure in the absence of the pathogen.

6.7 Resistance to other diseases and pests

Castro et al. (2013) reported the introgression of resistance to *Ceratocystis fimbriata*, the fungal agent of Ceratocystis or Macana canker, a disease affecting the open wounds produced by stumping in renewed coffee plantations or caused by stepping on the trees for support during harvesting in very sloped hills where coffee is cultivated on mountain areas.

Resistance to both root-knot (*Meloidogyne* spp.) and root-lesion (*Pratylenchus* spp.) nematodes has been observed in *C. canephora*, *C. liberica* and *C. excelsa*. In Central America and India, large-scale grafting of *C. arabica* scions into *C. canephora* rootstock cultivars, like Nemaya in Central America and Apoatã IAC 2258 in Brazil, has been the preferred method to use this resistance (Bertrand et al., 2002; Andreazi et al., 2015).

Progress has been achieved in the search for a tolerant variety against White Stem Borer (WSB), a major pest in India, causing annual losses estimated at US$17.5–26 million (Venkatesha, 2010), where Prakash et al. (2016) reported promising gene introgression results from the spontaneous amphidiploid *C. liberica* x *C. eugenioides* further crossed with the Timor Hybrid.

6.8 Adaptation to climate change

Many different models consider the possibility of adverse conditions for current coffee-growing areas if permanent changes in temperatures (median, highest and lowest) and rain patterns occur in the coming years (Camargo, 2010; Läderach et at., 2017). Such changes could affect not only the physiology of the plant for growth and flowering but also create better conditions for epidemics and fertiliser wash-off. Although a more integrated solution is required, involving the development and installation of resilient production systems, one possible component is a set of varieties that can tolerate dryer or wetter conditions, providing alternatives for the farmers to remain in the current coffee-producing zones.

Coffee genotypes that tolerate drought have a larger capacity to maintain a higher hydric potential in their tissues (DaMatta, 2004), while other genotypes hold the potential to improve their water-use efficiency (WUE), i.e. the ratio of water used in plant metabolism to water lost by transpiration, without reducing the growth rate or yield (Dias et al., 2007). Another mechanism is more efficient water extraction from the soil, either by reducing the osmotic potential or increasing the elasticity of tissues maintaining turgor under low water concentration (DaMatta et al., 2003). Studies on the drought-resistant Siriema variety (*C. arabica* x *C. racemosa* Lour.) indicate an early closing of stomata and a deeper rooting system (Dias et al., 2007). Finally, leaf venation patterns have been associated with tolerant genotypes in India (Mishra and Slater, 2012).

Excess water causes anoxia in root tissues, and several mechanisms have been identified in plants that are responsible for resistance to transitory low levels of oxygen, including the generation of glucose from starch and saccharose, and controlled acidification that results from the accumulation of hydrogen via the production of succinate (Arpagaus and Braendle, 2000; Bailey-Serres and Voesenek, 2008). These mechanisms remain to be studied in coffee plants.

6.9 Other traits of interest

The continuous pressure for higher efficiency and reduced production costs drives scientific interest towards introducing new features into coffee plants.

Improved photosynthesis has become one of those targets, in particular, when shade is required to provide adequate conditions for the plantations in terms of temperature and water retention; however, coffee plants seem to be performing very well under current increases in atmospheric CO_2 concentration (DaMatta et al., 2019). Sakiyama et al. (2017) described an interesting triple plagiotropic branch mutation on the Catuaí variety that results in an additional branch per node, changing the architecture of the plant and potentially increasing yield.

Another aspect is the suitability of plant architecture and physiology for mechanised and semi-mechanised harvesting technologies, selecting factors such as cherry detachment strength, defoliation, vegetative vigour after harvesting and efficiency of fruit harvesting, with interesting outcomes for reducing production costs (Dias et al., 2020).

7 CENICAFE's breeding programme: a case study for improved Arabica varieties

The Colombian National Coffee Research Center is a private institution created in 1938 by the National Federation of Coffee Growers of Colombia (Federación Nacional de Cafeteros - FNC), and funded with public money collected from the parafiscal contributions from green coffee exports. A National Coffee Congress, representing the interests and needs of the coffee producers in the country, annually defines the topics and amounts of funding for research. CENICAFE is a subsidiary institution of the Technical Management of FNC, and in 2020 employed 60 scientists organised into 12 departments, including 1 for plant breeding and 1 for regional experimentation, in charge of 8 experiment stations located across the country to carry out field tests and validations.

7.1 Proactive breeding and parallel advance of progenies

CENICAFE initiated a breeding programme for CLR resistance in 1952, with the introduction of genotypes and their agronomic evaluation during the 1960s. Before 1983, in the absence of the disease in Colombia, more than 55 000 plants were artificially inoculated, from 567 parents, at the CIFC using their collection of CLR races, a methodology that enabled the release of the Colombia variety 1 year before the disease arrived in the country (Castillo and Moreno, 1986).

This proactivity has continued in order to keep pace with the onset of new races in the country. A continuous schedule of crossings and advancement of progenies, until F5, enable a permanent evaluation of selected genotypes in multiple regional tests and supply the materials for the development of the multiline strategy.

7.2 Multiline/composite varieties

For a semi-perennial crop, the speed of change, either of environmental conditions, commercial aspects or pathogen and pest dynamics, poses a serious threat for the validity of a variety over time. The use of genetic diversity is key to compensate for variability in the surroundings. The suggested systems to introduce genetic diversity are numerous and include polygenic traits and the use of mixtures of lines and even species (Castillo and Moreno, 1986). For host-pathogen relationships, stability and effectiveness of resistance depend on the genetic balance of the pathogen populations due to reduced selection pressure and, therefore, very little adaptation to the offered plant resistance (Parlevliet and Zadoks, 1977). In this respect, monogenic lines have constantly failed to control epidemics in the long term (Castillo and Moreno, 1986), and the multilines concept provides a solution to the rational and long-term use of genetic resources. Multilines can also be dynamic (Alvarado, 2002) and allow replacement of the individual components that are not fulfilling the expectations of breeders, producers or consumers. Although an obvious and workable solution, multiline strategies have been difficult to implement, arguing high costs, being agronomically conservative, facilitating the onset of new pathogen races (Browning and Frey, 1969; Wolfe, 1985) and lacking phenotypic uniformity that is usual in monolines and desired by farmers.

Since 1982, CENICAFE has delivered 14 composite varieties that are currently planted on 82% of the 860 000 hectares that are responsible for the annual production of 14 million bags of green coffee. They vary between 4 and 35 components, and show durable resistance against CLR. The mechanisms by which genetic diversity using multiline varieties are effective may include a barrier effect on inoculum propagation among plants with different resistance levels, pathogen dissolution by increasing the distance among similar host plants, resistance induction with a non-compatible inoculum (Browning and Frey, 1969; Wolfe, 1985; Mundt, 2002), compensation for an average yield under different conditions as a result of environmental plasticity among genotypes (Reddy, 2012; Bowden et al., 2001; Roy and Kharkwal, 2004) and ecological stability that results in yields for the entire population greater than the average of the yields from individual lines (Creissen et al., 2013, 2016; Wolfe, 1985).

7.3 Seed production system

The Colombian National Federation of Coffee Growers has a certified seed production system for the varieties generated at CENICAFE, meeting the standards of the ISTA (International Seed Testing Association) and the ICA (Colombian Agricultural Institute). This has delivered around 80 tons of seeds

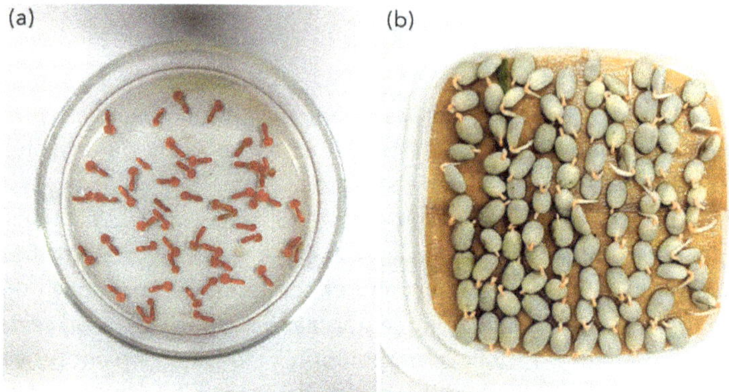

Figure 3 Seed viability tests. (a) Tetrazolium chloride staining of detached *C. arabica* embryos with the red colour indicating viable seeds. (b) Seed germination in humidity chambers with the percentage of radicle protrusion expressing viability.

per year since 1983, when the Colombia variety was released, representing 300 million new plants that annually rejuvenate coffee plantations.

The seed production system ensures an impact on the whole industry, materialising the technical change originated in the breeding programme and providing solutions to stay competitive through investment in research. Seeds must meet four requirements: (a) identity, assuring that all the components of the multiline are present; (b) appearance, presenting parchment beans with an amber colour; (c) viability, guaranteeing at least 75% germination and (d) health, with no more than 15 beans per kg displaying physical defects.

Embryo staining with 1% tetrazolium chloride allows the estimation of seed viability, confirmed by the determination of germination percentage after 25 days. For this test, 400 seeds without parchment are placed in boxes with wet paper (Fig. 3), and complemented by a vigour test, with the germination of 400 seeds on sand and checking for seedling growth parameters under controlled conditions.

In 2020, CENICAFE established 82.5 hectares of seed multiplication plots, located in six experiment stations across the country, to provide a continuous flow of seeds throughout the year, to ensure their availability when farmers require the seeds for the germination beds. A seed production system connected to the breeding programme facilitates reaching a large number of farmers with certified varieties, at very low prices, delivering 1 kg of seeds (approximately 4300 seeds) for less than US$10.

8 Conclusion

Cultivated varieties of *C. arabica* have a long history motivated mainly by the search for a solution to CLR, but most recently, guided by facing the challenges that new threats from pests and diseases, as well as novel requirements in adaptation to climate change, production costs and final customer expectative impose on the whole coffee business chain worldwide. The need to expand the narrow genetic background of the current cultivars, exploring the germplasm collections available, in addition to strengthening durable coffee-breeding programmes, and to support easy, cost-effective and certified bred-line distribution systems, are conditions that must be fulfilled in order to ensure sustainability for the producers and market. Science has provided tools to reach these goals, but they require adjustments and longstanding funding in order to have an impact on a perennial crop, with particular genetic features, such as the ones found in Arabica coffee.

9 Where to look for further information

- Centro de Café 'Alcides Carvalho' Instituto Agronômico (IAC), de Campinas, Brazil, http://www.iac.agricultura.sp.gov.br/areasdepesquisa/cafe/.
- Centro Nacional de Investigaciones de Café – CENICAFÉ, Federación Nacional de Cafeteros de Colombia, http://www.cenicafe.org.
- Coffee Research Institute, Kenya Agricultural and Livestock Research Organisation – KALRO, https://www.kalro.org/coffee/?q=node/25.
- Unidad de Agroforestería y Mejoramiento Genético de Café y Cacao (UAMCC), Centro Agronómico Tropical de Investigación y Enseñanza – CATIE, https://www.catie.ac.cr/unidad-agroforesteria-cafe-cacao-fe/.
- Institute français por le développement – IRD, https://www.ird.fr/projet-la-collection-du-centre-de-ressources-biologiques-crb-coffea.
- L'organisme français de recherche agronomique et de coopération internationale pour le développement – CIRAD, https://www.cirad.fr/nos-activites-notre-impact/filieres-agricoles-tropicales/cafe/contexte-et-enjeux.
- World Coffee Research, https://worldcoffeeresearch.org/.

10 References

Adkinsson, P. L. and Dyck, V. A. (1984). Variedades resistentes en los sistemas de manejo de plagas. In: Jennings, M. F. G. Y. and Manejo, P. R. (Eds), *de plantas resistentes a insectos*, Primera edición, México: Editorial Limusa, pp. 253-271.

Agwanda, C., Lashermes, P., Rouslot, P., Combes, M. and Charrier, A. (1997). Identification of RAPD markers for resistance to Coffee berry disease, Colletotrichum kahawae, in Arabica coffee. *Euphytica* 91, 241–248.

Allard, R. W., and Bradshaw, A. D. (1964). Implications of genotype-environmental interactions in applied plant breeding. *Crop Science* 4, 503-508.

Andrade, V. T., Goncalves, F. M. A., Nunes, J. A. R. and Botelho, C. E. (2016). Statistical modeling implications for coffee progenies selection. *Euphytica* 207(1), 177–189. https://doi.org/10.1007/s10681-015-1561-6.

Andreazi, E., Sera, G., Faria, R., Fonseca, I., Machado, A., Shigueoka, L., Carvalho, F. and Carducci, F. (2015). Behavior of 'IPR 100' and 'Apoatã IAC 2258' coffee cultivars under different infestation levels of meloidogyne paranaensis inoculum. *Australian Journal of Crop Science* 9, 1069–1074.

Alvarado, A. (2002). Mejoramiento de las características agronómicas dela Variedad Colombia mediante la modificación de su composición. *Avances Técnicos* 204. Manizales: Cenicafé, 8 p.

Alvarado, A. (2011). *El café y la roya: Estrategias de resistencia incompleta*. Manizales: Cenicafé, 96 p.

Alvarado, A. and Cortina, G. H. A. (1997). Comportamiento agronómico de progenies de híbridos triploides de Coffea arabica var. Caturra X (Caturra X Coffea canephora). *Cenicafé (Colombia)* 48(2), 73–91.

Al-Murish, T. M., Elshafei, A. A., Al-Doss, A. A. and Barakat, M. N. (2013). Genetic diversity of coffee (*Coffea arabica* L.) in Yemen via SRAP, TRAP and SSR markers. *Journal of Food, Agriculture and Environment* 11, 411–416.

Anthony, F. (1992). Les ressources génétiques des caféiers: collecte, gestion d'un conservatoire et évaluation de la diversité génétique, Collection Travaux & Documents Microfichés n°81, ORSTOM, Paris, 320 p.

Anthony, F., Astorga, C. and Berthaud, J. (1999). Los recursos genéticos: las bases de una solución genética a los problemas de la caficultura latinoamericana. In: Bertrand, B. and Rapidel, B. (Eds), *Desafíos de la Caficultura Centroamericana*. France, San José: IICA/PROMECAFE-CIRADIRD- CCCR, pp. 369–406.

Anthony, F., Bertrand, B., Quiros, O., Wilches, A., Lashermes, P., Berthaud, J. and Charrier, A. (2001). Genetic diversity of wild coffee (Coffea arabica L.) using molecular markers. *Euphytica* 118, 56–65.

Anthony, F., Astorga, C., Avendaño, J. and Dulloo, E. (2007). Conservation of coffee genetic resources in the CATIE field GeneBank. In: Engelmann, F., Dulloo, M. E., Astorga, C., Dussert, S. and Anthony, F. (Eds), *Complementary Strategies for Ex Situ Conservation of Coffee (Coffea arabica L.) Genetic Resources. A Case Study in CATIE, Costa Rica. Topical Reviews in Agricultural Biodiversity*. Roma, Italia: Biodiversity International, pp. 23–34.

Arcila, M. A., Bustillo, P. A. E. and Chaves, C. B. (1997). Residualidad de una aspersión de Beauveria bassiana en cafetales a libre exposición. In: *CONGRESO de la Sociedad Colombiana de Entomología, 24. Pereira, Colombia, Julio 16–18, Resúmenes*. Pereira: SOCOLEN, pp. 11–12.

Arpagaus, S. and Braendle, R. (2000). The significance of alpha-amylase under anoxia stress in tolerant rhizomes (Acorus calamus L.) and non-tolerant tubers (Solanum tuberosum l., var. Désirée). *Journal of Experimental Botany* 51(349), 1475–1477.

Barbosa, A. E., Albuquerque, E. V. S., Silva, M. C., Souza, D. S. L., Oliveira-Neto, O. B., Valencia, A., Rocha, T. L. and Grossi-de-Sa, M. F. (2010). α-amylase inhibitor-1 gene from Phaseolus vulgaris expressed in Coffea arabica plants inhibits α-amylases from the coffee berry borer pest. *BMC Biotechnology* 10, 44. https://doi.org/10.1186/1472-6750-10-44.

Barka, G. D. (2017). *Identification, molecular characterization and differential expression studies of genes activated during Coffea arabica L. - Hemileia vastatrix Berk. & Broome interactions*. Thesis PhD. Vicosa, MG. 149 p.

Bailey-Serres, J. and Voesenek, L. A. (2008). Flooding stress: aclimations and genetic diversity. *Annual Review of Plant Biology* 59, 313-339. https://doi.org/10.1146/annurev.arplant.59.032607.092752.

Bawin, Y., Ruttink, T., Staelens, A., Haegeman, A., Stoffelen, P., Mwanga Mwanga, J. I., Roldán-Ruiz, I., Honnay, O. and Janssens, S. B. (2021). Phylogenomic analysis clarifies the evolutionary origin of Coffea arabica L. *Journal of Systematics and Evolution* 59(5), 953-963. https://doi.org/10.1111/jse.12694.

Berecha, G., Aerts, R., Vandepitte, K., Van Glabeke, S., Muy, B., Roldán-Ruiz, I. and Honnay, O. (2014). Effects of forest management on mating patterns, pollen flow and intergenerational transfer of genetic diversity in wild Arabica coffee (Coffea arabica L.) from Afromontane rainforests. *Biological Journal of the Linnean Society* 112(1), 76-88.

Berry, V., Herrera, J. C. and Torres, J. (2021). Seed purity of the first two commercial coffee hybrids varieties (*Coffea arabica*) Star2 and Star3 in Ecuador. In *Proceedings of ASIC 28th International Conference on Coffee Sciences*. Montpellier France, 160.

Bertrand, B., Alpizar, E., Lara, SantaCreo, L., Hidalgo, M., Quijano, J., Montagnon, C., Georget, F. and Etienne, H. (2011). Performance of Coffea arabica F1 hybrids in agroforestry and full-sun cropping systems in comparison with American pure line cultivars. *Euphytica* 181, 147-158. https://doi.org/10.1007/s10681-011-0372-7

Berthaud, J. (1997). Strategies for conservation of genetic resources in relation with their utilization. *Euphytica* 96(1), 1-12.

Bertrand, B., Aguilar, V. G. J., Anthony, F., Etienne, E. and SantaCreo, P. (1997). *Comparación de híbridos F1 con variedades de Coffea arabica*, SIMPOSIO Latinoamericano sobre Caficultura, San José (Costa Rica).

Bertrand, B., Anzueto, F., Moran, M., Eskes, A. and Etienne, H. (2002). Creation and diffusion by somatic embryogenesis of the 'Nemaya' rootstock variety (Coffea canephora) in Central America. *Plantations, Recherche, Développement, Special Issue Research and Coffee Growing*, 105-107.

Bertrand, B., Etienne, H., Cilas, C., Charrier, A. and Baradat, P. (2005). Coffea arabica hybrid, fertility and bean weight. *Euphytica* 141(3), 255-262.

Bertrand, B., Vaast, P., Alpizar, E., Etienne, H., Davrieux, F. and Charmetant, P. (2006). Comparison of bean biochemical composition and beverage quality of Arabica hybrids involving Sudanese-Ethiopian origins with traditional varieties at various elevations in Central America. *Tree Physiology* 26(9), 1239-1248.

Bettencourt, A. J. (1973). *Consideracoes gerais sobre o "Hibrido de Timor"*. Sao Paulo: Instituto Agronómico, Circular N° 23, 20 p.

Bettencourt, E. and Konopka, J. (1988). *Directory of Germplasm Collections. 5.II Industrial Crops. Beet, Coffee, Oil Palm, Cotton and Rubber*. Rome, Italy: IBPGR. 64 p.

Bieysse, D., Mouen, J., et al. (2002). L'anthracnose des baies une menace potentielle pour la culture mondiale de l'Arabica. In: *Recherche et Caféicultures*. CIRAD, Montpellier, pp. 145-152.

Bowden, R., Shoyer, J., Roozeboom, K., Claasen, M., Evans, P., Gordon, B., Heer, B., Janssen, K., Long, J., Martin, J., Schlegel, A., Sears, R. and Witt, M. (2001). Performance of wheat variety blends in Kansas, Kansas State University Agric. *Extension Bulletin* 128, 6p.

Breitler, J. C., Dechamp, E., Campa, C., Zebral Rodrigues, L. A., Guyot, R., Marraccini, P. and Etienne, H. (2018).CRISPR/Cas9-mediated efficient targeted mutagenesis has the potential to accelerate the domestication of *Coffea canephora*. *Plant Cell, Tissue and Organ Culture* 134(3), 383-394 https://doi.org/10.1007/s11240-018-1429-2.

Bretting, P. K. and Widrlechner, M. P. (1995). Genetic markers and horticultural germplasm management. *Hortscience* 30(7), 1349-1356.

Brockhaus, R. and Oetmann, A. (1996). Aspects of the documentation of in situ conservation measures of genetic resources. *Plant Genetic Resources Newsletter* 108, 1-16.

Browning, J. A. and Frey, K. J. (1969). Multiline cultivars as a means of disease control. *Annual Review of Phytopathology* 7(1), 355-382.

Budar, F. and Palletier, G. (2001). Male sterility in plants: ocurrence, determinims, significance and use. *Comptes Rendus de l'Académie des Sciences - Series III - Sciences de la Vie* 324(6), 543-550.

Caixeta, E., Silva, R., Sousa, T., Silva, L., Nascimento, M., Barreiros, P., Oliveira, C. A., Pereira, A. and Zambolin, L. (2021). Genome-wide association study for morphological and yield component traits in *Coffea arabica*. *The 28th International Conference on Coffee Science*, Paris, Francia, 67 p.

Camargo, M. B. Pd. (2010). The impact of climatic variability and climate change on Arabic coffee crop in Brazil. *Bragantia* 69(1), 239-247. https://doi.org/10.1590/S0006-87052010000100030.

Carvalho, A. and Krug, C. A. (1949). Agentes de polinizacao da flor do cafeeiro (Coffea arabica L.). *Bragantia* 9(1-4), 11-24.

Carvalho, A. (1952). Taxonomia de Coffea arabica L. Caracteres morfológicos dos haploides. *Bragantia* 12(4-6), 201-212.

Carvalho, A. (1958). Advances in coffee production technology: recent advances in our knowledge of coffee trees 2. Genetics. *Coffee and Tea Industries and the Flavor Field* 81(11), 30-36.

Castillo, J. (1976). Tasa de polinización cruzada del café arábigo en la región de Chinchiná. *Cenicafé* 27(2), 78-88.

Castillo, J. and Moreno, G. (1981). Selección de cruzamientos derivados del híbrido de Timor en la obtención de variedades mejoradas de café para Colombia. *Cenicafé* 32(2), 37-53.

Castillo, J. and Moreno, G. (1986). *La variedad Colombia*. Chinchiná - Caldas: Cenicafé.

Castillo, J. and Moreno, G. (1988). *La variedad Colombia: Selección de un cultivar compuesto resistente a la roya del cafeto*. Chinchiná, Caldas, Colombia: Cenicafé.

Castro, B., Cortina, H., Roux, Y. and Wingfield, M. (2013). New coffee (*Coffea arabica*) genotypes derived from *Coffea canephora* exhibiting levels of resistance to leaf rust and Ceratocystis canker. *Tropical Plant Phatology* 38(6), 485-494.

Chaparro, A. P., Cristancho, M. A., Cortina, H. A. and Gaitán, A. L. (2004). Genetic variability of Coffea arabica L. accessions from Ethiopia evaluated with RAPDs. *Genetic Resources and Crop Evolution* 51(3), 291-297, http://doi.org/10.1023/B:GRES.0000024014.04471.9a.

Charrier, A. (1978). Etude de la structure et de la variabilité géné-tique des caféiers. Résultats des études et des expérimentationsréalisées au Cameroun, en Côte d'Ivoire et à Madagascar surl'espèce Coffea arabica L. collectée en Ethiopie par une misión ORSTOM. Bulletin IFCC 14, 99 p.

Charrier, A. and Eskes, A. (2004). Botany and genetics of coffee. In: Wintgens, J. N. (Ed.), *Coffee: Growing, Processing, Sustainable Production. A Guidebook for Growers, Processors, Traders, and Researchers*. Wiley VCH Verlag, Weinheim, pp. 24-56, 976 p.

Chevalier, A. (1947). Les caféiers du globe. II. Les vrais et faux caféiers. Nomenclature et systématique. In: Lechevalier, P. (Ed.), *Encyclopédie biologique 28, Fascicule III*. Paris, France, pp. 117-263.

Chinnappa, C. C. (1969). Octoploidy in Coffea arabica. *Genetica* 40(1), 508-516.

Cordeiro, A., Zambolim, L., Pétiard, V. and Spiral, J. (2000). Inter-specific protoplast fusion in coffea. *Coffee Biotechnology and Quality*, 151-156.

Creissen, H. E., Jorgensen, T. H. and Brown, J. K. (2013). Stabilization of yield in plant genotype mixtures through compensation rather than complementation. *Annals of Botany* 112(7), 1439-1447.

Creissen, H. E., Jorgensen, T. H. and Brown, J. K. (2016). Increased yield stability of field-grown winter barley (Hordeum vulgare L.) varietal mixtures through ecological processes. *Crop Protection* 85, 1-8.

DaMatta, F. M., Chaves, A. R. M., Pinheiro, H. A., Ducatti, C. and Loureiro, M. E. (2003). Drought tolerance of two field grown clones of Coffea canephora. *Plant Science* 164(1), 111-117.

DaMatta, F. M. (2004). Exploring drought tolerance in coffee: a physiological approach with some insights for plant breeding. *Brazilian Journal of Plant Physiology* 16(1), 1-6.

DaMatta, F. M., Rahn, E., Läderach, P., Ghini, R. and Ramalho, J. C. (2019). Why could the coffee crop endure climate change and global warming to a greater extent than previously estimated? *Climatic Change* 152(1), 167-178 https://doi.org/10.1007/s10584-018-2346-4.

Davis, A. P., Govaerts, R., Bridson, D. M. and Stoffelen, P. (2006). An annotated taxonomic conspectus of the genus Coffea (Rubiaceae). *Botanical Journal of the Linnean Society* 152(4), 465-512.

Davis, A. P., Tosh, J., Ruch, N. and Fay, M. F. (2011). Growing coffee: Psilanthus (Rubiaceae) subsumed on the basis of molecular and morphological data; implications for the size, morphology, distribution and evolutionary history of Coffea. *Botanical Journal of the Linnean Society* 167(4), 357-377. https://doi.org/10.1111/j.1095-8339.2011.01177.x.

Davis, A., Wilkinson, T., Kebebew, C., Williams, J., Baena, S., Wole, T. W. and Moat, J. (2018). *Coffee Atlas of Ethiopia, Kew, Richmond*. London: Royal Botanical Garden.

Davis, A. P., Chadburn, H., Moat, J., O´Sullivan, R., Hargreaves, S. and Lughadha, E. N. (2019). High extinction risk for wild coffee species and implications for coffee sector sustainability. *Science Advances* 5(1), eaav3473. https://doi.org/10.1126/sciadv.aav3473.

De Castro, R. D. and Marraccini, P. (2006). Cytology, biochemistry and molecular changes during coffee fruit development. *Brazilian Journal of Plant Physiology* 18(1), 175-199.

Dias, P. C., Araujo, W. L., Moraes, G. A., Barros, R. S. and DaMatta, F. M. (2007). Morphological and physiological responses of two coffee progenies to soil water availability. *Journal of Plant Physiology* 164(12), 1639-1647.

Dias, R., Dias, R., Botelho, C., Abrahão, J., Rezende, T., Carvalho, G. (2020). Genetic determination of characteristics related to semi-mechanized coffee harvests. Crop Breeding and Applied Biotechnology 20. https://doi.org/10.1590/1984-70332020v20n1n16.

Duarte, N. M. T. (1992). *Determinación de la atractividad de frutos de varios cultivares de café a la broca del fruto del cafeto Hypothenemus hampei F., utilizando el método de olfatometría a nivel de laboratorio. San Carlos de Guatemala*. Guatemala: Universidad de San Carlos. Facultad de Agronomía, 55 p.

Ducos, J. P., Lambot, C. and Pétiard, V. (2007). Bioreactors for coffee mass propagation by somatic embryogenesis. *International Journal of Plant Developmental Biology* 1(1), 1-12.

Dufour, M., Anthony, F., Bertrand, B. and Eskes, A. B. (1997). Identification de caféiers male-stériles de Coffea arabica au Catie, Costa Rica. *Plantations, Recherche, Developpement* 4, 401-407.

Dulloo, M. E., Ebert, A. W., Dussert, S., Gotor, E., Astorga, C., Vasquez, N., Rakotomalala, J. J., Rabemiafara, A., Eira, M., Bellachew, B., Omondi, C., Engelmann, F., Anthony, F., Watts, J., Qamar, Z. and Snook, L. (2009). Cost efficiency of cryopreservation as a long-term conservation method for coffee genetic resources. *Crop Science* 49(6), 2123-2138.

Dussert, S., Chabrillange, N., Engelmann, F., Anthony, F. and Hamon, S. (1997). Cryopreservation of coffee (Coffea arabica L) seeds: importance of the precooling temperature. *Cryo-Letters* 18(5), 269-276.

Dussert, S., Chabrillange, N., Engelmann, F., Anthony, F., Louarn, J. and Hamon, S. (1998). Cryopreservation of seeds of four coffee species (Coffea arabica, C. costatifructa, C. racemosa and C. sessiliflora): importance of water content and cooling rate. *Seed Science Research* 8(1), 9-15.

Dussert, S. and Engelmann, F. (2006). New determinants for tolerance of coffee (Coffea arabica L.) seeds to liquid nitrogen exposure. *Cryo Letters* 27(3), 169-178.

Dussert, S., Chabrillange, N., Vasquer, N., Engelmann, F., Anthony, F., Guyot, A. and Hamon, S. (2000). Beneficial effect of post-thawing osmoconditioning on the recovery of cryopreserved coffee (*Coffea arabica L.*) seeds. *Cryo Letters* 21(1), 47-52.

Ebert, A. (2008). Flujos de germoplasma facilitado por el Centro Agronómico de Investigación y Enseñanza Dentro y Fuera de Latinoamérica, Informe Especial Recursos Fitogenéticos. *Recursos Naturales y Ambiente* 53, 62-71.

Eira, M. T. S., Silva, E. A. Ad, De Castro, R. D., Dussert, S., Walters, C., Bewley, J. D. and Hilhorst, H. W. M. (2006). Coffee seed physiology. *Brazilian Journal of Plant Physiology* 18(1), 149-163, https://doi.org/10.1590/S1677-04202006000100011.

Ellis, R. H., Hong, T. D. and Roberts, E. H. (1990). An intermediate category of seed storage behaviour? I. Coffee. *Journal of Experimental Botany* 41(9), 1167-1174.

Ellis, R. H., Hong, T. D. and Roberts, E. H. (1991). An intermediate category of seed storage behaviour? II. Effects of provenance, immaturity, and imbibition on desiccation-tolerance in coffee. *Journal of Experimental Botany* 42(5), 653-657. https://doi.org/10.1093/jxb/42.5.653.

Engels, J. and Visser, L. (Eds.) (2007). *Guía para el manejo eficaz de un banco de germoplasma, Manuales de Bioversity para bancos de germoplasma N° 6*. Roma, Italia: Bioversity International.

Eskes, A. (1989). Resistance. In: Kushalappa, A. and Eskes, A. (Eds), *Coffee Rust: Epidemiology, Resistance and Management*. Boca Raton, FL: CRC Press, pp. 171-292.

Eskes, A. and Toma-Braghini, M. (1981). *FAO Plant Protection Bulletin* 2, 56-76.

Etienne, H., Bertrand, B., Georget, F., Lartaud, M., Montes, F., Dechamp, E., Verdeil, J. L. and Barry-Etienne, D. (2013). Development of coffee somatic and zygotic embryos to plants differs in the morphological, histochemical and hydration aspects. *Tree Physiology* 33(6), 640-653.

Etienne, H., Breton, D., Breitler, J. C., Bertrand, B., Déchamp, E., Awada, R., Marraccini, P., Léran, S., Alpizar, E., Campa, C., Courtel, P., Georget, F. and Ducos, J. P. (2018). Coffee somatic embryogenesis: how did research, experience gained and innovations promote the commercial propagation of elite clones From the two cultivated species? *Frontiers in Plant Science* 9, 1630. https://doi.org/10.3389/fpls.2018.01630.

Fanelli Carvalho, H., Galli, G., Ventorim Ferrão, L. F., Vieira Almeida Nonato, J., Padilha, L., Perez Maluf, M., Ribeiro de Resende, M. F., Guerreiro Filho, O. and Fritsche-Neto, R. (2020). The effect of bienniality on genomic prediction of yield in Arabica coffee. *Euphytica* 216(6), 101. https://doi.org/10.1007/s10681-020-02641-7.

FAO (2010). *The Second Report on the State of the World's Plant Genetic Resources for Food and Agriculture*. Rome, 399 p.

Ferh, W. R. (1987). *Principles of Cultivar Development. Theory and Technique*. New York: Macmillian Publishing Company. 536 p.

Ferreira, D., Demétrio, C., Manly, B., Machado, A. and Vencovsky, R. (2006). Statistical models in agriculture: biometrical methods for evaluating phenotypic stability in plant breeding. *Cerne, Labras* 12(4), 373–388.

Filho, O., Ramalho, M. A. P. and Andrade, V. T. (2018). Alcides Carvalho and the selection of Catuaí cultivar: interpreting the past and drawing lessons for the future. *Crop Breeding and Applied Biotechnology* 18(4), 460–466.

Garcia, F., López, C. and Posada, H. (2014). Identification and characterization of molecular markers linked to introgressions from Coffea Canephora Pierre ex Froehner in F5 coffee lines with resistance to coffee leaf rust. *The 24th International Conference on Coffee Science*, Armenia, Colombia, p. 247.

Georget, F., Courtel, P., García, E. M., Hidalgo, M., Alpizar, E., Breitler, J. C., Bertrand, B. and Etienne, H. (2017). Somatic embryogenesis-derived coffee plantlets can be efficiently propagated by horticultural rooted mini-cuttings: a boost for somatic embryogenesis. *Scientia Horticulturae* 216, 177–185.

Georget, F., Marie, L., Alpizar, E., Courtel, P., Bordeaux, M., Hidalgo, J. M., Marraccini, P., Breitler, J. C., Déchamp, E., Poncon, C., Etienne, H. and Bertrand, B. (2019). Starmaya: the first arabica F1 coffee hybrid produced using genetic male sterility. *Frontiers in Plant Science* 10, 1344. https://doi.org/10.3389/fpls.2019.01344.

Gichimu, B. M. and Omondi, C. O. (2010). Morphological characterization of five newly developed lines of Arabica coffee compared to commercial cultivars in Kenya. *International Journal of Plant Breeding and Genetics* 4(4), 238–246.

Gichuru, E. K., Agwanda, C. O., Combes, M. C., Mutitu, E. W., Ngugi, E. C. K., Bertrand, B. and Lashermes, P. (2008). Identification of molecular markers linked to a gene conferring resistance to Coffee berry disease (Colletotrichum kahawae) in Coffea arabica. *Plant Pathology* 57(6), 1117–1124.

Gimase, J. (2015). Coffee improvement by interspecific hybridization: a review. *Indian Journal of Agricultural Research* 3(3), 41–46.

Gimase, J. M., Thagana, W. M., Omondi, C. O., Cheserek, J. J., Gichimu, B. M., Gichuru, E. K., Ziiyomo, C. and Sneller, C. H. (2020). Genome-Wide Association Study identify the genetic loci conferring resistance to Coffee Berry disease (Colletotrichum kahawae) in Coffea arabica var. Rume Sudan. *Euphytica* 216(6), 86 https://doi.org /10.1007/s10681-020-02621-x.

Goncalves, W. and Pereira, A. (1998). Resistencia de cafeeiros a nematóides IV-reacao de cafeeiros derivados de Híbridos de Timor a meloidogyne exigua. *Nematologia Brasileira* 22, 39–50.

Goncalves, M., Rodrigues, M., Mexia, J. and Daehnhardt, E. (1977). Amelioration de la cafeiculture a Timor face a l' Hemileia vastatrix B. & Br, Colloque Scientifique International sur le Café, Abidjan (Costa de Marfil) 28-3 December. Paris (Francia): ASIC, pp. 515-522.

Guillaumet, J. and Halle, F. (1978). Échantillonnage du matériel. Coffea arabica récolté en éthiopie. p. 13-18. In: Charrier A. pg. 26/86. *Etude de la structure et de la variabilité génétique des caféiers: Résultats des études et des expérimentations réalisées au Cameroun, en Cote d'Ivoire et à Madagascar sur l'espèce Coffea arabica L. collectée en Ethiopie par une mission ORSTOM en 1966*. Paris: ORSTOM-IFCC. Bulletin IFCC N° 14, 100 p.

Harlan, J. R. and De Wet, J. M. J. (1971). Toward a rational classification of cultivated plants. *Taxon* 20(4), 509-517.

Hendre, P. and Aggarwal, R. (2007). DNA markers: development and application for genetic improvement of coffee. In: Varshney, R. K. and Tuberosa, R. (Eds), *Genomic Assisted Crop Improvement: Genomics Applications in Crops* (vol. 2) Germany: Springer-Verlag, pp. 399-434.

Hutchinson, J. (2012). *Impact of Honeypollen on Coffee Pollination in Jamaica, West Indies*. CA: Humboldt State University.

ICO (2020). Historical data on the global coffee trade, Available at: http://www.ico.org/historical/ 1990%20onwards/Excel/1a%20-%20Total%20production.xlsx.

Idárraga, S. M., Castro, A. M., Macea, E. P., Gaitán, A. L., Rivera, L. F., Cristancho, M. A. and Góngora, C. E. (2012). Sequences and transcriptional analysis of Coffea arabica var. Caturra and Coffea liberica plant responses to coffee berry borer Hypothenemus hampei (Coleoptera: Curculionidae: Scolytinae) attack. *Journal of Plant Interactions* 7(1), 56-70, https://doi.org/10.1080/17429145.2011.573151.

Jaramillo, A. (2010). Efecto de las abejas silvestres en la polinización del café (coffea arabica: Rubiaceae) en tres sistemas de producción en el departamento de Antioquia.

Jaramillo, A. (2018). *El clima de la caficultura colombiana*. Manizales: Cenicafé, 206 p.

Jiménez, W. and Castillo, J. (1976). Observaciones sobre la polinizacion de Coffea arabica L. en la zona cafetera central de Colombia. *Cenicafe* 27, 51-66.

Kathurima, C., Kenji, G., Muhoho, S., Boulanger, R. and Davrieux, F. (2010). Discrimination of Coffea arabica hybrids of the composite cultivar Ruiru 11 by sensorial evaluation and biochemical characterization. *Advance Journal of Food Science and Technology* 2(3), 148-154.

Kaul, M. (1988). *Male Sterility in Higher Plants*. Springer-Verlag, Berlin, 1005 p.

Koch, V. (1973). *Abondance de Hypothenemus hampei Ferr., scolyte des graines de cafe, en fonction de sa plant-hote et de son parasite Cephalonomia stephanoderis Betren, En Cote d'Ivoire*. Wageningen: Holanda, Veenman and Zonen, 84 p.

Krug, C. A. and Carvalho, A. (1951). The genetics of Coffea. *Advances in Genetics* 4, 127-158. https://doi.org/10.1016/s0065-2660(08)60233-3. PMID: 14943675.

Kushalappa, A. and Eskes, A. (1989). Advances in coffea rust research. *Annual Review Phytophatology* 27, 503-531.

Läderach, P., Ramirez-Villegas, J., Navarro-Racines, C., Zelaya, C., Martinez-Valle, A. and Jarvis, A. (2017). Climate change adaptation of coffee production in space and time. *Climatic Change* 141(1), 47-62, https://doi.org/10.1007/s10584-016-1788-9.

Labouisse, J. P., Bellachew, B., Kotecha, S. and Bertrand, B. (2008). Current status of coffee (Coffea arabica L.) genetic resources in Ethiopía: implications for conservation.

Genetic Resources and Crop Evolution 55(7), 1079–1093. https://doi.org/10.1007/s10722-008-9361-7.

Labouisse, J. P. and Adolphe, C. (2012). Conservation et gestion des ressources genetiques du caféier Arabica (*Coffea arabica* L.): un défi pour l'Ethiopie. *Cahiers Agricultures* 21(2–3), 98–105. https://doi.org/10.1684/agr.2012.0554.

Lashermes, P., Combes, M., Cros, J., Trouslot, P. and Charrier, A. (1995). Origin and genetic diversity of Coffea arabica L. based on DNA markers. In: *International Scientific Colloquium on Coffee*. Paris, France: ASIC, Orstom, Montpellier, pp. 528–536.

Lashermes, P., Trouslot, P., Anthony, F., Combes, M. C. and Charrier, A. (1996). Genetic diversity for RAPD markers between cultivated and wild accessions of Coffea arabica. *Euphytica* 87(1), 59–64. http://doi.org/10.1007/BF00022965.

Lashermes, P., Combes, M., Robert, J. and Charrier, A. (1999). Molecular characterisation and origin of the Coffea arabica L. genome. *Molecular Genetics and Genomics* 261(2), 259–266.

Lashermes, P., Combes, M., Ribas, A., Cenci, A., Mahé, L. and Etienne, H. (2010). Genetic and physical mapping of the SH3 region that confers resistance to leaf rust in coffee tree (*Coffea arabica L.*). *Tree Genetics and Genomes* 6, 973–980. http://doi.org/10.1007/s11295-010-0306-x.

Leon, J. (1962). *Especies y cultivares (Variedades) de café*. Turrialba: Inter-American institute for cooperation in agriculture.

Leroy, T., Henry, A. M., Royer, M., Altosar, I., Frutos, R., Duris, D. and Philippe, R. (2000). Genetically modified coffee plants expressing the Bacillus thuringiensis Cry1Ac gene for resistance to leaf minner. *Plant Cell Reports* 19(4), 382–385.

Mahé, L., Combes, M., Várzea, V. M. P., Guilhaumon, C. and Lashermes, P. (2008). Development of sequence characterized DNA markers linked to leaf rust (Hemileia vastatrix) resistance in coffee (Coffea arabica L.). *Molecular Breeding* 21(1), 105–113.

Marie, L., Abdallah, C., Campa, C., Courtel, P., Bordeaux, M., Navarini, L., Lonzarich, V., Bosselman, A. S., Turreira-García, N., Alpizar, E., Goerget, F., Breitler, J. C., Etienne, H.. and Bertrand, B. (2020). G × E interactions on yield and quality in *Coffea arabica*: new F1 hybrids outperform American cultivars. *Euphytica* 216(5), 78 https://doi.org/10.1007/s10681-020-02608-8.

Mazzafera, P., Eskes, A., Parvais, J. and Carvalho, A. (1989). Stérilité male détectée chez Coffea arabica et C. Canephora au Brésil. In: *13 Colloque Scientifique International sur le Café. Paipa (Colombia) 21–25 aout*. París (Francia): ASIC, Orstom, Montpellier, pp. 466–473.

Mendes, A. and Bacchi, O. (1940). Observacoes citológicas en Coffea. V. Uma variedade haploide ("Di-Haploide") de *C. arabica* L. *Boletim Técnico No. 77, Separata Do Jornal de Agronomia* 3(3), 183–206.

Mishra, M. K. and Slater, A. (2012). Recent advances in the genetic transformation of coffee. *Biotechnology Research International* 2012, 580857, https://doi.org/10.1155/2012/580857.

Moncada, M. D. P., Tovar, E., Montoya, J. C., González, A., Spindel, J.. and McCouch, S. (2016). A genetic linkage map of coffee (Coffea arabica L.) and QTL for yield, plant height, and bean size. *Tree Genetics and Genomes* 12(1), 5. https://doi.org/10.1007/s11295-015-0927-1.

Montagnon, C., Guyot, B., Cilas, C. and Leroy, T. (1998). Genetic parameters of several biochemical compounds from green coffee, Coffea canephora. *Plant Breeding* 117(6), 576–578. https://doi.org/10.1111/j.1439-0523.1998.tb02211.x.

Montagnon, C., Mahyoub, A., Solano, W. and Sheibani, F. (2021). Unveiling a unique genetic diversity of cultivated *Coffea arabica* L. in its main domestication center: Yemen. *Genetic Resources and Crop Evolution* 68(6), 2411–2422 https://doi.org/10.1007/s10722-021-01139-y.

Montoya, S. (1992). *Ciclo de vida de la broca del café (Hypothenemus hampei (Ferrari)) sobre frutos en diferentes estados de desarrollo*. Bogotá, Colombia: Universidad Nacional de Colombia, Facultad de Ciencias Agropecuarias, 78 p.

Moreno, G. (1977). Variabilidad de semillas anormales de café en una población F2 de Caturra x Híbrido de Timor. *Cenicafé* 28(2), 39–50.

Mundt, C. C. (2002). Use of multiline cultivars and cultivar mixtures for disease management. *Annual Review of Phytopathology* 40, 381–410.

Murashige, T. and Skoog, F. (1962). A revised medium for rapid growth and bioassays with tobacco tissue cultures. *Physiologia Plantarum* 15(3), 473–497.

Murashige, T. (1974). Plant propagation through tissue cultures. *Annual Review of Plant Physiology* 25(1), 135–166.

Narasimhaswamy, R. (1960). Arabica selection S.795 - its origin and performance - a study. *Indian Coffee* 24(5), 197–204.

Narasimhaswamy, R. (1962). Some thoughts on the origin of C. arabica L. Coffee. *Turrialba* 4, 1–5.

Noir, S., Anthony, F., Bertrand, B., Combes, M.-C. and Lashermes, P. (2003). Identification of major gene (Mex-1) from Coffea canephora conferring resistance to meloidogyne exigua in Coffea arabica. *Plant Pathology* 52(1), 97–103.

Omondi, C. O., Ayiecho, P. O., Mwang'ombe, A. W. and Hindorf, H. (2001). Resistance of Coffea arabica cv. Ruiru 11 tested with different isolates of Colletotrichum kahawae, the causal agent of Coffee Berry disease. *Euphytica* 121(1), 19–24.

Opile, W. R. and Agwanda, C. O. (1993). Propagation and distribution of cultivar Ruirú. *Kenya Coffee,* 58(11), 1496–1508.

Painter, R. (1951). *Insect Resistance in Crop Plant*. New York: Estados Unidos, Mc Millan, 23 p.

Parlevliet, J. E. and Zadoks, J. C. (1977). The integrated concept of disease resistance; a new view including horizontal and vertical resistance in plants. *Euphytica* 26(1), 5–21.

Pérez-Míguez, R., Castro-Puyana, M., Sánchez-López, E., Plaza, M. and Marina, M. L. (2020). Untargeted HILIC-MS-based metabolomics approach to evaluate coffee roasting process: contributing to an integrated metabolomics multiplatform. *Molecules* 25(4), 887. https://doi.org/10.3390/molecules25040887.

Philpott, S. M., Uno, S. and Maldonado, J. (2006). The importance of ants and high-shade management to coffee pollination and fruit weight in Chiapas, Mexico. *Biodiversity and Conservation* 15(1), 487–501.

Pinto, V. and Marques, D. (2005). Population variability of Hemileia vastatix vs. coffee durable resistance. In: Zambolim, L., Zambolim, E. and Pinto, V. (Eds), *Durable Resistance to Coffee Leaf Rust*. Vicosa - Minas Gerais, Brasil: Universidad Federal de Vicosa, pp. 53–74.

Portilla, M. (1996). Producción masiva y uso de parasitoides para el control de la broca del café en Colombia. *Cenicafé* 47(4), 215–230.

Prakash, N., Seetharam, H. and Kumar, T. (2016). Breeding for broad spectrum of host resistance in arabica: exploitation of C. librica and C. eugenioides hybrid derivatives. In: *Proceedings of the ASIC 26th International Conference on Coffee Sciences*.

Reddy, P. (2012). Variety mixture/cutivar mixtures/multilines. In: Reddy, P. (Ed.). *Recent Advances in Crop Protection*. Bangalore, Karnataka, India: Springer, pp. 201-221.

Reed, B., Engelmann, F., Dulloo, M. and Engels, J. (2004). *Technical Guidelines on Management of Field and In Vitro Germplasm Collections. Handbook for Genebanks* (vol.7). Rome, Italy: International Plant Genetics Resources Institute.

Rodrigues Jr., C. J., Bettencourt, A. J. and Rijo, L. (1975). Races of the pathogen and resistance to coffee rust. *Annual Review of Phytopathology* 13(1), 49-70.

Romero, J. V.and Cortina, H. A. (2004). Evaluación de germoplasma de café por antixenosis a *Hypothenemus hampei* (Ferrari) en condiciones controladas. *Cenicafé* 55(4), 341-346.

Raw, A. and Free, J. (1977). The pollination of coffee (*Coffea arabica L.*) by honeybees. *Tropical Agriculture of Jamaica* 54, 365-370.

Roy, D. and Kharkwal, M. (2004). Breeding for wider adaptability. In: Jain, H. K. and Kharkwal, M. C. (Eds), *Plant Breeding. Mendelian to Molecular Approaches*. New Delhi, India: Narosa Publishing House, pp. 573-584

Ruas, E., Teixeira, E., Vieira, T., Alves, A., Baião de Oliveira, A., Zambolim, L. and Sussumu, N. (2017). Marker-assisted selection provides Arabica coffee with genes from other Coffea species targeting on multiple resistance to rust and coffee berry disease. *Molecular Breeding* 37(6), 1-10.

Sabino, P. (1971). Observacoes sobre o gran de infestacao a desenvolvimiento da broca Hypothenemus hampei, en variedades de café. *Revista de agricultura* 46(1), 28-30.

Sakiyama, N. S., Picoli, E. AdT., Oliveira, A. C. Bd, Caixeta, E. T., Zambolim, L., Martinez, H. E. P. and Pereira, A. A. (2017). Triple plagiotropic branch in coffee: a new promising mutant? *Crop Breeding and Applied Biotechnology* 17(4), 408-411. https://doi.org /10.1590/1984-70332017v17n4n61.

Samson, N. P., Campa, C., Gal, L. L., Noirot, M., Thomas, G., Lokeswari, T. S. and de Kochko, A. (2006). Effect of primary culture medium composition on high frequency somatic embryogenesis in different Coffea species. *Plant Cell, Tissue and Organ Culture* 86(1), 37-45.

Sant'Ana, G., Pereira, P., Pot, D., Ivamoto, T., Domingues, S., Ferreira, V., Pagiatt, F., da Silva, R., Nogueira, M., Kitzberger, G., Scholz, S., de Oliveira, F., Sera, H., Padilha, L., Labouisse, J., Guyot, R., Charmetant, P., and Leroy, T. (2018). Genome-wide association study reveals candidate genes influencing lipids and diterpenes contents in *Coffea arabica L. Scientific Reports* 8(1), 465. https://doi.org/10.1038/s41598-017-18800-1.

Scalabrin, S., Toniutti, L., Di Gaspero, G., Scaglione, D., Magris, G., Vidotto, M., Pinosio, S., Cattonaro, F., Magni, F., Jurman, I., Cerutti, M., Suggi Liverani, F., Navarini, L., Del Terra, L., Pellegrino, G., Ruosi, M. R., Vitulo, N., Valle, G., Pallavicini, A., Graziosi, G., Klein, P. E., Bentley, N., Murray, S., Solano, W., Al Hakimi, A., Schilling, T., Montagnon, C., Morgante, M. and Bertrand, B. (2020). A single polyploidization event at the origin of the tetraploid genome of *Coffea arabica* is responsible for the extremely low genetic variation in wild and cultivated germplasm. *Scientific Reports* 10(1), 4642. https://doi.org/10.1038/s41598-020-61216-7.

Sera, G. H., Sera, T., Ito, D. S., Azevedo, J. Ad, Mata, J. Sd, Dói, D. S., Ribeiro Filho, C. and Kanayama, F. S. (2007). Resistance to leaf rust in Coffee carrying SH3 gene and other SH genes. *Brazilian Archives of Biology and Technology* 50(5), 753-757.

Sera, G., Sera, T., Ito, D., Filho, C., Villacorta, A., Kanayama, F., Alegre, C. and Grossi, L. (2010). Coffee Berry Borer Resistance in Coffee Genotypes. *Brazilian Archives of Biology and Technology* 53. https://doi.org/10.1590/S1516-89132010000200003.

Silva, M., Várzea, V., Guerra-Guimarães, L., Azinheira, H., Fernandez, D., Petitot, A., Bertrand, B., Lashermes, P. and Nicole, M. (2006). Coffee resistance to the main diseases: leaf rust and coffee berry disease. *Brazilian Journal of Plant Physiology* 18, 119-147.

Silva, R., Zambolim, L., Castro, I., Rodrigues, H., Cruz, C. and Caixeta, E. (2018). The Híbrido de Timor germplasm: identification of molecular diversity and resistance sources to coffee berry disease and leaf rust. *Euphytica* 214(153), 16.

Silvarolla, M. B., Cesar, M. J., Silva, A. J. and Mantovani, S. E. (2021). Development of decaffeinated clonal cultivar of Arabica coffee. *The 28th International Conference on Coffee Science*, Paris, Francia, 63 p.

Smith, R. (1985). A history of coffee. In: Clifford, M. N. and Wilson, K. C. (Eds). *Coffee: Botany, Biochemistry and Production of Beans and Beverage*, Croom Herm, Westport, London, pp. 1-12.

Sousa, T. V., Caixeta, E. T., Alkimim, E. R., Oliveira, A. C. B., Pereira, A. A., Sakiyama, N. S., Zambolim, L. and Resende, M. D. V. (2019). Early Selection enabled by the implementation of genomic selection in coffea arabica breeding. *Frontiers in Plant Science* 9, 1934 https://doi.org/10.3389/fpls.2018.01934.

Sybenga, J. (1960). Genetic and Cytology of coffee: a literatura review. *Bibliografia Genetica* 19, 217-316.

Talhinhas, P., Batista, D., Diniz, I., Vieira, A., Silva, D. N., Loureiro, A., Tavares, S., Pereira, A. P., Azinheira, H. G., Guerra-Guimarães, L., Várzea, V. and Silva, M. D. C. (2017). The coffee leaf rust pathogen Hemileia vastatrix: one and a half centuries around the tropics. *Molecular Plant Pathology* 18(8), 1039-1051, https://doi.org/10.1111/mpp.12564.

Teisson, C. and Alvard, D. (1995). A new concept of plant in vitro cultivation liquid medium: temporary immersion. In: Terzi, M., Cella, R. and Falavigna, A. (Eds), *Current Issues in Plant Molecular and Cellular Biology*. Dordrecht: Kluwer Academic Publishers, pp. 105-110.

Tran, H. T. M., Furtado, A., Vargas, C. A. C., Smyth, H., Slade Lee, L. and Henry, R. (2018) SNP in the Coffea arabica genome associated with coffee quality. *Tree Genetics and Genomes* 14(5), 72. https://doi.org/10.1007/s11295-018-1282-9.

Valencia-Lozano, E., Cabrera-Ponze, J. L., Gómez-Lim, M. A. and Ibarra, J. E. (2019). Development of an efficient protocol to obtain transgenic coffee, *Coffea arabica* L., expressing the Cry10Aa toxin of Bacillus thuringiensis. *International Journal of Molecular Sciences* 20(5334). doi:10.3390/ijms20215334.

Van Boxtel, J. and Berthouly, M. (1996). High Frequency somatic embryogenesis from coffee leaves. Factor influencing embryogenesis, and subsequent proliferation and regeneration in liquid medium. *Plant Cell, Tissue and Organ Culture* 44, 7-17.

Van der Graaff, N. (1981). *Selection for Resistance to Coffee Berry Disease in Arabica Coffee in Ethiopia*. Ph D. thesis. Wageningen the Netherlands: Agricultural University. 39 p.

Van der Graaff, N. A. (1982). The principles of scaling and the inheritance of resistance to coffee berry disease in Coffe arabica. *Euphytica* 31(3), 735-740.

Van der Vossen, H. A. M. and Walyaro, D. J. (1980). Breeding for resistance to coffee berry disease in Coffea arabica L. II. Inheritance of the resistance. *Euphytica* 29(3), 777-791.

Vega, F., Ebert, A. and Ming, R. (2008). Coffee germplasm resources, genomics, and breeding. In: *Plant Breeding Reviews*. Hoboken, NJ: John Wiley & Sons, Inc. 30, 415–417.

Venkatesha, M. (2010). *Sustainable Coffee Cultivation in India: Challenges and Management*, https://doi.org/10.13140/2.1.2825.7283.

Villagran, G. (1991). *Atractividad relativa y susceptibilidad de varias especies y cultivares de café Coffea spp. a la broca del fruto Hypothenemus hampei Ferr. en condiciones de laboratorio*. Guatemala: Universidad de San Carlos de Guatemala. Facultad de Agronomía, 43 p.

Villegas, A., Pérez, C., Posada, H., Garcia, F. and Rojas, A. (2014). Caracterization of Coffea arabica germoplasm by quality parameters. *24th International Conference on Coffee Science*, Armenia, Colombia, 251 p.

Vuillet, J. (1925). Degré de sensibilite des différents caféiers au Stephanoderes Coffeae. *Revue de Botanique Appliquée et d'Agriculture Coloniale* 5(48), 601–604.

Walyaro, D. (1983). *Considerations in Breeding for Improved Yield and Quality in Arabica Coffee (Coffea arabica L.)*. Wageningen, The Netherlands, 121 p.

Walters, C. (2015). Orthodoxy, recalcitrance and in-between: describing variation in seed storage characteristics using threshold responses to water loss. *Planta* 242(2), 397–406, https://doi.org/10.1007/s00425-015-2312-6.

Walyaro, D. J. and Van der Vossen, H. A. M. (1979). Early determination of yield potential in Arabica coffee by applying index selection. *Euphytica* 28(2), 465–472.

Wellman, F. (1961). Coffee: botany, cultivation, and utilization. *London Leonard Hill*, 488 p.

Wolfe, M. S. (1985). The current status and prospects of multiline cultivars and variety mixtures for disease resistance. *Annual Review of Phytopathology* 23(1), 251–273.

Chapter 7

Optimizing post-harvest practices in coffee cultivation

Carlos H. J. Brando and João Alberto P. Brando, P&A Marketing, Brazil

1 Introduction

Optimizing efficiency, quality, and sustainability in coffee processing is challenging. It should be done while ensuring that those who do it - growers, cooperatives, service providers, traders, etc. - can still make money. Optimizing processing efficiency includes using equipment at its maximum capacity, minimizing physical damage to coffee during processing which might

http://dx.doi.org/10.19103/AS.2021.0096.20

compromise quality, and ensuring the maximum amount of raw material is converted into a final, saleable product with minimum waste. Optimizing environmental impact means reducing input use where possible, minimizing emissions and effluents, and preventing pollution of the environment. Optimizing quality involves producing a product with the required sensory quality attributes that justify the price at which it is finally sold and, more specifically, precisely meeting the specifications for quality, volume, and price set out by the customer.

The objective of this chapter is to describe how coffee is processed, from picking coffee cherries to green coffee ready to be exported or roasted, and how to optimize each step in the processing chain.

2 Coffee types and processing

The components of a fresh coffee cherry are shown in Fig. 1, which also shows the basic process of transformation from coffee cherry to bean ready for roasting. The components of a coffee cherry are as follows:

- Skin or exocarp.
- Pulp or mesocarp.
- Mucilage: inner layer of sticky, sugar-rich pulp.
- Bean: known as the endosperm and referred to as green coffee once processed.
- Silver skin: thin epidermis covering the bean.
- Parchment: a papery hull encasing the bean, known as the endocarp. Coffee with the skin and pulp removed, but with the parchment remaining, is known as parchment coffee.

At a basic level, coffee processing involves the removal of successive layers of the coffee cherry to produce beans.

Figure 1 From fresh cherry to green coffee.

Coffee cherries can be processed in three different ways that each affects the quality or cup features of coffee. The differing final products are as follows:

- Natural coffee: produced by drying the green coffee beans with the pulp and mucilage retained around the beans.
- Pulped natural or honey coffee: obtained by drying parchment coffee without the pulp but with some or all of the mucilage.
- Washed coffee: produced by drying parchment coffee with the pulp and the mucilage removed.

Natural coffees are considered to have more body and sweetness, whereas washed coffees have more aroma and acidity. Pulped natural or honey coffees are seen to lie between the two, with those dried with all the mucilage retained being close to natural coffees, while those dried with little mucilage are considered close to washed coffees.

Natural Arabica coffees are widely produced in Brazil and Ethiopia although pulped natural coffees are growing in importance in Brazil as are washed coffees in Ethiopia. Washed coffees also predominantly use Arabica beans. Robusta coffees are mostly natural coffees, with some washed Robusta coffees produced in India and pulped natural Robusta coffees growing in importance in Brazil. Pulped natural coffees mainly use Arabica beans, with the vast majority coming from Brazil, followed by Central American countries and small quantities produced in several other countries.

While the natural and washed coffee systems have been used for centuries, the pulped natural or honey system was developed only in the 1980s in Brazil, where it is called Cereja Descascado or CD. The objective was to obtain natural coffee free from the astringency introduced by unripe and partially ripe cherries. The outcome was a product ideally suited for espresso coffee, a type of coffee that was then gaining market share around the world. Subsequent concerns about the time it took to pre-dry parchment coffee with all the mucilage attached led to processes for partial mechanical removal of mucilage, changing the organoleptic properties of the coffee. Central American growers, who use the word 'honey' for pulped natural coffees, created different names for the resulting coffees according to the amount of mucilage left attached to the parchment. Black honey coffee is dried with all the mucilage retained, while red and yellow coffee have more mucilage removed, with the colors associated with the way parchment looks according to the different amounts of mucilage attached.

The main processing steps for the different types of coffee are shown in Fig. 2 and are discussed in more detail in the following sections. Quality characteristics can be altered by processing steps up to and including drying. Most processing steps can also damage coffee quality if not properly undertaken.

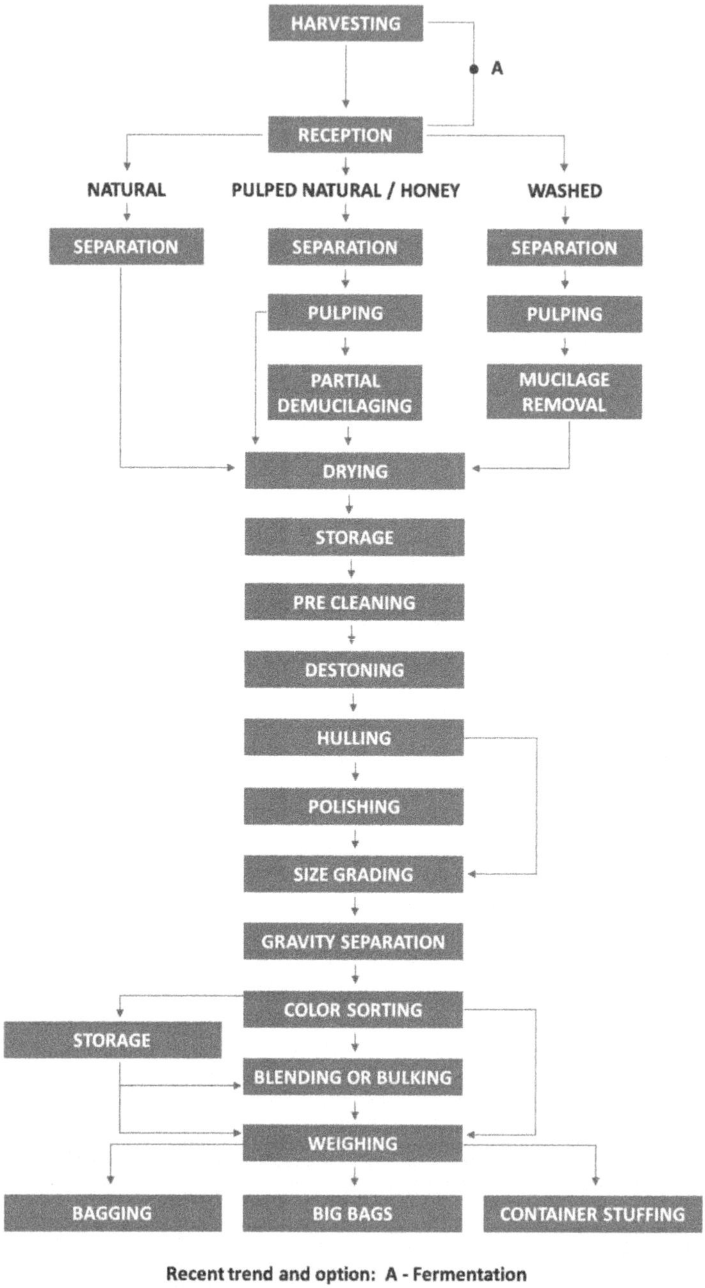

Figure 2 Processing from cherry to green coffee.

3 Harvesting, cleaning and separation processes

Ideally, harvesting should result in picking uniformly ripe cherries with the highest concentration of volatile compounds responsible for the aroma and flavor properties of coffee. Picking under-ripe or over-ripe cherries results in smaller volumes of top-quality coffee but may in many circumstances be the most profitable solution if modern processing techniques are used. There are two main harvesting techniques:

- Selective picking: only ripe cherries are picked as they come to maturity.
- Strip picking: coffee cherries are all picked at the same time or in a few picking rounds.

While it can be done manually, strip picking can be mechanized to improve efficiency. Since coffee cherries vary in maturity dates, harvesting at a single point will result in a percentage of under-ripe and over-ripe fruits. Selective picking by hand is more targeted but requires experienced pickers, involves several picking rounds, and is labor intensive.

The percentage of non-ripe cherries can vary from 3% to 5% to 10%, even with selective hand picking, and is higher with strip picking. This means some sort of separation process is required to separate ripe from non-ripe cherries because this affects both the final quality of the product and the effectiveness of subsequent processes such as drying.

In natural or dry-processed coffee, harvested cherries should be cleaned to remove impurities such as dirt and stones and separated to enable more uniform drying. This is traditionally done in water-intensive siphon tanks or channels and increasingly in mechanical siphons that consume very little water (Fig. 3). In larger processing facilities, a mechanical siphon may incorporate or bepreceded by a winnower to remove materials such as leaves and twigs. Siphoning separates coffee cherries into two fractions based on density:

- 'Floaters': over-ripe and partially dry cherries that are less dense.
- 'Sinkers': unripe, partially ripe, and fully ripe cherries with higher moisture content and density.

Since floaters have lower moisture content than sinkers, these two products should be dried separately to both obtain a more uniformly dried product and cut down total drying time with savings in labor and/or energy. Some producers of commercial rather than premium-priced coffees, Robusta coffees, in particular, choose to ignore this separation and to dry all coffee cherries together.

Separated out sinkers contain a mix of unripe, partially ripe, and fully ripe cherries, with the first two types having a lower overall quality and potentially

Figure 3 Mechanical siphon.

causing astringency in the cup. The solution for this is costly hand or color sorting of sinkers or, in the case of wet-processed coffee, the use of pulpers with an unripe cherry separator that separates these groups of cherries. Hand sorting is restricted to smaller, premium-quality growers. Color sorting of wet cherries, offered by most makers today, is restricted to and still little used only by larger processors, given the cost of equipment and level of expertise required for its operation.

4 Pulping processes

4.1 Pulped natural or honey coffee

In the pulped natural system, harvested cherries go through a process of removal of impurities and stones and separation of floaters and sinkers, using siphon tanks, water channels, or mechanical siphons. Floaters are usually dried, but over-ripe cherries may be pulped instead to remove the skin. Sinkers are pulped in a process that can remove non-ripe cherries before or after pulping to produce pulped natural coffee.

The parchment coffee coming out of pulping may go either straight into drying with all mucilage attached, to get pulped natural coffee with the maximum body and sweetness, or into a mucilage remover, for partial removal of the mucilage, with the corresponding cup approaching that of washed coffee as more mucilage is removed.

4.2 Washed coffee

In the washed coffee system, incoming fresh cherries go through the same process as natural or pulped natural coffee for the removal of impurities and stones and for the separation of floaters and sinkers, using siphon tanks, water channels, or mechanical siphons. Sinkers are pulped in one of the different types of machines available, with some machines designed to separate unripe cherries.

The mucilage attached to parchment coming out of pulping must be removed in one of three ways: fermentation with or without water, mechanically, in machines called mucilage removers or a combination of the two.

4.3 Optimizing wet milling

Separation of unripe and semi-ripe cherries *before* pulping has the advantage of being able to remove either both together, for top-quality coffee, or leaving the semi-ripe cherries mixed with the ripe cherries for less demanding clients (Fig. 4). Separation of unripe and under-ripe cherries *after* the cherries have

Figure 4 Vertical pulper preceded by unripe ('green') cherry separator.

been pulped does not allow this degree of fine tuning in producing top-quality washed and pulped natural coffees.

Two critical parameters to be controlled in the processing or pulped natural and washed coffees are the percentage of pulp mixed with parchment and the percentage of parchment lost with the pulp that is discarded. These two parameters should ideally approach zero. Physical damage to parchment should also be monitored at this stage.

Two major trends have, in recent years, produced major changes in pulping and demucilaging technology: environmental concerns about water usage and water contamination and the increasing percentage of non-ripe cherries harvested using techniques such as strip harvesting. Pulpers and mucilage removers have evolved more in the last three decades than in the previous century in order, first, to consume less water and, second, in the case of pulpers, to separate out unripe and partially ripe cherries from ripe cherries more efficiently. The pulpers available today range from disc pulpers to horizontal and vertical drum pulpers. These may include unripe cherry separators placed before or after the pulper itself. Horizontal mucilage removers are still used, but vertical machines are now more widely used. The most advanced machines are designed to consume as little water as possible, though zero water use is still synonymous with damage to coffee cherries and sub-optimal performance.

5 Mucilage removal processes: fermentation and mechanical removal

The mucilage attached to parchment coffee coming out of pulping must be removed in one of three ways:

- Fermentation with or without the use of water.
- Mechanically using mucilage removers.
- A combination of fermentation and mechanical processing.

The traditional way to remove mucilage is fermentation in water in tanks from 12 to 60 h, depending on temperature and amount of mucilage. This is the system that consumes the most water. Wet fermentation can be accelerated with the addition of enzymes and other substances though these may change coffee quality. Though much less common, dry fermentation is faster and consumes less water but is more difficult to control. In spite of its name, dry fermentation still consumes water to fill and empty the fermentation tank, which is then without water, while the parchment coffee is fermented. Fermentation may be followed by soaking in clean water. Natural fermentation helps develop acidity in the cup.

The mechanical removal of mucilage consumes much less water and enables continuous processing because there is no need to wait for fermentation to take place (Fig. 5). Another advantage of mechanical demucilaging is that there is a weight gain in the final product in comparison with fermentation, where weight gain varies according to local temperature and duration of fermentation. Fermentation for a short period followed by the mechanical removal of the remaining mucilage can combine some of the advantages of the two systems. Moreover, water consumption will remain high but not as high as in the fermentation alone method. This is because washing of fermented coffee is done in a mucilage remover that uses little water, instead of manually doing in the fermentation tank that has to be filled with clean water.

There has always been a debate about the quality of washed coffee that has been fermented or mechanically demucilaged. Experience suggests that the quality remains the same in both processes, but some experts suggest that, at higher altitudes, fermentation can develop more acidity than mechanical removal of mucilage. It is believed that partial fermentation, for 8–10 h, followed by mechanical demucilaging, yields the same quality as full fermentation, even at higher altitudes.

There have been many recent developments in the area of fermentation, not only in the fermentation of parchment, as described above for washed coffee, but also in the fermentation of recently harvested cherries, whether separated by density or not. Aerobic, anaerobic, and other types of cherry fermentation seek to develop different organoleptic features in the cup and to add value to

Figure 5 Mucilage remover.

coffee. Fermentation of cherries has been restricted to small volumes, not to say micro-lots, and has been done with the cherries alone in different conditions, e.g. higher temperatures, and with the addition of substances, including flavors. The cherries may be fermented before processing or after they are separated by density. It is still early to predict whether some of these forms of fermentation will be standardized to be used for large coffee lots.

6 Drying processes

The objective of drying is to lower the moisture content of parchment or cherry coffee to 12% to preserve coffee beans properly in storage and to hull them more easily. The moisture content above 12% increases the risks of quality and weight losses in storage. Moisture levels below 10% or 11% increase the breakage of beans at the hulling stage. The control of moisture in cherry or parchment coffee must be carried out with moisture meters in the case of both sun and mechanical drying in order to stop drying when the target moisture is reached.

Coffee should be dried neither too fast nor too slow to optimize quality. Fast drying can kill the coffee embryo, and slow drying may cause unwanted fermentation, with negative effects on quality in both cases. There is often a tendency to increase the temperature to shorten the drying time for economic reasons. The damage caused by overheating depends on temperature, coffee moisture content, and the duration of overheating. Damage from overheating can include parchment cracking, bending and bleaching of the beans, and widening of the center cut, as well as sour flavors, metallic taste, or a flat cup.

There are a wide range of drying techniques:

- Sun drying: using drying grounds or patios, drying racks or barbecues, or plastic sheds (Figs 6 and 7).
- Revolving mechanical driers:
    - Rotary: single or divided drum.
    - Vertical: batch, continuous, or multiple chambers with stirrers.
- Static mechanical driers: single or multiple chamber.

The drying techniques vary between producers. Arabica naturals are mostly sun dried in Ethiopia, while a combination of pre-drying under the sun followed by mechanical drying (typically using rotary driers) is the prevailing practice in Brazil. Wet-processed Arabicas are commonly pre-dried or fully dried in drying grounds or racks, with plastic sheds gaining in popularity. Pre-drying in large vertical driers is common, e.g. in Costa Rica. Rotary driers are mostly used for Arabica parchment though some countries rely on static driers, both single and multiple chamber types. Vertical multiple-chamber driers with stirrers

Figure 6 Drying ground in Central America.

Figure 7 Drying racks in Africa.

are now being introduced. Robusta naturals are mostly pre-dried under the sun by growers down to a moisture level that allows the cherries to be hulled. Drying of the resulting green beans is completed in large vertical driers though processors are now increasingly using rotary machines to complete the drying process in order to improve the quality. Brazilian Conilon naturals are dried using rotary driers. The advantages and disadvantages of different drying techniques are discussed below.

6.1 Coffee cherries (natural coffees)

Sun drying of natural coffees is mostly done on drying grounds with the surface lined with concrete, asphalt, or another material. Raised drying beds are also used. In both cases, the coffee cherries being dried should be stirred frequently to avoid unwanted fermentation, and they should be covered in case of rain or heavy dew after they are partially dry. Drying time can run from 12 to 21 days, depending on climate conditions. A solution used to deal with rain is to have the drying grounds or beds under a plastic roof. In the case of enclosed, green-house-like structures, fans can be used to remove inside air and accelerate drying.

The mechanical drying of Arabica naturals is preceded by pre-drying under the sun for a short period until the drier can receive 'wet' coffee or for the time the grower wants to use sun drying depending on climate, availability of labor and space, and the respective cost-benefit of the two options. Although there seems to be an agreement that the best drying machines for Arabica naturals are rotary driers, a few growers recently prefer static driers, mostly for commercial coffees. Drying time for cherry coffee is between 35 h and 45 h in efficient rotary driers, shorter for floaters and longer for sinkers, using temperatures under 45°C in the coffee itself (and *not* in the air used to dry the coffee). In the case of static driers loaded with cherries that have not been separated by density, the drying time is much longer.

The mechanical drying of Robusta naturals, after pre-drying under the sun, is also done with rotary driers in Brazil but mostly with vertical driers elsewhere. The usual practice in Asia and Africa is pre-drying of the cherries under the sun by growers, followed by hulling. The final drying of green beans is done by traders in large vertical driers, and rotary driers are starting to be used.

Drying of the whole Arabica coffee cherry with the sugar-rich pulp and mucilage attached to it tends to impart more body and sweetness to the cup of natural coffees that are known for these features. Although natural coffees coming from floaters are free from unripe or partially ripe cherries, this is not the case with sinkers because unripe, partially ripe, and fully ripe cherries are all denser ('heavier') and sink. These unripe and partially ripe cherries may cause astringency in the cup. The solution for this is costly hand or color sorting of sinkers or the use of the pulped natural/honey system that separates these unwanted cherries and can have cup features similar to naturals. Hand sorting is restricted to quality-conscious growers. Color sorting of wet cherries, offered by most makers today, still has a very restricted use because of its comparatively high costs, not to mention the care and personnel required to handle a sophisticated machine in a wet farming environment.

6.2 Parchment coffee (pulped natural and washed coffees)

Pre-drying and full drying of pulped natural and washed coffees may take place in drying grounds, with the surface lined with concrete, asphalt, or other material, or in raised beds, also called drying racks or barbecues (Fig. 7). Pre-drying under the sun, if required before mechanical drying, will take longer if more mucilage is attached to parchment. In the worst case, pre-drying may last up to 2 days. Parchment coffee may also go straight into rotary driers without pre-drying if a centrifuge is used to both drain excess water and feed coffee to the driers (Fig. 8).

The time it takes to complete drying of parchment under the sun is from 7 days to 15 days, taking longer depending on the amount of mucilage attached to parchment as well as climatic and weather conditions. Pre-drying of pulped natural parchment must be carried out with thin layers of parchment, less than 1-inch thick, with stirring required every 15 minutes. Layers may be thicker for washed coffee parchment. Handling of parchment in pre-drying and drying has

Figure 8 Centrifuge receiving parchment from the wet mill and draining (skin drying) and raising it to driers.

to be gentler than in the case of natural coffees in order to avoid damage to the parchment that is more fragile than the attached pulp in the case of naturals. The drier the parchment, the more careful turning and handling has to be because the risk of physical damage to the parchment and bean increases. Similar to the drying of naturals, a solution to deal with rain is to have drying grounds or beds under a plastic roof. In the case of enclosed, green-house-like structures, fans can be used to circulate air to accelerate drying and avoid build up of moisture. It is for this reason that the completion of drying is increasingly done in driers that minimize or eliminate this risk, particularly rotary driers.

The use of mechanical driers can reduce drying times from 7 days to 15 days to 24 h to 30 h at a maximum temperature of 40°C measured in the parchment coffee itself (and *not* in the air used for drying). Attempts to dry in a shorter time under the sun or in driers, by using temperatures higher than 40°C, are known to cause damage to coffee that ranges from bleaching in storage to unwanted cup features.

Mechanical completion of drying is more frequent in this case than in naturals, for the reasons explained above, with rotary driers being by far the machine of choice. Static driers may create conditions for unwanted fermentation with the risk increasing with more mucilage is attached to the parchment. Stirring of coffee in otherwise static driers may cause physical damage as is also the case with elevators that recycle coffee in vertical driers. Uniformity of drying is frequently a problem with static driers, even those that use multiple chambers and have the advantage of reusing unsaturated hot air. Round multi-chamber static parchment coffee driers are now offered with stirrers to improve the uniformity of drying. Vertical driers of all types are often associated with physical damage to parchment coffee because of the continuous recycling of coffee with elevators. They are not known for uniform drying either, especially in the case of incoming coffee with uneven moisture content. Revolving rotary driers are known for greater homogeneity of drying, even in the case of coffee with uneven moisture content. The practice of overheating coffee in hulling-polishing in order to correct the uneven parchment drying can shorten the shelf life of green coffee, whose color fades sooner.

6.3 Heat sources

Hot air to dry coffee can also be produced by different systems: indirect fire heat exchangers for one or several driers and radiators heated by hot water or steam from boilers. Direct fire furnaces, with hot air and smoke going through coffee, should be avoided because the fumes transmit unwanted aromas and flavors to the beans. Typical fuels used are coffee husk, wood, other solid fuels (e.g. sawdust), fuel oil, diesel, and gas. A heat exchanger for each drier is the norm around the world except for Central America, where a large heat exchanger is often used to supply hot air for up to six or eight driers. In Brazil, large drying facilities with more than eight (some with up to 40) rotary driers may use hot water or steam boiler and radiators to produce hot air.

6.4 Optimizing drying

It is important to note that the quality of Arabica coffee, which is *properly* dried under the sun or mechanically is the same. Contrary to what is normally believed, there may be greater chances of inadequate drying under the sun than in rotary driers machines, even more so if the digital systems are used to monitor and control mechanical drying. The risks of sun drying are not only limited to accidental wetting by rain but also include unwanted fermentation and overheating if labor is not available or not trained to stir the coffee frequently.

Sun drying can be optimized in several ways, from coffee handling to temperature control. Lack of labor, trained or not, coupled with its rising costs, is rendering labor-intensive sun drying more difficult and complicated, especially in mid-size to large farms and when the workers available are not trained, the latter in the case of smallholder farmers too.

Frequent stirring of coffee is required to avoid unwanted fermentation at the early stages of drying and overheating throughout the whole drying process. Another type of damage to be minimized and avoided, especially in the case of parchment drying, is physical damage from the use of inadequate tools and machines, hard handling, and undue pressure or weight on the beans during stirring and unloading of the drying facilities. In East Africa, parchment coffee is usually covered for a couple of hours to protect it from overheating at noon time. Coffee being sun dried requires protection from dew and rain, more often by covering coffee with tarpaulins or plastic or, much less often but increasingly, by drying under plastic shades.

Mechanical drying can also be optimized in several ways: control of temperature and duration, minimizing damage to coffee, drying uniformly, and fuel efficiency. Recent developments like digital control systems and the use of hot air that is wasted help to optimize drying. Fuel efficiency can be optimized by improving the design of heat exchangers that have been specially developed for coffee driers. Fossil fuel burners and boilers are not specifically for coffee-drying but have many other uses. Shortening drying time can also help with fuel efficiency providing it also does not cause damage to or uneven drying of coffee. Improving the uniformity of air distribution in the coffee mass is another way to save fuel. The digital drying control systems, especially the ones with temperature monitoring at three points - heat source, air entering the drier, and coffee mass - can minimize fuel consumption in a significant way (Fig. 9).

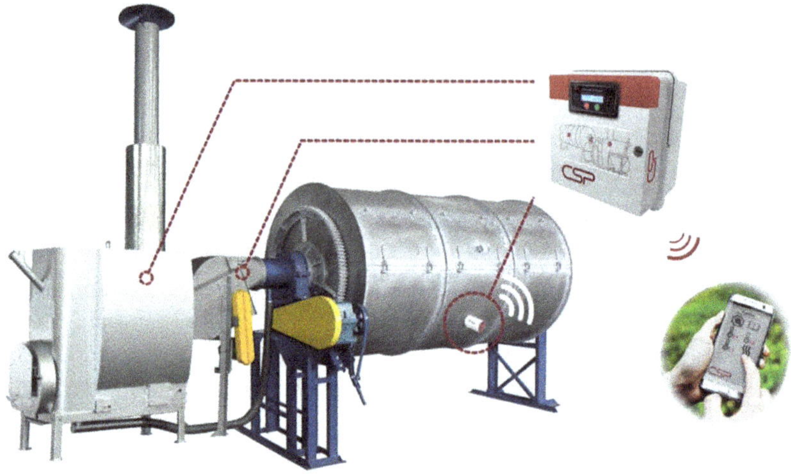

Figure 9 Rotary drier with indirect fire heat exchanger and digital three-point temperature control system.

One area that still deserves work to improve fuel efficiency is the reuse of air that leaves driers unsaturated, especially at the final stages of drying. This air has the potential to dry more coffee without adding to fuel costs. This is to some extent done in multiple-chamber static driers but does not occur in rotary and vertical driers. A different approach to save fuel in rotary driers is to use clean hot air otherwise lost around the chimney to pre-dry coffee in a silo above the drying drum.

7 Smallholder farmers

Coffee cherry processing by smallholder farmers should follow the steps described above with due consideration to the smaller volumes to be processed and the use of family labor. The type of harvesting and the labor available to separate cherries may not require the use of mechanical cherry separation even though there are small pulpers with unripe cherry separators available. The small driers available are either too large for smallholders or not economically viable. Smallholder coffee growers should use modern, water-saving small pulpers and mucilage removers and sun dry the coffee. In addition, they may use an unripe cherry separator, ferment coffee, and/or cover their drying grounds or beds with a plastic roof. This should allow them to produce as good quality coffee as in larger facilities if there is proper control at each step, especially fermentation, if used, and drying.

Given the greater economies of scale in processing larger volumes of coffee, collective processing of coffee cherries produced by a group of smallholder growers may be a good alternative and even make mechanical drying economically viable. This explains the current tendency toward central wet milling and drying facilities used by groups of smallholders. Another advantage of central milling is the ability to offer larger volumes of coffees with uniform features, which is often not the case when coffee from many smallholders is processed separately and blended later.

8 Environmental impacts of on-farm processing and drying

Environmental impacts differ along the coffee-processing chain. Water contamination and the production of wet and solid waste occur in pulping and mucilage removal. Air pollution may be caused by mechanical drying, whereas noise and dust can be produced in other areas of processing. Modern technology minimizes but not necessarily eliminates these adverse effects.

There are four important sources of environmental impacts in the processing of coffee cherries:

- Water consumption.
- Water contamination.
- Solid waste production.
- Fuel used in drying and the smoke it produces.

Pulped natural and washed coffee processing have the four impacts mentioned above, whereas the environmental impact of natural processing is limited to drying.

The removal of impurities and stones and the separation of floaters (over-ripe and partially dry cherries) and sinkers (unripe, under-ripe, and ripe cherries) may consume a lot of water if done in siphon tanks or water channels. The eco-friendly alternative is mechanical siphons that consume very little water.

The choice of pulpers to be used should consider low water consumption to the extent that it does not affect performance, i.e. loss of parchment with pulp, pulp mixed with parchment, and physical damage to coffee. The same holds for mechanical mucilage removal, with emphasis on the trade-off between water consumption and physical damage to parchment and green bean.

Fermentation of parchment, if used, is by far the greatest source of water use and contamination in wet processing. Dry fermentation consumes a little less water because fermentation tanks are usually loaded and unloaded with the help of water in either type of fermentation. Even though the pollution load that derives from the mucilage is basically the same in fermentation or mechanical removal of mucilage, in the latter case the volume of wastewater is much smaller and less costly to treat, if necessary, or dispose of.

Water contamination from wet processing can be spread over many smallholder farms or concentrated in small, mid-size, or large central wet mills. Both can be harmful to the environment. It is apparently easier to dispose off small volumes of contaminated water by infiltration into the ground, but this may contaminate the underground water table. On the other hand, the larger volumes concentrated in central mills can be treated cost-effectively with specialist equipment. The solid waste produced by the wet milling of pulped natural and washed coffee is the coffee pulp that can be used as a fertilizer after it decomposes.

Mechanical drying requires fuel to heat the air that dries coffee. The ideal fuel is dry parchment husk if coffee is hulled on the farm or nearby. Dry husk from natural cherries can also be used as a fuel, but it should be ideally returned to the field as a fertilizer. If the coffee husk is not available, the alternatives are wood, oil, or gas. The use of wood should be restricted ideally to coffee wood from pruning or to trees grown for commercial or industrial use. Smoke and gases coming out of the chimney of the heat exchangers that supply clean hot air to driers are least harmful when the fuel is natural gas. Although the smoke and gases from burning husk, wood, or oil are harmful to the environment, they tend to be more easily absorbed when drying takes place on farms than in urban areas.

Fuel efficiency can be optimized by improving the design of heat exchangers that have been specially developed for coffee driers. Fossil fuel burners and boilers are not coffee-drying specific and have many other uses. Shortening drying time can also help with fuel efficiency as long as it does not affect coffee quality. Improving the uniformity of air distribution in the coffee mass is another way to save fuel. Digital drying control systems, especially those with temperature monitoring at three points - heat source, air entering the drier, and coffee mass - can minimize fuel consumption significantly (Fig. 9).

One area that still deserves research to improve fuel efficiency is the reuse of unsaturated air that leaves driers, especially at the final stages of drying. This air has the potential to dry more coffee without adding to fuel costs. This to some extent is done in multiple-chamber static driers but does not occur in rotary and vertical driers. A different approach to save fuel in rotary driers is to use hot air otherwise lost around the chimney to pre-dry coffee in a silo above the drying drum. Filters and after-burners can be used to lower the emissions of heat exchangers independently of the fuel used. Not much can be done in the case of direct fire furnaces whose exhaust gases go through the mass of coffee being dried.

9 Storage of dry coffee

Dry cherry and parchment coffee should be stored for at least a few days before the removal of the cherry or parchment husk, i.e. before hulling. Ideally, coffee should be stored as cherry or parchment for as long as needed, with further processing taking place shortly before roasting or exporting to ensure maximum quality. Wet-milled coffee is primarily stored as parchment until processing for export or roasting. Arabica naturals are usually stored as cherries on farms but hulled for storage as green coffee outside farms. Robusta naturals are almost always hulled shortly after drying or even before drying is completed and stored as green coffee.

Storage of dry coffee may take place on farms, at cooperatives, or at traders' facilities, depending on the usual practices in different countries. Storage may be in bags of different sizes or in bulk in silos. In all cases, temperatures and moisture levels inside the storage areas if not in the ideal range - around 25°C and 60% moisture content - may negatively affect the quality and cause coffee weight loss or gain and, if the latter, require re-drying. Moisture contents below 12% not only mean a potential decrease in revenue (because of weight loss) but also increase the risk of physical damage in hulling and polishing. Properly designed warehouses and the way coffee is placed inside them can contribute to good air circulation and temperature control. Fans and/or equipment to increase air moisture can be used to inject or remove air at warehouses to keep the storage environment within the desired temperature and moisture levels.

10 Cleaning processes

Cleaning of dry cherries, parchment, or green coffee is performed in two stages:

- Sieving.
- Density separation.

These use two different machines:

- Pre-cleaners.
- Destoners.

Pre-cleaning machines use two or three screens with different sizes of holes to remove impurities (e.g. sand, twigs, and stones) that are larger or smaller than the coffee to be processed (Fig. 10). Impurities that are the same size as coffee, especially stones, leave the pre-cleaners along with coffee. The choice of pre-cleaning screens has to consider both the type of coffee to be processed - cherry, parchment, or green - and the type, shape, and size of impurities. The wrong choice of screens may jeopardize effective separation.

Destoners use an upward air flow current to make coffee float while the heavier stones remain in contact with the inclined vibrating deck (Fig. 11). While coffee flows downward by force of gravity along the inclined deck,

Figure 10 Pre-cleaner with dust suction at coffee inlet and outlet.

the stones are pushed upward by the longitudinal vibration of the deck. Proper adjustment of pressure and vibration are also critical to ensure proper separation. Magnets may be used to remove ferrous impurities. Magnets can be installed before the pre-cleaner, between the pre-cleaner and the destoner or after the destoner.

Overloading may cause poor performance and inadequate separation in pre-cleaners and destoners. Stones and other impurities that are not removed may damage other processing machines, specially hullers and hullers-polishers, and have a direct impact on the quality and price of the final product. Since lighter impurities can generate dust, pre-cleaners and destoners can cause air contamination in dry coffee processing if dust removal technology is not properly designed and built into the machines. Low-noise fans should be used, especially in destoners, to ensure a good working environment.

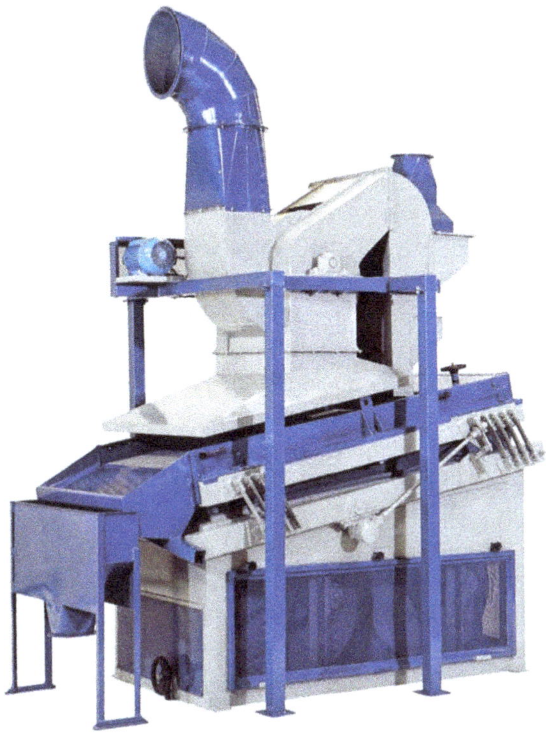

Figure 11 Destoner with dust suction hood.

11 Hulling and polishing processes

Hulling is the removal of the dry parchment or cherry husk. Polishing is the removal of the silver skin attached to the green bean. There are two major ways to hull coffee:

- Rubbing cherries or parchment against each other and against the metal parts of the machines.
- Forcing them with a cross-beater through a round screen with holes bigger than coffee beans but smaller than dry parchment or cherry.

Polishing, carried out by friction as the green beans rub against each other and against the machine, may be performed in the same machine that hulls coffee or at a different machine after the beans have been hulled. The husk and dust produced in the process must be aspirated and blown outside the mill into a silo that should retain the solid waste and discharge clean air into the environment.

Parchment is usually hulled and polished by friction in the same machine. Hullers-polishers have a single drum or a divided drum to hull first and then polish (Fig. 12). Parchment can also be hulled by a cross-beater/screen huller with polishing by friction in a separate machine if required. Unlike washed coffee, pulped natural/honey coffee may or may not be polished.

Arabica naturals are mostly hulled in cross-beater/screen hullers. This also holds for Robusta coffees in Brazil and a few other countries. Robusta naturals are mostly hulled by friction in machines that have a steel knife to tear the harder and thicker dry cherry husks. Naturals are not usually polished.

Hulling of parchment by friction causes a temperature increase that has to be controlled to avoid weight loss, damage to coffee, and bleaching (loss of color) in storage or shipping. It is not unusual for parchment coffee that is not evenly dried to be overheated in hulling and polishing, resulting in problems such as bleaching. Dry cherry hulling by friction is not known to cause these problems. Hulling with cross-beater/screen hullers does not heat coffee at all and is known as 'cold' hulling. Neither friction nor screen hulling of naturals removes the silver skin from, i.e. polishes, the green beans, which is not required for naturals.

Inadequate hulling may cause physical damage to green coffee ranging from bruising and cracking to breaking and even crushing. Friction hullers are known to be more aggressive than cross-beater/screen hullers, and proper operation is critical to avoid damage in both cases. Adequate adjustment is also important to ensure that green beans are not lost with the husk that is

Figure 12 Two-stage friction huller for parchment coffee.

Figure 13 Cross-beater huller preceded by destoner and followed by devices to separate coffee.

discarded or that the husk remains mixed with green coffee. Cross-beater/screen hullers are often equipped with devices to sort out and repass unhulled coffee and to separate green beans by density (Fig. 13). Hulling performance can be optimized by choosing the right machine and running it properly

The environmental impact of hulling is mostly associated with dust that can be released if the husk aspiration and storage system is not properly designed to retain solid waste and only to let out clean air. In the case of parchment hulling by friction, it is important to avoid parchment husk particles remaining with coffee and generating dust in the processing stages after hulling.

12 Size grading

Coffee is separated by size and shape to enable improved density and color separation, to meet national standards or clients' requirements, and to allow more uniform roasting. Green coffee is size graded in a set of flat vibrating screens placed one above the other and with decreasing hole sizes. Flat beans

Figure 14 Upward-flow size grader with 16 screens.

are separated by screens with round holes, while peaberries (round-shaped beans) are separated, if required, by screens with slotted holes. Coffee flows downward, by gravity, in the inclined flat screens. There are now upward-flow size graders that are more precise because of a longer contact time between beans and screens (Fig. 14).

The choice of sizes (and shapes) provided by grading machines, using interchangeable screens, is related to a producing country's grade description system (e.g. Kenya's AA or Colombia's Supremo), clients' requirements, or both. Flat-screen graders in the market today use from three to nine different screens. Even in the case of countries that export ungraded coffee, the use of a grader is recommended both to remove unwanted over-sized and small and broken beans and to facilitate density and color separation. The most common deficiency in size grading is excess capacity that causes smaller beans to remain mixed with larger ones. A set of sample screens should be used from time to time to verify the efficiency of grading and to adjust the machine if necessary.

13 Gravity separation

The purpose of gravity separation is to separate green beans by density because many defective beans have a low density, and there is a positive correlation between high-density beans and coffee quality. Separation is achieved with the help of coffee flotation produced by an upward-moving air flow. Lower density ('light') beans float while higher density ('heavy') beans remain in contact with the deck. The inclination of the deck to the front and sideways causes good denser beans to be discharged at the higher outlet(s) in front of the machine while less dense defective beans leave the machine through the lower front outlet(s). There is usually a center fraction with a mix of beans of different densities that leaves the machine through the central outlet(s) and should be repassed.

Some gravity separators (Fig. 15) now make it possible to adjust not only the pressure but also the direction of air flow. Metallic decks with different types of perforations and dents are used by some makers instead of the traditional wire mesh. There may be up to six different adjustments in a gravity separator: feeding rate, transversal and longitudinal slopes, air pressure and direction, and intensity of deck vibration. This requires the closest attention from knowledgeable operators and is a step that is critical to optimize quality. Negative pressure gravity separators, that aspirate rather than blow air through the deck, have fewer adjustment possibilities and are more difficult to operate.

Modern high-efficiency gravity separators should always precede color sorters because they are a lower-cost way to remove many defects that are associated with color and density, e.g. black beans, over-fermented beans, and even stinkers. This reduces reliance on color sorters with considerable

Figure 15 Gravity separator with dust suction hood.

savings in investment and maintenance costs. Gravity separators may be a source of air contamination if incoming coffee is not aspirated or separators do not have a dust hood. The latter is essential if coffee has not been properly cleaned or there are parchment husk particles mixed with the coffee. Another environmental impact of gravity separation is the noise coming from the fans used to create the air flow. Low-noise fans should be used to improve the working environment.

14 Micro-lot processing

Micro-lot processing is discussed here because it typically comprises the dry processing steps just reviewed, from cleaning to gravity separation, using small capacity machines starting, e.g. at 300 kg/h. Depending on the quality and cleanliness of incoming parchment or cherry coffee and the client's requirements, cleaning and/or gravity separation may be skipped, but ideally a micro-lot processing mill should include cleaning, hulling and polishing, size grading, and gravity separation (Fig. 16).

Small capacity machines used should have the same features as larger machines because micro-lots are supplied primarily to the specialty coffee market, which usually has stricter quality requirements than other coffee markets. If the quality of incoming lots is high and defects are few, color sorting

Figure 16 Micro-lot mill with (A) cold huller, (B) huller-polisher, (C) size grader, and gravity separator (behind/not shown).

is not required and a small gravity separator may suffice, if needed. Because micro-lots are mostly single-origin, often from a single farm, blending is not usually required. Micro-lot processing can be done in compact machines that combine several processing steps in one single set-up, e.g. hulling, polishing, repassing, and size grading.

Export mills, even small ones, cannot handle micro-lots in their main processing line(s) because the small size of the lots means they pass through the machinery too quickly, without time to adjust to a new lot, resulting in too much downtime between micro-lots. It is for this reason that coffee export mills have to rely on separate lines for micro-lots using small machines. The other users of micro-lot mills are local traders, cooperatives, and mid-to-large-size farmers.

15 Color sorting

Since color is a good indicator of quality defects, green coffee beans are color sorted by comparing their color to acceptable colors using digital micro-processing technology (Fig. 17). Off-color beans, associated with defects, are rejected by a compressed-air ejection system. Modern machines can 'learn' what to accept or reject from green coffee samples. Color sorting machines have a feeding system that creates uniform layers or streams of beans that go through an optical system connected to a processing unit that identifies the beans to be ejected. The most common feeding systems are gravity chutes and belt conveyors. Processing units have evolved from simple analogue circuits to advanced digital microprocessors. The sorting techniques range from mono- to trichromatic sorting, with UV sorting also available.

Figure 17 Color sorters.

Different color sorting techniques include the following:

- Monochromatic sorting: this can only separate beans that are lighter or darker than the average: white, pale, dark, or black. They are mostly used for Robustas, where color defects are easier to identify.
- Bichromatic sorters: these measure two wavelengths and are more sensitive to more subtle color differences. They can separate lighter and darker beans and, additionally, sort unripe, pale green, yellowish, brownish, and other defective beans. They are primarily used for Arabicas, where defects are associated with a wider range of colors and tones.
- Trichromatic sorters: these improve on biochromatic color separation and add the capabilities to sort by shape or size and to detect impurities like stones. Cracks or holes can also be identified in beans of acceptable colors.
- Ultraviolet (UV) sorting: this is mostly used to separate moldy or stinker beans out of freshly dried coffee. It loses accuracy after the coffee is stored because the fluorescence of the beans falls sharply.

Laser and X-ray are other technologies available for coffee sorting whose use is likely to increase depending on the type of application and relative costs.

Color sorting optimization must be analyzed from three different perspectives:

- First, as far as possible, defects should be removed at an earlier stage by mechanical machines, e.g. gravity separators, that proportionally cost less per ton of coffee processed than color sorters.
- Second, the color sorter solution, from mono- to trichromatic and including laser and X-ray options, should result from a cost-benefit analysis considering the type of coffee to be processed, the defects to be removed, and market requirements to be met.
- Third, repassing should be considered to improve the capacity and efficiency of sorting.

Sorting by size and shape using trichromatic technology needs to be compared with mechanical size graders that also separate coffee beans by size and shape. This comparison should include not only just costs but also at which stage of the processing flow this operation should be performed. Since size grading improves the efficiency and capacity of gravity separation and color sorting, it is best performed at an earlier stage.

16 Coffee storage in dry mills

Coffee received in dry mills is dry cherry, parchment, and/or green coffee. Storage before or after processing, and before blending or bulking, is done

Figure 18 Coffee stored in big-bags.

in regular bags, big-bags (Fig. 18), or in bulk in silos, with bags typically made from jute/hessian. The use of big-bags or bulk storage minimizes handling and labor but requires a much larger investment in silos and transport equipment. The decision to choose bags, big-bags, or bulk depends on the cost of labor, land, civil works, transport equipment, and silos besides the size of the mill.

Natural coffee is mostly stored as green coffee because green beans occupy one-third of the space of dry cherry coffee, and unpolished natural coffee beans lose less color in storage than polished washed coffee beans. The storage of green naturals is mostly done after full processing and before blending for export or roasting. Different grades and types of green coffee are stored to be blended later, shortly before shipment, according to importers' specifications or country's standards. Wet-milled coffees, especially if polished, are shipped shortly after processing because their green beans lose color in storage. Large parchment millers now have some sort of buffer storage for a short period of time after processing and before shipment for logistical reasons.

17 Blending and bulking

Blending is the mixing of different coffee sizes, cup qualities, and/or origins to obtain a homogeneous lot with the characteristics and volumes required by the country's (export) or the client's specifications. Bulking is the mixing of coffees of the same quality produced by different growers or coming from different regions to obtain a homogeneous lot at the volume required by the client.

The principles and techniques of blending and bulking are the same. Coffee must be brought together in agreed proportions and mixed to obtain homogeneous lots. This is usually done with the help of a round silo fed by a high-capacity elevator that both receive coffee in the pre-determined proportions and recycles it constantly to ensure uniform mixing. As mills get larger, more blending silos and elevators are used. High-capacity export mills may mix different coffees previously fed into a battery of silos that unload onto a belt conveyor through an adjustable volume feeding system, e.g. a rotary valve or a variable speed belt, installed at the outlet of each silo. The rate of feeding from individual silos onto the belt is selected according to the proportions required in the final lot. Blending or bulking requires having the coffee bags fed in the right proportion into the round silos or adjusting the rotary valve or variable speed belt according to the required proportions. Frequent sampling and checking of the final product should be performed to ensure that coffee is evenly mixed.

18 Weighing and packing for shipment

The usual ways to weigh coffee to be bagged are traditional mechanical platform scales, for small lots and where labor is available, or automatic mechanical or electronic batch scales for larger volumes and processing lines. Bags are closed by sewing them with a sewing machine that may be hand-held or attached to the structure of a belt conveyor. Big-bags can be either fed from automatic scales in continuous-flow mode or weighed in specific big-bag floor-level platform scales with load cells (Fig. 19). Finally, coffee to be transported in containers is weighed on automatic scales in continuous-flow mode or using flow scales.

Automatic batch scales have a weighing bucket under an overhead silo. In the mechanical version, the bucket is connected by a lever to a counterweight where the desired weight is set, e.g. 60 kg. The main flow into the scale is interrupted shortly before this weight is reached, and a much smaller restricted flow continues for a short time to reach the target weight. Electronic batch scales use load cells to weigh coffee and interrupt the flow into the bag when the required weight is reached. Electronic scales are much easier to connect to digital systems that may control the full processing operation or parts of it. Big-bags with capacities of 600–1200 kg of coffee are usually weighed at floor-level platforms equipped with load cells. The empty big-bag placed on the platform is fed up to the point it reaches the desired weight, as indicated by the load cell. Alternatively, coffee may be fed into a big-bag from a mechanical or electronic batch scale working as a flow scale.

Bulk container loaders receive pre-weighed coffee from overhead silos or elevators and use a high-capacity fan to blow coffee beans into a container with a liner (Fig. 20). The bulk loaders may have a dust extraction system or leave the dust inside the container. Mechanical or electronic bucket scales used as flow

Figure 19 Big-bag scale.

scales or flow scales themselves allow a pre-set weight of coffee to go through before it is fed in bulk into containers.

Optimal weighing is achieved by a combination of good-to-high precision equipment and proper operation and control that together enable coffee to be delivered within the range of error specified in contracts. The choice of packing for shipment - conventional bags, hermetic bags, big-bags, or containers in bulk - depends on logistic requirements and the clients' ability to receive coffee shipped in the different ways available. There is a strong tendency to use containers in bulk for commercial coffees and also other types of coffee required in large volumes. Different sizes and materials of bags are increasingly used for specialty coffees and micro-lots. Big-bags are more used for intermediate pre- or post-processing storage; they are also placed in containers for shipment but in much smaller scale than regular bags. The environmental impact of weighing and packing is restricted to dust generation during container stuffing in bulk, depending on the type of equipment used. Some countries recommend the use of smaller bags, under the usual 60–69 kg bags, to facilitate handling and to decrease labor effort.

Figure 20 Bulk container loader.

19 Processing lines

The individual pieces of equipment described before are usually combined into processing lines designed by experienced engineers (Fig. 21). Although separation, pulping, and demucilaging may feed each other by gravity in small- and mid-size lines, the usual way to combine machines into a processing line is with the help of transport equipment, usually bucket elevators and belt conveyors. Silos may be used for buffer or temporary storage.

The design of a coffee-processing line can be optimized by using as little floor area, transport equipment, and electric power as possible. The level of mechanization depends on the relative costs of labor and machinery. The choice of machines that deliver the required capacity and quality with the least power/energy requirement is a critical decision because of its impacts on processing costs over the useful life of the equipment.

A growing trend in the design of coffee-processing lines, especially larger ones, is to use central dust aspiration and disposal systems in order to keep the working environment with as little dust as possible. The alternative is to use specific dust aspiration and filtering systems for machines or group of machines that produce the most dust. Automation of parts of the processing line or for the full line is increasing for larger lines. The control of processing efficiency enabled by automation can be complemented by automatic sampling at strategic points to control coffee quality.

Figure 21 Dry processing line with central dust aspiration in Southeast Asia.

20 Where to look for further information

Borém, F. (Ed.). 2014. *Handbook of Coffee Post-Harvest Technology*. Specialty Coffee Association, Chelmsford, UK.

Farah, A. (Ed.). 2019. *Coffee: Production, Quality and Chemistry*. Royal Society of Chemistry, London, UK.

Folmer, B. (Ed.). 2016. *The Craft and Science of Coffee*. Elsevier Academic Press, Switzerland.

Hameed, A., Hussain, S. A., Ijaz, M. U., Ullah, S., Pasha, I. and Suleria, H. A. R. 2018. Farm to consumer: Factors affecting the organoleptic characteristics of coffee II: Postharvest processing factors. *Comprehensive Reviews in Food Science and Food Safety* 17(5): 1184–1237.

Illy, A. and Viani, R. (Ed.). 2005. *Espresso Coffee, the Science of Quality*. Elsevier Academic Press, Cambridge, MA.

Lashermes, P. (Ed.). 2018. *Achieving Sustainable Cultivation of Coffee*. Burleigh Dodds Science Publishing, Cambridge, UK.

Sivets, M. and Desrosier, N. W. (Ed.). 1979. *Coffee Technology*. Avi Publishing Co, Westport, CT.

Taveira, J. H. D. S., Sttela, D. V. F. D. R., Pedro, D. O., Gerson, S. G. and Eder, P. I. 2015. Post-harvest effects on beverage quality and physiological performance of coffee beans. *African Journal of Agricultural Research* 10(12): 1457–1466.

Wintgens, J. N. (Ed.). 2012. *Coffee: Growth, Processing, Sustainable Production* (2nd edn.). Wiley-VCH Press, Weinheim, Germany.

Part 2

Sustainable pest and disease management

Chapter 8

Insect pests affecting coffee: understanding agroecosystems and alternative methods of control

Luis Fernando Aristizábal-Aristizábal, Independent Consultant, USA

1 Introduction

Originating in Africa, *Coffee arabica* L. and *Coffee canephora* L. (Gentianales: Rubiaceae) are the two most popular species of coffee produced commercially around the world. *C. arabica* represents about 70% of worldwide production and is recognized for its lower caffeine content and sweet taste, while *C. canephora* has higher caffeine content and a more bitter taste. There are about 10 million hectares of planted coffee globally in about 80 countries located in tropical and subtropical regions. Global coffee production was estimated to be 170 977 million 60 kg bags (the basic unit used in the industry) during the coffee

http://dx.doi.org/10.19103/AS.2021.0096.21

season 2018/19 (ICO, 2019). An estimated 20 million people are estimated to depend on coffee production. This means insect pests and diseases can have a major impact on their livelihoods as well as the economies of many countries.

The aim of this chapter is to review some of the most important insect pests affecting coffee plantations worldwide. In each case, the chapter discusses the damage they cause, insect biology and ecology, and how this understanding can be used to develop alternative more environmentally friendly and sustainable methods of pest management. The impact of climate change on the coffee agroecosystem and especially on insect pests are discussed. The chapter includes a case study on improving integrated pest management of coffee berry borer (CBB) in commercial coffee farms in Hawaii. This case study looks at improving cultural control practices for CBB in coffee farms in Ka'u District, Hawaii.

2 Entomofauna associated with coffee agroecosystems

Insects associated with coffee agroecosystems are key to the stability of agroecosystems and thus the sustainability of coffee production. The presence of phytophagous (plant feeding), detritivorous (feeding on plant detritus), and beneficial insects (such as pollinators or parasitoid species) is fundamental for ecosystem homeostasis, allowing a balance in insect populations to be achieved. When a number of biotic and abiotic factors interact simultaneously, the equilibrium of the natural ecosystem is maintained. However, in agroecosystems such as monocultures of coffee crops, this equilibrium is easily broken due to the low diversity of plants and crop uniformity. Populations of insects become unbalanced and phytophagous insect populations grow sufficiently to become a pest.

An example of diversified coffee systems is the description given by my grandfather of coffee farms in Colombia before 1950s. He described more than 25 different plants (trees, tropical fruits, and vegetables) associated with the coffee plantation. Trees and other plants included avocado, mango, guava, cocoa, guanabana, tangerines, oranges, lemons, sapota, banana, sugar cane, cassava, corn, onion, tomato, and bean. In addition, the farm had ornamental plants and incorporated animals in the farm. In this agroecosystem, with a high diversity of plants supporting a range of insects and other species, it was difficult for any insect pest to reach high enough infestation levels to cause significant damage, given factors such as natural enemies and inter-species competition.

There are some coffee farms that still have a high diversity of plants. Some organic coffee farms are, e.g. examples of high diversity. However, the demand for higher volumes of coffee production, more productivity per hectare, the introduction of coffee plants resistant to disease, technological innovations, and other factors have made coffee monoculture the common denominator in coffee-producing countries. As a result, phytophagous insects have become significant insect pests in coffee agroecosystems. Moreover, the indiscriminate use of insecticides, fungicides, and herbicides, plus the impact of climate change, has created further imbalances, contributing in some cases to an increased number of insect pests and infestation levels. In addition, invasive exotic insect pests remain a risk for coffee countries in which those insects have not yet been reported.

Phytophagous insects are found in all parts of the coffee plant. Mealybugs (*Dysmicoccus* sp.) and beetles (*Phyllophaga* sp.) are, e.g. commonly observed feeding in the root system. Stem borers such as the black stem borer (*Apate monachus*) and grasshoppers (*Gryllotalpa* sp.) cause damage to stems and foliage, especially in young coffee plants or stumped trees. Other borers such as black twig borer (*Xylosandrus compactus*) and brown twig borer (*Xylosandrus morigerus*) affect branches. Coffee leaves are attacked by both chewing and sucking insects. The coffee left miner (*Leucoptera coffeella*), Lepidoptera defoliators, and grasshoppers cause damage to leaves as chewing insects, while the citrus mealybug (*Planococcus citri*) and coffee green scale (*Coccus viridis*) are sucking insect pests damaging foliage. In addition, mites such as red spider mites (*Olygonychus yothersi*) and broad mites (*Polyphagotarsonemus latus*) (arachnids rather than insects) also cause significant damage to leaves. Coffee berries or cherries are attacked by the CBB (*Hypothenemus hampei*). Insect pests affect processed and stored coffee as well.

Besides phytophagous insect species affecting coffee plants, there are many beneficial insects associated with coffee agroecosystems, including predators, parasitoids, and pollinators. Generalist predators such as dragonflies, lacewings, ladybirds, wasps, and ants are commonly observed in coffee plantations. Ecto-parasitoids and endo-parasitoids also play a role in the regulation of insect pests. Honey bees and other pollinators are present in coffee plantations, most commonly during coffee blooms. Finally, several detritivorous insect species such as beetles, flies, and butterflies are frequently found feeding on dead plant residues in coffee agroecosystems.

3 Key insect pests of coffee crops

There are more than 850 species of insects that feed on coffee plants around the world. However, only a few of these species cause significant damage to plants and losses in coffee yields. Le Pelley (1973) reported 38 significant species, Barrera (2008) reported 121, Constantino et al. (2013) reported 136, and Bustillo (2015) reported 40 (Tables 1 and 2). These authors have also reported beneficial insects (predators, parasitoids, and pollinators) and microorganisms (entomopathogenic fungi and entomopathogenic nematodes), all of which interact in coffee agroecosystems (Table 1). This chapter discusses six major insect pests that significantly affect coffee plants and production around the world. These pests are:

1 Coffee berry borer,
2 Coffee leaf miner,

Table 1 Number of insect pest (species or genus) affecting different structures of coffee trees (root system, stem, branches, leaves, and berries or beans) reported in different regions.

	Region	Genus /					Affected part of the plant			Natural	**
Reference	Country	Species	Order	Family	Root	Stem	Branches	Leaves	Berries	Enemies *	Microorganisms
Le Pelley (1973)	Worldwide	38	7	20	3	3	2	26	4	30	2
Cardenas & Posada (2001)	Colombia	90	10	38	13	7	1	65	4	30	6
Bittenbender and Smith (2008)	Hawaii	7	4	6	1	0	1	4	1	2	1
Barrera (2008)	America	121	6	28	33	12	1	73	2	54	20
Constantino et al. (2013)	Colombia	136	11	42	26	8	1	89	12	41	10
Mugo et al. (2013)	Kenya	5	4	4	0	0	0	2	2	0	1
Bustillo (2015)	Worldwide	40	4	14	7	6	2	18	7	47	6
Magina et al. (2016)	Tanzania	10	4	9	0	2	0	6	2	9	3

Source: Natural enemies* including predators and parasitoids. Microorganism** affecting insect pests including entomopathogenic fungi, entomopathogenic nematodes, bacteria, and virus.

3 Antestia bug,
4 Twig borer,
5 Root mealybug, and
6 Stem borers.

In each case, the chapter describes the damage caused, insect's life cycle, ecology, and biology, and how this understanding can be used to develop alternative pest management strategies (Fig. 1):

Figure 1 Damages caused by insect pests and mites on coffee trees: (a) damage caused by coffee leaf miner, (b) damage caused by black twig borer, (c) damage caused by red mite, and (d) damage caused by Broad mite.

4 Coffee berry borer

(Coleoptera: Curculionidae: Scolytinae) *Hypothenemus hampei* (Ferrari).

Originating in Central Africa, the CBB is considered the most economically damaging and serious insect pest affecting coffee trees around the world (Baker, 1999; Kucel et al., 2009; Vega et al., 2009). This small but aggressive pest is found in almost all coffee-producing countries located in tropical and sub-tropical regions in the world, except Australia, China, and Nepal (Johnson et al., 2020). Direct damage caused in berries affects the quality and proportion of marketable coffee. Infestations higher than 5% in the field can cause yield loses since the quality of processed coffee (parchment or green beans) is affected. High CBB infestations (30-90%) are observed in poorly managed and abandoned coffee plantations. Economic losses include a lower proportion of marketable coffee, reduction of quality, and thus price premium, as well as an increase in production cost due to CBB control (Fig. 2).

4.1 Life cycle, ecology and behaviour

The mated female abandons the coffee berry in which she was born and searches for a new berry to colonize as her new home. This is the beginning of the CBB life cycle. The CBB female opens a hole and enters inside the berry. CBB attacks all berries (from green developing berries until cherries and raising berries). CBB attacks younger berries (45–120 days old) but oviposition does not start until berries become hard (Bustillo et al., 1998; Baker, 1999). If the green developing berry becomes hard (> 20% of dry weight), which occurs when the berry is around 150 days old, the CBB makes tunnels and galleries in which she starts oviposition. Oviposition can thus be delayed for many weeks on young berries. CBB oviposition can occur in berries 150 days old until berries reach maturity (240 days old) (Baker, 1999). The CBB may also abandon a berry and search for a new one. Overall the CBB lays about 40-120 eggs in her life span (Damon, 2000; Bustillo et al., 1998). During the first 15-20 days, the CBB lays between 1 and 3 eggs per day. She then takes a break for some days and then starts oviposition again, but with a smaller number of eggs.

Eggs hatch in 4-8 days. Larvae pass through two instars, which take about 18-28 days. Larvae feed actively, making galleries inside the berries which affect the quality of the bean. Pupation occurs inside the berries and takes about 12-18 days. The duration of the complete life cycle of CBB depends on the temperature and the maturity of infested berries. Overall, the life cycle takes between 21 days (27°C) and 63 days (19°C). The emerging teneral adults (light brown in colour) take about 1-3 days to harden their exoskeleton (becoming dark brown). Adults then reach sexual maturity. Mating occurs between siblings with an overall sex ratio of (1:10) male:female. Males are smaller than females

Figure 2 Damages caused by coffee berry borer: (a) healthy uninfested berries, (b) CBB female starting the attack in a green berry, (c) damage inside the berry, and (d) damage on processed coffee.

due to sexual dimorphism. Only females fly and colonize coffee berries, while males have no functional wings. The longevity of females is about 100–155 days, while it is about 45–70 days for males. Several generations occur per year depending on local conditions and available coffee berries. In Colombia (2–4), Hawaii (3–5), Ethiopia (2), Kenya (3), and Tanzania (4) generations occur per year, respectively.

CBB is typically observed in 'hot spots' with high infestation levels. CBB is commonly found in field borders, close to coffee-processing facilities, close

Table 2 List of most relevant insect pests and mites reported in coffee crops around the world including distribution, structure of plant that is affected, and developmental stage that feeds on coffee plants.

Insect and mite pests	Geographic	Distribution			Developmental stage
Taxon (scientific name)	Africa	America	Asia	Structure of plant affected	feeding on coffee plant
Antestia bugs					
(Hemiptera: Pentatomidae)					
Antestiopsis orbitalis bechuana (Kirkaldy)	x			Berries, buds, leaves	Nymph, adult
Antestiopsis orbilatils ghesquierei (Carayon)	x			Berries, buds, leaves	Nymph, adult
Antestiopsis intricata (Guesquiere & Carayon)	x			Berries, buds, leaves	Nymph, adult
Antestiopsis thunbergii (Gmeln)	x			Berries, buds, leaves	Nymph, adult
Antestiopsis facetoides (Greathaed)	x			Berries, buds, leaves	Nymph, adult
Antestiopsis clymeneis (Kirkaldy)	x			Berries, buds, leaves	Nymph, adult
Antestiopsis cruciata (F.)			x	Berries, buds, leaves	Nymph, adult
Aphids					
(Hemiptera: Aphididae)					
Toxoptera auranti (Boyer de Fonscolombe)	x	x	x	Leaves, buds tissues	Nymph, adult
Aphis gossypii (Glover)	x	x		Leaves, new tissues	Nymph, adult
Mysuz persicae (Sulzer)		x		Leaves, new tissues	Nymph, adult
Cicadas					
(Homoptera: Cicadidae)					
Quesada gigas (Oliver)		x		Roots	Nymph
Quesada sodalis (Walker)		x		Roots	Nymph
Fidicina mannifera (Fabricius)		x		Roots	Nymph

Fidicina pullata (Berg.)		x		Roots	Nymph
Carinta fasciculata (Germar)		x		Roots	Nymph
Coffee berry borer					
(Coleoptera: Curculiondae, Scolytinae)					
Hypotenmeus hampei (Ferrari)	x	x	x	Berries	Larva, Adult
Coffee berry moth					
(Lepidoptera: Pyralidae)					
Prophantis smaragdina (Butler)	x			Berries	Larvae
Coffee leaf miners					
(Lepidoptera: Lyonetiidae)					
Leucoptera coffeella (Guérin-Méneville)			x	Leaves	Larvae
Leucoptera meyricki (Guesquiere)	x			Leaves	Larvae
Leucoptera coma (Guesquiere)	x			Leaves	Larvae
Leucoptera caffeina (Washbourn)		x		Leaves	Larvae
Cutworms and armyworms					
(Lepidoptera: Noctuidae)					
Agrotis spp.		x		Stems & small plants	Larva
Agrotis ipsilon (Hufnagel)		x		Stems & small plants	Larva
Agrotis repleta (Walker)		x		Stems & small plants	Larva
Helicovepava virescens (Fabricius)		x		Berries	Larva
Pseudoplusia includens (Walker)		x		Leaves	Larva
Spodoptera eridania (Stoll)		x		Stems & small plants	Larva
Spodoptera frugiperda (Smith)		x		Stems & small plants	Larva
Trichoplusia ni (Hubner)		x		Stems & small plants	Larva

(Continued)

Table 2 (*Continued*)

Insect and mite pests	Geographic	Distribution			Developmental stage
Taxon (scientific name)	Africa	America	Asia	Structure of plant affected	feeding on coffee plant
Foliage defoliators					
(Hymenoptera: Formicidae)					
Acromyrmex coronatus (F.)		x		Leaves	Larva
Acromyrmex octuspinosus (Wheeler)		x		Leaves	Larva
Atta cephalotes (L.)		x		Leaves	Larva
Atta fervens (Say)		x		Leaves	Larva
Atta insularis (Guérin-Méneville)		x		Leaves	Larva
Atta mexicana (Smith)		x		Leaves	Larva
(Lepidoptera: Arctiidae)					
Estigmene acrea (Drury)		x		Leaves	Larva
(Lepidoptera: Elachistidae)					
Stenoma cercropia (Meyrick)		x		Leaves	Larva
(Lepidoptera Geometridae)					
Apicia sp.		x		Leaves	Larva
Ascotis selenaria (Dennis & Schiff)	x				
Glena bisulca (Rindge)		x		Leaves	Larva
Oxydia vesulia (Cramer)		x		Leaves	Larva
Oxydia hispata (Cramer)		x		Leaves	Larva
Oxydia trychiata		x		Leaves	Larva
Oxydia saturniata (Guenée)		x		Leaves	Larva
(Lepidoptera: Megalopygidae)					

Megalopyge lanata (Stoll)	x	Leaves	Larva
Podalia sp.	x	Leaves	Larva
(Lepidoptera: Noctuidae)			
Trichoplusia ni (Hübner)	x	Leaves	Larva
(Lepidoptera Psychidae)			
Oiketicus geyeri (Berg)	x	Leaves	Larva
Oiketicus kirbyi (Lucas)	x	Leaves	Larva
(Lepidoptera: Saturniidae)			
Automeris sp.	x	Leaves	Larva
Automeris complicata (Walker)	x	Leaves	Larva
Automeris ilustris (Walter)	x	Leaves	Larva
Fruit flies			
(Diptera: Tephritidae)			
Anastrepha fraterculos (Wiedemann)	x	Berries/cherry	Larva
Ceratitis capitata (Wiedemann)	x	Berries/cherry	Larva
Grasshopper			
(Orthoptera: Gryllotalpidae)			
Gryllotalpa sp.	x	Root, stems	Nymph, adult
(Orthoptera: Tettigonidae)			
Idiarthron subquadratum (Saussure & Pictet)	x	Stems, leaves, berries	Nymph, adult
Root mealybugs			
(Hemiptera: Pseudococcidae)	x	Roots	Nymph, adult female
Brevicoccus sp.	x	Roots	Nymph, adult female
Dysmicoccus sp.	x	Roots	Nymph, adult female

(Continued)

Table 2 (*Continued*)

Insect and mite pests	Geographic	Distribution			Developmental stage
Taxon (scientific name)	Africa	America	Asia	Structure of plant affected	feeding on coffee plant
Dysmicoccus bispinosus (Beardsley)		x		Roots	Nymph, adult female
Dysmicoccus brevipes (Cockerell)	x	x	x	Roots	Nymph, adult female
Dysmicoccus texensis (Tinsley)		x		Roots	Nymph, adult female
Geococcus sp.		x		Roots	Nymph, adult female
Geococcus coffeae (Green)		x		Roots	Nymph, adult female
Geococcus radicum (Green)		x		Roots	Nymph, adult female
Neachavesia caldasiae (Balachowsky)		x		Roots	Nymph, adult female
Rhizoecus sp.	x	x	x	Roots	Nymph, adult female
Rhizoecus andensis (Hambleton)		x		Roots	Nymph, adult female
Rhizococus coffeae (Laing)		x		Roots	Nymph, adult female
Planococcus citri (Risso)		x	x	Roots and aerial structures	Nymph, adult female
Planococcus fungicola (Watson & Cox)	x			Roots	Nymph, adult female
Planococcus minor (Maskell)		x	x	Roots and aerial structures	Nymph, adult female
Planococcus radicum (Watson & Cox)	x			Roots	Nymph, adult female
Pseudococcus elisae (Borchsenius)		x		Roots	Nymph, adult female
Pseudococcus jackbeardsleyi (Gimpel & Miler)		x		Roots	Nymph, adult female
Pseudococcus longispinus (Targioni-Tozzetti)		x		Roots	Nymph, adult female
Rhizoeccus campestris (Hambleton)		x		Roots	Nymph, adult female
Rhizoeccus kondonis (Kuwana)		x		Roots	Nymph, adult female

(Hemiptera: Putoidae)			
Puto sp.	x	Roots	Nymph, adult female
Puto barberi (Cockerell)	x	Roots	Nymph, adult female
Puto antioqensis (Murillo)	x	Roots	Nymph, adult female
Root beetles			
(Coleoptera: Scarabaeidae/Melolonthidae)			
Anomala sp.	x		
Anomala cinta (Say)	x	Roots	Larva
Anomala caucana (Ohaus)	x	Roots	Larva
Callistethus cupricollis (Chevrolat)	x	Roots	Larva
Cyclocephala stictica (Burmeister)	x	Roots	Larva
Cyclocephala signata (Fabricus)	x	Roots	Larva
Cyclocephala mafaffa (Burmeister)	x	Roots	Larva
Ceraspis innotata (Blanchard)	x	Roots	Larva
Phyllophaga sp.	x	Roots	Larva
Phyllophaga latipes (Bates)	x	Roots	Larva
Phyllophaga menetriesii (Blanchard.)	x	Roots	Larva
Phyllophaga obsoleta (Blanchard)	x	Roots	Larva
Phyllophaga sanjosecola (Saylor)	x	Roots	Larva
Phyllophaga vicina (Moser)	x	Roots	Larva
Plectris fassli (Moser)	x	Roots	Larva
Plectris pavida (Burmeister)	x	Roots	Larva

(*Continued*)

Table 2 (*Continued*)

Insect and mite pests	Geographic	Distribution			Developmental stage
Taxon (scientific name)	Africa	America	Asia	Structure of plant affected	feeding on coffee plant
Scales					
(Hemiptera: Coccidae)					
Coccus sp.		x		Aerial structures	Nymph, adult female
Coccus viridis (Green)	x	x	x	Aerial structures	Nymph, adult female
Parasaissetia nigra (Nietner)		x		Aerial structures	Nymph, adult female
Saissetia spp.		x		Aerial structures	Nymph, adult female
Saissetia coffeae (Walker)	x	x		Aerial structures	Nymph, adult female
Saissetia olae (Oliver)		x		Aerial structures	Nymph, adult female
Toumeyella coffea (Kondo)		x		Root system	Nymph, adult female
(Hemiptera: Diaspididae)					
Chrysomphalus dictyospermi (Morgan)		x		Aerial structures	Nymph, adult female
Lepidoshaphes beckii (Newman)		x		Aerial structures	Nymph, adult female
Selenaspidus articulatus (Morgan)		x		Aerial structures	Nymph, adult female
(Hemiptera: Monophlebidae)					
Icerya purchasis Maskell		x		Aerial structures	Nymph, adult female
(Hemiptera: Ortheziidae)					
Insignorthezia insignis Browne		x		Aerial structures	Nymph, adult female

Stem borers					
(Coleoptera: Bostrichidae)					
Apate monachus (Fabricius)	x	x	x	Stem	Larva, Adult
(Coleoptera: Cerambycidae)					
Acalolepta cervina (Hope)			x	Stem	Larva, Adult
Bixadus sierricola (White)	x			Stem	Larva, Adult
Dirphya nigricornis (Oliver)	x			Stem	Larva, Adult
Monochamus leuconotus (Pascoe)	x			Stem	Larva, Adult
Plagiohammus sp.		x		Stem	Larva, Adult
Plagiohammus colombiensis sp. nov.		x		Stem	Larva, Adult
Xylotrechus quadripes Chevrolat			x	Stem	Larva, Adult
(Lepidoptera: Tineidae)					
Opogona sacchari (Bojer)			x	New shoots/stem	Larva
Coffee bean weevil					
(Coleoptera: Anthribidae)					
Araecerus fasciculatus (De Geer)		x		Berries/parchment	Larva, Adult
Sucking bugs					
(Hemiptera: Cicadellidae)					
Agalla sp		x		Leaves	Nymph, Adult
Clinonella declivata (Osborn)		x		Leaves	Nymph, Adult
Graphocephala sp.		x		Leaves	Nymph, Adult
(Hemiptera: Cercopidae)					
Clastoptera biguttata Melichar		x		Peduncle of berries	Nymph, Adult

(Continued)

Table 2 (*Continued*)

Insect and mite pests Taxon (scientific name)	Geographic Africa	Distribution America	Asia	Structure of plant affected	Developmental stage feeding on coffee plant
(Hemiptera: Membracidae)					
Ennya chrysura (Fairmaire)		x		Leaves	Nymph, Adult
(Hemiptera: Ortheziidae)					
Orthezia insignis Browne		x		Leaves	Nymph, Adult
Orthezia praelonga Douglas		x		Leaves	Nymph, Adult
(Hemiptera: Miridae)					
Monalonion velezangeli (Carvalho & Costa)		x		Leaves, flowers	Nymph, Adult
Termites					
(Isoptera: Rhinotermitidae)					
Comatermes perfectus (Hagen)		x		Stem	Larva, Adult
Thrips					
(Thysanoptera: Thripidae)					
Selenothrips rubrocintus (Giard)		x		Leaves	Nymph, Adlult
Heliothrips heamorrhoidalis (Bouché)		x		Leaves	Nymph, Adlult
Twig borers					
(Coleoptera: Curculionidae)					
Xylosandrus compactus Eichhhoff	x	x	x	Young stems, branches	Larva, Adult
Xylosandrus morigerus (Blandford)	x	x	x	Young stems, branches	Larva, Adult
Weevils					
(Coleoptera: Curculionidae)					
Compsus sp.		x		Leaves, roots	Adult, Larva

Macrostylus boconoensis Bordón		x		Leaves, roots	Adult, Larva
Epicarus capetilensis Sharp		x		Leaves, roots	Adult, Larva
White flies					
(Hemiptera: Aleyrodidae)					
Aleurothrixus floccosus Maskell	x	x	x		
Aleurocanthus woglumi Ashby	x	x	x		
Spider mites					
(Acari: Tetranychidae)					
Olygonychus coffeae (Nietner)	x	x	x	Leaves	Nymph, Adult
Olygonychus ilicis (McGregor)		x		Leaves	Nymph, Adult
Olygonychus punicae (Hirst)		x		Leaves	Nymph, Adult
Olygonychus yothersi (McGregor)		x		Leaves	Nymph, Adult
(Acari: Tarsonemidae)					
Polyphagotarsonemus latus (Banks)		x		Leaves	Nymph, Adult

to streams of water, under shady trees, within depressions in the landscape, and close to roads (Bustillo et al., 1998). Those 'hot spots' are evidence of the gregarious spatial distribution of CBB (Aristizábal, 2018). Female emergency from infested berries occurs when the endosperm is almost consumed and weather conditions are appropriate. A high emergence of CBB females is observed when rainfall occurs following 2-4 weeks of dry conditions (Bustillo et al., 1998; Baker, 1999; Damon, 2000). CBB females are attracted to kairomones (complex substances and alcoholic compounds) produced and released by developing berries and ripe berries (Bustillo et al., 1998).

4.2 Damage caused by coffee berry borer

Direct damage is caused to coffee berries by CBB female and larval stages. Tunnels and galleries are made in berries by CBB during colonization, feeding, and oviposition (Bustillo et al., 1998; Baker, 1999; Damon, 2000). Larval stages also make galleries inside berries (endosperm) while they are feeding. Infested berries lose both weight and quality. High infestations (30–90%) may result in 20–70% yield losses. Early attack of developing green berries may cause abortion of berries. In Hawaii, e.g. harvested cherries with more than 10% CBB damage may affect the selling prices for processing in wet mills (Aristizábal, 2018). In Colombia, processed coffee with more than 5.5% of CBB damage is penalized at the point it is sold (Bustillo et al., 1998). In addition to the reduction in quality and price of processed coffee, the cost of production is increased by the need to control CBB. Moreover, the use of insecticides can result in risks to human health and the environment. Overall, global losses from CBB are estimated at about US$500 million per year (Vega et al., 2015).

4.3 Alternative pest management approaches

Integrated pest management (IPM) of CBB includes cultural control practices, use of biological control agents, post-harvest practices, and selected use of chemical controls. An IPM programme for CBB starts with a sampling plan for monitoring CBB populations and activity. In addition, traps baited with alcohol are used for monitoring CBB flight periods and as a mass trapping method for reducing CBB populations. In general, CBB infestation levels higher than 3% or 5% are considered as a threshold to spray insecticides.

The life cycle of the CBB is linked to flowering periods, which then influences the timing of pest management. In Colombia, e.g. two flowerings (major and minor) occur each year: in Brazil and Central America only one major flowering occurs; in Mexico two or more high-intensity flowerings occur in a three-month period (Bustillo et al., 1998; Olivera et al., 2018). Insecticides or mycoinsecticides are therefore used during the early season (45–150 days after flowering). During this period, green developing berries are not hard enough to allow oviposition. This means the colonizer CBB females wait in a tunnel in infested berries until the endosperm reaches the hardness needed for oviposition, making the female vulnerable to insecticides. Monitoring CBB populations by using traps containing a mixture (3:1) of methanol and ethanol to lure females is effective in detecting periods with high CBB flight activity (Aristizábal et al., 2015, 2017a). High flight activity of CBB corresponds with the early coffee season (45–90 days after flowering). CBB may emerge from the remains of berries that were not collected at the end of the harvesting season or which fell on the ground are were not collected. Traps based on alcohols can be used to regulate CBB populations, but this does not apply to all coffee regions.

The entomopathogen *Beauveria bassiana* (Basl. Criv.) Vuill. (Hypocreales: Cordycipitaceae) is the most important biological control agent used against CBB (Bustillo et al., 1998; Jaramillo et al., 2006; Vega et al., 2015). Different *B. bassiana* strains have been isolated from infected CBB adults in the field. Mass production of *B. bassiana* has been developed in several countries to support the control of CBB. *B. bassiana* strains with high pathogenicity and adapted to local conditions are used in Colombia, Ecuador, Central America, India, and Africa (Benavides et al., 2012; Wraight et al., 2018; Hollingsworth et al., 2020). There are several formulations (liquid and granular) of *B. bassiana* with different concertation of spores. In some countries such as Ecuador and India, coffee farmers' associations produce their own cheaper and adapted *B. bassiana* strains while, in other locations (e.g. Hawaii), only a fully tested commercial *B. bassiana* formulation is approved to control CBB (Aristizábal et al., 2017b; Aristizábal, 2018). The use of entomopathogenic nematodes has also been evaluated for the regulation of CBB populations with a focus on infested berries on the ground (Vega et al., 2015; Aristizábal et al., 2016). Native strains of *Heterorhabiditis* sp. (Rhabditida: Heterorhabditidae) and *Steinernema* sp. (Rhabditida: Steinernematidae) are used in Colombia, Hawaii, and in some countries in Central America (Aristizábal et al., 2017b). However, difficulties in mass rearing of those entomopathogenic nematodes have limited use by coffee farmers.

Since the CBB is hidden most of the time, the control of this pest by spraying insecticides or myco-insecticides (*B. bassiana*) is challenging. Other alternative methods of control are needed. Cultural control practices have been shown to be highly effective in the regulation of CBB populations. The cultural control of CBB is focused on the frequent harvesting of berries during the harvesting season plus picking cherries or raisin berries outside the harvesting season. In some locations, this means strip picking of all berries (green – cherries – raisin) at the end of the harvesting season to remove habitat for the CBB. Such cultural control practices reduce about 80% of total CBB populations, since infested berries, the source of the pest, are removed from coffee trees (Bustillo et al., 1998; Aristizábal et al., 2011a; Aristizábal et al., 2012). In addition, preventing berries from falling on the ground during harvesting, and removing those that have, helps to minimize pest populations, since fallen berries harbour many CBB individuals (Johnson et al., 2019; Constantino et al., 2021).

Post-harvest controls are practices that prevent the escape and dispersal of CBB females from infested berries that are collected during the harvest season. Post-harvest control practices need to be conducted from the field (at the moment of harvest) by tying up burlap bags for storing berries until the coffee is ready to be processed in wet and dry mills (Bustillo et al., 1998; Aristizábal et al., 2016). Install traps based on a mixture (3:1) (methanol:ethanol) alcohols help to capture CBB females escaping from processing facilities (Aristizábal, 2018). Transparent plastic smeared with grease is used as a physical barrier in poppers (place where cherries are peeled). This physical barrier catches CBB females escaping from poppers (Aristizábal et al., 2012). In the dry mill, high temperatures (50–60°C) reached by the dryer machine kill all stages of CBB presented in parchment coffee. When the coffee is dried under sunlight, an enclosed deck with a transparent plastic cover and cement floor is recommended (Bustillo et al., 1998; Aristizábal, 2018).

Several natural enemies (predators and parasitoids) are reported to predate CBB. Three parasitoids originated from Africa. *Cephalonomia stephanoderis*, *Prorops nasuta* (Hymenoptera: Bethylidae), and *Phymastichus coffea* (Hymenoptera: Eulophidae) were introduced into America for control of CBB (Bustillo et al., 1998; Baker, 1999; Aristizábal et al., 2016). In addition, several ant (Hymenoptera: Formicidae) species from the genus *Wasmannia*, *Pheidole*, *Crematogaster*, *Dorymyrmex*, and *Solenopsis* have been reported as predating different CBB stages in many coffee-producing countries. The predator *Cathartus quadricollis* (Coleoptera: Silvanidae) has been reported in America and Hawaii, while *Karnyothrips flavipes* (Thysanoptera: Phlaeothripidae) has been reported in Africa (Follet et al., 2016; Vega et al., 2015; Bril et al., 2021).

An additional practice that both helps in the reduction of CBB populations and in the renovation of coffee plantations is stumping, usually involving blocks of trees. In this practice, old coffee trees are cut about 1.5 feet high from the ground. This means they cannot produce berries for around 15 months, depriving the CBB of food and the opportunity to reproduce. New suckers start to grow on stumped coffee trees and 3–4 verticals are selected for the renovation of the new tree. During the first year after stumping, the coffee tree only grows foliage, and, during the second year, some flowers appear. New developing berries should have low CBB populations (< 2%) during the first year of harvest since CBB needs time to recolonize the coffee field (Bustillo et al., 1998; Aristizábal et al., 2017b).

5 Coffee leaf miners

(Lepidoptera: Lyonetiidae) species are as follows:

- *Leucoptera coffeella* (Guérin-Méneville),
- *Leucoptera meyricki* (Guesquiere),
- *Leucoptera coma* (Guesquiere),
- *Leucoptera caffeina* (Washborun).

All these species of coffee leaf miner (CLM) are considered a monophagous pest since they only feed on coffee leaves. The species *L. meyricki* and *L. coma* are reported in Africa, while *L. caffeina* and *L. coffeella* are reported in Asia and America, respectively. All CLM species are considered serious and economically important insect pests. In Brazil, for example, a high infestation (70–95%) of *L. coffeella* causes high defoliation, reducing the photosynthetic activity of leaves and reducing coffee production by between 68% and 80%.

5.1 Life cycle, ecology and behaviour

Adult CLM is microlepidoptera with nocturnal habits. The moths are white pale colour, 2.5 mm long and 6.5 mm in wingspan. Adult CLM feeds on honeydew produced by aphids and other insects, while larvae feed only on coffee leaves (Dantas et al., 2020). CLM female and male mate normally during nighttime. The female pre-oviposition period lasts 3.6 days at 20°C. During the night, females lay between 3 eggs and 6 eggs on the upper side of coffee leaves. Females lay between 30 eggs and 80 eggs during their life span (Barrera, 2008; Bustillo, 2015; Dantas et al., 2020) . The larvae hatch at 4–6 days after oviposition. Four larval instars (developmental stages) occur during development. Larvae penetrate and make tunnels in leaves during feeding. At the fourth developmental stage, larvae reach about 3–4 mm long before they pupate. At 20°C the larvae stage takes between 16 days and 26 days.

In the axial part of leaves, larvae weave a cocoon for the pupation stage which typically lasts about 14 days if the temperature is constant. The cocoon is formed in an 'X' shape. The complete life cycle of the CLM from egg to adult ranges between 25 days and 75 days depending on temperature. The sex ratio is 1:1 (male:female) with sexual dimorphism between male and female.

CLM is considered a seasonal species, most commonly observed in warm areas during dry periods and climatic condition known as 'El Nino' in tropical regions. In Colombia, for example, high infestations of *L. coffeella* are observed in coffee plantations located at low elevation (<1300 metres above sea level (masl), in younger coffee trees (<18 months old)), where fertilization is deficient, on bare soil without green cover, and where herbicides are applied in high amounts. Temperature is a major factor affecting the population dynamic of CLM. At 18°C about 6 generations occur per year, while at 22°C, 10 generations can be produced per year. The impact of climate change is clearly observed in Colombia and other locations. Before 2008, *L. coffeella* was reported in coffee plantations located at low elevations (<1300 masl) in Colombia (Cardenas and Posada, 2001). However, recently CLM was observed in coffee farms located between 1500 masl and 1700 masl due to higher temperatures (Benavides et al., 2013). Rain in combination with natural enemies can suppress CLM populations without the need for insecticides in coffee regions with a high diversity of plants and corresponding populations of beneficial insects, as it occurs in Mexico and Colombia (Barrera, 2008; Bustillo, 2015).

5.2 Damage caused by coffee leaf miner

The larval developmental stages are directly responsible for damaging coffee leaves where larvae live and feed. Galleries or tunnels (mines) observed in coffee leaves are lesions caused by feeding behaviours of CLM larvae. After emergence (eclosion), larvae rapidly penetrate the upper epidermis of the leaf and feed on palisade parenchyma tissue. Those galleries have an irregular shape, are a pale-yellow colour, and can occur on 80% of the coffee leaf surface.

Coffee leaves with CLR damage (galleries) turn into brownish colour until they become necrotic and finally drop, causing defoliation (Dantas et al., 2020). Photosynthesis is reduced due to necrosis and defoliation, resulting in yield loss. A high level of defoliation affects the formation of flowers and the development of berries. In Brazil, for example, coffee production has been reduced by 68–80% when CLM infestation was high (>70%). In Ecuador, coffee yields have been reduced by over 50% due to infestations of *L. coffeella* that occurs during flowering periods.

5.3 Alternative pest management approaches

There are a high number of natural enemy species coexisting with *L. coffeella* (55 at one count), including predators, ecto-parasitoids, and endo-parasitoids (Waller et al., 2007). In Colombia, for example, 8 predators (families Vespidae and Chrysopidae) and 15 parasitoid species (families Eulophidae, Braconidae) have been reported (Cardenas and Posada, 2001, Constantino et al., 2013). In Mexico, several species of parasitoids and predators have been found to control CLM. In countries such as Colombia and Mexico, populations of *L. coffeella* are kept under control, thanks to the existing diversity of natural enemies in the coffee agroecosystem. However, when the natural balance of agroecosystem is disrupted, the population of beneficial insects (pollinators, predators, and parasitoids) is reduced and there can be a resurgence of insect pests such as CLM. This is exactly what is happening in coffee plantations in Brazil, the world's largest coffee-producing country. High infestations of CLM have been reported in Brazil due to high-density, monocultural coffee plantations, as well as overuse of insecticides and herbicides.

Alternatives for the management of CLM include cultural, biological, and selected chemical controls, resistant cultivars, and effective monitoring to make informed decisions when it is best to spray insecticides in a targeted and more effective way. Agronomic practices such as pruning to stimulate vigorous growth, balanced fertilization, adequate control of weeds, and regulation of shade help to keep coffee trees healthy and productive. Green covers with flowers between rows attract natural enemies that help in the regulation of CLM. Tropical and fruit trees cultivated within coffee plantations also provide shelter to natural enemies, increasing the biodiversity that benefits balanced agroecosystems. High-density planting should be avoided. In Brazil traps with a synthetic female pheromone are used both to monitor males and regulate CLM populations through mating disruption (Bacca et al., 2008, 2012). The coffee field is saturated with traps to lure males who then fail to find (and mate with females) and/or are trapped, breaking the reproduction cycle (Bacca et al., 2012).

In terms of biological control, populations of CLM are regulated by several parasitoids and predators. In Colombia Rueda et al. (2016) reported levels of natural parasitism of CLM of 58–89%. In Central America, Barrera (2008) reported parasitism between 2% and 70%. Parasitoids commonly found in America include *Closterocerus coffeellae*, *Closterocerus lividus*, *Pnigalio sarasolai*, *Zagrammosoma multilineatum*, *Neocgrysocharis* sp., *Aprotucetus* sp., *Horismenus* n. sp., and *Apleurotropis* n. sp. (Hymenoptera: Eulophidae), Allobracum *primus*, and *Sitropius letifer* (Hymenoptera: Braconidae). Predators that feed on CLR larvae include *Polistis* sp. and *Polibia* sp. (Hymenoptera: Vespidae), individual ants from the Formicidae family and *Chrysoperla* sp. (Neuroptera: Chrysopidae). Due to the high diversity of natural enemies parasitizing or predating CLM at various stages of growth, conservation biological control is highly recommended. This means avoiding the use of broad-spectrum and residual contact synthetic insecticides. The use of such insecticides should be restricted to certain conditions, e.g. when most females are still outside leaves (Benavides et al., 2013).

The selection of coffee cultivars resistant to CLM is an alternative method to manage this insect pest (Pantoja-Gomez et al., 2019; Dantas et al., 2020). Genetically modified coffee plants expressing the *Bacillus thuringiensis CryAc* gene have shown resistance to CLM under field conditions in French Guinea (Perthus et al., 2005). Intercropping coffee with other crops or fruit trees reduces the habitat favoured by CLM species while promoting beneficial insects. In Africa, e.g. *L. meyricki* is less abundant in plantations with shade trees. In contrast, *L. caffeina* is more common where coffee crops are grown under shade (Mugo et al., 2013). Appropriate planting can take account of these conditions.

6 Antestia bug

(Hemiptera: Pentatomidae) species include:

- *Antestiopsis orbitalis* ssp. *bechuana* (Kirkaldy),
- *Antestiopsis orbitalis* ssp. *ghesquierei* (Carayon),
- *Antestiopsis intricata* (Guesquiere and Carayon),
- *Antestiopsis thunbergii* (Gmeln),
- *Antestiopsis facetoides* Greathaed.

The genus *Antestiopsis* (Hemiptera: Pentatomidae) known as the antestia bug or variegated coffee bug has several species that are considered major insect pests in Africa. *A. thunbergii* and *A. orbitalis* ssp. *bechuana* are more commonly observed in eastern and southern Africa, while *A. orbitalis ghesquierei*, *A. intricata*, and *A. facetoides* are common in western Africa. *A. thunbergii* has a preference for *C. arabica* planted at high elevations (1600-2100 masl), while *A. intricate* and *A. facetoides* are commonly observed at low and middle elevations (1000-1600 masl). Antestia bugs are distributed both in Africa and in some coffee-producing countries in Asia, including India, Sir Lanka, Myanmar, and China, causing considerable losses in yield and quality (Azrag et al., 2018; Bigirimana et al., 2018).

Low populations of antestia bugs (1-2) per coffee tree are the threshold for spraying insecticides. Since antestia bugs attack developing green berries, flower buds, and growing tips, infestations cause a loss of coffee quality with significant crop losses (15-38%) reported in African countries. In Kenya, *Antestiopsis* sp. are considered as a major insect pest on coffee farms on which coffee is grown under shade trees (Mugo et al., 2013). *Antestiopsis* sp. also are important insect pest in Ethiopia.

6.1 Life cycle, ecology and behaviour

The lifespan of adult females of *Antestiopsis* sp. is about 1 year during which they can lay about 500 eggs. Oviposition occurs on the underside of coffee leaves, on which clusters of 10-12 eggs are laid by the female. The life cycle of *A. intricate* has been evaluated under laboratory conditions at 27°C and 60% relative humidity by Mendesil and Abebe (2004). Results showed that on average 12 eggs were laid per day with 324 eggs per female laid overall during the study. Nymphal instars emerge about 6-8 days after oviposition.

The developmental stages involve five instars. All five nymphal stages take an average of 42.9 days at 27°C from eggs to adults, *A. intricate* took on an average of 50.3 days with females and males living on an average of 187 and 135 days, respectively (Mendesil and Abebe, 2004).

The major host plants for antestia bugs belong to the Rubiaceae. Antestia bugs are found on *Coffee canephora* and *C. liberica*, but most commonly on *C. arabica*. Different host plants for antestia bugs include *Citrus sinensis* (L.) (Rutaceae), *Bidens pilosa* L. (Asteraceae), *Zea mays* L (Poaceae), *Ricinus communis* L (Euphorbiacea), and *Musa* spp. (Musaceae). Other plants from Annonaceae, Asteraceae, Malvaceae, and Solanaceae families that are associated with coffee agroecosystems are considered temporary host plants for antestia bugs.

6.2 Damage caused by Antestia bugs

Both adults and nymphs suck on flower buds, developing green berries, cherries, growing foliage tissues, and green shoots. Early attacks by antestia bugs on flower buds and small berries prevent the full development of fruits cause abortion in those parts of the plant and thus reduce coffee yields. Coffee losses have reported at up to 45% in East Africa (Babin et al., 2018). Affected flower buds turn dark (black or brown) into colours. In addition, major damage in infested berries is caused by the fungus *Asbya* spp, which is carried and transferred by antestia bugs. This fungus affects the coffee seed and consequently the quality of processed coffee. Microbes such as those from the family Enterobacteriaceae enter coffee berries through wounds caused by antestia bugs. Bacteria cause the defect known as potato taste defect (PTD) in coffee, a potato-like aroma found in processed coffee (green beans, roasted coffee, and in the coffee drink) (Gueule et al., 2015). A population between 1 and 3 antestia bugs per tree is considered the threshold for taking control measures before the population increases and causes economic losses.

6.3 Alternatives management approaches

Research suggests that shaded coffee agroecosystems in Kenya suffer from higher levels of infestation and damage caused by antestia bugs than in unshaded agroecosystems (Mugo et al., 2013). Effective management of shade trees and regular pruning of coffee trees make conditions less suitable for antestia bugs (Bigirimana et al., 2018). Pruning to create more open coffee bushes exposes them to more sunlight and aeration, favouring the presence of natural enemies (predators and parasitoids) (Bustillo, 2015).

Botanical insecticides are used for the control of antestia bugs. Pyrethrum and neem extract have been shown to effectively control adults and nymphal stages with minimal risk to humans and the agroecosystem. Synthetic

Table 3 Natural enemies (predators and parasitoids) of key insect pests of coffee crops. Ecological role and hot stage attacked by the natural enemy species.

Key insect/mite pests Affecting coffee trees	Natural enemy	Ecological role Parasitoid/predator	Host stage Attacked
Coffee berry borer	**(Coleptera: Silvanidae)**		
Hypothenemus hampei	*Cathartus quadricollis* (Guérin-Méneville)	Predator	All stages
	(Coleptera: Laemophoeidae)		
	Leptophloeus sp.	Predator	All stages
	(Hymenoptera: Bethylidae)		
	Cephalonomia stephanoderis (Betrem)	Ecto-parasitoid	Larvae, pupae
	Cephalonomia hyalinipennis (Ashmead)	Ecto-parasitoid	Larvae, pupae
	Prorops nasuta (Waterston)	Ecto-parasitoid	Larvae, pupae
	(Hymenoptera: Braconidae)		
	Heterospillus coffeicola (Schimideknecht)	Predator	Eggs, larvae
	(Hymenoptera Eulophidae)		
	Phymastichus coffea (La Salle)	Endo-parasitoid	Adults
	(Hymenoptera: Formicidae)		
	Azteca instabilis (F. Smith)	Predator	All stages
	Crematogaster sp.	Predator	All stages
	Crematogaster curvispinosa (Mayr)	Predator	All stages
	Dolichoderus bituberculatus (Mayr)	Predator	All stages
	Dorymyrmex sp.	Predator	All stages
	Pheidole radoszkowskii (Mayr)	Predator	All stages
	Solenopsis geminata	Predator	All stages
	Wasmannia sp.	Predator	All stages
	(Thysanoptera: Phlaeothripidae)		

(Continued)

Table 3 (*Continued*)

Key insect/mite pests Affecting coffee trees	Natural enemy	Ecological role Parasitoid/predator	Host stage Attacked
	Karnyothips flavipes (Jones)	Predator	Eggs, Larvae
Coffee leaf miners	**(Neuroptera: Chrysopidae)**		
(Lepidoptera: Lyonetiidae)	*Chrysoperla externa* (Hagen)	Predator	Larvae
Leucoptera coffeellum (Guérin-Méneville)			
Leucoptera meyricki (Guesquiere)	**(Hymenoptera: Eulophidae)**		
Leucoptera coma (Guesquiere)	*Aprostucetus* sp.	Ecto-parasitoid	Larvae
Leucoptera caffeina (Washbourn)	*Cirrospilus* sp.	Ecto-parasitoid	Larvae
	Chrysocharis livida (Ashmead)	Endo-parasitoid	Larvae
	Closterocerus coffeellae (Ihering)	Endo-parasitoid	Larvae
	Closterocerus lividus (Ashmead)	Endo-parasitoid	Larvae
	Horismenus cupreus (Ashmead)	Ecto-parasitoid	Larvae
	Pnigalio sarasolai (De Santis)	Ecto-parasitoid	Larvae
	Tetrastichus sp.	Ecto-parasitoid	Larvae
	Zagrammosoma multilineatum (Ashmead)	Ecto-parasitoid	Larvae
	Zagrammosoma variegatum (Masi)	Ecto-parasitoid	Larvae
	Zagrammosoma zebralineatum (De Santis)	Ecto-parasitoid	Larvae
	(Hymenoptera: Encyrtidae)		
	Ageniaspsis sp.	Parasitoid	Larvae
	(Hymenoptera: Braconidae)		
	Apanteles bordagei (Giard)	Parasitoid	Larvae
	Allobracom primus (Muesebeck)	Endo-parasitoid	Larvae

(*Continued*)

Table 3 (*Continued*)

Key insect/mite pests Affecting coffee trees	Natural enemy	Ecological role Parasitoid/predator	Host stage Attacked
	Centistidae striata (Penteado-Dias)	Parasitoid	Larvae
	Stiropius letifer (Mann)	Parasitoid	Larvae
	Stiropius reticulatum (Penteado-Dias)	Parasitoid	Larvae
	Mirax insularis (Muesebeck)	Parasitoid	Larvae
	(Hymenoptera: Vespidae)		
	Polistes sp	Predator	Larvae
	Polybia sp.	Predator	Larvae
Antestia bugs	**(Diptera: Tachinidae)**		
(Hemiptera: Pentatomidae)	*Bogosia rubens* (Villenueve)	Parasitoid	Adults
Antestiopsis orbitalis bechuana (Kirkaldy)			
Antestiopsis orbilatils ghesquierei Carayon	**(Hymenoptera: Platygastroidea)**		
Antestiopsis intricata (Guesquiere & Carayon)	*Telenomus sechellenis* (Kieffer)	Parasitoid	Eggs
Antestiopsis thunbergii (Gmeln)	*Trissolcus mopsus* (Nixon)	Parasitoid	Eggs
Antestiopsis facetoides Greathaed	*Trissolcus suranus* (Nixon)	Parasitoid	Eggs
Antestiopsis clymeneis (Kirkaldy)	*Gryon fulviventris* (Crawford)	Parasitoid	Eggs
Antestiopsis cruciata (F.)			
	(Hymenoptera: Braconidae)		
	Aridelus sp.	Parasitoid	Nymphal stages
	(Hymenoptera: Eulophidae)		
	Pediobius sp.	Parasitoid	Eggs
	(Hymenoptera: Eupelmidae)		
	Anastatus antestiae (Ferrière)	Parasitoid	Eggs

(*Continued*)

Table 3 (*Continued*)

Key insect/mite pests Affecting coffee trees	Natural enemy	Ecological role Parasitoid/predator	Host stage Attacked
	(Hymenoptera: Pteromalidae)		
	Acroclisoides africanus (Ferrière)	Parasitoid	Eggs
	(Strepsiptera: Corioxenidae)		
	Corioxenus antestiae (Blair)	Parasitoid	Nymph, adults
Twig borers	**(Hymenoptera: Bethylidae)**		
(Coleoptera: Curculionidae)	*Prorops nasuta* (Waterston)	Ecto-parasitoid	Larvae, pupae
Xylosandrus compactus (Eichhhoff)			
Xylosandrus morigerus (Blandford)	**Hymeniptera Eulophidae)**		
	Tetrastichus xyleborus (Kalshoven)	Parasitoid	Larva
	Hymenoptera: Eupelmidae)		
	Eupelmus sp.	Ecto-Parasitoid	Larva
	(Hymenoptera: Formicidae)		
	Crematogaster sp.	Predator	All stages
	Pheidole sp.	Predator	All stages
	Solenopsis sp.	Predator	All stages
	Pseudomyrmes sp.	Predator	All stages
	Leptothorax sp.	Predator	All stages
	(Coleoptera: Cleridae)		
	Callimerus sp.	Predator	All stages
	(Coleoptera: Silvanidae)		
	Cryptamorpha desjardinsi (Guérin-Méneville)	Predator	All stages

(*Continued*)

Table 3 (*Continued*)

Key insect/mite pests Affecting coffee trees	Natural enemy	Ecological role Parasitoid/predator	Host stage Attacked
Root mealybugs			
(Hemiptera: Pseudococcidae)	**(Hymenoptera: Encyrtidae)**		
Dysmicoccus sp.	*Leptomastix dactylopii* (Howard)	Parasitoids	Nymph, adults
Dysmicoccus brevipes (Cockerell)	*Aenasius colombiensis* (Compere)	Endo-parasitoid	Nymph, adults
Planococcus citri (Risso)	*Aenasius cariocus* (Compere)	Endo-parasitoid	Nymph, adults
	Anagyrus ananatis (Gahan)	Parasitoids	Nymph, adults
(Hemiptera: Putoidae)	*Hambletonia pseudococcina* (Compere)	Endo-parasitoid	Nymph, adults
Puto sp.			
Puto barberi (Cockerell)	**(Coleptera: Coccinellidae)**		
	Cryptolaemus montrouzieri (Mulsant)	Predator	All stages
	Scymnus sp.	Predator	All stages

pyrethroids used for control of this insect pest show good results (>90% field efficacy). However, this group of insecticides can cause toxicity, e.g. to fish when sources of water are contaminated. Synthetic insecticides such as organophosphates are still used in some countries but restricted in others due to negative effects on human health and the environment. The effectiveness of new neonicotinoid systemic insecticides (imidacloprid) against antestia bugs has been demonstrated on coffee plantations in Rwanda. Spraying should occur before the harvesting season to prevent residues in processed coffee.

Antestia bugs have several indigenous natural enemies such as egg parasitoids which can be used in population control (Table 3). *Telenomus sechellensis*, *Gryon fulviventre*, *Trissolcus mopsus*, *Anastatus antestia*, and *Pediobius* sp. are some of the egg parasitoids found on antestia bugs. On eggs of *A. intricate*, high egg parasitism rates (from different species) were reported (40–80%) in Cameroon, showing a seasonality variation over the year. These results show the value of conservation biological control as a strategy to promote natural enemies in the coffee agroecosystem. Minimal use of high-spectrum synthetic insecticides, green cover, and plant diversification enhance the survival of beneficial insects (parasitoids, predators, and pollinators).

7 Twig borer

(Coleoptera: Curculionidae: Scolytinae) species are as follows:

- *Xylosandrus morigerus* (Blandford) (Brown twig borer),
- *Xylosandrus compactus* Eichhoff (Black twig borer).

Both twig borer species are ambrosia beetles that live in a symbiotic relationship with fungi. *X. morigerus*, e.g. feeds on *Ambrosiaemyces zeylanicus* Trotter (Hyphomycetes), while *X. compactus* feeds on *Ambrosiella xylebori* and *Fusarium solani* (Mart) (Hyphomycetes). Normally ambrosia beetles attack sick, weak, or stressed trees. However, the genus *Xylosandrus* attacks both weak and healthy host plants. *X. compactus* has been reported as a serious insect pest affecting avocado and orchid cultivation in Florida and this species is an important pest in coffee plantations in Hawaii.

7.1 Life cycle, ecology and behaviour

A sex ratio of 1:9 male to female is observed under field conditions. Females mate with sibling males inside the gallery within the plant where they were born. Mated females then leave their brood chamber to find and colonize a new host plant. The flight activity of females emerging and searching for new hosts normally occurs in the afternoon. Smaller flightless males generally remain in their brood gallery.

The adult female opens a small hole (1 mm diameter) to enter branches or twigs, in which she and her progeny are protected. In a new gallery created within the branch or twig, a colonizer female starts laying eggs 4-7 days after boring inside. Eggs are laid in clusters (8-15) and hatch 3-5 days later, depending on temperature. Under field conditions, the mean number of eggs per branch ranges between 2 and 23 (Greco and Wright, 2015). Larvae pass through three instars and take about 7-11 days to reach pupation. Pupae take about 7-9 days before adults emerge. The total life cycle from eggs to adults takes between 20 and 40 days, depending on temperature. At 23-27°C the life cycle of *X. compactus* takes 28.5 days on average. After adults emerge from pupae, hardening of the exoskeleton takes 2 days and sexual maturity is reached after 6 days.

According to Greco and Wright (2013), *X. compactus* has an aggregated spatial distribution. Twig borer preference for coffee varieties is influenced by the hardness of coffee branches, texture of tissues, water content, and quality of sap. Semiochemicals produced by host plants also affect aggregated patterns of distribution observed in twig borers and other ambrosia beetles. In addition, the presence of alternative host plants providing shade to coffee or the proximity of forested areas may also influence the behaviour and population dynamics of twig borers. Entry holes made by twig borers in coffee trees are observed on lateral and vertical branches and berries. Lateral branches are most commonly attacked. The number of galleries with adults or progeny ranges between 1 and 14 per branch, while the number of females per branch ranges between 1 and 41 with 4.2 on average (Greco and Wright, 2015).

The symbiotic relationship between twig borers and the fungi on which they feed is crucial for their survival. The fungus provides nutrition to adults and larvae by predigesting tissue. Twig borers and other ambrosia beetles also transport fungi from one host plant to another. The survival of fungi depends on the inoculation of the host plant by beetles inside galleries.

7.2 Damage caused by twig borers

A large number of plants (>200 host species) are attacked by twig borers including avocado, cedar, coffee, cocoa, cypress, macadamia nut, magnolia, landscape ornamentals, shrubs, orchids, and native trees. Twig borers have a preference for *Coffee canephora*, but have recently become an important pest on *C. arabica*. As noted, twig borers' selection and preference for coffee varieties are influenced by factors such as hardness of coffee branches, texture of tissues, and water content. The condition of coffee plants is important because healthy trees with balanced nutrition and a good supply of water are less vulnerable than weak and stressed trees with poor nutrition.

Adults and larvae both cause damage to coffee trees. Galleries and tunnels are made by colonizer females and larvae. Recently infested branches become light green, discoloured, weak, and withered since the normal movement of nutrients and water inside the branch is disrupted both by the physical damage caused and by the growth of the fungus associated with infestation. After a few days, the affected area around the hole made by the female turns black or brown. Infested branches turn dry and leaves fall off. As a consequence, berries on affected branches do not fill properly. If such berries are collected during harvesting, they become floats during the wet mill process with no market value.

While infestation by the CBB can affect one or more coffee seeds, a branch infected by a twig borer can affect most or all the berries on the branch. Over three infested branches per tree is a threshold for control measures. In Mexico, Barrera (2002) found that infested branches with 1, 2, or 3 effective holes resulted in losses of 6, 12, and 18 berries, respectively.

7.3 Alternative management approaches

Since black twig borer and brown twig bores are cryptic species, adults and their progeny are protected inside brood chambers made by the colonizer female in coffee branches. This makes it difficult to control this insect pest with insecticides. Synthetic contact insecticides cannot reach pest populations inside infested branches. This puts the emphasis on cultural control practices. Cutting and destruction of infested branches from coffee trees can reduce twig borer populations by up to 90%. Infested branches need to be cut 5-8 cm below the entry hole. Infested branches should be cut into small pieces and burned. Promoting healthy coffee trees helps plants resist attack or recover from previous infestations.

A sequential sampling plan for monitoring *X. compactus* on coffee plantations has been suggested by Greco and Wright (2013). A minimum of 20 coffee trees should be sampled per coffee lot to determine the number of infested branches per tree. Control measures (cultural practices or spraying with insecticides) should be taken when more than three infested branches per tree are observed. However, the use of insecticides is only effective when female adults are searching for new host branches. It is essential to determine periods of high flight activity of females during the coffee season by using contact insecticides effectively before damage and reproduction occur. In Hawaii, e.g. high flight activity and attacks by black twig borer females in commercial coffee farms were observed during late spring and summer. In Hawaii, traps based on alcohol (3:1 - methanol - ethanol) used for monitoring coffee berry borer also captured *X. compactus*. Those traps should be used for monitoring twig borer populations to establish seasonal trends.

In coffee plantations in Hawaii, the use of a repellent based on verbenone and limonene was reported as a potential semiochemical in the regulation of *X. compactus*. Some products based on clay also have acted as repellents for black twig borer, reducing the number of infested branches per coffee tree, the number of holes per infested branch, and the overall pest population (L. F. A. unpublished).

A list of natural enemies, including parasitoids and predators of *X. compactus* and *X. morigerus*, is shown in Table 3. Microbial control of both *Xylosandrus* sp. includes the use of the entomopathogenic fungus *B. bassiana*, which is also commonly used in coffee farms to control CBB.

8 Root mealybug

(Hemiptera: Pseodococcidae) species are as follows:

- *Dysmicoccus bispinosus* (Beardsley), *D. brevipes* (Cockerell), *D. texensis* (Tinsley),
- *Geococcus coffeae* Green, *G. radicum* Green,
- *Neochavesia caldasiae* (Balachowky),
- *Rhizoecus andensis* Hambleton, *Rhizoecus coffeae* Laing,

- *Pseudococcus elisae* (Borchsenius), *P. jackbeardsleyi* Gimpel & Miler, *P. longispinus* (Targioni-Tozzetti),
- *Rhizoecus campestris* (Hambleton), *Rhizoecus kondonis* Kuwana.

(Hemiptera: Putoidae): *Puto barberi* (Cockerell), *Puto antioquensis* (Murillo)

Root mealybugs affecting coffee trees are reported in all tropical regions where coffee crops are produced and are, in many cases, endemic in some locations. In Colombia, for example, *Puto barberi*, *Pseudococcus jackbeardsleyi*, *Dysmicoccus brevipes*, and *D. texensis* are the most common and destructive insect pests affecting coffee crops. In Costa Rica and El Salvador, *D. brevipes* is economically important, while in Guatemala *Dysmicoccus bispinosus* is the most damaging species. In Africa, *Planococus fungicola* and *P. radicum* affect coffee, while *P. minor* is reported in coffee-producing countries in Asia and recently in Brazil. Root mealybugs species are polyphagous insect pests that have several host plants commonly observed in the coffee agroecosystem including citrus, sugarcane, cassava, banana, and cacao. These pests affect the coffee tree root system and the neck of the stem (Tables 2).

8.1 Life cycle, ecology and behaviour

Most root mealybug species pass through three nymphal stages. Hundreds of eggs (200–600) are laid by females inside an ovisac (a layer of cotton-like wax or egg sac of crystalline wax hairs). Several species of mealybugs are ovoviviparous, which means the hatching of eggs occurs inside the ovisac. The first nymphal stage (named crawlers) emerge from the female ovisac. Crawlers search for a place to settle and feed on the root system. They are sucking insects that feed on root sap. Adults and developing nymphal stages are covered by a white waxy layer (cotton-like wax) when sucking on roots. In many cases, nymphal males reach maturity with two moults (first and second instar) and a formation of a cocoon for the third instar, while female nymphs moult three times (third instar) to reach maturity. Depending on the temperature, mealybug species and the host plant, the life cycle from egg to adult ranges between 30 and 140 days. The age of coffee trees, weather conditions, and soil conditions (pH, moisture, organic matter, and physical structure) influence mealybug development.

Sugary substances (honeydew) excreted by mealybugs attract some ant species. A symbiotic association between mealybugs and ants allows dispersal in the field and protection of mealybugs from natural enemies. Some of the ant genera associated with mealybugs include *Acropyga* sp., *Brachymyrmes* sp., *Crematogaster* sp., *Heteroponera* sp., *Paratrechina* sp., *Mycocepurus* sp., *Pheidole* sp., *Solenopsis* sp., and *Wasmania* sp.

8.2 Damage caused by root mealybugs

Infected root systems cause weakening of coffee trees. High infestations can kill young trees. However, productive adult coffee trees rarely die from mealybug infestation. Damage to coffee trees includes high extraction of nutrients (e.g. potassium) by the pest, deterioration of the root system (e.g. loss of root hairs), atrophy, and necrosis of roots. In addition, open wounds caused by pests sucking roots allow entry by soil-plant pathogens such as *Ceratocystis* spp. and *Roselinia* spp. Coffee plants in nurseries are also affected by root mealybugs. Commercial nurseries that sell new coffee trees for planting are a major factor in the dispersal of root mealybugs. In Colombia, yield losses in infested coffee trees are caused by *Planococcus citri*, can reach up to 30%. Infected coffee trees are also more susceptible to foliage plant pathogens such as *Cercospora* spp.

8.3 Alternatives management approaches

The management of root mealybugs includes monitoring, cultural control practices, use of resistant varieties, and chemical control. Sampling is particularly recommended for new coffee fields with nursery plants. The presence of ants is a potential indicator of root mealybugs. Yellowish seedlings in nurseries are symptoms of potential attack by mealybugs. These seedlings should be removed and inspected for infestation. In new coffee fields, additional coffee trees are planted to act as indicators of the presence of root mealybugs. Those plants are removed and sampled (destructive sampling) during the first year of establishment of a new coffee field. The presence of just one infested coffee plant with root mealybugs is enough to establish an immediate control programme.

Some insecticides used for the control of root mealybugs include organophosphates (chlorpyriphos, fenitrothion, and dimethoate), pyrethroids (cypermethrin), and neonicotinoids (thiamethoxam and imidacloprid). However, frequent use of these insecticides can negatively impact natural enemies and other beneficials. Controlling ants in coffee fields helps in the regulation of mealybugs. The use of the entomopathogens fungi *B. bassiana* (Bals.- Criv) and the entomopathogenic nematode *Steinernema carpocapsae* (Weiser) have showed some promising results against root mealybugs species in some studies but low effectiveness in others. Further laboratory and field trials are needed to fully understand the role of those biological agents in the regulation of root mealybug populations.

The management of weeds and other host plants can reduce favourable conditions for root mealybug establishment. In addition, balanced nutrition improves the health of coffee trees and their root systems. Some natural enemies (predators or parasitoids) are reported to attack some root mealybug species. For example, the parasitoid *Leptomastix dactylopii*

Howard and the predator *Cryptolaemus montrouzieri* Mulsant are reported as natural enemies of *P. citri*. Two predators *C. montrouzieri* and *Scymnus* spp. and two parasitoids *Aenasius colombiensis* Compere and *Anagyrus ananatis* Gahan are reported to attack *D. brevipes*. In general, natural enemies of roots mealybugs are limited given the cryptic behaviours of these pests (Table 3).

9 Stem borers

(Coleoptera: Bostrichidae) species are as follows:

- *Apate monachus* Fabricius (Black stem borer).

(Coleoptera: Cerambycidae) species are as follows:

- *Acalolepta cervine* (Hope) (coffee brown beetle),
- *Bixadus sierricola* (White) (West African coffee bore),
- *Dirphya nigricornis* (Oliver) (yellow-headed borer),
- *Monochamus leuconotus* (Pascoe) (white coffee borer),
- *Plagiohammus* spp.,
- *Plagiohammus colombiensis* sp. nov.,
- *Plagiohammus maculosus* (Bates),
- *Plagiohammus mexicanus* Breuning,
- *Plagiohammus spinipennis* (Thomson),
- *Xylotrechus quadripes* Chevrolat.

Overall, stem borers are reported as insect pests affecting coffee trees in all major regions in which coffee is produced. The black stem borer, *Apate monachus*, is an economically important pest in Africa. *A. monachus* was reported in Tanzania affecting coffee crops. It is commonly seen in coffee crops associated with shade trees or close to the forest. In West Africa, *Bixadus sierrricola* affects commercial coffee plantations, while in eastern Africa, the species *Monochamus leuconotus* and *Dirphya nigricornis* are most commonly reported to attack coffee trees. In Asian countries, *Xylotrechus quadripes* and *Acalolepta cervine* are major pests affecting coffee crops. However, *X. quadripes* is considered one of the most serious insect pests in Asia (China, India, Sir Lanka, Thailand, and Viet Nam). High economic losses due to the destruction of coffee trees caused by *X. quad*ripes are reported in India. In America, the species *Plagiohammus mexicanus* and *P. spinipennis* were reported to affect coffee trees in Mexico, while *P. maculosus* is more frequently reported in Central America (Barrera, 2008). In Colombia, the new species *P. colombiensis* was reported to affect coffee tree stems and roots (Constantino et al., 2013).

9.1 Life cycle, ecology, and behaviour

The black stem borer (*A. monachus*) is a wood beetle, which had been reporting attacking ornamental and fruit trees including apple, avocado, carob, lemon, palms, pear, peach, pomegranate, and coffee among other wood trees (Bonsignore, 2012; Kilambo et al., 2005). Adults caused damage to the stem of coffee trees. However, the oviposition takes place on dead wood or wilting trees in which better conditions exit for larval development. The life cycle is completed in the forest. During summer, high temperatures favour the attack by *A. monachus* adults, Therefore, inspection and monitoring are recommended for early detection of this insect pest.

Female species of stem borers from the family Cerambycids known as longhorns' beetles lay eggs on the stem of coffee trees (cervices). After the larvae hatch, they bore into the stem making tunnels. Tunnels are down inside the stem until the root (20 cm long) and up on the stem (100 cm long). Larval stages take between 2 and 3 years. The life cycle of *P. maculosus* from eggs to adults took a duration of 18 months (Barrera et al., 2004). *Plagiohammus* spp. Is most commonly observed at high elevations (>1000 m) and in dry locations. A high infestation of *Plagiohammus* spp. is observed on abandoned or feral coffee trees.

9.2 Damage caused by stem borers

Adults of *A. monachus* feeds on the truck and branches making tunnels about 20 cm long upwards in the stem. Large tunnels bored by adults may cause the death of branches and in some cases the tree. An accumulation of fine dust is observed on the bottom of trees and holes on the trunk are an indication of *A. monachus* presence.

Coffee stem borers (Cerambycidae) cause severe damage to coffee trees. Tunnels made by the larvae on young trees (<2 years old) kill the trees. Coffee production is highly reduced in coffee trees that survive the attack of stem borers. Yellowing and wilting are observed on leaves of infested trees. The attack by coffee stem bores causes weakening of trees making them more susceptible to diseases and termites attack. Infested trees are recognized by the presence of sawdust mounds at the base of trees.

9.3 Alternative management approaches

Cultural practices including sanitation, cleaning, pruning, and burning dead stems and branches are recommended. Infested coffee trees and other infested trees need to be destroyed. Application of insecticide on the truck and branches helps to kill adults outside the stems. The management of coffee stem borers (Cerambycid) is a focus on prevention, since the control of larvae inside the stems is difficult to reach. Infested coffee trees need to be removed and destroyed to prevent dispersal and re-infestation on other trees. Preventive application of insecticides is recommended in locations in which those stem bores appear yearly. For example, in Mexico and Central America, adults of *Plagiohammus* spp. are most commonly observed during the rainy season (April–June). Only, a braconid species *Sclerodermus* spp. was reported as a parasitoid of *A. monachus*. Few is known about the natural enemies of stem borers. The use of entomopathogenic fungi *B. bassiana* and *Metarhizium robertsii* may play an important role as a biological control against adults of *A. monachus* and stem borers.

10 Natural enemies for key insect pests in coffee agroecosystems

A wide diversity of vegetation in coffee agroecosystems promotes natural enemies (predators and parasitoids) that play an important role in the regulation of insect pests. Table 3 lists the main natural enemies of some key insect pests in coffee agroecosystems, including the CBB, CLM, antestia bug, twig borer, and root mealybug. In total 64 species are reported as natural enemies for those five key insect pests. However, the overuse of highly toxic and non-selective insecticides, fungicides and herbicides, coffee monocultures with a high density of coffee trees per area, and deforestation are some of the factors that reduce the number and diversity of natural enemies in coffee agroecosystems. There is therefore a need to re-evaluate integrated pest management of insect pests in coffee agroecosystems with the participation of coffee growers.

Monitoring key insect pests and diseases is the first step to establishing an IPM programme. The use of biological control agents, the preservation of natural enemies, the use of entomopathogenic fungi and nematodes, the use of botanical extracts, and the more selective use of new more targeted insecticides are recommended. In addition, cultural control practices such as sanitation and pruning help to regulate insect pests without affecting the abundance of natural enemies. Green cover, some shade trees, and border shrubs support the presence of beneficial insects. Fruit and timber trees associated with coffee plantations provide habitat for natural enemies. However, shading needs to be carefully managed since it can create conditions favourable to some pests and plant pathogens such as *Hemileia vastatrix* (Berk & Brome) which causes coffee leaf rust, one of the most serious diseases in coffee crops.

Conservation biological control practices favour the survival of natural enemies in agricultural areas. Since coffee agroecosystems are perennial systems, supporting natural enemies by providing sources of food (nectar and pollen) should be a common practice by coffee growers. Plantings tropical fruit trees and other plants on islands or on borders of coffee fields helps to re-establish the habitat of natural enemies and the diversity of beneficial insects and microorganisms is increased. Conservation biological control practices include collection of berries infested with CBB during harvesting, storage in cages for promotion, and release of parasitoids (*P. nasuta* and *C. stephanoderis*), as reported in Colombia by Aristizábal et al. (2011b). Results showed that parasitism of *P. nasuta* (8.3%) was higher in comparison with coffee fields without screened enclosures (4.1%). Those results suggest that the use of parasitoids is compatible with cultural control practices such as sanitation picking. Green cover plants such as *Commelina elegans, C. diffusa* (Commelinaceae), *Polygonium nepalense* (Poluganaceae), and *Oxalis latifolia* (Oxalidacea) that produce flowers and have low competition with coffee trees, supply pollen, and nectar to several beneficial insects, including parasitoids of CBB.

11 Climate change impacts on insect pests in coffee agroecosystems

Climate change is already affecting tropical and sub-tropical regions in which coffee and many other crops are produced (Groenen, 2018). Coffee agroecosystems face multiple impacts from increases in temperature which affect where coffee can be grown and insect population dynamics. Changes in rainfall patterns can affect the physiology of coffee trees, especially flowering patterns and therefore harvest seasons. An increase in temperatures, longer dry periods, and more extreme weather events will limit coffee production in some regions. It is predicted that there will be a reduction in coffee fields areas located at low elevations due to high temperatures, longer dry periods, reduction of rainfall, and low humidity. In contrast, highland mountains areas with temperate and cold climates are expected to become new areas for planting coffee. Donatti et al. (2019) have reviewed the potential impact of climate change on coffee growth and yields in Central America and Mexico through to 2050. Overall, the suitability of a country to grow coffee and coffee yields are expected to decrease by between 10–40% and 4–82%, respectively, depending on each coffee-producing country, respectively. In terms of insect pests, it is estimated that increases in temperature, while they might restrict a pest's life cycle, may increase the number of generations per year, making them harder to control. This reflects similar concerns about the impact of increasing temperature in promoting coffee plant pathogens such as coffee leaf rust (Talhinhas et al., 2017)

Jaramillo et al. (2013) have reviewed the impact on traditional coffee production systems in East Africa from climate change over the last 80 years.

One aspect is the effect of climate change on CBB populations in shaded and non-shaded coffee production systems. CBB infestation was 12.2-fold times high in unshaded compared to shaded coffee systems. In addition, the overall number of CBB females was 18.2 times higher in coffee monoculture exposed to full sunlight in comparison with coffee grown under shade (Jaramillo et al., 2013). Those results show how an insect pest population can increase significantly with only a 2°C difference between shaded and unshaded coffee agroecosystems. The introduction of shade trees in coffee agroecosystems could mitigate the negative impact of temperature increases caused by climate change (Jaramillo et al., 2013). These benefits could extend beyond limiting pest numbers to promoting beneficials, regulating temperature, conserving moisture, and providing additional, diversified sources of income for smallholders. Lemma and Megersa (2021) has suggested measures such as improving plant diversity, soil and water conservation measures, and breeding plants more resistant to high temperature and reduced water availability, as measures to mitigate the negative impacts of climate's change on coffee plantations in East Africa.

In addition, climate change is expected not just to impact just coffee agroecosystems and insect pest populations but also to affect coffee quality. In a systematic review study, Ahmed et al. (2021) reported that an increase in altitude is linked with positive improved sensory attributes of coffee quality, while increased light exposure (sun radiation) is linked with a reduction of coffee quality. The effect of CBB (*H. hampei* Ferrari) damage on coffee flavour profiles reported by Walker et al. (2018) is an example of how damage by an insect pest can affect the quality of coffee. Populations of CBB are expected to increase as a result of climate change, increasing damage to coffee berries, and thus affecting coffee quality. Finally, an interesting study conducted by Echeverria-Beirute et al. (2018), in Costa Rica, explored the impact of management of coffee leaf rust CLR (*Hemileia vastatrix* Berk. Et Br.) on coffee beverage quality. According to Echeverria-Beirute et al. (2018), the sensory and chemical analysis showed that 14% of quality attributes and 11% of chemical volatile compounds of coffee beverage were negatively affected by coffee genotype and CLR management. Those results suggest that new CLR epidemics caused by the impact of climate change (Avelino et al., 2015) will increase the use of fungicides in coffee agroecosystems affecting coffee quality and the environment.

12 Case study: improving integrated pest management of coffee berry borer in Hawaii

Coffee production from Hawaii represents a small proportion (3000 ha) of all coffee planted globally (>10 000 000 ha). However, due to the unique location in the middle of the Pacific Ocean, the volcanic origin of the soils, the high variety of micro-climates, and resulting high coffee quality, Hawaiian coffees are recognized as one of the most valuable specialty coffees around the world.

Some insect pests such as scales, mealybugs, black twig borer, grasshoppers, and some mites such as red mite and broad mite have been reported as affecting coffee trees across the Hawaiian Islands (Table 4) (Nelson et al., 2005; Bittenbender and Smith, 2008) . However, it was only when the CBB (*H. hampei*) was reported in Hawaii in 2010. that coffee production and the quality of processed coffee began to be significantly affected by an insect pest (Burbano et al., 2011; Aristizábal et al., 2016). In addition, the recent discovery of coffee leaf rust (*Hemileia vastatrix*), the most devastating global disease affecting coffee crops, in late 2020 in Maui and the Big Island, is one of the major challenges that the coffee industry in Hawaii is facing.

An initiative to establish an IPM programme to manage the CBB has been developed by coffee farmer associations and cooperatives and by several federal and state agencies in Hawaii. The IPM programme developed for CBB includes pest monitoring, spraying of *B. bassiana* (commercial formulation), cultural control practices, use of natural enemies (flat bark beetles), and selective use of insecticides. A recent, ongoing initiative is the introduction of *Phymastichus coffee* La Salle, an endo-parasitoid of CBB adults.

One aspect of IPM is cultural control, including good sanitation practices. A key indicator of good practice is the number of cherries and raisin berries left on average per tree after a harvesting round is conducted (Bustillo et al., 1998; Aristizábal et al., 2016, 2017b):

- An 'Excellent' sanitation practice is when less than 5 raisin berries are left per tree,
- A 'Good' practice is when less than 10 raisins are left per tree,
- A 'Bad' practice is when more than 10 raisin berries are left per tree.

Figure 3 (a and b) shows both variables - the CBB infestation and the efficacy of harvesting practices - over a number of years in several coffee farms in Kona and Ka'u Districts in Hawaii. They reflect the progress achieved by coffee farmers between 2015 and 2020. When farmers improved the harvesting of cherries and raisin berries, a cultural control practice known as a 'sanitation pick', CBB infestation levels were significantly reduced.

This case study compares CBB infestation before and after coffee farmers received training in how to improve cultural control practices (covering the 2018 coffee season with the participation of 25 coffee farmers from Ka'u District, Hawaii).

Table 4 Most common insect and mite pests found in coffee farms in Hawaii

Insect and mite pests	Structure affected	Stage	Observations
Coffee berry borer	Berry	Adult, larvae	Raisin berries major source of pest
Hypotenmeus hampei (Ferrari)			A major insect pest
(Coleptera: Curculionidae)			
Black twin borer	Branches	Adult, larvae	Most common during spring and summer
Xylosandrus morigerus (Blandford)			
(Coleptera: Curculionidae)			
Banana moth	New verticals stemps	Larvae	Attack recently stumped trees
Opogana sacchari (Bojer)			
(Lepidoptera: Tineidae)			
Green scale	Aerial structures	Nymph, adult female	Most common during summer and dry periods
Coccus viridis (Green)			Associated with ants
(Hemiptera: Coccidae)			
Little fire ant	Aerial structures		Interfere with coffee pickers during harvesting
Wasmannia auropuntata (Roger)			
(Hymenoptera: Formicidae)			
Mediterranean fruit fly	Berry	Larvae	Most common on coffee cherry
Ceratitis capitata (Wiedemann)			
(Diptera: Tephritidae)			
Grasshoppers			Most damage in new growth coffee and nursery
Conocephalus saltator (Saissure)	Aerial structures	Nymph, adult	
Elimea punctifera (Walker)	Aerial structures	Nymph, adult	
(Orthoptera: Tettigoniidae)			
Coffee bean weevil			
Araecerus fasciculatus (De Geer)	Processed beans	Larvae, adults	Most common on stored processed coffee
(Coleptera: Curculionidae)			
Red mite	Leaves	Nymph, adult	Most common during summer and dry periods
Olygonychus yothersi (McGregor)			
(Acari: Tetranychidae)			
Broad mite	Leaves	Nymph, adult	New tissue leaves
Polyphagotarsonemus latus (Banks)			In nurseries and crops
(Acari: Tarsonemidae)			

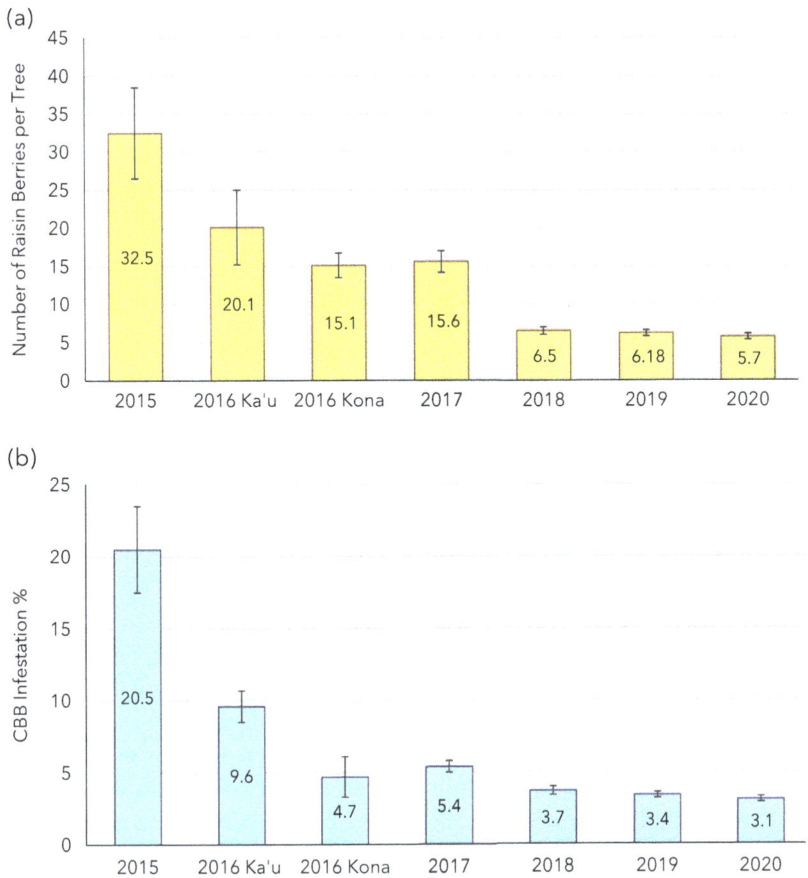

Figure 3 (a) Efficacy of harvestings practice (number of raisin berries left after a harvesting round) and (b) coffee berry borer infestation levels in commercial coffee farms from Hawaii (2015–2020).

12.1 Improving cultural control practices to reduce coffee berry borer in coffee farms in Ka'u District, Hawaii

This study was funded by the 2017 Socially Disadvantaged Group Grant (SDGG) from the US Department of Agriculture (USDA) obtained by the Ka'u Coffee Grower's Cooperative (KCGC). The overall goal of this study was to provide training to coffee growers from Ka'u District on how to establish an IPM programme for the CBB with a focus on cultural control practices. Activities related with this project were conducted from February until October (2018). In total, 92 individual coffee farm visits were conducted to assess levels of CBB infestation and give specific recommendations for control of the pest. The project targeted 25 coffee growers growing the crop on an estimated 120 ha overall.

The first step was to determine the CBB population in each commercial coffee plantation. A starting point was the '30 tree sampling plan' developed in Colombia (Bustillo et al., 1998), adapted for Hawaii (Kawabata et al., 2017), and modified by Aristizábal et al. (2018). Infestation rates were contrasted between:

- the initial situation (at beginning of the study in February-March 2018): 'Before training', and
- the situation over the rest of the coffee season (April-October 2018): 'After training'.

During coffee farm visits growers were instructed in how regulate CBB populations using two specific strategies:

1 Collecting all raisin berries, which are the major reservoir of CBB inside coffee plantations, and
2 Applying *B. bassiana* (bio-insecticide) when high activity (>25%) of CBB in AB was observed and infestation rates were higher than (3%).

The goal was to reduce CBB infestations rates lower than 3%.

The major reservoir of CBB populations are infested raisin berries which may harbour up to 120 CBB individuals per berry (Bustillo et al., 1998; Vega et al., 2015; Aristizábal et al., 2017a). Removing them can reduce infestation rates to the target of 3% or lower (Bustillo et al., 1998; Aristizábal et al., 2011a). A cultural control measure known as a 'sanitation pick' is the best way to achieve this (Damon, 2000; Jaramillo et al., 2006). The control is based on collection of cherries and raisin berries, evaluated by number of the raisin berries left on the tree after a single harvesting round (Bustillo et al., 1998; Aristizábal et al., 2011a).

In the 'Before training' period, from February to March 2018, only 8% of the coffee fields scored as 'Excellent' in their sanitation practices with a low number of raisin berries left per tree, while in 76% of fields the sanitation practice was 'Bad'. However, the efficacy of these sanitation practices was improved following additional coffee farm visits. In the 'After training' period (April-October 2018) 82.8% of fields showed 'Good' or 'Excellent' practice with only 17.28% of fields classified as 'Bad'. Overall, the number of raisin berries per tree was 15.62 ± 1.3 (n = 25) (mean ± SEM), ranging from 1.5 to 28.2 at the beginning of the first coffee farm training visit during (February and March), indicating 'Bad' harvesting practices. Following additional (n = 48) coffee farms visits (April-October), the efficacy of harvesting practice ranged between 0.5-20.7 dropping to 6.35 ± 0.6 raisins berries per tree, which is considered a 'Good' practice (Fig. 4 a).

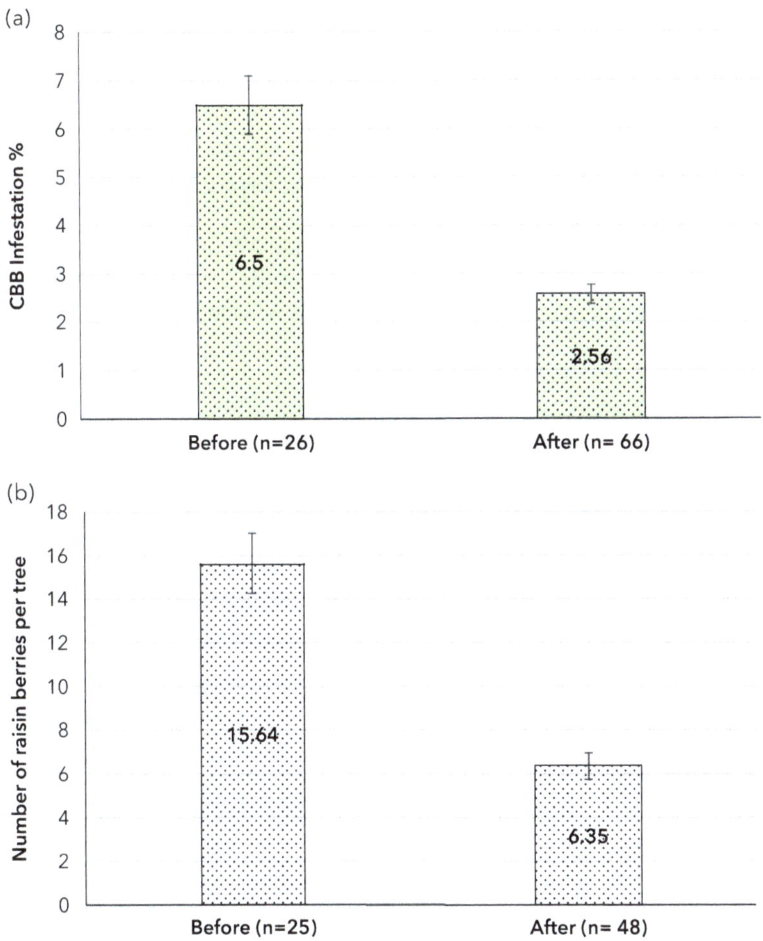

Figure 4 Case study: (a) number of raisin berries left per tree and (b) Coffee berry borer infestation rate (%) in average (SEM) at the beginning of the training, February-March 2018 (before) and following additional coffee farms visits during April-October 2018 (After) in commercial coffee farms in Ka'u District, Hawaii. Number of evaluations (n) conducted for each variable.

Coffee infestation rate for first farm visit was in average 6.53% ± 0.6 (n = 26) (mean ± SEM), ranging between 1.2% and 14.5%. Infestation dropped to an average 2.56% ± 0.2 (n = 66) ranging between 0.75% and 7.5% following additional visits and recommendation given to coffee growers (Fig. 4b). CBB infestation rates were classified into three groups:

1 Good (lower than 3%), which means that CBB infestation is low and does not represent a major risk for the coffee plantation,

2 Alert (3.1–5%), which means that control measures need to be conducted to prevent damage to coffee yields, and
3 Danger (>5.1%), which means that damage to coffee yields has already occurred and could increase if control action is not taken.

Based on this classification, for the first coffee farm visits (total 26):

- 11.6% of coffee plantations showed less than 3% CBB infestation: 'Good',
- 34.6% with infestation rates ranging between 3.1 and 5%: 'Alert', and
- 53.8 % of coffee fields with more than 5.1% infestation rate: 'Danger'.

Following additional (66) coffee farm visits, this situation changed significantly. Results showed:

- 65.1% of the coffee fields with less than 3% CBB infestation,
- 22.7% ranging between 3.1 and 5%, and
- only 12.1% of the fields with more than 5.1% CBB infestation rates.

A positive correlation ($R^2 = 0.81$) between the number of raisin berries left in the trees and the CBB infestation rates was observed, providing strong evidence of the impact of a 'sanitation pick' on CBB regulation. In most cases, CBB infestation rates higher than 5% were found when more than 10 raisin berries per tree was observed. Those results showed how important the 'sanitation pick' practice is in the control of CBB populations. For this reason, it is recommended in all coffee farm visits that all cherries and raisin berries should be collected before conducting any spraying with *B. bassiana*.

Most coffee growers from Ka'u have relied for control of CBB on calendar (monthly) applications of *B. bassiana*. This approach has had a positive impact on the reduction of CBB population in comparison with coffee growers that do not apply *B. bassiana* or other insecticides. However, combining a 'sanitation pick' with spraying *B. bassiana* when CBB is high has proved particularly effective with CBB populations lower than 3% on Ka'u coffee farms. This can be compared to previous years when CBB infestation rates in coffee fields from Ka'u District were high at 20.5% on average during 2015 with a pick of 66.4% (Aristizábal et al., 2016) and 9.6% on average during 2016 with a pick of 34% (Aristizábal et al., 2017a).

A study of 'sanitation pick' practices in 2015 in coffee fields from Ka'u District showed between (20–45) raisin berries left per tree after harvesting, as 'Bad' practice (Aristizábal et al., 2016). A similar situation was observed in coffee farms from Kona District during 2016, in which the average of raisin berries left per tree was 13.9, ranging between 2.5 and 31.4 (Aristizábal et al., 2017b). At this time 30.3% of evaluations in 36 coffee fields were classified as 'Good' or 'Excellent' and 69.7% as 'Bad' (Aristizábal et al., 2017b). The improvement after training was reflected in the later low infestation rates (2.45% on average) observed during April-October in all coffee fields.

In a small coffee farm (1 ha) in Colombia, sanitation (harvesting) practices were monitored during 2002-2004. Results showed 6.2 raisin berries left per tree on average resulting in an average 2.2% CBB infestation rate over 2 years (Aristizábal et al., 2011a). Similarly, the average of raisin berries left per tree was 6.3 (April-October) in (25) coffee fields in Ka'u District, resulting in 2.5% CBB average infestation rates. In another case study in Hawaii (Luis F. Aristizábal, unpublished), coffee farms in which the management of CBB was focused on cultural control practices resulted in an average 4.6 berries left per tree during the 2019 and 2020 coffee seasons while the CBB infestation level averaged 1.8% during the same period. Both the examples from Colombia and Hawaii suggest that removing raisin berries from coffee trees significantly reduces CBB populations at a low cost over the full coffee season with good quality processed coffee obtained.

Applications of *B. bassiana* or other authorized insecticide at the right time (early coffee season), when more activity of CBB in AB position is observed, contribute in regulating CBB populations and in preventing damage to processed coffee. The studies were conducted in Colombia by Bustillo et al. (1998), Bustillo et al. (1999), Vera et al. (2011), Benavides et al. (2012), and in India by Haraprasad et al. (2001), in Hawaii by Aristizábal et al. (2017a), Greco et al. (2018), and Hollingsworth et al. (2020) have reported a reduction of CBB populations when the entomopathogenic fungus *B. bassiana* is applied on time and in 'hot spots', Some studies have addressed the impact of weather conditions such as the altitude (coffee farms located at high elevation), in which relative humidity is high and temperatures low, favouring CBB control using *B. bassiana* (Mascarin and Jaronski, 2016; Wraight et al., 2018). However opposite situation was observed in coffee farms located at low elevation, in which low relative humidity and high temperatures affect the efficacy of *B. bassiana*. Finally, a rational use of *B. bassiana* sprays in combination with sanitation pick could be an excellent approach for controlling CBB in an IPM programme at low cost with minimal risk to the environment and human health.

13 Summary and future trends

Coffee agroecosystems are located in tropical and sub-tropical regions with a high biodiversity of arthropods. Herbivorous insects that feed on coffee trees are abundant, but only few of them are key insect pests. In addition, high numbers of beneficial insects including predators, parasitoids, and pollinators are natural inhabitants of coffee agroecosystems. Those beneficial insects provide natural regulation of mites and herbivorous insects. However, monoculture of coffee with a high density of trees per area, low diversity of vegetation, and high use of pesticides disrupt the natural balance between beneficial and herbivorous insects. This situation is one of the reasons why some herbivores become primary or secondary insect pests.

This chapter has reviewed aspects of the biology, ecology, life cycle, behaviour, damage, and alternative control measures for six key insect pests: CBB, CLM, antestia bug, twig borers, root mealybug and stem borers. The chapter has also discussed natural enemies of these pests and their conservation, as well as the impact of climate changes on coffee agroecosystems and pest dynamics. The chapter has also included case studies on improving integrated pest management of CBB in commercial coffee farms in Hawaii; these discuss cultural control practices and the use of *B. bassiana* (bio-pesticide). Results showed strong evidence that a frequent and effective 'sanitation pick' of cherries and raisin berries (cultural control practice) is the most economic and effective strategy of control CBB under Hawaiian conditions.

The use of insect molecular engineering is one of the new trends in the search for new alternatives for controlling insect pests. The improvement of native locally adapted and highly virulent strains of entomopathogenic fungi and entomopathogenic nematodes used in the control of insect pests is another trend that is promising. New repellent products, using vegetables and mineral oils, and new generations of synthetic insecticides are also in development. Many of those new products are less toxic and pose less risk to human health and the environment. The tendency of coffee marketing is toward specialty coffees which are sold at a premium price. Some specialty coffees require certification such as fair trade, organic, eco-friendly, and so on. This shift reinforced the necessity of establishing IPM programmes with a focus on cultural control practices, use of biological control agents, and use of insecticides with a low impact on the environment. High precision agriculture is another innovation that may be applied in the prevention and control of insect pests. Technologies include monitoring CBB populations by using mobile application technologies connected to weather stations (Johnson et al., 2020). This technology can support coffee farmers to take decisions for controlling CBB in real time, based on infestation levels, flight activity of the pest, and visualization of specific 'hot spots' in maps support by Global Positioning System (GPS) technologies.

14 Where to look for further information

Overall, the most important coffee producer countries have official government institutions or agencies and privacy farmers associations supporting research and extension services on coffee production. For example, the National

Federation of Coffee Growers of Colombia, support the National Coffee Research Center (Cenicafé), which is one of the most important and recognized research center in coffee production around the world. Cenicafé has several serial publications including Technical Advances and Cenicafé Journal (Spanish: Avances Técnicos y Revista Cenicafé) respectively, in which is documented basic and applied research in all aspects of coffee production from selection of coffee seeds until evaluation the quality of coffee cupping.

The Coffee Institute of Costa Rica (ICAFE), the Honduran Coffee Institute (IHCAFE), and the National Coffee Association (Anacafé) in Guatemala are some of the institutions that have supported coffee growers in those countries with research, training, and divulgation in many aspects related with coffee production, coffee agroecosystems, and coffee marketing. In addition, the Tropical Agricultural Research and Higher Education Center (CATIE) in Costa Rica, the Zamorano Pan-America Agricultural School 'El Zamorano' in Honduras, and 'El Colegio de la Frontera Sur, (Ecosur)' in Mexico are some of the educational institutions that are supporting research and education in coffee agroecosystems in Central America.

In Africa, the International Center of Insect Physiology and Ecology (ICIPE), the Ethiopia Institute of Agricultural Research (EIAR), the Agricultural Research for Development (CIRAD), and CABI-Bioscience have supported basic and applied research on coffee production including IPM strategies.

15 References

Ahmed, S., Brinkley, S., Smith, E., Sela, A., Theisen, M., Thibodeau, C., Warne, T., Anderson, E., Van Dusen, N., Giuliano, P., Ionescu, K. E. and Cash, S. B. (2021). Climate change and coffee quality: Systematic review on the effects of environmental and management variation on secondary metabolites and sensory attributes of Coffea arabica and coffea Canephora. *Front. Plant Sci.* 12, 708013. doi: 10.3389/fpls.2021.708013.

Aristizábal, L. F. (2018). *Controlling the Coffee Berry Borer Through Integrated Pest Management: A Practical Manual for Coffee Growers and Field Workers in Hawaii.* Kailua-Kona, HI: Steuben Press, p. 79.

Aristizábal, L. F., Bustillo, A. E. and Arthurs, S. P. (2016). Integrated pest management of coffee berry borer: Strategies from Latin America that could be useful for coffee farmers in Hawaii. *Insects* 7(1), 6. doi: 10.3390/insects7010006.

Aristizábal, L. F., Jimenez, M., Bustillo, A. E. and Arthurs, S. P. (2011a). Monitoring cultural practices for coffee berry borer *Hypothenemus hampei* (Coleoptera: Curculionidae: Scolytinae) management in a small coffee farm in Colombia. *Fla. Entomol.* 94(3), 685–687. doi: 10.1653/024.094.0334.

Aristizábal, L. F., Jimenez, M., Bustillo, A. E. and Arthurs, S. P. (2011b). Introduction of parasitoids of *Hypothenemus hampei* (Coleoptera: Curculionidae: Scolytinae) on small coffee plantations in Colombia through farmer participatory methods development. *Fla. Entomol.* 94(3), 690–693.

Aristizábal, L. F., Lara, O. and Arthurs, S. P. (2012). Implementing an integrated pest management program for coffee berry borer in a specialty coffee plantation in Colombia. *J. Integr. Pest Manag.* 3(1), G1–G5. doi: 10.1603/IPM11006.

Aristizábal, L. F. M., Jiménez, A. E. Bustillo, H. I. Trujillo and Arthurs, S. P. (2015). Monitoring coffee berry borer, *Hypothenemus hampei* (Coleoptera: Curculionidae), populations with alcohol-baited funnel traps in coffee farms in Colombia. *Fla. Entomol.* 98: 381-383. doi: 10.1603/IPM11006.

Aristizábal, L. F., Shriner, S., Hollingsworth, R. and Arthurs, S. P. (2017a). Flight activity and field infestation relationships for coffee berry borer *Hypothenemus hampei* (Ferrari) in commercial coffee plantations in Kona and Kau districts, Hawaii. *J. Econ. Entomol.* 110(6), 2421-2427. doi: 10.1093/jee/tox215.

Aristizábal, L. F., Johnson, M., Shriner, S., Hollingsworth, R., Manoukis, N. C., Myers, R., Bayman, P. and Arthurs, S. P. (2017b). Integrated pest management of coffee berry borer in Hawaii and Puerto Rico: Current status and prospects. *Insects* 8(4), 123. doi: 10.3390/insects8040123.

Aristizábal, L. F., Shriner, S., Hollingsworth, R., Mascarin, G. M., Chaves, B., Matsumoto, T. and Arthurs, S. P. (2018). Field sampling strategies for coffee berry borer (Coleoptera: Curculionidae: Scolytinae) infesting berries in coffee farms in Hawaii. *Int. J. Trop. Insect Sci.* 38(4), 418–426. doi: 10.1017/S174275841800022X.

Avelino, J., Cristancho, M., Georgion, S., Imbach, P., Aguilar, L., Bornemann, G., Laderach, P., Anzueto, F., Hruska, A. J. and Morales, C. (2015). The coffee rust crises in Colombia and Central America: Impacts, plausible causes and proposed solutions. *Food Sec.* 7(2), 303–321.

Azrag, A. G. A., Pirk, C. W. W., Yusuf, A. A., Pinard, F., Niassy, S., Mosomtai, G. and Babin, R. (2018). Prediction of insect pest distribution as influenced by elevation: Combining field observations and temperature-dependent development models for the coffee stink bug, *Antestiopsis thunbergii* (Gmelin). *PLoS ONE* 13(6), e0199569. doi: 10.1371/journal.pone.0199569.

Babin, R., Mbondji Mbondji, P., Mendesil, E., Mugo, H. M., Lee, J. H., Serracin, M., Rukazambuga, N. D. T. M. and, Miller, T. A. (2018). The Antestia bug complex in Africa and Asia. In Mc Pherson, J. E. (Ed.), *Invasive Stink Bugs and Related Species (Pentatomidae) Biology, Higher Systematics, Semichemistry and Management*. Boca Raton, FL: CRC Press, Taylor & Francis Group, pp. 465–493.

Bacca, T., Lima, E. R., Picanço, M. C., Guedes, R. N. C. and Viana, J. H. M. (2008). Sampling plan for the coffee leaf miner *Leucoptera coffeella* with sex pheromone traps. *J. Appl. Entomol.* 132(6), 430–438.

Bacca, T., Saraiva, R. M. and Lima, E. R. (2012). Captura de *Leucoptera coffeella* (Lepidoptera: Lyonetiidae) en trampas con feromona sexual y su intensidad de dano. *Rev. Colomb. Entomol.* 38(1), 42–49.

Baker, P. S. (1999). The coffee berry borer in Colombia. Final Report of the DFID - Cenicafé CABI Bioscience IPM for coffee project. *Chinchiná Cen. Nac. Investig. Café* 154.

Barrera, J. F. (Ed.). (2002). Tres plagas del cafe. en Chiapas. El Colegio de la Frontera Sur. *Mexico*, p. 198.

Barrera, J. F. (2008). Coffee pests and their management. In Capinera, J. L. (Ed.), *Encyclopedia of Entomology*. New York: Springer, pp. 961-998.

Barrera, J. F., Herrera, J., Villalobos, J. and Gomez, B. (2004). El barrenador del tallo y la raiz del café. Una plaga silenciosa. El Colegio de la Frontera Sur, ECOSUR. *Folleto Tecnicao* 9, 8.

Benavides, P., Gil, Z. N., Constantino, L. M., VIllegas, C. and Giraldo, J. M. (2013). Plagas del café: Broca, minador, cochinillas harinosas, arañita roja y monalonion pp-Tomo 2. In *CENICAFÉ. Manual del Cafetero Colombiano*. Chinchiná, p. 349.

Benavides, P., Góngora, C. and Bustillo, A. (2012). IPM program to control coffee berry borer *Hypothenemus hampei*, with emphasis on highly pathogenic mixed strains of *Beauveria bassiana*, to overcome insecticide resistance in Colombia. In Perveen, F. (Ed.), *Insecticides-Advances in Integrated Pest Management* (1st ed.). Rijeka: InTech Europe, pp. 512–540.

Bigirimana, J., Gerard, A., Mota-Sanchez, D. and Gut, L. J. (2018). Options for managing *Antestiopsis thunbergii* (Hemiptera: Pentatomidae) and the relationship of bug density to the occurrence of potato taste defect in coffee. *Fla. Entomol.* 101(4), 580–586.

Bittenbender, H. C. and Smith,V. E. (2008). Growing coffee in Hawaii. College of tropical Agriculture and human Resources. *University of Hawaii at Manoa* (rev. ed.), p. 40.

Bonsignore, C. P. (2012). *Apate Monachus* (Fabricius, 1775), a Bostrichid pest of pomegranate and carob trees in nurseries - Short Communication. *Plant Prot. Sci.* 48(2), 94–97.

Bril, E., Follett, P. A. and Kawabata, A. M. (2021). Feeding habits, movement, and reproduction of the predatory flat bark beetles *Cathartus quadricollis* (Coleptera: Silvanidae) and *Leptophloeus* sp. (Coleoptera: Laemophloeidae) in Hawaii coffee and macadamia nut. *Int. J. Trop. Insect Sci.* 4, 285–294.

Burbano, E., Wright, M., Bright, D. E. and Vega, F. E. (2011). New record for the coffee berry borer, *Hypothenemus hampei*, in Hawaii. *J. Insect Sci.* 11, 117. doi: 10.1673/031.011.11701.

Bustillo, A. E., Bernal, M. G., Benavides, P. and Chaves, B. (1999). Dynamics of *Beauveria bassiana* and *Metarhizium anisopliae* infecting *Hypothenemus hampei* (Coleoptera: Scolytidae) populations emerging from fallen coffee berries. *Fla. Entomol.* 82(4), 491–498. doi: 10.2307/3496468.

Bustillo, A. E., Cardenas, M. R., Villalba, D., Orozco, J., Benavides, M. P. and Posada, F. J. (1998). *Manejo Integrado de la Broca del Café Hypothenemus Hampei (Ferrari) en Colombia. Cenicafé*. Chinchiná: Centro Nacional de Investigaciones de Café, p. 134.

Bustillo, P. A. E. (2015). Coffee insect pests, part II: Damages caused by arthropods. In Gaitan, A. L., Cristancho, M. A., Castro-Caicedo, B. L., Rivillas, C. A. and Cadena-Gomez, G. (Eds.), *Compendium of Coffee Diseases and Pests*. St. Paul, MN: The American Psychopathological Society Press, pp. 46–60.

Cardenas, M. R. and Posada, F. F. J. (2001). Los insectos y otros habitants de cafetales y platanales. Comite. Departamental de Cafeteros del Quindio. *Armenia, Colombia*, p. 250.

Constantino, C. H. L. M., Gil, P. Z. N., Benavides, M. P., Martiniez, H., Giraldon, J. M. and Villegars, G. C. (2013). Tomo 2. Otros habintantes naturales del cafetal. In *CENICAFÉ. Manual del Cafetero Colombiano*. Chinchiná, pp. 262–306.

Constantino, L. M., Gil, Z. N., Montoya, E. C. and Benavides, P. (2021). Coffee berry borer (*Hypothenemus hampei*) emergence from ground fruits across varying altitudes and clime cycles, and the effect on coffee tree infestation. *Neotrop. Entomol.* 50(3), 374–387 doi: 10.1007/s13744-021-00863-5.

Damon, A. A. (2000). A review of the biology and control of the coffee berry borer, *Hypothenemus hampei* (Coleoptera: Scolytidae). *Bull. Entomol. Res.* 90(6), 453–465. doi: 10.1017/s0007485300000584.

Dantas, J., Motta, I., Vidal, L., Bilio, J., Pupe, J. M., Veiga, A., Carvalho, C., Lopes, R. B., Benito, N., Lima, T., Silva, L. P., Pujol-Luz, J. R. and Albuquerque, E. V. S. (2020). A comprehensive review of the coffee leaf miner *Leucoptera coffeella* (Lepidoptera: Lyonetiidae), with special regard to neotropical impacts, pest management and control. *Insects* 11. doi: 10.20944/preprints202010.0629.v1.

Donatti, C. I., Harvey, C. A., Martinez-Rodriguez, M. R., Vignola, R. and Rodriguez, C. M. (2019). Vulnerability of smallholder farmers to climate change in Central America and Mexico: Current knowledge and research gaps. *Clim. Dev.* 11(3), 264–286. doi: 10.1080/17565529.2018.1442796.

Echeverria-Beirute, F., Murray, S. C., Klein, P., Kerth, C., Miller, R. and Bertrand, B. (2018). Rust and thinning management effect on cup quality and plant performance for two cultivars of *Coffea arabica* L. *J. Agric. Food Chem.* 66(21), 5281–5292. doi: 10.1021/acs.jafc.7b03180.

Follett, P. A., Kawabata, A., Nelson, R., Asmus, G., Burt, J., Goschke, K., Ewing, C., Gaertner, J., Brill, E. and Geib, S. (2016). Predation by at bark beetles (Coleoptera: Silvanidae and Laemophloeidae) on coffee berry borer (Coleoptera: Curculionidae) in Hawaii coffee. *Biol. Control* 101, 152–158. doi: 10.1016/j.biocontrol.2016.07.002.

Greco, E. B. and Wright, M. G. (2013). Dispersion and sequential sampling plan for *Xylosandrus compactus* (Coleoptera: Curculionidae) infesting Hawaii coffee plantations. *Environ. Entomol.* 42(2), 277–282. doi: 10.1603/EN12182.

Greco, E. B. and Wright, M. G. (2015). Ecology, biology, and management of *Xylosandrus compactus* (Coleoptera: Curculionidae: Scolytinae) with emphasis on coffee in Hawaii. *J. Integ. Pest Manag.* 6(1), 7. doi: 10.1093/jipm/pmv007.

Greco, E. B., Wright, M. G., Burgueño, J. and Jaronski, S. T. (2018). Efficacy of *Beauveria bassiana* applications on coffee berry borer across an elevation gradient in Hawaii. *Biocontrol Sci. Technol.* 28(11), 995–1013 doi: 10.1080/09583157.2018.1493088.

Groenen, D. (2018). The effects of climate change on the pests and diseases of coffee crops in Mesoamerica. *J. Climatol. Weather Forecasting* 6(3), 239. doi: 10.4172/2332-2594.1000239.

Gueule, D., Fourny, G., Ageron, E., Le Flèche-Matéos, A., Vandenbogaert, M., Grimont, P. A. D. and Cilas, C. (2015). *Pantoea coffeiphila* sp. nov. cause of 'potato taste' of arabica coffee from African Great Lakes region. *Int. J. Syst. Evol. Microbiol.* 65(1), 23–29.

Haraprasad, N., Niranjana, S. R., Prakash, H. S., Shetty, H. S. and Wahab, S. (2001). *Beauveria bassiana* a potential mycopesticide for the efficient control of coffee berry borer, *Hypothenemus hampei* (Ferrari) in India. *Biocontrol Sci. Technol.* 11(2), 251–260. doi: 10.1080/09583150120035675.

Hollingsworth, R. G., Aristizábal, L. F., Shriner, S., Mascarin, G. M., Moral, RdA. and Arthurs, S. P. (2020) Incorporating *Beauveria bassiana* into an integrated pest management plan for coffee berry borer in Hawaii. *Front. Sustain. Food Syst.* 4, 22. doi: 10.3389/fsufs.2020.00022.

International Coffee Organization (ICO). (2019). Trade statistics. Available at: http://www.ico.org/trade_statistics.asp (accessed 21 April 2021).

Jaramillo, J., Borgemeister, C. and Baker, P. (2006). Coffee berry borer *Hypothenemus hampei* (Coleoptera: Curculionidae): Searching for sustainable control strategies. *Bull. Entomol. Res.* 96(3), 223–233. doi: 10.1079/BER2006434.

Jaramillo, J., Setamou, M., Muchugu, E., Chabi-Olaye, A., Jaramillo, A., Mukabana, J., Maina, J., Gathara, S. and Borgemeister, Ch. (2013). Climate change or urbanization? Impacts on a traditional coffee production system in East Africa over the last 80 years. *PLoS ONE* 8(1), e51815. doi: 10.1371/journal.pone.0051815.

Johnson, M. A., Fortna, S., Hollingsworth, R. G. and Manoukis, N. C. (2019). Postharvest population reservoirs of coffee berry borer (Coleoptera: Curculionidae) on Hawai'I island. *J. Econ. Entomol.* 112(6), 2833–2841. doi: 10.1093/jee/toz219.

Johnson, M. A., Ruiz-Diaz, C. P., Manoukis, N. C. and Verle, J. C. R. (2020). Coffee berry borer (*Hypothenemus hampei*), a global pest of coffee: Perspectives from historical and recent invasions, and future priorities. *Insects* 11(12), 882. doi: 10.3390/insects11120882.

Kawabata, A. M., Nakamoto, S. T. and Curtiss, R. T. (2017). *Recommendations for coffee berry borer integrated pest management in Hawai'i 2016*. University of Hawaii, Mänoa Publication. Available at: http://www.ctahr.hawaii.edu/ oc/freepubs/pdf/IP-33.pdf.

Kilambo, D. L., Ng'Homa, N., Mahamed, R., Teri, J. M., Oduor, G., Phiri, N., Flori, A. and Pinard, F. (2005). Coffee disease survey in Tanzania. In *ASIC. 20th International Conference (Vol. 2004) on Coffee Science*, 11–15 October 2004. Bangalore, India, pp. 1263–1266.

Kucel, P., Kangire, A. and Egonyu, J. P. (2009). *Status and Current Research Strategies for Management of the Coffee Berry Borer (Hypothenemus Hampei Ferr) in Africa*. Mukono: Coffee Research Centre, National Crop Resources Research Institute (NaCRRI), p. 9.

Lemma, T. D. and Megersa, H. G. (2021). Impact of climate change on East African coffee production and its mitigation strategies. *World J. Agric. Sci.* 17(2), 81–89.

Le Pelley, R. H. (1973). *Las Plagas Del Cafe*. Barcelona, SA: Editorial Labor, p. 693.

Magina, F. L., Kilambo, D., Maerere, A. and Teri, J. (2016). Innovative strategies for control of coffee insect pests in Tanzania: A review. *Huria J. Open Univ. Tanzan.* 22, 63–72.

Mascarin, G. M. and Jaronski, S. T. (2016). The production and uses of *Beauveria bassiana* as a microbial insecticide. *World J. Microbiol. Biotechnol.* 32(11), 177. doi: 10.1007/s11274-016-2131-3.

Mendesil, E. and Abebe, M. (2004). Biology of Antestia bug, *Antestiopsis intricata* (Ghesquire & Carayon) (Hemiptera: Pentatomidae) on Coffea arabica. *L. J. Coffee Res* 32, 30–39.

Mugo, H. M., Kimemia, J. K. and Mwangi, J. M. (2013). Severity of antestia bugs, *Antestiopsis* spp. And other key insect pests under shaded coffee in Kenya. *Int. J. Sci. Nat.* 4(2), 324–327.

Nelson, S., Smith, V. E. and Wright, M. (2005). *Insect Pests IP-21. Banana Moth as a Pest of Coffee*. Manoa: Cooperative Extension Service, College of Tropical Agriculture and Human Resources, University of Hawaii, p. 4.

Olivera, C. M., Santos, M. J., Amabile, R. F., Frizzas, M. R. and Bartholo, G. F. (2018). Coffee berry borer in conilon coffee in the Brazilian cerrado: An ancient pest in a new environment. *Bull. Entomol. Res.* 108(1), 101–107. doi: 10.1017/S0007485317000530.

Pantoja-Gomez, L. M., Corrêa, A. S., de Oliveira, L. O. and Guedes, R. N. C. (2019). Common origin of Brazilian and Colombian populations of the neotropical coffee leaf miner, *Leucoptera coffeella* (Lepidoptera: Lyonetiidae). *J. Econ. Entomol.* 112(2), 924–931. doi: 10.1093/jee/toy416.

Perthus, B., Pradon, J. L., Montagnon, C., Dufour, M. and Leroy, T. (2005). Stable resistance against the leaf miner *Leucoptera coffeella* expressed by genetically transformed Coffea canephora in a pluriannual field experiment in French Guiana. *Euphytica* 144(3), 321–329.

Rueda, G. D., Constantino, L. M., Montoya, E. C., Ortega, O. E., Gil, Z. N. and Benavides, M. P. (2016). Diagnostico de *Leucoptra coffeella* (Lepidoptera: Lyonetiidae) y sus parasitodes en el departamento de Antioquia, Colombia. *Rev. Colombina Entomol.* 42(1), 4–11.

Talhinhas, P., Batista, D., Diniz, I., Vieira, A., Silva, D. N., Loureiro, A., Tavares, S., Pereira, A. P., Azinheira, H. G., Guerra-Guimarães, L. and Várzea, V. (2017). The coffee leaf rust pathogen *Hemileia vastatrix*: One and a half centuries around the tropics. *Mol. Plant Pathol.* 18(8), 1039–1051. doi: 10.1111/mpp.

Vega, F. E., Infante, F., Castillo, A. and Jaramillo, J. (2009). The coffee berry borer, *Hypothenemus hampei* ((Ferrari) (Coleoptera: Curculionidae): A short review, with recent finding and future research directions. *Terr. Arthr. Rev.* 2, 129–147. doi: 10.1163/187498209X12525675.

Vega, F. E., Infante, F. and Johnson, A. J. (2015). The genus *Hypothenemus*, with emphasis on *H. Hampei*, the coffee Berry Borerin bark beetles. In Vega, F. E. and Hofstetter, R. W. (Eds.), *Biology and Ecology of Native and Invasive Species* (1st ed.). London: Elsevier, pp. 427–494. doi: 10.1016/B978-0-12-417156-5.00011-3.

Vera, J. T.,Montoya, E. C., Benavides, P. and Góngora, C. E. (2011). Evaluation of *Beauveria bassiana* (Ascomycota: Hypocreales) as a control of the coffee berry borer *Hypothenemus hampei* (Coleoptera: Curculionidae: Scolytinae) emerging from fallen infested coffee berries on the ground. *Biocontrol Sci. Technol.* 21(1), 1–14. doi: 10.1080/09583157.2010.517605.

Walker, H. E., Lehman, K. A., Wall, M. M. and Siderhurst, M. S. (2018). Analyis of volatile profiles of green Hawaiian coffee beans damaged by the coffee berry borer (*Hypothenemus hampei*). *J. Sci. Food Agric.* 99(4), 1954–1960. doi: 10.1002/jsfa.9393.

Waller, J. M., Bigger, M. and Hillocks, R. J. (2007). *Coffee Pests, Diseases and Their Management.* Wallingford: CAB International, p. 434.

Wraight, S. P., Galaini-Wraight, S., Howes, R. L., Castrillo, L. A., Carruthers, R. I., Smith, R. H., Matsumoto, T. K. and Keith, L. M. (2018). Prevalence of naturally-occurring strains of *Beauveria bassiana* in populations of coffee berry borer *Hypothenemus hampei* on Hawai'I Island, with observations on coffee plant-*H. hampei-B. bassiana* interactions. *J. Invertebr. Pathol.* 156, 54–72. doi: 10.1016/j.jip.2018.07.008.

Chapter 9

Diseases affecting coffee: an overview

S. Daivasikamani and A. P. Ranjini, Central Coffee Research Institute, India

1 Introduction

In India, coffee is mainly cultivated in the hills of the states of Karnataka, Kerala and Tamil Nadu and to a lesser extent in Andhra Pradesh, Odisha, West Bengal, Maharashtra and some northeastern states. The two main varieties grown are arabica (*Coffea arabica* L.) and robusta (*C. canephora* Pierre ex Froehner). Coffee is commercially cultivated in India over an area of 459 895 ha consisting of 375 542 farm holdings producing 319 500 MT coffee. The average productivity of *C. arabica* and *C. canephora* is 470 kg ha^{-1} and 1047 kg ha^{-1}, respectively, with a total average productivity of 767 kg ha^{-1}. India exports coffee to more than 45 countries around the world. In 2019, India exported 239 630 MT of coffee worth US$545.07 million. Consumption of coffee in the country is 108 000 MT (Anon, 2019).

Arabica coffee is more susceptible to fungal disease than robusta coffee. Leaf rust, blackrot, anthracnose, root diseases, collar rot, brown eye spot, stem necrosis and leaf spot are important diseases that need good plant protection measures to maintain production (Anon, 1998, 2014; Daivasikamani et al.,

http://dx.doi.org/10.19103/AS.2021.0096.22

2016). Integrated disease management (IDM) is used to minimize pesticide use and disturbance to the environment and maintain pathogen population levels below economic damage levels (Muller, 2009; Gaitan et al., 2016). Below is a list of key coffee diseases:

Diseases affecting adult coffee plants:	
Leaf rust:	*Hemileia vastatrix B.* & Br.
Black rot:	*Koleroga noxia* Donk.
Anthracnose:	*Colletotrichum gloeosporioides* Penz.
Twig die-back:	*Colletotrichum gloeosporioides* Penz.
Stalk rot of berries:	*Colletotrichum gloeosporioides* Penz.
Brown blight of leaves:	*Colletotrichum gloeosporioides* Penz.
Root diseases:	
Brown root disease:	*Fomes noxius* Corner
Red root disease:	*Poria hypolateritia* Berk.
Santavery root disease:	*Fusarium oxysporum* f. sp. *coffeae*
Coffee trunk canker:	*Ceratocystis fimbriata* Ell. & Halst
Berry blotch:	*Cercospora coffeicola* B. & Cke.
Diseases affecting seedlings:	
Collar rot:	*Rhizoctonia solani* Kuhn.
Brown eye spot:	*Cercospora coffeicola* B. & Cke.
Stem necrosis and leaf spot:	*Myrothecium roridum* Tode ex Fries
Minor diseases of coffee:	
Grey blight of leaves:	*Pestalotiopsis clavispora* (Atk.) Steyt.
Black leaf:	*Cylindrocladium illicicola* (Hawley) Boedj & Reitsma
Target leaf spot:	*Myrothecium advena* Sacc.
Tip blight:	*Myrothecium roridum* Tode ex. Fr.
Sclerotium disease:	*Sclerotium coffeicola* Stahel
Coffee blight:	*Phoma costarricensis* Echandi
Black root disease:	*Rosellinia bunodes* (B. & Br.) Sacc. and *R. arcuate* Petch.

These are discussed in more detail in the following sections.

2 Diseases affecting adult plants: coffee leaf rust

Leaf rust commonly known as 'coffee leaf rust' (CLR), 'orange rust' or 'La Roya' is a major disease affecting coffee. CLR is caused by the fungus *Hemileia vastatrix* Berk. & Br. (Large, 1940). The fungus infects the foliage of the genus *Coffea* though there are reports that the fungus can infect developing berries and young shoots (Anon, 1996; Madhu et al., 2020).

Coffee rust has been reported in over 50 coffee-growing countries. The disease was first noticed in Lake Victoria in Africa during the year 1861 and subsequently found in Sri Lanka and India in 1868–1869 and in Latin American countries from 1970 (Rangaswamy, 1972; Kushalappa, 1989a). Rust affects berry yield in many ways, by reducing photosynthetic area, defoliation as well as reducing plant vigour (Anon, 1957). Depending on the severity of infection, fewer flowers are formed, flowers and fruit fall prematurely and remaining berries do not fully develop. In severely affected areas, the pathogen can cause losses of 30–50% and even up to 90% (Muthappa, 1975; Vallega and Chiarappa, 1964; Avelino et al., 2015).

The causal organism of coffee leaf rust is *Hemileia vastatrix* B. & Br. Its taxonomy is:

Kingdom:	Fungi
Division:	Basidiomycota
Class:	Pucciniomycetes
Order:	Uredinales
Family:	Pucciniaceae
Genus:	*Hemileia*
Species:	*vastatrix*

The pathogen is host-specific and capable of producing cultivar-specific races or strains. There are over 50 races of *H. vastatrix* reported worldwide, of which 35 races exist in India (Anon, 1965, 2014; Kushalappa, 1989a; Mayne, 1932, 1939; Olivera and Rodrigues, 1961; Ramachandran et al., 1979; Rodrigues et al., 1975; Sreenivasan, 1989; Varzea and Marques, 2005). There have been various attempts to understand the relatively rapid emergence of new races, including via cryptosexual reproduction (Varzea and Marques, 2005; Carvalho et al., 2011).

H. vastatrix produces uredinial, telial and basidial stages in coffee. Uredospores are the infection propagule through which the rust pathogen spreads and completes its life cycle (Bawden et al., 1971; Bock, 1962; Gopalakrishnan, 1951). Uredinospores germinate under favourable conditions to form germ tubes and produce appressorium and vesicles. Infection hyphae enter into substomatal cavity to form uredinal sori and spores on lesions within 20–22 days of infection. A single lesion produces spores for several weeks (Thirumalachar and Narasimhan, 1947; Vishveshwara and Nagaraj, 1960; Ward, 1882). Small discoloured pale yellow spots on the lower surface of leaves appear first. Within 2 weeks, the spots enlarge, turn yellow and produce a yellow-orange powdery mass of uredospores (see Figs 1 and 2). The spots gradually increase in size, coalesce and cover the entire surface of the leaf.

Figure 1 Leaf rust: affected leaves with rust spots.

Figure 2 Leaf rust: Uredosorus with urediniospores.

Diseased leaves fall prematurely and induce die-back (Rayner, 1961; Wellman, 1961).

Factors affecting disease development include temperature (with an estimated optimum of 23°C), rainfall levels (affecting leaf wetness), wind (which helps to spread the disease) level of shade and planting density (Kushalappa, 1989a; Muthappa, 1975; Zambolim, 2016). There is a range of chemical, cultural and biological methods of control (Rutherford and Phiri, 2006). Copper-based fungicides, especially Bordeaux mixture, have been used successfully to control leaf rust in the field for many years (Mayne et al., 1933; Muthappa, 1979a; Ramanathan et al., 1951). An alkaline Bordeaux mixture has been found to have beneficial effects such as good foliage retention, slow weathering of copper from spray deposits and prolonged persistence (Muthappa, 1979b; Muthappa and Bakre, 1976; Zambolim, 2016; Anon, 2014).

Since the 1970s, a number of fungicides have been developed, including oxycarboxin, carboxin, pyracarbolid, triadimefon, propiconazole, hexaconozole and epoxyconazole (Anon, 1976; Daivasikamani and Govindarajan, 1989; Muthappa and Kannan, 1981; Muthappa et al., 1977). Triadimefon (Bayleton 25 WP) and oxycarboxin (Plantvax 20 EC) have been tested extensively in the field for their efficacy in rust control (Muthappa, 1980, 1981). Combined use of these fungicides and Bordeaux mixture has also been found to be effective (Souza, 2011).

Cultural control techniques include growing rust-tolerant arabica coffee cultivars, for example, in India, varieties such as Chandragiri (Sln.13), Sln.5B, Sln.6 and Sln.9 (Ramachandran et al., 1979; Sreenivasan, 1989). Although progress has been made in identifying resistance genes, achieving durable disease resistance is challenging, given pathogen evolution and variability (Prakash et al., 2004; Alvarado, 2005; Cabral, 2015). Other measures include growing arabica coffee under optimal shade conditions (using top and middle canopies) (Anon, 1976) as well as improving plant nutrition. Pruning coffee bushes after harvest helps to reduce initial inoculum on the host (Kushalappa, 1989b; Mayne, 1937).

Looking at biological methods of control, attempts to use the hyperparasite *Verticillium hemileiae* on coffee leaf rust have not generally been effective, in part because of its sensitivity to humidity levels and failure to establish fully at the initial stage of rust disease development (Eskes, 1989; Silveira and Rodrigues, 1972). There has been some success with using *Bacillus* spp. and *Pseudomonas* spp. as bioprotectants to boost plant resistance (Daivasikamani and Naika, 2009; Haddad et al., 2009).

3 Diseases affecting adult plants: black rot

Black rot (also known as Koleroga) is the second most important disease of coffee in India, attacking both robusta and particularly arabica coffee (Anon, 2014). The disease has also been reported in over 30 other countries (Anon, 1988). In India, the severity of the disease is linked to rainfall patterns, with a particular prevalence in southern coffee regions in the states of Karnataka, Kerala and Tamil Nadu, whilst drier coffee regions can be disease free (Thomas, 1924; Mathew, 1954). In severely diseased areas, losses can reach 10–20% in a year, with die-back then affecting yields in subsequent years (Duruz, 1953; Flores et al., 1955; Mathew, 1954; Sanchez de Leon, 1977).

The causal organism is *Koleroga noxia* Donk (synonyms=*Pellicularia koleroga* Cke.=*Corticium koleroga* (Cke.) Hohnel=*Ceratobasidium anceps* (Bres. & Syd.) Jackson). Its taxonomy is as follows:

Kingdom:	Fungi
Division:	Basidiomycota
Class:	Basidiomycotina
Order:	Polyporales
Family:	Thelephoraceae
Genus:	*Koleroga*
Species:	*noxia*

Out of 80 different wild hosts of this pathogen recorded from various countries, 37 different plant species have been identified from India alone as collateral hosts of *K. noxia*.

Studies of *K. noxia* isolates from different locations show divergence in morphological characteristics and virulence indicating different strains among isolates (Sudhakar et al., 1988, 1995). Pellicle and sclerotial stages are both observed in the disease cycle of *K. noxia*. In the pellicle stage, the fungus forms a continuous mycelial mat on the leaf surface with dense, whitish powdery spore-bearing patches. The pellicle form which is entirely superficial can be peeled off (Coleman et al., 1923). Basidiospores produced from this stage help secondary spread. In the sclerotial stage, the leaves become brownish and later black. The fluffy, pulverulent appearance of the film changes into dot-like hyphal clumps. The micro-sclerotia can remain dormant for 8–10 months (Sudhakar and Bhat, 1993).

Primary spread of the fungus is mostly by contact from leaf to leaf and bush to bush through vegetative mycelium. Affected leaves become detached and are carried by wind to other plants to cause further infection. Secondary spread through the basidiospores can occur either by wind or rain dispersal. Close examination reveals the presence of characteristic threads of mycelia running along twigs, petioles and on the lower surface of leaves (see Fig. 3). Defoliation and berry drop from the infected branches occur in the advanced stage of the disease (Coleman et al., 1923; Mathew, 1954). High levels of humidity promote disease spread (Narasimhan, 1933; Mathew, 1954).

Black rot can be effectively controlled through cultural methods that include good sanitation procedures, good drainage, circulation of air and sunlight by regulating shade, pruning and removal of infected leaves and berries and then prophylactic application of Bordeaux mixture (Coleman et al., 1923; Mathew, 1954; Mayne et al., 1933). Several fungicides have been evaluated, including carbendazim, carboxin and captafol (Anon, 1976; Sudhakar et al., 1992). Carbendazim was found to be particularly effective as well as combining

Figure 3 Black rot: affected leaves showing rotting of leaves and fungal mycelia.

fungicides to optimize efficacy and minimize the development of resistance (Anon, 2014; Giri et al., 2019).

4 Diseases affecting adult plants: anthracnose

Anthracnose is found worldwide, although it is more abundant in the tropics and subtropics than in temperate regions. It has been reported in most coffee-growing countries, particularly Brazil, Vietnam, India and African countries where it causes considerable losses, particularly in arabica coffee.

The pathogen is a generally weak bark-inhabiting fungus but which becomes virulent under certain conditions, affecting leaves, twigs and berries (Narasimhan and Mayne, 1933). The main causal organism is *Colletotrichum gloeosporioides* Penz though other species can also cause infection (Nguyen et al., 2010; Livramento Freitas-Lopes et al., 2013). Its taxonomy is:

Kingdom:	Fungi
Division:	Basidiomycota
Class:	Deutermycotina
Order:	Melanoconiales
Family:	Melanoconiaceae
Genus:	*Colletotrichum*
Species:	*gloeosporioides*

The fungus causes three different diseases of coffee (Muthappa, 1973a, 1976), such as twig die-back, stalk rot and brown blight, which are discussed in the following sections.

4.1 Twig die-back or summer die-back

Predisposing factors include inadequate overhead shade, prolonged drought and soil moisture stress (Ramaiah and Chokkanna, 1964). In India, for example, the disease occurs during dry weather from October to May (Ananth, 1961; Chokkanna, 1962). Under South-West monsoon conditions, the disease occurs at varying intensities across a wide cross-section of coffee areas, varying from zone to zone, estate to estate and even from block to block in the same estate (D'Souza, 1971; Ramaiah and Chokkanna, 1964; Muthappa, 1976). Symptoms include yellowing and necrosis of nodes and internodes towards the twig tip, resulting in die-back (see Fig. 4). Floral buds on infected branches fail to open. New leaves formed are small, crinkled, chlorotic, thick and leathery. Internodes are short and give a fan shape appearance (Mayne, 1931b).

There are a range of control measures (Mayne, 1931b; Muthappa, 1973a; Ramaiah and Chokkanna, 1964; Ramanathan et al., 1951; Sharma, 1967). These include pruning of affected plants typically in the Spring; maintaining adequate overhead shade to protect the bush from sun scalding and mulching of leaf litter around the plants to conserve soil moisture during dry weather; application of balanced nutrients to maintain plant vigour; spraying 0.5% Bordeaux mixture or carbendazim 0.03% (Bavistin 50 WP, 120 g/200 L of water) during February-March (Pre-blossom), April-May (Pre-monsoon) and September-October (Post-monsoon). There has also been research into novel techniques such as zincoxide delivered as nanoparticles (Mosquera-Sanchez et al., 2020).

Figure 4 Anthracnose: die-back affected coffee twig.

4.2 Stalk rot

In India, stalk rot of berries and leaves generally occurs on arabica and also robusta coffee plants during July–August under South-West monsoon conditions (Muthappa, 1976; Daivasikamani et al., 2017).

Necrosis of nodes and internodes occurs at the junction of brown and green wood towards the apex followed by berry drop and defoliation (see Fig. 5). Even if some berries remain, they show premature ripening towards the end of September or October. Conditions favouring infection include low temperature, high relative humidity (95–100%), surface wetness of plants due to rain or mist, wounds on the stalk of berries and leaves and excess soil moisture.

Disease management measures include preventing water logging and good drainage around the root zone; maintaining overhead shade during monsoon periods; removal of mulch to expose the soil around the plants; pre-monsoon spraying of Bordeaux mixture at 0.5% during May–June giving a good coverage of branches and stalks of berries; spraying carbendazim at 0.03% (120 g/200 L of water) in areas of stalk rot to prevent further spread of the disease (Ranjini and Daivasikamani, 2015).

4.3 Brown blight

This disease is generally noticed on leaves during hot weather. Injury caused by sun scalding or other types of wounds on leaves predispose them to infection by *C. gloeosporioides* (Muthappa, 1975, 1976; Muller, 2009). The disease is

Figure 5 Anthracnose: stalk rot of berries.

Figure 6 Anthracnose: brown blight of leaves.

characterized by round necrotic spots measuring up to 25 mm in diameter appearing on leaves. Two or more such spots may coalesce and the entire leaf or a portion of it may look blighted. Necrotic spots are brown in colour hence the name brown blight (see Fig. 6). Fructifications (acervuli) of the fungus are visible as black dots on the spots on upper or lower surface of leaf.

Favourable factors for disease development include coffee plants exposed to sunlight, poor overhead canopy of shade, injury caused by sun scalding, sucking pests or any other type of wounds on leaves also predispose coffee bushes to the disease. Methods of disease management include maintaining good overhead shade to avoid sun scalding of leaves and spraying 0.5% Bordeaux mixture to protect the leaves from the rust fungus also appears to give adequate protection against brown blight.

5 Root diseases: brown root disease and red root disease

Brown, red, black and santavery root disease are the different types of root diseases reported on coffee in India (Nirmala Kannan, 1986, 1996; Daivasikamani et al., 2009; Sudha et al., 2016). Of the four, red and brown root diseases not only affect both arabica and robusta coffee but also several shade trees commonly grown on coffee estates. Black root disease is rare and does not cause economic damage at present. Santavery root disease of coffee can cause severe damage to arabica coffee plants. Large-scale tree felling and growing coffee under the shade such of single species such as Silver oak has resulted in the increased incidence of brown and red root diseases in recent years. This section focuses on brown root disease and red root disease.

5.1 Brown root disease

This disease is also known as 'stump rot' and is mostly associated with the rotting stumps of shade trees in the vicinity of the affected coffee plants. The pathogen infects shade trees grown in coffee plantations such as *Grevilla robusta* and *Erythrina lithosperma*. The disease spreads by root contact between infected stumps and coffee plants. The causal organism is *Fomes noxius* Corner. Its taxonomy is as follows:

Kingdom:	Fungi
Division:	Basidiomycota
Class:	Basidiomycetes
Order:	Polyporales
Family:	Polyporaceae
Genus:	*Fomes*
Species:	*noxius*

Low soil pH and organic content favour the multiplication of the pathogen and its spread, as does debris from infected coffee plants if not completely removed.

Affected plants show gradual yellowing of leaves, wilting, defoliation followed by death of the bush. The lower stem becomes spongy and soft. The root system develops thick brown encrustations with fungus mycelium appearing as brown woolly hyphae or in brown crusts. The brown fungal encrustation gives the name brown root disease (see Fig. 7). Affected roots are very brittle and readily snap off. The interior of roots shows dark brown to black wavy lines consisting of tangled fungus hyphae (Mayne, 1931a; Nirmala Kannan, 1986).

5.2 Red root disease

This root disease is also associated with infected stumps of shade trees in the vicinity of coffee plants (Nirmala Kannan, 1996). Shade tree species vulnerable to infection include *Grevilla robusta*, *Syzygium jambolana*, *Pterocarpous marsupium*, *Dalbergia latifolia*, *Cedrellia tona*, *Ficus tsiela*, *Ficus jakela*, *Ficus glomerata*, *Acrocarpus fraxinifolius*, *Artocarpus integrifolia* and *Erythrina lithosperma*. The disease spreads by means of root contact of infected stumps

Figure 7 Brown root disease: brown encrustation of fungal mycelia on root.

with coffee plants. The causal organism is *Poria hypolateritia* Berk. Its taxonomy is as follows:

Kingdom:	Fungi
Division:	Basidiomycota
Class:	Basidomycotina
Order:	Polyporales
Family:	Polyporaceae
Genus:	*Poria*
Species:	*hypolateritia*

Diseased plants show gradual yellowing of foliage, wilting of branches, defoliation followed by death of the bush identical to brown root disease. The root system of dying or dead coffee plants shows red encrustation from fungal mycelia (see Fig. 8). The affected root appears deep red in colour when washed in water which is a good diagnostic symptom of this disease (Venkata Ram and Muthappa, 1975).

5.3 Disease management of brown and red root disease

These diseases can be contained by adopting a number of measures (Nirmala Kannan, 1996; Venkata Ram and Muthappa, 1975; Ranjini et al., 2017a). Cultural measures include the following:

Figure 8 Red root disease: red encrustation of fungal mycelia.

- Uproot and destroy affected shade trees and coffee plants to reduce the available inoculum for further spread.
- Remove fallen logs and felled shade tree stumps from the coffee plantation as soon as possible instead of allowing decomposition.
- Isolate the affected area and one row of apparently healthy adjoining plants with trenches 60 cm deep and 15 cm wide. Soil taken out from the trenches should be placed inside the infected area. The trenches should be cleaned every 3 months.
- Apply 2–3 kg of agricultural lime and turn the soil in the pits where affected coffee plants and shade trees were uprooted. Also apply agricultural lime to the affected areas inside the trench and take up cover digging during the dry period to expose the pathogen to sunlight.
- Planting in the affected areas should be done only after exposing the pits to sunlight for 6–8 months.
- Application of well-composted organic manure at 10–15 kg per plant can also help in suppression of pathogens.

These can be supplemented by chemical and biological measures:chemical measures include soil drenching of carbendazim solution (8 g/L of water)

@ 2–3 L/plant at the collar region of the plant twice a year (pre- and post-monsoon). Biological measures include soil application of 3–5 kg of the mixture of bioagent *Trichoderma* in compost (*Trichoderma* sp. 125 g and compost 3 kg) to the affected coffee plants in a drip circle (15–20 cm radius and 3–5 cm depth) during September–October and May–June (Anon, 1996).

6 Root diseases: black root disease and santavery root disease

The causal organism of black root disease in coffee is *Rosellinia bunodes* (B. & Br.) Sacc. and *Rosellinia arcuate* Petch. They are classified as follows:

- Class: Ascomycotina.
- Order: Sphaeriales.
- Family: Xylariaceae.

Though the disease has been reported on coffee, its spread and severity is limited, unlike brown and red root diseases (Muthappa, 1977). Symptoms are similar to brown root disease. A black fungal rhizomorph or black woolly mycelium is seen on infected plants. Transverse sections of roots show thread-like black lines or dots on the root surface. A fan-shaped black fungal mat with pellet-like fructifications can also be seen on the stem near ground level.

6.1 Santavery root disease

The organism responsible for this disease is a soil-inhabiting saprophytic fungus *Fusarium oxysporum* f. sp. *coffeae* (Nataraj, 1973). Its taxonomy is as follows:

Kingdom:	Fungi
Division:	Basidiomycota
Class:	Deuteromyctes
Order:	Moniliales
Family:	Tuberculariaceae
Genus:	*Fusarium*
Species:	*oxysporum*

Favourable factors for disease development include prolonged low or high soil temperature, poor soil physical condition, moisture stress from low rainfall,

Figure 9 Santavery root disease: internal root discoloration.

inadequate shade and wounds on the roots due to wind damage (Nataraj, 1973). Symptoms include sudden wilting, yellowing of leaves followed by defoliation and death of aerial plant parts. Roots in transverse section show brown to pinkish discoloration (see Fig. 9). Scraping the stem bark near ground level also shows internal discoloration (Govindarajan and Subramanian, 1968; Rehuman and Subramanian, 1967).

There are a number of measures to contain the disease (Nirmala Kannan, 1986, 1996). Cultural measures include the following:

- uprooting and destroying affected plants;
- maintaining the vigour of coffee bushes by applying organic manure and agricultural lime to maintain soil health;
- protecting coffee plants from exposure to high winds by developing proper wind belts;
- maintaining adequate overhead shade;
- replanting diseased areas with grafted plants (e.g. disease-resistant robusta root stock and arabica scions).

Chemical measures include drenching the soil with carbendazim (0.4%–8 g/L of water) at 2–3 L solution in a drip circle around affected plants in the initial stage of infection. Biological measures include treating affected plants with bio-control agent *Trichoderma viride or T. harzianum* along with farmyard manure

(FYM) or compost (*Trichoderma* 125 g and 3 kg FYM), incorporated in the soil in a drip circle when sufficient soil moisture exists.

7 Coffee trunk canker and berry blotch

Coffee trunk canker (CTC) is also known as stem canker, tree canker, stem rot and bark disease. It is caused by the fungus *Ceratocystis fimbriata* Ell. & Halst. Since the 1950s, this has been one of the most important diseases of arabica coffee in Colombia, Venezuela, Guatemala and Costa Rica where it continues to cause serious losses (Marin et al., 2003; Van Wyck et al., 2010). It was reported in India for the first time in Karnataka state during the late 1990s. The CTC pathogen infects only arabica coffee and causes considerable crop loss (Nirmala et al., 1998).

As noted, the causal organism is *Ceratocystis fimbriata* Ell. & Halst. Itstaxonomy is as follows:

Kingdom:	Fungi
Division:	Ascomycota
Class:	Ascomyctes
Order:	Microascales
Family:	Ophiostomataceae
Genus:	*Ceratocystis*
Species:	*fimbriata*

Bark wounds provide entry for the pathogen into the host. Wet weather, high humidity and excess soil moisture promote the disease. The CTC fungus produces abundant conidia and ascospores under wet weather conditions during the monsoon period. Spores are transmitted from diseased to healthy coffee plants. The earliest visible symptoms are a decline in foliage formation and a faded appearance of bushes with pale green leaves. A close examination reveals irregular lesions on the main stem. Brown to black discoloration of the internal tissues can be observed at a depth of 5–10 mm (Bhat et al., 2000; Sudhakar et al., 2009). At the advanced stage, leaves turn yellow and branches start wilting, leading to death of the plant (see Fig. 10).

In terms of disease management, it is essential to avoid bark injuries while carrying out cultural operations. Severely infected plants must be uprooted and destroyed. Plants at the initial stage of infection should be stumped (to 25 cm above ground) and the cut end protected with Bordeaux paste (copper sulphate 75 g and spray lime 150 g in 1L of water). Pruning knives and other implements should be disinfected with carbendazim solution.

Figure 10 Coffee trunk canker.

7.1 Berry blotch

Berry blotch is also known as berry spot and cherry blotch and affects only the berries on coffee plants. The causal organism is *Cercospora coffeicola* B. & Cke. Its taxonomy is as follows:

Kingdom:	Fungi
Division:	Ascomycota
Class:	Deuteromycotina
Order:	Moniliales
Family:	Dematiaceae
Genus:	*Cercospora*
Species:	*coffeicola*

Figure 11 Berry blotch.

The diseased tissue of the fruit skin becomes dry making fruits difficult to pulp and thus lowering quality considerably (Siddique, 1970; Swaminathan et al., 1960). The disease is a significant problem in India and Latin America (Wellman, 1961; Ariel, 2018).

The disease affects developing berries exposed to direct sunlight due to lack of good overhead shade (Mouen Bedimo et al., 2008). High humidity and low temperatures favour the production of conidia by the pathogen. Symptoms include dark brown, irregular, slightly sunken necrotic spots on the exposed surface of green berries (Nelson, 2008). Necrotic spots enlarge in size and cover a major portion of berry surface. The skin of the affected berries shows a purple halo around the necrotic spots (see Fig. 11). Affected tissue turns brown to black, shrivels, dries up and sticks fast to the fruit parchment. Disease management options include maintaining good overhead shade to avoid scorching of berries and, if infection is observed, spraying the developing berries with 1.0% Bordeaux mixture.

8 Diseases affecting seedlings

Coffee is mainly propagated through seeds. Seedlings are raised in two stages under pandal shade (a identified place provided by high density polyethylene shade mats drawn over poles fixed at specified distance to form a shed with filtered overhead shade for raising coffee seedlings). Seeds are sown on the raised germination beds containing sieved jungle soil, compost and sand (6:2:1 ratio). The germinated seedlings at 'topee' or 'button' stage are transplanted to the poly bags containing same mixture used as in the case of germination bed. The success of establishing new plantations in a perennial crop like coffee mainly depends on the use of healthy seedlings with good vigour for field planting. For this purpose, it is most essential to protect the seedlings in the primary nursery bed and after transplantation to the poly bags from attacks of pests and diseases. Coffee seedlings at the nursery stage are susceptible to three important fungal diseases *viz.*,collar rot, brown eye spot and stem necrosis and leaf spot (Anon, 2014; Ranjini et al., 2017b).

8.1 Collar rot (damping-off disease)

'Collar rot' or 'damping-off' disease occurs on 1–4-month-old tender coffee seedlings in the nursery. Both *C. arabica* and *C. canephora* are susceptible to the pathogen. Under conditions favouring the pathogen, seedling loss may go up to 10–20% (Venkata Subbaiah and Muthappa, 1981; Wellman, 1953).

The causal organism is *Rhizoctonia solani* Kuhn. Its taxonomy is as follows:

Kingdom:	Fungi
Division:	Basidiomycota
Class:	Deutermycotina
Order:	Mycelia sterelia
Family:	Ceratobasidiaceae
Genus:	*Rhizoctonia*
Species:	*solani*

The soil-inhabiting fungus has a wide host range including coffee (Vekataramaiah, 1968). The pathogen is capable of surviving in soil for many months in the form of sclerotia. Under favourable conditions, sclerotia germinate into mycelia and infect coffee seedlings.

Disease development is promoted by excessive soil moisture around nursery beds due to factors such as poor drainage, over watering and overcrowding of seedlings as well as high rainfall levels and hot, humid weather conditions (Anon, 1957).

There are two phases of the disease:

1. **Pre-emergence stage:** The pathogen invades the embryo and endosperm of coffee seeds before germination in the nursery seedbeds; consequently, seeds start rotting, disintegrate and fail to germinate.
2. **Post-emergence stage:** Seedlings show brown to black discoloration at the collar region of the stem leading to rotting of the tissues (see Fig. 12). The tip of the seedling wilts, collapses and dies.

Figure 12 Collar rot of coffee seedlings.

There are a number of ways of managing the disease (Anon, 2014; Govindarajan, 1971; Venkata Subbaiah and Muthappa, 1981; Vekataramaiah, 1968; Sudha et al., 2019). Cultural measures include the following:

- changing the site of seedbeds in the nursery every 2–3 years;
- using sufficiently dry soil, compost and sand in the recommended proportion (6:2:1 ratio) after sieving for preparation of seedbeds or filling of poly bags for transplanting of seedlings;
- avoiding overcrowding of seeds while sowing and excessive watering;
- controlling light using a 50% polythene shade net.

Chemical control measures include using fungicide-treated seeds (carbendazim 1g/kg) and spraying carbendazim at 0.05% (1g/L) or Dithane M-45 at 0.4% (5 g/L) to prevent spread of the disease.

8.2 Brown eye spot

Brown eye spot or leaf spot disease has been reported in all coffee-growing countries including India (Siddique, 1970). Affected leaves drop and seedlings become unfit for field planting. The pathogen responsible is *Cercospora coffeicola* Berk. & Cke (Sridhar and Subramanian, 1969). Its taxonomy is as follows:

Kingdom:	Fungi
Division:	Ascomycota
Class:	Deuteromycetes
Order:	Moniliales
Family:	Deniatiaceae
Genus:	*Cercospora*
Species:	*coffeicola*

Commonly occurring weeds in coffee plantations such as *Chinopodium ambrosoides, Bidens pilosa* and *Ricinus communis* act as collateral hosts for the pathogen (Sridhar and Subramanian, 1966). The pathogen can survive in soil as long as 35 weeks.

Conditions favouring disease development include high humidity, exposure to sunlight due to inadequate overhead shade either in the nursery or in the field. Under favourable conditions, the pathogen produces conidia and dispersal takes place through wind and rain splash. Conidia germinate and penetrate through sun-injured tissues/stomata, leading to disease development.

Affected leaves show circular necrotic spots with a dark brown margin and light brown or pale centre. Around the necrotic spot, the colour of the leaves changes from green to yellow (see Fig. 13). Necrotic spots increase in size and the central portion turns light grey. The central portion of the spot then disintegrates and collapses, leaving a hole in the middle of the spot. Affected leaves turn yellow and result in pre-mature defoliation (Subramanian and Sridhar, 1966).

Figure 13 Brown eye spot disease on coffee seedlings.

Cultural management techniques include using properly dried soil, well-composted cattle manure and sand in recommended proportions (6:2:1 ratio) in either seedbeds or for filling poly bags for transplanting, as well as ensuring suitable shade using a 50% polythene shade net or other techniques such as coconut leaf thatch. In terms of chemical control, the use of Bordeaux mixture has been discontinued because it caused stem wasting (Sheriff et al., 1960). Alternative fungicides have been developed and found effective, including Captan, Dithane M-45, Ziride, Ferbam, carbendazim and captafol (Govindarajan, 1971; Hanumantha et al., 1994; Muthappa and Nataraj, 1979; Nataraj, 1976; Nataraj and Muthappa, 1980). Seedlings can, for example, be protected in the nursery by spraying with either Dithane M-45 @ 0.4% (5 g/L) or carbendazim @ 0.05% (1 g/L) once in 30 days till the seedlings are transplanted to the field.

8.3 Stem necrosis and leaf spot

Stem necrosis and leaf spot of coffee seedlings caused by *Myrothecium roridum* has recently become a more significant disease of coffee seedlings, causing severe damage (Sudhakar et al., 2011; Ranjini and Naika, 2018). The

Figure 14 Stem necrosis on coffee seedlings.

causal organism is *Myrothecium roridum* Tode ex Fries. Its taxonomy is as follows:

Kingdom:	Fungi
Division:	Ascomycota
Class:	Ascomycetes
Order:	Hypocreales
Family:	Insertaesedis
Genus:	*Myrothecium*
Species:	*roridum*

The pathogen is a soil-inhabiting fungus that can survive on plant debris for 8–10 months. Under favourable environmental conditions, it multiplies and becomes actively parasitic on coffee and other plants. The pathogen forms spore-bearing black dots which form numerous conidia on infected regions. Favourable factors for disease development include hot, humid weather conditions, particularly if seedlings are wet for long periods.

Diagnostic symptoms are as follows:

- **Stem:** The infected plant shows water-soaked brown to grey discoloration on the stem just above the soil (see Fig. 14). The infected region later shows cushion-shaped black fruiting bodies surrounded by white mycelia. Affected seedlings gradually start wilting and then die.

Figure 15 Leaf spot of coffee seedling.

- **Leaves:** Infected leaves initially show water-soaked circular necrotic spots which gradually spread (see Fig. 15). These spots then become brown with concentric rings. In severe infections, two to three such spots coalesce to form irregular necrotic patches and may even cover the entire lamina of the leaf. Black fruiting bodies occur on the surface of the affected leaf. Occasionally, the centre of the lesion cracks irregularly, ruptures and some of the affected tissues drop down during rainfall.

Cultural management options include changing nursery sites every 3 years as well as regularly removing and destroying infected seedlings. Seedlings can be protected by spraying propiconazole at 0.02% (0.8 mL/L of water) or Dithane M-45 at 0.4% (5 g/L) at monthly intervals from May to August.

9 Minor diseases of coffee

Diseases described below are of minor importance at present since they occur on a small scale, that is, below an economic threshold level. These diseases need not require dedicated control measures at present. The main diseases are as follows:

Grey blight of leaves:	*Pestalotiopsis clavispora* (Atk.) Steyt.
Black leaf:	*Cylindrocladium illicicola* (Hawley) Boedj & Reitsma
Target leaf spot:	*Myrothecium advena* Sacc.
Tip blight:	*Myrothecium roridum* Tode ex. Fr.
Sclerotium disease:	*Sclerotium coffeicola* Stahel
Coffee blight:	*Phoma costarricensis* Echandi
Black root disease:	*Rosellinia bunodes* (B. & Br.) Sacc. and *R. arcuate* Petch.

These are discussed in the following sections.

9.1 Grey blight of leaves

The disease has been recorded on mature leaves of robusta coffee (Muthappa et al., 1976). The causal organism is *Pestalotiopsis clavispora* (Atk.) Steyt. A brown or grey necrotic spot up to 1–2 cm in diameter is observed on mature leaves. These spots enlarge and coalesce with leaves appearing brown or grey. Black fructification of the fungus on both sides of leaves in concentric rings is irregularly distributed on necrotic spots.

9.2 Black leaf

In India, this disease occurs during the South-West monsoon from July to September (Muthappa and Krishna, 1978). The causal organism is *Cylindrocladium illicicola* (Hawley) Boedj & Reitsma. Small circular water-soaked necrotic black spots appear on the leaf blade mostly on the margins. Spots enlarge in size, and the leaf turns black and rots, hence the name black leaf.

9.3 Target leaf spot

In India, this disease occurs during the South-West monsoon from June to September (Anon, 2014). The causal organism is *Myrothecium advena* Sacc. Brown circular or irregular necrotic spots are formed on leaves. Concentric circles visible on the upper surface of leaves give a characteristic 'target board' appearance, hence the name target leaf spot. On the lower surface, black fructifications (sporodochia) with white margin are seen. Severe infection may lead to defoliation. Mostly leaves near or touching the soil are affected.

9.4 Tip blight

The causal organism is *Myrothecium roridum* Tode ex. Fr. Five-to six-month-old plants show infection in internodes and on leaves. The symptoms observed are water-soaked brown to grey discolouration of the internodes from the tip followed by defoliation and death similar to die-back. On necrotic tissue, cushion-shaped fructifications with shiny black centre and white margins can be seen (Nirmala Kannan and Muthappa, 1982).

9.5 Sclerotium disease

The causal organism is *Sclerotium coffeicola* Stahel. Small circular necrotic spots appear on leaves. These spots gradually increase in size up to 25 mm in diameter and show concentric rings of light and dark brown shades. Rhizomorphs and mustard-like sclerotia are seen on necrotic spots.

9.6 Coffee blight

The causal organism is *Phoma costarricensis* Echandi. Severe infection leads to 30–35% defoliation. Wet weather, high humidity, low temperature and wounds caused by insects favour development of disease, particularly at higher elevations (about 1500 m) (Rajendran et al., 1983; Nirmala Kannan and Hanumantha, 1986). The affected leaves show dark brown to black necrotic spots resembling blight. Necrotic spots are nearly round in shape with a diameter up to 25 mm. Two or more such spots might coalesce and part of the leaf becomes necrotic.

9.7 Black root disease

The causal organism is *Rosellinia bunodes* (B. & Br.) Sacc. and *R. arcuate* Petch. Diseased plants show gradual yellowing of foliage, wilting of branches, defoliation followed by death of the coffee bush. Black woolly fungal mycelia or rhizomorphs can be seen on tap roots of affected plants (Muthappa, 1977). Fan-shaped black fungal mats with pellet-like fructifications can be seen on stems near ground level. The fructifications are the main way to diagnose this disease.

Figure 16 Flowering shoot parasite: *Cuscuta* on coffee branch.

Figure 17 Flowering root parasite: *Balanophora* on roots of coffee.

10 Flowering parasites of coffee

Some primitive angiospermic flowering plants occasionally grow as parasites on coffee plants. Important flowering parasites of coffee are shoot and root parasites.

10.1 Shoot parasite

The causal organism is *Cuscuta reflexa* Roxb. Commonly known as 'dodder' or 'strangle vine', the plant is observed on coffee plants both in nurseries and in the field (Muthappa, 1973b). A dense strand of leafless yellow vines twining round plants can be observed (see Fig. 16). During dry weather from November to January, they produce white flowers and form seeds. Primary infection comes from seeds and secondary spread was from neighbouring infected plants. This parasite gradually weakens coffee plants by drawing nourishment through haustoria, resulting in reduced host vigour. Complete removal of the parasite from infected coffee plants and burning assures good control. While removing the parasite, care should be taken not to leave the piece of vine or its haustoria on the plant as they will continue to grow again. Fence hedges and neighbouring plants should be kept free of this parasite to prevent secondary spread.

10.2 Root parasite

The causal organism is *Balanophora indica* (Arn.) Griff. The root parasite belongs to the family Balanophoraceae and occurs on roots of coffee growing at elevations above 1000 m. Affected plants gradually decline in vigour (Sreenivasan, 1939). The parasite occurs in the form of root stock composed of tuberous outgrowths on the lateral roots of coffee plants. In the early stages of development, the tubers are small round bodies hidden in the soil. Towards the end of monsoon, the tubers become bigger in size and appear above ground as a pale brown mass of tubers. Flower heads begin to appear by about November. When the tubers are dug out, they can be seen attached to the roots of coffee (see Fig. 17). The affected coffee bushes do not show clear-cut symptoms of attack externally except for a general unhealthy appearance and low productivity. Control involves digging the soil to collect the tubers. All tubers should be collected before flowering and destroyed.

11 Summary

Effective and economic disease control measures in coffee plantation depend on good management, proper diagnosis of the disease in the initial stage followed by selection and effective implementation of the appropriate control measures. Use of any single component in an IDM programme will not be as effective as the combination of cultural, chemical and, where possible, biological methods of control. Good diagnosis is likely to become more important as climate change alters the pattern and severity of individual diseases.

Major coffee diseases like leaf rust, black rot, anthracnose and root diseases can be effectively controlled by cultural measures that include removal and destruction of affected plant parts, shade regulation, pruning, good field and bush sanitation, maintaining coffee bush vigour through good nutrition and prophylactic application of Bordeaux mixture. Judicious, needs-based application of the triazole group of fungicides such as hexaconazole, propiconazole or expoxiconazole, and integrating them with prophylactic Bordeaux mixture, affords better control of major diseases like coffee leaf rust. Soil-borne red root disease and brown root disease become more significant. They can be effectively managed by isolating affected plants, soil sanitation, destruction of the affected coffee plants and shade tree stumps and soil application of biocontrol agents like *Trichoderma*.

12 Where to look for further information

12.1 Books

- Wellman, F. L. (1961). *Coffee: Botany, Cultivation and Utilization*. New York, Leonard Hill (Books) Limited. Inter Science Publishers Inc.
- Anon. (1998). *A Compendium on Pests and Diseases of Coffee and Their Management in India*. Coffee Board Research Department, Central Coffee Research Institute, Chikmagalur, India.
- Anon. (2014). *Coffee Guide - A Manual of Coffee Cultivation*. Coffee Board of India.

12.2 Journals/Magazines

- *Journal of Coffee Research*.
- *Journal of Plantation Crops*.
- *Tea, Coffee and Cocoa*.
- *Indian Coffee*.
- *Kissan World*.

12.3 Conferences

- Association Scientific International Colloquium on Coffee (ASIC).

12.4 International organizations for collaborative research on coffee

- Centro de Investigacao de Ferrugens do Cafeeiro (CIFC), Oeiras, Portugal.
- Centre for Agriculture and Bioscience International (CABI), UK.
- Agricultural Research Centre for International Development (CIRAD), France.
- Institut de Recherche pour le Development (IRD), France.
- Natural Resource Institute (NRI), UK.
- World Coffee Research (WCR), Texas, USA.
- International Coffee Organization (ICO), UK.

13 References

Alvarado, G. (2005). Evolution of *Hemileia vastatrix* virulence in Colombia. In: Zambolim, L., Zambolim, E. M., Varzea, V. M. P. (Eds), *Durable Resistance to Coffee Leaf Rust*, Brazil, UFV-Viçosa, 99-115.

Ananth: K. C. (1961). A note on the incidence of die-back and investigations carried out by Coffee Board Res. Dept. *Indian Coffee* 25: 224-229.

Anon. (1957). Tenth annual detailed technical report. *Coffee Board Res. Dept., India*, 1956-57: 96-99.

Anon. (1965). *Progress Report from 1960 to 1965. Centro de Investigacao Das*, Ferrugens Do Caffeiro (CIFC) Oeiras, Portugal.

Anon. (1976). Twenty ninth annual detailed technical report. *Coffee Board Res. Dept., India*, 1975–76:94.

Anon. (1988). *Distribution of Plant Diseases (Koleroga noxia) on Coffee, Citrus etc.* (4th edn.), Chikmagalur, Karnataka, India: Coffee Board Research Department, Central Coffee Research Institute.

Anon. (1996). Forty ninth annual detailed technical report. *Coffee Board Res. Dept., India*, 1995–96:141.

Anon. (1998). *A Compendium on Pests and Diseases of Coffee and Their Management in India*, Chikmagalur, Karnataka, India: Coffee Board Research Department, Central Coffee Research Institute, 67 P.

Anon. (2014). *Coffee Guide - A Manual of Coffee Cultivation*, Chikmagalur, Karnataka, India: Coffee Board of India.

Anon. (2019). *Database on Coffee (July, 2019)*, Bengaluru: Coffee Board of India, p. 131. Available at: http://www.indiacoffee.org.

Ariel, C. (2018). Impact of berry blotch disease on coffee quality and value in Colombia. APS-ICPP Conference, Boston, MA.

Avelino, J., Cristancho, M., Georgiou, S., Imbach, P., Aguilar, L., Bornemann, G., Läderach, P., Anzueto, F., Hruska, A. J. and Morales, C. (2015). The coffee rust crises in Colombia and Central America (2008–2013): impacts, plausible causes and proposed solutions. *Food Sec.* 7(2): 303–321.

Bawden, J., Gregory, P. H. and Johnson, C. G. (1971). Possible wind transport of coffee leaf rust across the Atlantic Ocean. *Nature* 229(5285): 500–501.

Bhat, S. S., Hanumantha, B. T., Daivasikamani, S. and Nandagopal, N. (2000). A new destructive disease of arabica coffee in India incited by *Ceratocystis fimbriata*. Plantation Crops Research and Development in the New Millennium. In: *PLACROSYM- XIV*, Hyderabad, India,pp. 544–547.

Bock, K. P. (1962). Dispersal of uredospores of *Hemileia vastatrix*. *Trans. Br. Mycol. Soc.* 45: 63–74.

Cabral, P. (2015). Genetic diversity and structure of *Hemileia vastatrix* populations on *Coffea* spp. *Plant Path.* 12411: 1–6.

Carvalho, C. R., Fernandes, R. C., Carvalho, G. M., Barreto, R. W. and Evans, H. C. (2011). Cryptosexuality and the genetic diversity paradox in coffee rust, *Hemileia vastatrix*. *PLoS ONE* 6(11): e26387.

Chokkanna, N. G. (1962). A note on the incidence of die-back and investigations carried out by Coffee Board Res. Dept. *Indian Coffee* 26: 294–301.

Coleman, L. C., Venkata Rao, M. K. and Narasimhan, M. J. (1923). Black rot or Koleroga of coffee in Mysore. *Dep. Agric. Mysore State* 5: 1–12.

D'Souza, G. I. (1971). Studies on die-back of arabica coffee in South India. Role of pest and parasites. *Turrialba* 21: 146–148.

Daivasikamani, S. and Govindarajan, T. S. (1989). Field efficacy of a new systemic fungicide PP 523 5% SC (hexaconazole) for the control of coffee leaf rust. *J. Coffee Res.* 19: 65–70.

Daivasikamani, S. and Naika, R. (2009). Biological control of coffee leaf rust pathogen, *Hemileia vastatrix* Berkeley and Broome using *Bacillus subtilis* and *Pseudomonas fluorescens*. *J.Biopesticides* 2(1): 94–98.

Daivasikamani, S., Ranjini, A. P. and Raghuramulu, Y. (2016). Diseases of coffee and their management. In: Trivedi, P. C. (Ed.) *Plant Pathogens and Their Management*: India, Avaishkar Publishers, pp. 19-50.

Daivasikamani, S., Ranjini, A. P., Madhu, S.,Giri., Santhoshreddy, M., Sudha, M., Anand, C. G. and Raghuramulu, Y. (2017). Coffee diseases during monsoon season and their management. *Indian Coffee* LXXXI(5 and 6): 20-22.

Daivasikamani, S., Sudhakar, S. B., Hanumantha, B. T. and Sudha, M. (2009). Root diseases in coffee and their management. *Indian Coffee* 73(8): 13-15.

Duruz, W. P. (1953). Spray controls coffee blight in Nicargua. *Foreign Agric.* 17: 204-205.

Eskes, A. B. (1989). Natural enemies and biological control. In: Kushalappa, A. C. and Eskes, A. B. (Eds), *Coffee Rust Epidemiology, Resistance and Management*, FL: CRC Press. Inc., pp. 161-169.

Flores, M. A., Le-Beaes, F. J. and Straube, E. (1955). El. Koleroga en el Cafe. Guatemala Instituto Agropeculario Nacional, *Bol. Agric.* 4: 1-14.

Gaitan, A. L., Cristancho, M. A., Castro Caicedo, B. L., Rivillas, C. A. and Gomez, G. C. (2016). *Compendium of Coffee Diseases and Pests*, St Paul, MN: American Phytopathological Society, 79 P.

Giri, M. S., Ranjini, A. P., Santoshreddy, M., Sudha, M., Daivasikamani, S. and Raghuramulu, Y. (2019). Effectiveness of fungicides against *Koleroga noxia* Donk, the black rot pathogen of coffee. *J. Mycopath. Res.* 52(2): 95-99.

Gopalakrishnan, K. S. (1951). Notes on the morphology of the genus *Hemilea*. *Mycologia* 43(3): 271-283.

Govindarajan, T. S. (1971). Uses of some newer fungicides for the control of coffee diseases. *Indian Coffee* 35: 308-311.

Govindarajan, T. S. and Subramanian, S. (1968). Fusarium wilt of coffee in India. *Indian Coffee* 32: 270-271.

Haddad, F., Maffia, L. A., Mizubuti, E. S. G. and Teixeira, H. (2009). Biological control of coffee leaf rust by antagonistic bacteria under field conditions. *Biol. Control* 49(2): 114-119.

Hanumantha, B. T., Kamanna, B. C., Kannan, N. and Govindarajan, T. S. (1994). Effect of chlorothalonil and captafol on brown-eye-spot. *J. Coffee Res.* 24: 15-21.

Kannan, N. (1986). Root diseases of coffee. *Indian Coffee* 50: 21-24.

Kannan, N. (1996). Technical report on diseases affecting coffee in India - a review (Part III). *Indian Coffee* 60: 21-24,32.

Kannan, N. and Hanumantha, B. T. (1986). Effect of fungicides on the control of coffee blight. *ProcPlacrosym* 7: 245-247.

Kannan, N. and Muthappa, B. N. (1982). Tip blight disease of young coffee plants in the nursery. *J. Coffee Res.* 12: 38-41.

Kushalappa, A. C. (1989a). Introduction. In: Kushalappa, A. C. and Eskes, A. B. (Eds). *Coffee Rust Epidemiology, Resistance and Management*, FL: CRC Press. Inc., pp. 1-11.

Kushalappa, A. C. (1989b). Rust management and epidemiological approach and chemical control. In:Kushalappa, A. C. and Eskes, A. B. (Eds), *Coffee Rust Epidemiology, Resistance and Management*, FL: CRC Press. Inc., pp. 81-133.

Large, E. C. (1940). *The Advance of Fungi*, London: Johathen, Cape.

Livramento Freitas-Lopes, R., Maciel-Zambolim, E., Zambolim, L., Lelis, D. T., Caixeta, E. T., Lopes, U. P. and Pereira, O. L. (2013). *Colletotrichum boninense* causing anthracnose on coffee trees in Brazil. *Plant Dis.* 97(9): 1255-1265.

Madhu, S. G., Ranjini, A. P., Santoshreddy, M., Sudha, M. and Daivasikamani, S. (2020). First report of rust on Robusta coffee berries (*Coffea canephora* Pierre Ex A. Froehner) caused by *Hemileia vastatrix* Berk. & Br. From India. *J. Plant Disease Science* 15(1): 104–105.

Marin, M., Castro, B., Gaitan, A., Preisig, O., Wingfield, B. D. and Wingfield, M. J. (2003). Relationship of *Ceratocystis fimbriata* isolates from Colombian coffee-growing regions based on moelcular data and pathogenicity. *J. Phytopathol.* 151(7–8): 395–405.

Mathew, K. T. (1954). Studies on black rot of coffee. I. The disease in South India and some general considerations. *Indian Acad., Sci. Proc. Sect. B.* 39: 131–170.

Mayne, W. W. (1931a). Annual report of the coffee scientific officer 1930-1931: root disease - progress report. *The Mysore Coffee Expt. Stn. Bull.* 5: 6–9.

Mayne, W. W. (1931b). Die-back of coffee. *Planter's Chron.* 26: 492–495.

Mayne, W. W. (1932). Physiologic specialization of *Hemileia vastatrix* B. & Br. *Nature* 129: 150–152.

Mayne, W. W. (1937). Factors affecting spray success in control of coffee leaf disease. *Mysore Coffee Expt. Stn. Bull.* 15: 1–46.

Mayne, W. W. (1939). Annual report of the coffee scientific Officer 1938–39. *Mysore Coffee Expt. Stn. Bull.* 19: 1–16.

Mayne, W. W., Narasimhan, M. J. and Srinivasan, K. H. (1933). Spraying of coffee in South India. *Mysore Coffee Expt. Stn. Bull.* 9: 1–63.

Mosquera-Sanchez, L. P., Arciniegas-Grijalba, P. A., Patino-Portelab, M. C., Guerra-Sierra, B. E., Munoz-Florez, J. E. and Rodriguez,-Paez, J. E. (2020). Antifungal effect of zinc-oxide nanoparticles on *Colletotrichum* spp., causal agent of anthracnose in coffee crops. *Biocatal. Agric. Biotech.* 25: 1–9.

Mouen Bedimo, J. A., Njiayouom, I., Bieysse, D., Ndoumbè Nkeng, M., Cilas, C. and Nottéghem, J. L. (2008). Effect of shade on arabica coffee berry disease development: toward an agroforestry system to reduce disease impact. *Phytopathology* 98(12): 1320–1325.

Muller, R. (2009). Coffee diseases. In: Wintgens, J. (Ed.), *Coffee: Growing, Processing, Sustainable Production*, Weinheim, Germany: Wiley-VCH Press.

Muthappa, B. N. (1973a). *Colletotrichum coffeanum* - A fungus in diversity. *Planters' Chron.* 68: 459–461.

Muthappa, B. N. (1973b). Histopathology of a new parasite on coffee. *J. Coffee Res.* 3: 34–36.

Muthappa, B. N. (1975). Progress and status of coffee disease research in India. *Pestic. Annu.* 115–121.

Muthappa, B. N. (1976). A note on the fungus *Colletotrichum* on coffee in South India. *Indian Coffee* 40: 68–69.

Muthappa, B. N. (1977). *Rosellinia bunodes* on *Coffea* spp. *J. Coffee Res.* 7: 109–110.

Muthappa, B. N. (1979a). Bordeaux mixture, past, present status in coffee plantations. *Indian Coffee* 43: 7–9.

Muthappa, B. N. (1979b). Effect of neutral and alkaline bordeaux mixture on yield of S.795 arabica. *Indian Coffee* 43: 265–267.

Muthappa, B. N. (1980). Plantvax for coffee rust control in India. *Indian Coffee* 44: 145–152.

Muthappa, B. N. (1981). Field efficacy of bayleton 25 EC for control of coffee leaf rust. *J. Coffee Res.* 11: 4–6.

Muthappa, B. N. and Bakre, S. G. (1976). Efficacy of bordeaux mixture of varying pH on coffee leaf rust control. *Indian Coffee* 40: 301–304.

Muthappa, B. N. and Kannan, N. (1981). Efficacy of Vitavax 75 WP for the control of leaf rust. *J. Coffee Res.* 11(2): 47–48.

Muthappa, B. N. and Krishna, V. R. (1978). Black leaf - A new disease of coffee caused by *Cylindrocladium illicicola*. *J. Coffee Res.* 8: 26–28.

Muthappa, B. N. and Nataraj, T. (1979). Control of Brown-eye-spot disease of coffee by bavistin. *Proc Placrosym* 2: 135–139.

Muthappa, B. N., Govindarajan, T. S., Nirmala Kumari, K., Krishna, V. R. and Nataraj, T. (1977). Chemotheropeutic effect of oxycarboxin 20 EC on coffee leaf rust. *Pesticides* 11: 46–49.

Muthappa, B. N., Krishna, V. R. and Nirmala Kumari, K. (1976). Grey blight of coffee. *FAO Plant Prot. Bull.* 24: 54–55.

Narasimhan, M. J. (1933). Black rot of coffee in Mysore. *Phytopathology* 23: 875–886.

Narasimhan, M. J. and Mayne, W. W. (1933). Report on the disease situation in coffee areas in 1933. *Planters' Chron.* 28: 492–495.

Nataraj, T. (1973). Preliminary studies on pathogenicity of *Fusarium oxysporum* on arabica coffee plants as a possible cause of coffee root rot. *J. Coffee Res.* 3: 58–60.

Nataraj, T. (1976). Effect of systemic fungicides on brown-eye-spot disease. *J. Coffee Res.* 6: 81–84.

Nataraj, T. and Muthappa, B. N. (1980). Control of brown-eye-spot: efficacy of Bayleton 25 WP. *Indian Coffee* 44: 203–205.

Nelson, S. (2008). *Cercospora Leaf Spot and Berry Blotch of Coffee*. Hawaii: College of Tropical Agriculture and Human Resources - University of Hawaii at Manoa.

Nguyen, P. T. H., Vinnere Pettersson, O., Olsson, P. and Liljeroth, E. (2010). Identification of *Colletotrichum* species associated with anthracnose disease of coffee in Vietnam. *Eur. J. Plant Pathol.* 127(1): 73–87.

Nirmala, K., Bhat, S. S., Hanumantha, B. T., Daivasikamani, S. and Naidu, R. (1998). A new coffee stem disease in India. International Symposium on Ecology of Fungi, Goa, India (vol. 102).

Olivera, B. and Rodrigues, C. J. Jr. (1961). O problem das ferrugens do caffeiro. *Rev. Cafe. Port.* 7: 5–50.

Prakash, N. S., Marques, D. V., Varzea, V. M., Silva, M. C., Combes, M. C. and Lashermes, P. (2004). Introgression molecular analysis of a leaf resistance gene. *Theor. Appl. Genet.* 109(6): 1311–1317.

Rajendran, C., Ahmed, A. and Rao, M. (1983). Coffee blight - a new disease of coffee in India. *J. Coffee Res.* 13: 35–39.

Ramachandran, M., Sreenivasan, M. S. and Vishweshwara, S. (1979). Preliminary survey of physiological races of coffee leaf rust, *Hemileia vastatrix* in South India. *J. Coffee. Res.* 9: 51–60.

Ramaiah, P. K. and Chokkanna, N. G. (1964). Observations on die-back of coffee. I. Percentage of incidence of die-back in different liaison zones. *Indian Coffee* 28: 195–204.

Ramanathan, V., Bal, D. V. and Thomas, K. M. (1951). A review of the past experimental work on coffee in South India with suggestions for the future. *Indian Coffee Board Res. Dept.*:1–56.

Rangaswamy, G. (1972). *Diseases of Crop Plants in India*, New Delhi: Prentice Hall of India Pvt. Ltd.

Ranjini, A. P. and Daivasikamani, S. (2015). Stalk rot disease in coffee and its management. *Indian Coffee* LXXIX(7): 14-15.

Ranjini, A. P. and Naika, R. (2018). Leaf spot and stem necrosis disease of coffee seedlings caused by *Myrothecium roridum* Tode ex Fr. in India. *J. Mycopathol. Res.* 56(1): 5-10.

Ranjini, A. P., Santhosh, N. R., Daivasikamani, S. and Raghuramulu, Y. (2017a). Efficacy of fungicides against *Poria hypolateritia* the red root pathogen of coffee. *J. Mycopathol. Res.* 55(1): 61-65.

Ranjini, A. P., Uma, M. S., Daivasikamani, S. and Raghuramulu, Y. (2017b). Coffee nursery: pest and disease management. *Indian Coffee* LXXXI(11 and 12): 22-24.

Rayner, R. W. (1961). Germination and penetration studies on coffee leaf rust (*Hemileia vastatrix* B. & Br.). *Ann. Appl. Biol.* 49: 497-505.

Rehuman, M. U. and Subramanian, S. (1967). A new Fusarium Wilt of coffee (*Coffea arabica* L.) in South India. *Plant Dis. Reptr.* 51: 758-759.

Rodrigues, C. J. Jr., Bettencourt, A. J. and Rijo, L. (1975). Races of the pathogen and resistance to coffee rust. *Annu. Rev. Phytopathol.* 13(1): 49-70.

Rutherford, M. and Phiri, N. (2006). *Pests and Diseases of Coffee in East Africa*, Wallingford (Oxford), UK: CAB International.

Sanchezde Leon, A. (1977). Neuvas investigacious sobre el control del Koleroga en el Cafe. *Rev. Cafetalera (Guatem.)* 167: 21-25.

Sharma, M. S. (1967). A review of work on the "new malady of coffee". *Indian Coffee* 31: 3-7.

Sheriff, M., Nataraj, T. and George, K. V. (1960). Stem wasting or Kondli disease of coffee. *Indian Coffee* 24: 160-167.

Siddique, M. A. (1970). Incidence and development and symptoms of *Cercospora* disease of coffee in Malawi. *Trans. Br. Mycol. Soc.* 54: 415-421.

Silveira, H. L. and Rodrigues, C. J. Jr. (1972). Bursting of rust uredospores caused by *Verticillium hemileiae* Bour. culture filtrates. *Agronomia Lusit.* 33: 391-396.

Souza, A. (2011). Chemical applications to manage coffee leaf rust. *Austral. Plant Path.* 40: 293-300.

Sreenivasan, K. H. (1939). A coffee root parasite *Balanophora* sp. *Mysore Coffee Expt. Stn. Bull.* 5: 1-4.

Sreenivasan, M. S. (1989). Breeding coffee for leaf rust resistance in India. In: Kushalappa, A. C. and Eskes, A. B. (Eds), *Coffee Rust Epidemiology, Resistance and Management*, FL: CRC Press. Inc., pp. 316-323.

Sridhar, T. S. and Subramanian, S. (1966). Studies on brown-eye-spot disease (*Cercospora coffeicola* B. & Cke.) of coffee (*Coffea arabica* L.) III. Mode of survival of the pathogen. *Rev. Path. Veg.* 2: 141-146.

Sridhar, T. S. and Subramanian, S. (1969). Studies on brown-eye-spot disease of coffee. *Indian Coffee* 33: 97-98.

Subramanian, S. and Sridhar, T. S. (1966). Studies on brown-eye-spot disease I. Examination of ethylene from infected leaves. *Rev. Path. Veg.* 2: 127-132.

Sudha, M., Daivasikamani, S. and Jayarama (2016). Root diseases. In: Gaitan, A. L., Cristancho, M. A., Castro Caicedo, B. L., Rivillas, C. A. and Cadena Gomez, G. (Eds), *Compendium of Coffee Diseases and Pests*, St. Paul, MN: The American Phytopathological Society, pp. 15-17.

Sudha, M., Giri, M. S., Santoshreddy, M., Ranjini, A. P., Daivasikamani, S. and Surya Prakash Rao, N. (2019). Biological management of *Rhizoctonia solani* Kuhn. Collar rot causing pathogen in coffee, *National Symposium on Enhancing Farm Income Through Sustainable Plant Health Management*, India, pp. 122-123.

Sudhakar, S. B., Hanumantha, B. T., Daivasikamani, S. and Jayarama. (2009). Coffee trunk canker disease in India and its management. *Indian Coffee*, LXXIII(10): 26-30.

Sudhakar, S. B., Hanumantha, B. T., Daivasikamani, S., Kiran Kumar, K. C., Shyam, S., Sudha, M. and Jayarama. (2011). Occurrence of stem necrosis and leaf spot disease of coffee seedlings. *Indian Coffee*, LXXV(7&8): 29--32.

Sudhakar, S. B. and Bhat, S. S. (1993). Studies on the viability of Koleroga *noxia* on coffee. *J. Coffee Res.* 23: 93-101.

Sudhakar, S. B., Bhat, S. S. and Govindarajan, T. S. (1992). Field efficacy of bavistin 50 WP and Foltaf 80 WP in comparison with bordeaux mixture for control of *Koleroga noxia* on coffee. *Proc. Placrosym* 10: 152-154.

Sudhakar, S. B., Bhat, S. S. and Govindarajan, T. S. (1995). Physical and chemical factors affecting growth of three isolates of *Koleroga noxia* on coffee. *J. Plan. Crops* 23: 78-83.

Sudhakar, S. B., Govindarajan, T. S., Ramaiah, P. K. and Bhat, S. S. (1988). A comparison of *Koleroga noxia* on *Coffea arabica*. *Proc. Placrosym* 8: 257-263.

Swaminathan, B., Ranga Setty, H. T. and Parthasarathy, S. (1960). Studies on "Berry blotch" disease of coffee. *Indian Coffee* 24: 246-250.

Thirumalachar, M. J. and Narasimhan, M. J. (1947). Studies on the morphology and parasitism of *Hemileia* species on Rubiaceae in Mysore. *Ann. Bot.* 11(1): 77-89.

Thomas, K. M. (1924). Some coffee diseases of South India and their control. *Planters Chron.* 19: 697-704.

Vallega, J. and Chiarappa, L. (1964). Plant disease losses as they occur worldwide. *Phytopathology* 54: 1305-1308.

Van Wyck, M., Wingfield, B. D., Marin, M. and Wingfield, M. J. (2010). *Ceratocystis fimbriata* species infecting coffee, cacao, citrus and native trees in Colombia. *Fungal Divers.* 40(1): 103-117.

Varzea, V. and Marques, D. (2005). Population variability of *Hemileia vastatrix*. In: Zambolim, L., Zambolim, E. M., Varzea, V. M. P. (Eds), *Durable Resistance to Coffee Leaf Rust*, Brazil: UFV-Viçosa.

Vekataramaiah, G. H. (1968). Sclerotia and rhizomorph formation by *Rhizoctonia solani* Kuhn in presence of fungicide. *Indian Phytopathol.* 25: 73-78.

Venkata Ram, C. S. and Muthappa, B. N. (1975). Red root rot of coffee. *Indian Coffee* 39:108.

Venkata Subbaiah, P. and Muthappa, B. N. (1981). Seed dressing chemicals for coffee against *Rhizoctonia solani*. *J. Coffee Res.*11: 1-3.

Vishveshwara, S. and Nagaraj, T. R. (1960). Nuclear status in basidiospores of *Hemileia vastatrix*. *Indian Coffee* 24: 118-119,124.

Ward, H. M. (1882). Researches on the life history of *Hemileia vastatrix*, the fungus of the coffee leaf disease. *J. Linn. Soc. Lond. Bot.* 19: 229-235.

Wellman, F. L. (1953). Some important diseases of coffee. *Plant Diseases, US Dept. Agric*: 891-896.

Wellman, F. L. (1961). *Coffee: Botany, Cultivation and Utilization*,London: Leonard Hill Ltd., pp. 271-272.

Zambolim, L. (2016). Current status and management of coffee leaf rust in Brazil. *Trop. Plant Pathol.* 41(1): 1-8.

Chapter 10

Ecological perspectives on the coffee leaf rust

Zachary Hajian-Forooshani and John Vandermeer, University of Michigan, USA

1 Introduction

The yellow blotches spotted on coffee leaves in Ceylon (Sri Lanka) in 1869 turned out to be a symptom of what would become one of the most devastating diseases of one of the most important crops in the world. British authorities rushed one of their most brilliant pathologists, Henry Marshall Ward, to one of their most intensively planned agroecosystems in one of their most profitable colonies to solve what was correctly seen as the tip of a disease iceberg. A keen observer of natural history, Ward determined the cause to be a fungus and meticulously worked out the details of its natural history, most of which remain the central core of our biological understanding yet today. As much as the current authors regard Ward as a hero, British authorities of the nineteenth century regarded his work as worthless, since he failed to find a solution to the problem. Providing detailed information on the underlying biology and ecology of the disease was regarded by the plant pathologists of the day as rather unimportant if such information did not lead to immediate solutions, reflective of nineteenth-century British imperial attitudes about their empire.

The rolling appearance of the coffee leaf rust (CLR), coupled with its devastating impact on local coffee economies, as told brilliantly by historian Stuart McCook in his recent book *Coffee Is Not Forever* (2019), is remarkable for its qualitative predictability: if you have coffee, CLR will come; if the CLR comes, your economy will be devastated. From the mid-nineteenth century in

http://dx.doi.org/10.19103/AS.2021.0096.24

Ceylon, to the beginning of the twentieth century in East Africa, to the middle of the twentieth century in West Africa, to the end of the twentieth century in the Americas, much like other species invasions, a plant invades a new area and prospers initially (coffee), followed by one or more natural enemies (CLR), which stifle its initial prosperity. The only difference here is the interest generated among the individuals of the species *Homo sapiens* due to its propensity for drug addiction.

The Americas represent something special. As documented in a great deal by McCook, the so-called Great Rust occurred largely without warning in 2012/2013. Having arrived in Brazil in 1970, coffee farmers throughout the Americas, cognizant of what had happened in Sri Lanka and elsewhere, panicked. Warnings with admonitions to do everything from pruning coffee bushes, to increase fertilization, to reducing or eliminating shade, and to prophylactic spraying of fungicides were common throughout the area for the next 30 or 40 years. Yet as time wore on, farmers began to realize that CLR was here to stay, although it never did cause the sort of devastation the nineteenth-century epidemic in Sri Lanka caused and was effectively treated as just one more of the yield-reducing nuisances they had learned to live with. Major problems with the disease were periodically reported in isolated areas throughout the Americas, but all seemed local and temporary. Then quite suddenly, the Great Rust hit them with devastating force. Walking through a coffee field in December of 2012, one's clothes would become yellow with the incredible abundance of CLR spores seeking new victims (Lopez-Bautista, personal communication).

The history of research on the disease (see McCook, 2019) seems oriented much like the original British establishment that judged Ward a failure. Genetics and chemistry seem to be the reigning discourses where the ultimate solution will be found. Ward's careful consideration of biological and ecological factors is acknowledged, to be sure, but a reading of the disease's history cannot help but suggest that a sort of technocentric or instrumentalist, as the philosophers say (Dear, 2008), point of view has prevailed. The search continues for genetic solutions, which is to say resistant varieties, as well as chemical ones, which is to say fungicides. Less common, albeit important, is an approach the philosophers would call intelligibility (Dear, 2008), seeking the underlying mechanisms as to why, for example, open landscapes seem to favor the rust (Avelino et al., 2012), or fungicides are effective only when fine-tuned at a local level (de Souza and Zambolim, 2011). The technocrat (instrumentalist) might say, if moisture favors CLR, reduce moisture. The ecologist, interested in the intelligibility of nature, would ask what it is about the natural history of CLR that results in its rapid growth when moisture is high. This latter approach, the ecological, we feel should represent an important complement to the more direct search for technological solutions.

In this chapter, we attempt to, at least in part, refocus an ecological approach to the problem of CLR. First, we provide a brief overview of the research, which has stemmed from a more technocentric approach to understanding and subsequently controlling CLR. We intentionally keep this section brief but try to highlight areas that are of potential ecological and evolutionary interest within this thread of research. This technocentric approach has been the main line of inquiry regarding CLR, and there are various resources available that provide in-depth information for ongoing research on these topics for interested readers (see the section on Where to look for further information). Next, we synthesize and discuss the sparsely available information about the community ecology and natural enemies of CLR around the world. We then present two case studies about CLR that we feel represent a more ecological (i.e. intelligible) approach to understanding the pathogen.

2 Technocentrism and control of the coffee leaf rust

The prevailing wisdom on the control of CLR has long considered fungicides and resistant coffee varieties to be the two main factors to be explored, with the tacit assumption that these two strategies will be the most effective (Talhinhas et al., 2017). There is a long history of developing and deploying fungicides with mixed results due to a variety of reasons ranging from the relatively fast evolution of resistance (McCook and Vandermeer, 2015), incorrect application technique (Belan et al., 2015), the economic costs to producers (Narayana, 2014), and environmental health concerns (Loland and Singh, 2004). While fungicides have been shown to be an important component of pathogen management under certain circumstances (Virginio Filho, 2017), it is almost certain that application impacts non-target organisms that occur in coffee agroecosystems communities, such as mycoparasitic fungi that attack CLR (Vandermeer et al., 2009; Pico Rosado, 2014). The extent to which fungicide application may have unintended consequences such as inhibiting biological control of CLR by fungal mycoparasites is not fully appreciated, but preliminary evidence suggests this may be the case (Carrión and Rico-Gray, 2002; Pico Rosado, 2014). Apart from potentially interfering with the biological control of CLR, there are examples of fungicide applications benefiting other devastating pests of coffee, such as reports from Kenya that a popular triazol used to control CLR, that resulted in promoting the Coffee Berry Disease (CBD) (Kushalappa and Eskes, 1989, 100). A consideration of the ecology of the agroecosystem will be a prerequisite to understanding how non-target effects of fungicide applications impact other pests, as well as biological control agents in coffee agroecosystems.

In addition to the fungicides, the development of resistant coffee varieties has been thought to be an effective way of controlling CLR (Silva et al.,

2006; McCook and Vandermeer, 2015). The history of this approach is long and complex, beginning with coffee's evolution under domestication. As a tetraploid, *Coffea arabica* is likely of hybrid origin from two diploid species, one perhaps similar to the other commercially important *Coffea* species, *C. canephora* (popularly known as Robusta coffee and the main source for instant coffee today). Prehistoric Arab traders brought *C. arabica* to Yemen from where it made its way, through international commerce, to south India, Sri Lanka, and Java in the 1690s. Most modern varieties stem from two sources, the variety 'Typica' transferred from Java to the botanical garden in Amsterdam by the Dutch in 1706 and the variety 'Bourbon' taken from Yemen to Reunion Island and then on to South America by the French. Thus, only two original accessions, historically, provide a rather narrow genetic basis for further evolution under domestication (Ferwerda, 1976). One of the most popular modern varieties, extremely susceptible to CLR and planted extensively before the arrival of CLR, is 'Caturra,' derived as a single dwarf mutant in Brazil from Bourbon stock.

By the 1930s, it was acknowledged that a variety of coffee had implications for the severity of CLR infection and that there was also substantial genetic diversity of the CLR, which could be classified into different strains or races. The CIFC rapidly identified several dozen CLR strains of varying virulence, and today more than 50 races of the pathogen are recognized (Talhinhas et al., 2017). With the discovery that both the variety of coffee and the strain of CLR impact the outcome of CLR infection, numerous efforts spawned focusing on the breeding of resistant varieties and exist to this day.

Unfortunately, as early as resistant breeding programs arose, so did the observation that CLR can eventually overcome resistance (Mayne, 1932). Although breeding programs have undoubtedly become more sophisticated through time, the loss of resistance for most and if not all varieties seems inevitable due to the relatively rapid evolution of CLR in response to selective pressures from varieties. The recent breakdown in resistance of the Lempira variety in Honduras in 2017, following almost 20 years of widespread use, illustrates how rapidly resistance can be overcome and the dangers of relying on a small number of varieties (Morales and Yonis, 2018; Avelino and Records, 2018). Although widely documented, the evolution of resistance presented somewhat of a paradox for biologists working on the CLR as the pathogen was long thought to clonally reproduce (Gouveia et al., 2005; Carvalho et al., 2011), suggesting that the genetic variability needed to evolve the ability overcome resistance should be relatively difficult to attain. There have been a number of hypotheses proposed to explain CLR genetic variability, from cryptosexuality, based on evidence of meiosis in urediniospores (Carvalho et al., 2011), to hybridization of different ancestral lineages of CLR that attack different species of *Coffea* (Silva et al., 2018). While an active area of research, there seems to be an emerging picture that mechanisms of reproduction are likely not homogenous everywhere CLR is present (Ramírez-Camejo et al., 2022). This complex picture of the population genetics of the CLR emerging around the world suggests the need for a regionally specific approach to understanding the dynamics of different CLR strains with different coffee varieties.

3 The community ecology of coffee leaf rust

The community of organisms associated with CLR has largely been considered a question of natural history curiosity rather than one of practical importance, something illustrated by the relatively small body of literature on this topic when compared to classical or technocentric research briefly described above. Rust fungi, evolutionarily a very ancient group, are known to have communities of organisms that regularly consume them (Anikster and Wahl, 1979; Henk et al., 2011). Currently little is known about the general ecology of these assemblages of natural enemies of rust fungi and what, if any, impact they have on the dynamics of the pathogens. Better understanding these communities of natural enemies will likely have practical implications for our understanding of how rust fungi are controlled in both 'natural' and agricultural settings. The context in which organisms are regulated in their native environment undoubtedly lends insight into the dynamics of how they escape control when exported elsewhere, something that has been long appreciated with classical programs in biological control. Understanding how and why pests and pathogens are released from control by natural enemies is an area of active research (Keane and Crawley, 2002; Liu and Stiling, 2006; Horrocks et al., 2020) but with regard to CLR, little is known. We suggest that understanding the dynamics of CLR in the regions where it has been and remains 'under control' can potentially shed light on how exactly it has escaped control historically in regions such as Sri Lanka and more recently in Central America, and where it might escape control in the future.

A growing body of work conducted in coffee's native range in Ethiopia sheds some light on the community ecology of CLR in its native range. With regard to CLR's natural enemies, Zewdie et al. (2021) collected 3 years of data on CLR and its fungal mycoparasite, *L. lecanii*, and showed that there is a signature of top-down control of CLR. It was found that *L. lecanii* was most abundant in the wet season and that its intensity in the wet season was negatively associated with the growth of CLR in the subsequent dry season, which is when CLR typically reaches its peak. Another recent study by Stüber et al. (2021) found that arboreal nesting *Crematogaster* sp. ants throughout parts of the coffee-producing region of Ethiopia were associated with lower amounts of CLR. These studies provide to our knowledge the first reports of the community ecology of coffee and CLR in its native range.

Similar relationships with CLR, *L. lecanii*, and arboreal ants are known from Mexico and have been extensively studied, revealing a complex network of direct and indirect ecological interactions (see Vandermeer et al., 2009, 2010, 2019; Jackson et al., 2012b). The similarities between these systems that have vastly different ecological and evolutionary histories suggest that generalities in interaction structures may exist and that interaction networks may assemble in introduced areas of coffee production that mirror those in the native range, as suggested earlier in a different context (Perfecto and Vandermeer, 2015a). Although similar work on the community ecology of CLR is sparse around the world, the little information that is available suggests potentially informative connections that span across novel ecological communities assembled in coffee agroecosystems.

While much remains to be understood about the community ecology of coffee agroecosystems in coffee's native range, much also remains to be discovered in the many regions around the world where coffee has been introduced and cultivated. As noted above, each introduction into a new region results in the assembly of a novel community with its own distinct assemblage of organisms and subsequent ecological dynamics. Simply documenting these communities could help piece together gaps in our understanding and serve as a basis to form hypotheses about the potential for control by natural enemies. Coffee researchers generally acknowledge the existence of these complex communities, but for a variety of reasons they are rarely reported in the literature. For example, *Cecidomyidae* consuming CLR in Latin America was not reported until 2016 (Hajian-Forooshani et al., 2016), although the authors have heard anecdotal reports of researchers having observed it for years in the region. It is likely the case that many more such ecological interactions are observed regularly but go unreported. Although potentially very important observations, creating incentives to report this type of information in the scientific or technical literature will likely be difficult, as documentation of natural history is seemingly less and less valued in the ecological literature, and without direct practical application, this information is similarly not highly valued in the agronomic literature.

As more information is accumulated about the organisms associated with CLR and their potential impact on the pathogen's dynamics, determining how environmental factors and management interact with this community of natural enemies will be exceedingly important. Given that CLR is a resource for its natural enemies, the first step is to understand how these environmental factors impact the dynamics of CLR itself since that will constrain the natural enemy community. In short, although regionally variable, some generalities have emerged in our understanding of environmental factors and management impacts on CLR. For example, the negative relationship between attitude and CLR has been widely documented (Bock, 1962; Avelino et al., 2006; Belachew et al., 2020; Zewdie et al., 2021). Various climatic factors such as participation and temperature are known to impact the development of the CLR as well (Brown et al., 1995; Avelino et al., 2015). Due to the interaction of temperature and precipitation on CLR, management factors that impact these variables have been proposed to be important for explaining realized CLR dynamics although there remains considerable controversy surrounding the topic. One of the most well-known examples of management in coffee agroecosystems is that of shade trees, which have shown variable effects in different systems, with some finding positive impacts (López-Bravo et al., 2012), others with negative impacts on CLR dynamics (Soto-Pinto et al., 2002; Zewdie et al., 2020), and some finding an interaction of shade effects depending on context (Pico Rosado, 2014). Shade trees in coffee agroecosystems present a variety of plausible pathways to impact CLR, from increasing germination through microclimatic modifications (López-Bravo et al., 2012), to decreasing wind speeds in coffee farms (Orozco and Jaramillo, 1978), to modifying rain-drop kinetics below shade trees (Avelino et al., 2020).

While a nuanced and growing literature carefully teases apart these effects on the CLR itself, there is some information emerging as to how these same factors may impact natural enemies, and in particular, attention has been paid to fungal mycoparasites such as *L. lecanii*. Recently Zewdie et al. (2021) found that higher levels of shade are associated with an increase in *L. lecanii*, and Pico Rosado (2014) found similar results with shade favoring *L. lecanii*. Pico Rosado's (2014) results suggest that natural enemy control by *L. lecanii* is more effective than the fungicide application and that fungicides inhibit *L. lecanii* ability to grow in the field. Regarding other natural enemies, there is currently little known about how environmental factors or management may impact their distribution or ability to control CLR. Basic ecological theory suggests that in systems where there are multiple natural enemies consuming the same resource there will likely be competition for that resource. In Puerto Rico, gastropods have been documented as consuming CLR both in the field and laboratory settings, and the invasive snail, *Bradybaena similaris*, was also shown to consume *L. lecanii* infected lesions of CLR. The extent to which resource competition and intra-guild predation between natural enemies in the field shapes the ecological dynamics of CLR needs to be explored in detail although preliminary observations suggest they are important factors for structuring the natural enemy communities in Puerto Rico (Personal observations).

4 The natural enemy community of coffee leaf rust

In this section we focus on summarizing the relatively small body of literature on the communities of organisms that are known to interact with CLR in various parts of the world.

4.1 Cecidomyidae

One of the first notes on what we might call the community ecology of CLR's natural enemies were published by H. F. Barnes when describing the species of *Cecidomyidae* associated with coffee (Barnes, 1939). Barnes provides a taxonomic description of both the larvae and adult of *Mycodiplosis hemileiae*, which lives among and feeds on CLR urediniospores in what is today Tanzania and Madagascar. Barnes also published notes from A. H. Ritchie on the ecology of *M. hemileiae* dated June 4, 1935, from the Boos' Plantation, Kaboia District, Bukoba, Tanganyika Territory. Ritchie provides a brief description of the life cycle, noting patterns in pupation and population densities that reach as high as 25 to 30 larvae per leaf. He also gives some insight into the community of organisms indirectly associated with CLR through interactions with *M. hemileiae*, describing a fungus that attacks larvae just prior to pupation and arrests the development resulting in the desiccation of the larvae. This unknown fungus is said to sporulate in the pupal webbing, and a sign of past attack is noted by the fawn to brownish color in the larvae. Mention of this fungus attacking *M. hemileiae* is absent from all subsequent literature and has not been observed by the authors in Mexico or Puerto Rico. In addition to *M. hemileiae* being attacked by fungi, Barnes describes another *Cecidomyidae*, *Lestodiplosis sp.*, which he suggests is predaceous on *M. hemileiae*. Predation of *M. hemileiae* by *Lestodiplosis* sp. was not mentioned by Richie in his notes.

After Richie's initial observations of *M. hemileiae* in Tanzania and Madagascar, two *Cecidomyidae, Lestodiplosis sp.* and *Mycodiplosis sp.* were reported to be commonly found in CLR pustules eating urediniospores in Kenya (Crowe, 1963), although it is unclear if these two species are the same as observed by Richie. It seems unlikely that *Lestodiplosis* is a fungus feeder given that most reports of species in the genus describe it to be an arthropod predator. In addition to noting these two natural enemies of CLR, Crowe makes note of two parasitoids that attack both species of *Cecidomyidae* larvae, *Leptacis kivuensis* Risbec, and *Synopeas* sp. (Hymenoptera: *Platygasteridae*) and suggests that these parasitoids may be vectors of long-range dispersal of CLR urediniospores. Crowe describes observations of the searching behavior of parasitoids within and around pustules with CLR and documents the accumulation of CLR urediniospores on their bodies. He found an average of 37 spores per individual and estimated the population of the parasitoids to be approximately 20 000 per acre. The next mention of *Cecidomyidae* associated with CLR came again from Kenya by D.L. Milne in 1975, where he notes relatively low densities of *M. hemileiae* (seven or eight per leaf), completely clearing leaves of CLR spores (Milne, 1975). Milne speculated that under certain conditions the larvae may play a role in controlling CLR stating:

> The coffee rust fungus unfortunately produces spores on such a prolific scale that even under conditions favorable for infestation by this midge, it is doubted whether the insect could ever be considered as more than a contributory factor to rust control. However in areas where rust infections are light, it is possible that M. hemileiae could play quite an important role in reducing the spore load.

Given the potential for a *Cecidomyidae* community to contribute to the control of CLR, it is somewhat surprising that no subsequent work was done to understand the impact of these organisms on CLR dynamics. It took over 40 years until *Cecidomyidae* associated with the CLR were mentioned again in the scientific literature when they were noted in Southern Mexico and Puerto Rico (Hajian-Forooshani et al., 2016), although CLR researchers seem aware of its existence and potential to consume CLR. While the realized distribution of *Cecidomyidae* that consume CLR is unclear, it seems likely that in areas where they have been described (Mexico and Puerto Rico) they have likely been co-introduced with CLR. Analyses have shown that *Mycodiplosis* genera overwhelmingly tend to be rust species-specific (Henk et al., 2011). This suggests that although there are likely native *Mycodiplosis* or other *Cecidomyidae* in both regions feeding on native rust fungi, the organisms found feeding on CLR are likely non-native. Work on both the molecular identification and population genetics of CLR-feeding *Cecidomyidae* could help elucidate their origins and potential as biological control agents.

4.2 Thrips

In Richie's notes on the community ecology of the CLR published by Barnes (1939), there is brief mention of a number of thrips associated with coffee, but he makes particular note of *Physothrips xanthoceros* feeding in CLR pustules. Since Richie's observations, various thrips associated with and potentially feeding on CLR have been mentioned sporadically throughout the literature of the twentieth century (see references in Kushalappa and Eskes p. 163–164) but largely dismissed, seemingly in the absence of detailed observational or experimental work. Knowledge of the ranges and occurrences of thrips associated with CLR mainly comes from taxonomic descriptions. Interestingly some of these thrip species seem to be relatively widespread and associated with CLR. Two African species *Craspedothrips antennatus* and *C. xanthocerus* have been found in Uganda, Kenya, Tanzania, and Angola (Mound et al., 2012). Additionally another thrip, *Megaphysothrips subramanii*, was recently observed in Timor Leste and was said to be found exclusively in pustules of CLR and not exploring any other portions of the coffee leaf (Mound, 2018). *M. subramanii* was found to be covered in CLR urediniospores, and it is suggested that *M. subramanii* consumes CLR to complete its life cycle while not causing feeding damage to coffee leaves themselves.

4.3 Mites

Coffee is known to host a community of mites that are likely facilitated by the presence of acadomatia on some coffee leaves. While these mites have a variety of life histories, some of them are associated with CLR. Flechtmann (1976) first suggested that the mite *Ricoseius loxocheles* feeds on CLR, and recently a detailed study of its biology and behavior was reported by Oliveira et al. (2014), including the important observation that for the females to oviposit, they need to be fed a diet that includes CLR urediniospores. Once eggs are ovipoisted in pustules of CLR urediniospores, they are covered in the CLR urediniospores and take about 9 days to develop from egg to adult. While CLR is seemingly important for the life-cycle of this mite, it has also been shown that *R. loxocheles* consumes resources other than CLR in a laboratory setting, which include cattail pollen (*Typha sp.*) and clover rust (*Puccinia oxalidis*) (Ajila et al., 2018). While the feeding behavior of these mites on CLR has been established, it has also been noted that they frequently carry spores on their bodies. The implications of this behavior on the dynamics of the CLR is currently unknown but potentially important for dispersing the pathogen. Personal observations suggest that they frequently carry spores to new areas on leaves, potentially increasing the rate of spread of the CLR within a plant.

It is unclear how widespread *R. loxocheles* is in coffee-producing regions around the world. While the only published data we are aware of comes from Brazil, we have personally made observations of mites associated with CLR both in Southern Mexico and throughout the central coffee-producing region of Puerto Rico. While we have not identified the species of mite, the description of behavior from both Flechtmann (1976) and Oliveira et al. (2014) suggests it is likely *R. loxocheles*. The behavior of laying eggs and aggregating in pustules of CLR urediniospores as well as coating their bodies in urediniospores has been observed numerous times in both regions. In Puerto Rico, these mites are widespread across the island and quite commonly found on CLR-infected leaves (personal observations). While seemingly the same mites occur in Southern Mexico they are less common than in Puerto Rico.

4.4 Gastropods

Although gastropods are not widely known to consume rust fungi (see Ramsell and Paul, 1990), both field surveys and laboratory experiments have recently shown gastropods to consume CLR in Puerto Rico (Hajian-Forooshani et al., 2020). Bright orange excrement left behind on leaves was found to be associated with leaves cleaned of urediniospores and were identified as being from gastropods consuming CLR. Laboratory experiments conducted with *Bradybaena similaris* showed that the snail consumes CLR as well as the mycoparasitic fungus *L. lecanii*, approximately halving the percentage of the coffee leaf covered in CLR urediniospores in 24 h as well as reducing the *L. lecanii* coverage. *B. similaris*, as well as other unidentified gastropods (see Hajian-Forooshani et al., 2020 for photographs) are widespread across the coffee-producing region in Puerto Rico (Personal observations), but it is currently unknown if or to what extent they may be altering the dynamics of CLR on the island.

Notably, *B. similaris* is a non-native species that has been introduced to Puerto Rico as well as a number of other coffee-producing regions and is often considered a pest in various forms of agriculture (Idris and Abdullah, 1997). Many questions remain unanswered regarding the impact that *B. similaris* and other gastropods may have in other parts of the world. Given its widespread distribution, it seems likely that they may consume CLR in other parts of the world, although thus far no other reports are in the literature. Additionally, and of potential importance, it is unclear if spores are viable after passing through the digestive system of these gastropods.

4.5 Mycoparasitic fungi

The first report of mycoparasitic fungi attacking CLR came from Stanleyville Congo in 1929 (Steyaert, 1930). Steyaert notes that when strolling through coffee plantations near the railroad, they were examining CLR on plants (he notes that coffee rust was rare in the region) and noticed a blackish (*noirâtre*) mass in the center of CLR lesion. Upon examination under the microscope, they discovered that the mycelia were not in the host plant tissue but surrounding the urediniospores of CLR. Steyaert denotes a new species, *Cladosporium hemileiae n.sp.* (now *Digitopodium hemileiae* see Heuchert et al., 2005), and notes that this first record of a mycoparasite of CLR could very well contribute to efforts to control the pathogen (Steyaert, 1930). Since this initial documentation, there has been growing appreciation of a highly diverse and widespread community of CLR fungal mycoparasites occurring across the globe which likely has the potential to impact CLR dynamics (Pirozynski, 1977; Lim and Nik, 1983; Shaw, 1988; Kushalappa and Eskes, 1989; Carrión and Rico-Gray, 2002; James et al., 2016; Gómez-De La Cruz et al., 2018; Colmán et al., 2021).

Given that over 15 fungal mycoparasites of CLR have been documented (see references above), understanding the extent to which they may contribute to the control of CLR has been a point of inquiry for many researchers and practitioners. Within coffee's native range, the well-known mycoparasite *L. lecanii* is widespread (Zewdie et al., 2021), as in the Central African Republic (Kushalappa and Eskes, 1989 p.163). Outside of coffee's native range, many fungal mycoparasites have been documented. Carrión and Rico-Gray (2002) note that the region in Mexico where they documented six mycoparasitic fungi does not use fungicides for CLR due to its low prevalence in the region. They speculate that this community of fungal antagonists may play a role in controlling CLR in the region. Furthermore, Carrión (1988) compared the attack by *L. lecanii* in two states in Mexico, Veracruz and Chiapas, and found that Veracruz, which had less CLR, had 21.2% of attacks by *L. lecanii* as compared to Chiapas which only had 2.9% of attack. A similar comparative study was done with Mexico and Puerto Rico and found similarly that regions with lower CLR had a higher attack rate by *L. lecanii* (Hajian-Forooshani et al., 2016). Aside from observational studies, there is mixed evidence that this mycoparasitic community of fungi could be manipulated to control CLR in the field. Laboratory experiments show that CLR urediniospores can be inoculated with *L. lecanii*, but with mixed results (Carrión and Ruíz-Belin, 1988). Some field trials have successfully demonstrated the efficacy of *L. lecanii* (Carrión, 1988), while others showed mixed to no results from field applications of *L. lecanii* and *Vertucillium leptobactrum* (Kushalappa and Eskes, 1989 p. 165–167).

Mixed results in field and laboratory manipulations may be a result of isolating this mycoparasitic community from the context of the interactions in which it is typically embedded. Research in Southern Mexico emphasized the fact that *L. lecanii* is a natural enemy not only of CLR but also of the green coffee scale (*Coccus viridis*), suggesting the potential for complicated indirect effects, where the scale insect might provide a reservoir of *L. lecanii* that would attack the coffee rust (Vandermeer et al., 2009, 2012; Jackson et al., 2012b; Galvão and Bettiol, 2014). The community of alternate hosts of *L. lecanii* is widely acknowledged (Jackson et al., 2012a) and understanding the dynamics of host utilization under different biotic and abiotic conditions will likely shed light on the potential for mycoparasites to control CLR. In addition to insects as alternative hosts for *L. lecanii*, the mycoparasite, *Digitopodium tectonae*, discovered in 1930 by Seyert, has been reported attacking an alternative rust host *Olivea tectonae* in Brazil (Colmán et al., 2021).

It is clear that a complicating factor in most field studies of CLR mycoparasites is a lack of certainty in their identification. For example, Carrión and Rico-Gray (2002) document five fungal mycoparasites of coffee and note that all of them are white to some extent when cultured. One notable expectation from the white-colored fungal mycoparasites is *Cladosporium hemileiae* described as olive brown by Kushalappa and Eskes (1989 p.163) and blackish by Steyaert (1930). James et al. (2016), highlighting the inadequacy of visual identification of mycoparasites in the field, used molecular methods to identify fungal communities from samples of CLR lesions that contained apparent mycoparasites in Mexico and Puerto Rico, finding 15 species of fungi belonging to clades that contained at least one species of known mycoparasite. In James et al. (2016) sampling material was simply collected from CLR lesions that had white fungal growth on them. These results suggest that the true identities of fungal mycoparasites are likely obfuscated from simple field identification, and what often looks like a single white mycoparasite is in fact harboring a cryptic community of which we know relatively little. Microscopic structural and morphological characterizations of these fungal mycoparasites may also offer little hope of elucidating the true diversity given what seems to be the convergent evolution of structures (Colmán et al., 2021). While documentation of the fungal communities associated with *Coffea* is ongoing (see Vega et al., 2010; James et al., 2016), it is clear that much remains unknown about these communities and their potential impacts on the dynamics of CLR around the globe. Approaches that prioritize not only identification but also elucidation of the ecology of these fungal communities will be particularly informative.

5 Case studies of ecological complexity and the coffee leaf rust

5.1 The regionally distinct ecological communities of Puerto Rico and Mexico

Almost all forms of agriculture involve the construction of a novel ecosystem to some extent (Perfecto and Vandermeer, 2015a). One of the most common ways to construct an agroecosystem involves the transplanting of a plant or animal away from the context in which it evolved and placing it into a new ecosystem. Coffee provides an example of one such system where the plants evolved in Eastern and Central Africa and were subsequently transplanted for cultivation throughout the tropics and sub-tropics, where it resides in new ecological contexts. From an ecologist's perspective, any attempt to understand the dynamics of agroecosystems should be an attempt to understand the dynamics of how new interactions develop and play out through time in novel ecological settings. An appreciation of this is seen in both the ecological and agronomic literature, although cross-fertilization between these two bodies of work is rare. The ecologist Donald Strong illustrated how quickly new interactions develop in agroecosystems through a series of papers in the 1970s which analyzed the accumulation of herbivores through time in several agroecosystems (Strong, 1974, 1979; Strong et al., 1977). He failed to find a signature of the time since introduction of different crops on the number of herbivores reported feeding on them. Rather, the area in cultivation was the main predictor of the number of novel herbivores. These studies highlighted just how quickly agroecosystems can assemble themselves when introduced species (crops) interact with the surrounding biodiversity in which they are embedded. In addition to Strong's work, there is a growing understanding of how rapidly new interactions can form in ecological time (Agosta, 2006; Agosta and Klemens, 2008; Agosta et al., 2010; Nylin et al., 2018). We suggest that the dramatic differences in the dynamics of agroecosystems around the world may have more to do with the ecological context in which they are embedded than is traditionally appreciated.

The apparent difference in the dynamics of the CLR in Mexico and Puerto Rico provides an example of the idiosyncratic dynamics of this pathogen in different parts of the world. These two locations were met with vastly different fates in recent times, where Mexico (along with other counties in Mesoamerica) experienced the epidemic of 'the Great Rust,' but Puerto Rico has been largely unscathed by island-wide outbreaks. This simple observation raises the obvious question of what is responsible for these vastly different outcomes? One may be inclined to focus on possible distinctiveness in abiotic conditions, searching for correlations in rainfall, humidity, and temperature to explain the apparent differences. While these factors have been extensively studied (Avelino et al., 2015), there is mixed evidence for their effect and no definitive smoking gun that explains why we see such drastic differences in the recent dynamics of the pathogen in one region but not the other. In fact, one recent study by Bebber et al. (2016) suggests that climate factors alone are insufficient to explain CLR dynamics. While we acknowledge that approaches that prioritize investigation of abiotic factors are important, we argue that a community ecological perspective that focuses on differences in the ecological communities of natural enemies between different regions may lend additional insight into the variable dynamics of CLR around the world.

Strong's original analyses illustrated how distinct agroecosystems can be assembled depending on the local ecological communities. Given that tropical crops such as coffee tend to be cultivated in some of the most biodiverse regions of the world, there is potential for a large number of novel interactions to arise quickly in areas of cultivation. There is now extensive documentation of the ecological communities in coffee agroecosystems in Mexico and Puerto Rico (Perfecto and Vandermeer, 2015b; Vandermeer et al., 2019), including natural enemies of CLR. This documentation permits study and speculation on how these uniquely assembled ecological communities and interaction networks may be impacting the regional dynamics of CLR (Fig. 1).

In a comparative study between coffee farms in Puerto Rico and Southern Mexico, it was found that the attack rates (or the proportion of each natural enemy to CLR lesions on leaves) was 3.8% in Mexico and 32.2% in Puerto Rico for *Mycodiplosis sp.*, and for *L. lecanii*, 8.7% in Mexico and 61% in Puerto Rico (Hajian-Forooshani et al., 2016). Early work in Mexico suggested that the attack of *L. lecanii* on CLR may be mediated by indirect interactions with a dominant arboreal ant *Azteca sericeasur* (Vandermeer et al., 2009; Jackson et al., 2012b). Another coffee pest, the green coffee scale insect (*Coccus viridis*), has a mutualistic association with this arboreal ant and can reach high local

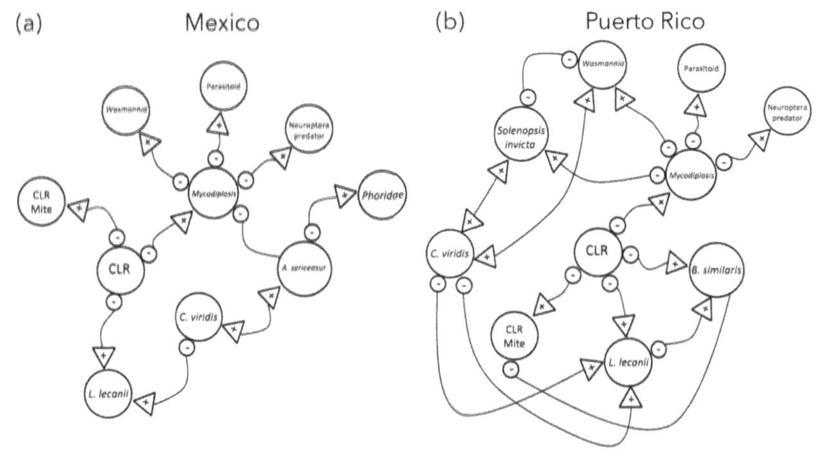

Figure 1 The regionally distinct communities associated with the coffee leaf rust in Mexico and Puerto Rico. Links between organisms denote interactions, with triangles showing positive effects and circles negative effects. Note that interaction networks are constructed from personal observations by the authors in both field and laboratory settings.

population densities when under ant protection (Vandermeer et al., 2010). As with many other pathogens, when a host population becomes sufficiently large, epidemic outbreaks become more probable, and this is the case with the green coffee scale and *L. lecanii* (Vandermeer and Perfecto, 2019). Studies in Mexico demonstrate that in sites where the *A. sericeasur* nests are present, *L. lecanii* attack on CLR is more common the following year, likely a consequence of *L. lecanii* building up on the green coffee scale populations and spilling over to attack CLR (Jackson et al., 2012a). Additionally, laboratory experiments suggest that *L. lecanii* can be dispersed by *A. seriseasur* (Jackson et al., 2012b). Thus, through various indirect and complex interactions, a native ant in Mesoamerica has a positive impact on biological control of a pathogen that is presumably from eastern Africa. An additional complication is that although *A. sericeasur* has a positive impact on *L. lecanii* it seems to have a negative impact on the abundance of *M. hemileiae* (Hajian-Forooshani et al., 2016). It was found that coffee plants around trees where *A. sericeasur* was nesting had fewer *M. hemileiae* but interestingly no difference in the amount of CLR on plants. These results illustrate how the ecological contexts in which crops and their natural enemies are embedded influence them differentially and may have the potential to shape pathogen dynamics, such as those of CLR.

Coffee farms in both Mexico and Puerto Rico largely share the same suite of natural enemies, although there are some exceptions (see Hajian-Forooshani et al., 2020), yet the relative abundances of the various organisms are dramatically different. Years of personal observations in both systems from

2013 to 2021 suggest that *L. lecanii* and the CLR-dwelling mites *R. loxocheles* tend to be far more common in Puerto Rico than in Mexico while *M. hemileiae* tends to be more common in Mexico. It is likely that these realized differences in abundances emerge from their interactions not only with their resource(s) (including CLR) but also from the plethora of interactions they take part in with the associated biodiversity within the coffee agroecosystems. From our experience, many interactions have been noted in the field although not yet formally reported in the literature, which includes parasitism of *M. hemileiae* by unidentified parasitoids in Mexico and Puerto Rico, predation of *M. hemileiae* by a neuropteran larvae in Mexico, and predation of *M. hemileiae* by a number of ants in Mexico and Puerto Rico. While the extent to which these interactions impact the population dynamics of these potential biological control agents is currently unknown, it seems plausible that the accumulated impact of these diverse interactions could have consequences for the dynamics of CLR.

5.2 Theoretical perspectives on space and environmental forces

Classic ecological models highlight how consideration of space can fundamentally alter our understanding and intuition about the dynamics of ecological systems (Tilman and Kareiva, 1997), and this is especially true for disease ecology, where space can take a central role in governing disease dynamics. Relatively little has been done to understand the spatial ecology of CLR outside of understanding its modes of transmission, although there is a growing appreciation that space plays a central role in CLR dynamics (Rosas et al., 2021; de Carvalho Alves et al., 2009). Given that coffee plants often have characteristic spatial arrangements and that several management factors are understood to interact with transmission (e.g. shade), focusing on the spatial ecology of the CLR can potentially lead to insights on how to better manage coffee agroecosystems with regard to the pathogen. Here we highlight some theoretical approaches to understanding the dynamics of CLR and highlight findings about the spatial ecology of CLR. We suggest that additional theoretical work on understanding the dynamics of CLR in space may be a fruitful path forward to understanding how the pathogen operates locally, within landscapes, and at a regional scale.

When considering CLR, some ecologists have abstracted the pathogen as operating at two distinct but interacting spatial scales. Perhaps most evidently, the pathogen spreads at a local scale from plant to plant within a given area, but in addition to this neighbor-to-neighbor spread, it can disperse large distances in wind currents. Avelino et al. (2012) highlighted how larger-scale regional transmission can be related to the management of the surrounding landscape of coffee production. They found that the amount of pasture in the surrounding landscape was associated with higher CLR within farms. Exploring the often non-intuitive implications of both the local and regional scale transmissions interacting has been a focus of recent theoretical explorations of CLR.

Vandermeer and Rohani (2014) used an understanding of the two scales of pathogen transmission to model the CLR. The basic idea behind their dynamical model is that there is an interaction between pathogen spore reservoirs at regional and local scales. CLR is modeled as building up within farms through the local transmission dynamics of rain splash and plant-to-plant contact, and as the reservoir of pathogen spores builds on a farm it feeds into the regional atmospheric environment's reservoir of spores which can be transmitted to larger distances and 'rain down' to infect other farms in the region. Using a modified patch-occupancy approach, they model both the portion of bushes infected within a farm and the proportion of farms infected within a region. Their results suggest three qualitative possibilities exist for the dynamics of the CLR, which depend on the underlying parameters of the model, which include: the bush-to-bush infection rate, the recovery rate of bushes, and the infection rate of farms. Depending on these parameters they find that the regional dynamics CLR can either be persistently absent, persistently epidemic, or a combination of absent and at intermediate infection levels depending on the initial conditions of the system. Furthermore, they explore how socioeconomic drivers that influence the number of farms within a region (e.g. the international market price of coffee) impact the overall dynamics of the CLR and find that evidence of tipping-point behavior regardless of different dispersal environments (i.e. high or low dispersal environments). This tipping-point or catastrophic transition behavior seems to robustly emerge from the interaction of the two scales of transmission as was illustrated by Vandermeer et al. (2015) who incorporated the additional effect of shade on modifying local dispersal of the CLR.

While the above studies employed a spatially implicit approach to understanding how multiscale transmission impacts CLR dynamics, there have also been spatially explicit approaches to understanding CLR dynamics. Vandermeer et al. (2018) first used spatial networks to empirically infer the local transmission dynamics of CLR in Mexico. Hajian-Forooshani and Vandermeer (2021) employed spatial networks to explore the dynamic process of how spatial pattern is created in coffee agroecosystems and subsequently how that spatial pattern influences the pathogen dynamics. Beginning with the abstraction of a local area of coffee production with plants arranged in an approximate lattice formation, they allowed for stochastic death of plants, followed by imperfect replanting in a local neighborhood of the recently killed plant. The simple model of stochastic plant death and replanting created a range of spatial patterns that resembled the range of observed empirical spatial distributions of coffee plots on a farm in Mexico. They found that the empirical plots show a similar time-dependence of the spatial organization as observed in the model, where newly planted areas on the farm are better approximated by a lattice-like spatial distribution, and older areas have more random-like distributions of plants. To understand the impact of different spatial patterning of plants on the dynamics of the CLR they modeled pathogen spread on the spatial patterns that emerged from the death and replanting model. By exploring a wide range of parameter combinations, they found that more lattice-like spatial patterns are susceptible to small changes in the spatial scale of CLR transmission, resulting in pathogen epidemics. It is suggested that these small changes in transmission scale can emerge from changes to management practices that interact with transmission, such as the amount of shade in a system (Fig. 2).

Another recent spatially explicit modeling approach by Beasley et al. (2020) focused on the regional dynamics of CLR. They use a stochastic cellular automata approach to model the dynamics of CLR across heterogeneous landscapes. By incorporating different habitat types in the landscape to simulate the known impact of pasture land on CLR dynamics (Avelino et al., 2012), they explore how regional landscape composition interacts with the local dynamics of the pathogen. They found that the clustering of coffee plants at a local scale is one of the most important factors in the model, but this interacts with the spatial distribution of deforestation in the landscape. Areas in the model where deforestation is more clustered resulted in more severe epidemics than when deforestation was more uniform across the landscape. This larger-scale and spatially explicit approach to modeling the CLR highlight the importance of not only local dynamics but also how both coffee-producing and non-producing areas together have implications for CLR.

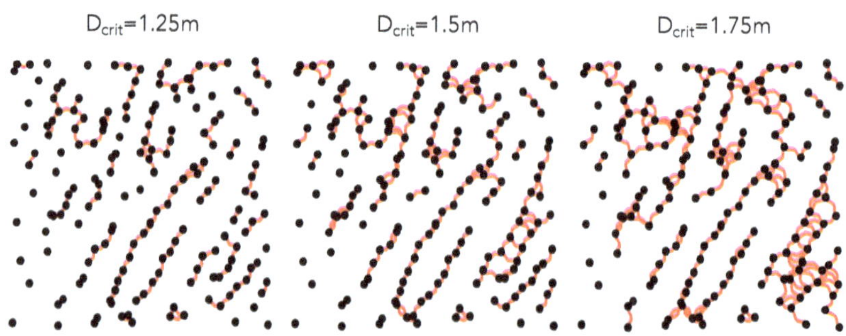

Figure 2 The network approach adopted by Vandermeer et al. (2018) and Hajian-Forooshani and Vandermeer, (2021) to spatially explicitly model CLR. Dots show the distribution of coffee plants from a 20 m × 20 m plot in southern Mexico. The lines connecting the dots are stipulated by D_{crit} (denoting critical distance) and illustrate theoretical connections between plants and thus tendency for local pathogen spread. This network approach has been used to estimate CLR transmission distance (D_{crit}) in empirical data (see Vandermeer et al., 2018 for details) and provides information such as the number of 'outside infections' necessary to infect an area and the speed of pathogen spread across an area (see Hajian-Forooshani and Vandermeer, 2021 for details).

6 Where to look for further information

The scope of this chapter has been made intentionally narrow to focus on what the authors consider to be potentially fruitful areas of research that they hope to bring attention to the larger community of coffee researchers as well as other agroecologists. In this section we briefly recommend resources for more in-depth presentations on various aspects of CLR research and point readers in the direction of various researchers and research groups working with CLR which the authors find their approaches particularly informative.

Kushalappa and Eskes provided a comprehensive reference text on CLR with their 1989 monograph *Coffee Rust: Epidemiology, Resistance, and Management*. Various aspects of the biology of CLR can be found there as well as a wealth of information regarding control strategies among other topics. For a more recent and succinct treatment of CLR, readers should be aware of *Prevention and control of coffee leaf rust: Handbook of best practices for extension agents and facilitators* by Elias de Melo Virginio Filho and Carlos Astorga Domain (2019) from CATIE, which details more recent information regarding the ecology of the CLR (including some information on fungal mycoparasites) and best practices for management. For an excellent analysis of the history of the coffee rust, see Stuart McCook's 2019 book *Coffee Is Not Forever*. The ongoing work of Jacques Avelino and his research group has been cited throughout this chapter, and his bibliography offers several insights into CLR's ecology. Ongoing work in Ethiopian coffee agroecosystems by Kristoffer Hylander and Ayco Tack's research groups at Stockholm University provides a unique ecological perspective on CLR in its native range. The authors feel that continued work by these two groups in Ethiopian coffee could be particularly informative in developing a more ecological approach to understanding CLR.

7 References

Agosta, S. J. (2006). On ecological fitting, plant-insect associations, herbivore host shifts, and host plant selection. *Oikos* 114(3), 556-565.

Agosta, S. J., Janz, N. and Brooks, D. R. (2010). How specialists can be generalists: Resolving the "parasite paradox" and implications for emerging infectious disease. *Zoologia* 27(2), 151-162.

Agosta, S. J. and Klemens, J. A. (2008). Ecological fitting by phenotypically flexible genotypes: Implications for species associations, community assembly and evolution. *Ecology Letters* 11(11), 1123-1134.

Anikster, Y. and Wahl, I. (1979). Coevolution of the rust fungi on Gramineae and Liliaceae and their hosts. *Annual Review of Phytopathology* 17(1), 367-403.

Avelino, J., Cristancho, M., Georgiou, S., Imbach, P., Aguilar, L., Bornemann, G., Läderach, P., Anzueto, F., Hruska, A. J. and Morales, C. (2015). The coffee rust crises in Colombia and Central America (2008-2013): Impacts, plausible causes and proposed solutions. *Food Security* 7(2), 303-321.

Avelino, J. and Records, A. 2018. *Coffee Leaf Rust: A Persistent Threat to the Livelihoods of the People Who Produce Your Morning Cup*. Boston: APS Publishing.

Avelino, J., Romero-Gurdián, A., Cruz-Cuellar, H. F. and Declerck, F. A. (2012). Landscape context and scale differentially impact coffee leaf rust, coffee berry borer, and coffee root-knot nematodes. *Ecological Applications: A Publication of the Ecological Society of America* 22(2), 584-596.

Avelino, J., Ten Hoopen, G. M. and DeClerck, F. 2011. *Ecological Mechanisms for Pest and Disease Control in Coffee and Cacao Agroecosystems of the Neotropics*. London: Earthscan Publications, pp. 91-117.

Avelino, J., Vilchez, S., Segura-Escobar, M. B., Brenes-Loaiza, M. A., Virginio Filho, EdM. and Casanoves, F. (2020). Shade tree Chloroleucon eurycyclum promotes coffee leaf rust by reducing uredospore wash-off by rain. *Crop Protection* 129, 105038.

Avelino, J., Zelaya, H., Merlo, A., Pineda, A., Ordóñez, M. and Savary, S. (2006). The intensity of a coffee rust epidemic is dependent on production situations. *Ecological Modelling* 197(3-4), 431-447.

Barnes, H. F. (1939). Gall Midges (Cecidomyiidae) associated with coffee. *Revue de Zoologie et de Botanique Africaines* 32, 324-336.

Beasley, E. M., Aristizábal, N., Bueno, E. and White, E. R. (2020). The spore of the beans: spatially explicit models predict coffee rust spread in fragmented landscapes. *bioRxiv*.

Bebber, D. P., Castillo, Á. D. and Gurr, S. J. (2016). Modelling coffee leaf rust risk in Colombia with climate reanalysis data. *Philosophical Transactions of the Royal Society of London. Series B, Biological Sciences* 371(1709), 20150458.

Belachew, K., Senbeta, G. A., Garedew, W., Barreto, R. W. and Del Ponte, E. M. (2020). Altitude is the main driver of coffee leaf rust epidemics: A large-scale survey in Ethiopia. *Tropical Plant Pathology* 45(5), 511-521.

Belan, L. L., de Jesus Junior, W. C., de Souza, A. F., Zambolim, L., Tomaz, M. A., Alves, F. R., Ferrão, M. A. G. and do Amaral, J. F. T. (2015). Monitoring of leaf rust in conilon coffee clones to improve fungicide use. *Australasian Plant Pathology* 44(1), 5-12.

Bock, K. R. (1962). Seasonal periodicity of coffee leaf rust and factors affecting the severity of outbreaks in Kenya Colony. *Transactions of the British Mycological Society* 45(3), 289-300.

Brown, J. S., Kenny, M. K., Whan, J. H. and Merriman, P. R. (1995). The effect of temperature on the development of epidemics of coffee leaf rust in Papua New Guinea. *Crop Protection* 14(8), 671-676.

Carrión, G. (1988). Estudio sobre el control biológico de la roya del cafeto mediante Verticillium lecanii. *Micologia Neotroprical Aplicada* I, 79-86.

Carrión, G. and Ruiz-Belin, F. (1988). Inoculación en el laboratorio de Verticillium lecanii sobre la roya del cafe (Hemileia vastatrix). *Revista Mexicana de Micologia* 4, 317-332.

Carrión, L. G. and Rico-Gray, V. (2002). Mycoparasites on the coffee rust in Mexico. *Fungal Diversity* 11, 49-60.

Carvalho, C. R., Fernandes, R. C., Carvalho, G. M. A., Barreto, R. W. and Evans, H. C. (2011). Cryptosexuality and the genetic diversity paradox in coffee rust, Hemileia vastatrix. *PLoS ONE* 6(11), e26387.

Colmán, A. A., Evans, H. C., Salcedo-Sarmiento, S. S., Braun, U., Belachew-Bekele, K. and Barreto, R. W. (2021). A fungus-eat-fungus world: Digitopodium, with particular reference to mycoparasites of the coffee leaf rust, Hemileia vastatrix. *IMA Fungus* 12(1), 1.

Crowe, T. J. (1963). Possible insect vectors of the uredospores of Hemileia vastatrix in Kenya. *Transactions of the British Mycological Society* 46(1), 24.

de Carvalho Alves, M., da Silva, F. M., Pozza, E. A. and de Oliveira, M. S. (2009). Modeling spatial variability and pattern of rust and brown eye spot in coffee agroecosystem. *Journal of Pest Science* 82(2), 137–148.

De Filho, E. V. and Domian, C. A. (2019). Prevention and control of coffee leaf rust: handbook of best practices for extension agents and facilitators. *Tropical Agricultural Research and Higher Education Center (CATIE)*, Turrialba, Costa Rica.

de Souza, A. F. and Zambolim, L. (2011). Chemical approaches to manage coffee leaf rust in drip irrigated trees. *Australasian Plant Pathology* 40(3), 293–300.

Dear, P. 2008. *The Intelligibility of Nature: How Science Makes Sense of the World*. Chicago, IL: University of Chicago Press.

Ferwerda, F. P. (1976). Coffees: Coffea spp. (Rubiaceae). Evolution of Crop Plants. NW Simmonds, ed. 1976, 257–260.

Flechtmann, C. H. W. (1976). Observações sobre dois ácaros (Mesostigmata, Acari) de vida livre. *Anais da Sociedade Entomológica do Brasil* 5(1), 95–96.

Galvão, J. A. H. and Bettiol, W. (2014). Effects of UV-B radiation on Lecanicillium spp., biological control agents of the coffee leaf rust pathogen. *Tropical Plant Pathology* 39(5), 392–400.

Gómez-De La Cruz, I., Pérez-Portilla, E., Escamilla-Prado, E., Martínez-Bolaños, M., Carrión-Villarnovo, G. L. L. and Hernández-Leal, T. I. (2018). Selection in vitro of mycoparasites with potential for biological control on coffee leaf rust (*Hemileia vastatrix*). *Revista Mexicana de Fitopatología* 36(1), 172–183.

Gouveia, M. M. C., Ribeiro, A., Várzea, V. M. and Rodrigues Jr., C. J. (2005). Genetic diversity in Hemileia vastatrix based on RAPD markers. *Mycologia* 97(2), 396–404.

Hajian-Forooshani, Z., Rivera Salinas, I. S., Jiménez-Soto, E., Perfecto, I. and Vandermeer, J. (2016). Impact of regionally distinct agroecosystem communities on the potential for autonomous control of the coffee leaf rust. Journal of *Environmental Entomology* 45(6), 1521–1526.

Hajian-Forooshani, Z. and Vandermeer, J. (2021). Emergent spatial structure and pathogen epidemics: the influence of management and stochasticity in agroecosystems. *Ecological Complexity* 45, 100872.

Hajian-Forooshani, Z., Vandermeer, J. and Perfecto, I. (2020). Insights from excrement: Invasive gastropods shift diet to consume the coffee leaf rust and its mycoparasite. *Ecology* 101(5), e02966.

Henk, D. A., Farr, D. F. and Aime, M. C. (2011). Mycodiplosis (Diptera) infestation of rust fungi is frequent, wide spread and possibly host specific. *Fungal Ecology* 4(4), 284–289.

Heuchert, B., Braun, U. and Schubert, K. (2005). Morphotaxonomic revision of fungicolous Cladosporium species (Hyphomycetes). *Schlechtendalia* 13, 1–78.

Horrocks, K. J., Ward, D. and Suckling, D. M. (2020). Can natural enemies of current insect pests provide biotic resistance to future pests? *Agricultural and Forest Entomology* 22(1), 20–29.

Idris, A. B. and Abdullah, M. (1997). The phorid fly, *Megaselia scalaris* (Loew), as a candidate for managing molluscicide-resistant round snail, *Bradybaena similaris* (Ferussas). *Resistant Pest Management* 9, 28–29.

Jackson, D. W., Zemenick, K. and Huerta, G. (2012a). Occurrence in the soil and dispersal of Lecanicillium lecanii, a fungal pathogen of the green coffee scale (*Coccus viridis*) and coffee rust (*Hemileia vastatrix*). *Tropical and Subtropical Agroecosystems* 15(2).

Jackson, D., Skillman, J. and Vandermeer, J. (2012b). Indirect biological control of the coffee leaf rust, *Hemileia vastatrix*, by the entomogenous fungus Lecanicillium lecanii in a complex coffee agroecosystem. *Biological Control* 61(1), 89–97.

James, T. Y., Marino, J. A., Perfecto, I. and Vandermeer, J. (2016). Identification of putative coffee rust mycoparasites via single-molecule DNA sequencing of infected pustules. *Applied and Environmental Microbiology* 82(2), 631–639.

Keane, R. M. and Crawley, M. J. (2002). Exotic plant invasions and the enemy release hypothesis. *Trends in Ecology and Evolution* 17(4), 164–170.

Kushalappa, A. C. and Eskes, A. B. 1989. *Coffee Rust: Epidemiology, Resistance, and Management*. Boca Raton, FL: CRC Press.

Lim, T. K. and Nik, W. Z. W. (1983). Mycoparasitism of the coffee rust pathogen, *Hemileia vastatrix*, by *Verticillium psalliotae* in Malaysia. *Pertanika* 6(2), 23–25.

Liu, H. and Stiling, P. (2006). Testing the enemy release hypothesis: A review and meta-analysis. *Biological Invasions* 8(7), 1535–1545.

Loland, J. Ø. and Singh, B. R. (2004). Copper contamination of soil and vegetation in coffee orchards after long-term use of Cu fungicides. *Nutrient Cycling in Agroecosystems* 69(3), 203–211.

López-Bravo, D. F., Virginio-Filho, EdM. and Avelino, J. (2012). Shade is conducive to coffee rust as compared to full sun exposure under standardized fruit load conditions. *Crop Protection* 38, 21–29.

Mayne, W. W. (1932). Physiological specialisation of *Hemileia vastatrix* B. and Br. *Nature* 129(3257), 510–510.

McCook, S. 2019. *Coffee Is Not Forever: A Global History of the Coffee Leaf Rust*. Ohio: Ohio University Press.

McCook, S. and Vandermeer, J. (2015). The big rust and the red queen: long-term perspectives on coffee rust research. *Phytopathology* 105(9), 1164–1173.

Milne, D. L. (1975). Coffee rust consumed by a helpful insect Mycodiplosis hemileiae. *Citrus and Sub-tropical Fruit Journal*, 11–12.

Morales, R. and Yonis, A. (2018). *Virulence Genes of a New Race of Rust (Hemileia Vastatrix) Affecting Lempira Coffee and Other Resistant/Suscepti Varieties in Honduras* (No. IAEA-CN--263).

Mound, L. A. (2018). Euphysothrips: an old world genus of Thripidae (Thysanoptera) associated with rust fungi (Pucciniales). *Zootaxa* 4532(3), 447–450.

Mound, L. A., Masumoto, M. and Okajima, S. (2012). The Palaeotropical genus Craspedothrips, with new species from Africa and Malaysia (Thysanoptera, Thripinae). *Zootaxa* 3478(1), 49–61.

Narayana, M. R. (2014). Economic analysis of leaf rust management by chemical controls: Evidence and implications for household coffee farmers in India. *Indian Journal of Agricultural Economics* 69, 459–473.

Nylin, S., Agosta, S., Bensch, S., Boeger, W. A., Braga, M. P., Brooks, D. R., Forister, M. L., Hambäck, P. A., Hoberg, E. P., Nyman, T., Schäpers, A., Stigall, A. L., Wheat, C. W., Österling, M. and Janz, N. (2018). Embracing colonizations: A new paradigm for species association dynamics. *Trends in Ecology and Evolution* 33(1), 4–14.

Oliveira, C. M., Ferreira, J. A., Oliveira, R. M., Santos, F. O. and Pallini, A. (2014). Ricoseius loxocheles, a phytoseiid mite that feeds on coffee leaf rust. *Experimental and Applied Acarology* 64(2), 223–233.

Orozco, F. J. and Jaramillo-Robledo, A. (1978). Efecto del déficit de humedad en el suelo sobre la temperatura del suelo y de las hojas en plantas de *Coffea Canephora* y *Coffea arabica*. *Cenicafé (Co)* 29(4), 121-134.

Perfecto, I. and Vandermeer, J. (2015a). Structural constraints on novel ecosystems in agriculture: the rapid emergence of stereotypic modules. *Perspectives in Plant Ecology, Evolution and Systematics* 17(6), 522-530.

Perfecto, I. and Vandermeer, J. (2015b). *Coffee Agroecology: A New Approach to Understanding Agricultural Biodiversity, Ecosystem Services and Sustainable Development*. London: Routledge.

Pico Rosado, J. T. 2014. *Efecto de la sombra del café y el manejo sobre la incidencia, severidad, cantidad de inóculo y dispersión de Hemileia vastatrix en Turrialba, Costa Rica*.

Pirozynski, K. A. (1977). Notes on hyperparasitic Sphaeriales, Hypocreales and 'hypocreoid Dothideales'. *Kew Bulletin*, 595-610.

Ramírez-Camejo, L. A., Eamvijarn, A., Díaz-Valderrama, J. R., Karlsen-Ayala, E., Koch, R. A., Johnson, E., Pruvot-Woehl, S., Mejía, L. C., Montagnon, C., Maldonado-Fuentes, C. and Aime, M. C. (2022). Global analysis of *Hemileia vastatrix* populations shows clonal reproduction for the coffee leaf rust pathogen throughout most of its range. *Phytopathology* 112(3), 643-652.

Ramsell, J. and Paul, N. D. (1990). Preferential grazing by molluscs of plants infected by rust fungi. *Oikos* 58(2), 145-150.

Rosas, J. T. F., de Assis Silva, S., de Almeida, S. L. H., Medauar, C. C., Moraes, W. B. and de Souza Lima, J. S. (2021). Spatial and temporal behavior of coffee rust in *C. canephora* and its effects on crop yield. *European Journal of Plant Pathology* 161(3), 677-692.

Shaw, D. E. (1988). Verticillium lecanii a hyperparasite on the coffee rust pathogen in Papua New Guinea. *Australasian Plant Pathology* 17(1), 2-3.

Silva, D. N., Várzea, V., Paulo, O. S. and Batista, D. (2018). Population genomic footprints of host adaptation, introgression and recombination in coffee leaf rust. *Molecular Plant Pathology* 19(7), 1742-1753.

Silva, MdC., Várzea, V., Guerra-Guimarães, L., Azinheira, H. G., Fernandez, D., Petitot, A. S., Bertrand, B., Lashermes, P. and Nicole, M. (2006). Coffee resistance to the main diseases: Leaf rust and coffee berry disease. *Brazilian Journal of Plant Physiology* 18(1), 119-147.

Soto-Pinto, L., Perfecto, I. and Caballero-Nieto, J. (2002). Shade over coffee: its effects on berry borer, leaf rust and spontaneous herbs in Chiapas, Mexico. *Agroforestry Systems* 55(1), 37-45.

Steyaert, R. L. (1930). CLADOSPORIUM HEMILEIAE N. SPEC. Un Parasite de l'HEMILEIA VASTATRIX Berk. et Br. Bulletin de la Société Royale de Botanique de Belgique/ Bulletin van de Koninklijke Belgische Botanische Vereniging 63(Fasc. 1), 46-48.

Strong, D. R. (1974). Rapid asymptotic species accumulation in phytophagous insect communities: The pests of cacao. *Science* 185(4156), 1064-1066.

Strong, D. R., Jr. (1979). Biogeographic dynamics of insect-host plant communities. *Annual Review of Entomology* 24(1), 89-119.

Strong, D. R., Jr., McCoy, E. D. and Rey, J. R. (1977). Time and the number of herbivore species: The pests of sugarcane. *Ecology* 58(1), 167-175.

Stüber, M., Tack, A. J. M., Zewdie, B., Mendesil, E., Shimales, T., Ayalew, B., Nemomissa, S., Sjögren, J., Vesterinen, E., Wezel, A. and Hylander, K. (2021). Multi-scale mosaics in top-down pest control by ants from natural coffee forests to plantations. *Ecology* 102(7), e03376.

Talhinhas, P., Batista, D., Diniz, I., Vieira, A., Silva, D. N., Loureiro, A., Tavares, S., Pereira, A. P., Azinheira, H. G., Guerra-Guimarães, L., Várzea, V. and Silva, M. D. C. (2017). The coffee leaf rust pathogen *Hemileia vastatrix*: One and a half centuries around the tropics. *Molecular Plant Pathology* 18(8), 1039–1051.

Tilman, D. and Kareiva, P. M. (Eds.). 1997. *Spatial Ecology: The Role of Space in Population Dynamics and Interspecific Interactions*. Princeton, NJ: Princeton University Press.

Vacacela Ajila, H. E., Ferreira, J. A. M., Colares, F., Oliveira, C. M., Bernardo, A. M. G., Venzon, M. and Pallini, A. (2018). Ricoseius loxocheles (Acari: Phytoseiidae) is not a predator of false spider mite on coffee crops: what does it eat? *Experimental and Applied Acarology* 74(1), 1–11.

Vandermeer, J., Armbrecht, I., de la Mora, A., Ennis, K. K., Fitch, G., Gonthier, D. J., Hajian-Forooshani, Z., Hsieh, H., Iverson, A., Jackson, D., Jha, S., Jiménez-Soto, E., Lopez-Bautista, G., Larsen, A., Li, K., Liere, H., MacDonald, A., Marin, L., Mathis, K. A., Monagan, I., Morris, J. R., Ong, T., Pardee, G. L., Rivera-Salinas, I. S., Vaiyda, C., Williams-Guillen, K., Yitbarek, S., Uno, S., Zemenick, A., Philpott, S. M. and Perfecto, I. (2019). The community ecology of herbivore regulation in an agroecosystem: Lessons from complex systems. *BioScience* 69(12), 974–996.

Vandermeer, J., Hajian-Forooshani, Z. and Perfecto, I. (2018). The dynamics of the coffee rust disease: An epidemiological approach using network theory. *European Journal of Plant Pathology* 150(4), 1001–1010.

Vandermeer, J. and Perfecto, I. (2019). Hysteresis and critical transitions in a coffee agroecosystem. *Proceedings of the National Academy of Sciences of the United States of America* 116(30), 15074–15079.

Vandermeer, J., Perfecto, I. and Liere, H. (2009). Evidence for hyperparasitism of coffee rust (Hemileia vastatrix) by the entomogenous fungus, Lecanicillium lecanii, through a complex ecological web. *Plant Pathology* 58(4), 636–641.

Vandermeer, J., Perfecto, I. and Philpott, S. (2010). Ecological complexity and pest control in organic coffee production: uncovering an autonomous ecosystem service. *BioScience*, 60(7), 527–537.

Vandermeer, J. and Rohani, P. (2014). The interaction of regional and local in the dynamics of the coffee rust disease. *arXiv Preprint ArXiv:1407.8247*.

Vandermeer, J., Rohani, P. and Perfecto, I. (2015). Local dynamics of the coffee rust disease and the potential effect of shade. *arXiv Preprint ArXiv:1510.05849*.

Vega, F. E., Simpkins, A., Aime, M. C., Posada, F., Peterson, S. W., Rehner, S. A., Infante, F., Castillo, A. and Arnold, A. E. (2010). Fungal endophyte diversity in coffee plants from Colombia, Hawai'i. *Fungal Ecology* 3(3), 122–138.

Virginio Filho, E. de M. 2017. Cafetales sanos, productivos y ambientalmente amigables: Guia para trabajo con familias productoras. *CATIE, Independent Pamphlet*.

Zewdie, B., Tack, A. J. M., Adugna, G., Nemomissa, S. and Hylander, K. (2020). Patterns and drivers of fungal disease communities on arabica coffee along a management gradient. *Basic and Applied Ecology* 47, 95–106.

Zewdie, B., Tack, A. J. M., Ayalew, B., Adugna, G., Nemomissa, S. and Hylander, K. (2021). Temporal dynamics and biocontrol potential of a hyperparasite on coffee leaf rust across a landscape in arabica coffee's native range. *Agriculture, Ecosystems and Environment* 311, 107297.

Chapter 11

Coffee wilt disease

Julie Flood, CABI, UK

1 Introduction

During the 20th century, a devastating disease called Coffee Wilt Disease (CWD) spread across Africa, reducing yields, destroying millions of coffee trees and costing millions of USD in lost earnings for producers (Flood, 2009). The disease has also contributed to a decline in the revenues of several African nations. CWD reached epidemic proportions twice in the 20th century - through the 1930s, 1940s, 1950s - and was responsible for the failure of the commercialisation of excelsa coffee. However, following the systematic introduction of sanitation methods (uprooting and burning affected bushes *in situ*) together with comprehensive breeding programmes in affected African countries, CWD appeared to decline in incidence and was considered a minor disease (Flood, 2009). Yet the disease re-emerged initially as a series of sporadic outbreaks on robusta coffee in remote areas of Central Africa during the 1970s. Later, incidence increased to epidemic proportions in the last decades of the 20th century and into the 21st century (Flood, 2009), affecting robusta coffee in DRC, Uganda and Tanzania. In addition, the disease has become an increasingly important constraint to production of arabica coffee in Ethiopia (Girma et al., 2001, 2009a).

http://dx.doi.org/10.19103/AS.2021.0096.25

CWD, also called Tracheomycosis, is a vascular wilt disease. Initially, several fungi were implicated as the causal agent, but in 1939, Steyaert (1948) isolated a species of *Fusarium* called *F. xylarioides*. In 1956, Saccas reported that the pathogen produced perithecia on the bark of affected trees. The perithecia represent the sexual stage *Gibberella xylarioides* (Heim and Saccas, 1950). In Ethiopia, symptoms similar to those of CWD in West and Central Africa were documented on *Coffea arabica* for the first time in the Keffa province by Stewart (1957) and later by Lejeune (1958). Based on comparative studies of isolates from arabica taken from different origins and different *Coffea* species, the disease was confirmed to be caused by *F. xylarioides* (Kranz and Mogk, 1973). Unlike many other coffee diseases, CWD kills trees. The pathogen evolved in Africa and the disease remains, to date, confined to the African continent. There is an unconfirmed report of the disease in the Americas – in El Salvador (Abrego, 1965) – but is likely a mis-identification (CABI, 2020). There is only one report of the disease in South Africa and Swaziland (Anon., 1989), which may indicate that *F. xylarioides* is not common in Southern Africa, but further surveys are needed to confirm absence or presence and extent of distribution. It is also highly likely that there has been considerable confusion with other *Fusarium* species (CABI, 2020).

This chapter discusses the history, impact and spread of this disease as well as its symptoms, host range and management. The chapter also provides a summary of the outputs of the Regional Coffee Wilt Programme (RCWP), which examined many different aspects of CWD and its management following its re-emergence as a significant threat to sustainable coffee production. Future research trends include investigating host specificity in the pathogen as well as underlying resistance mechanisms in the host and the role of alternative hosts as a means to reduce field inoculum. The ecology of the fungus also needs further investigation to allow for a greater focus on more agroecological management practices.

2 History and impact

CWD was first observed in 1927 in a plantation of *Coffee excelsa*, near Bangui in the Central African Republic (CAR), then known as Oubangui-Chari (Figueres, 1940). The disease developed slowly but by 1942 it had become a serious problem throughout the country and, by 1945, had destroyed most of the country's excelsa plantations (Saccas, 1951). It is generally considered that CWD was chiefly responsible for the permanent collapse of excelsa as a commercial crop. It is likely that initially, the pathogen was present in wild coffee species and had co-evolved with these (possibly with limited symptoms) but with the planting of large areas of excelsa and its apparent greater susceptibility, then disease symptoms became more apparent. Once the pathogen had become established in the excelsa crop, it subsequently spread rapidly through plantations due to agricultural practices. Later Saccas (1956) also found the disease on *C. canephora* and *C. neo-Arnoldiana* in parts of CAR; all species of *Coffea* including wild relatives are susceptible to the pathogen to some degree. In 1939, Steyaert (1948) discovered it on excelsa in plantations in DRC (Democratic Republic of Congo, formerly Zaire and previously the Belgian Congo) close to the border with Sudan at Aba (Fraselle, 1950). Initially, it was considered a minor problem but, as with CAR, it later reached epidemic proportions (Fraselle et al., 1953) when it spread to robusta. Subsequently, it was found to have extended to Haut-Uélé, North Kasaï and Katanga (Kalonji-Mbuyi et al., 2009). Between 1938 and 1945 the disease became established on excelsa in Cameroon, causing complete destruction of plantations in the east of that country. Then in 1947, it was discovered on *C. canephora* in Côte d'Ivoire and major losses occurred there into the1950s, with more than 50% of the coffee-producing areas being destroyed in both Côte d'Ivoire and then DRC too. The Kouillou line of robusta was completed destroyed in Cote d'Ivoire (Flood, 2009). In 1958, the disease was reported in Guinea and spread quickly to most of the coffee areas, causing coffee production to fall by nearly 50% (Chiarappa, 1969).

3 Host range

The pathogen attacks all wild and cultivated species of *Coffea*. *Fusarium xylarioides* has also been reported to be pathogenic to cotton seedlings (IAC 20 cultivar) under laboratory conditions, less so under glasshouse conditions (Pizzinatto and Menten, 1991). Interestingly, it has been detected in the banana cultivar Kayinja syn 'pisang awak' in Uganda (Serani et al., 2007), which is often intercropped with coffee for the production of banana juice and banana beer (Rutherford et al., 2009). Serani et al. (2007) raised the possibility of solanaceous weeds being reservoirs of infection in coffee fields following the report of the pathogen being isolated from rotting tomatoes from fruit markets in Nigeria (Onesirosan and Fatunla, 1976). The report of pathogenicity to cotton also raises the possibility of other *Malvaceae* being alternative hosts.

4 Symptoms

The pathogen invades the coffee host and colonises the vascular system. Colonization of the vascular system induces host responses which disrupt water conduction, and this is manifested as wilting and desiccation of the leaves which turn inwards (Fig. 1). This is followed by defoliation and dieback of the affected branches (Fraselle, 1950). See also Fig. 2. Coffee berries ripen prematurely (Fig. 3). Swelling of the trunk can be seen as well as a series of vertical or spiralling cracks in the bark (Fig. 4). Small blackish-brown perithecia

Figure 1 Initial symptoms on infected tree - leaves turn inwards © Julie Flood, CABI.

Figure 2 Defoliation and die-back follow © Noah Phiri, CABI.

(sexual stage) may be seen in these cracks. Internally, a diagnostic blue-black coloration of the wood is seen (Flood, 1997). See also Fig. 5. Symptoms appear at any stage of crop growth and the rate of symptom development is variable. Symptoms may be pronounced on one part of the tree, a likely consequence of initial infection occurring in one of the main stems but the symptoms gradually extend to the whole plant. Juvenile plants may be killed in a few weeks from the initial observation of symptoms while mature trees may take 3-15 months to die.

Figure 3 Premature ripening of coffee berries © Julie Flood, CABI.

Figure 4 Swelling at the base of the trunk with tiny, black perithecia (sexual stage) in the cracks © Julie Flood, CABI.

5 Ecology of the fungus

The pathogen is considered to be a soil-inhabiting fungus, but its life cycle and epidemiology are not fully understood. The fungus can be readily isolated from stems, twigs and roots of diseased coffee plants, but isolation from the soil has proven difficult. This may suggest that the fungus is unable to compete well with other soil-borne microbes and remains within infected plant debris in soil or, potentially, within alternative hosts. Host infection, colonization and symptom development are thought to be similar to other vascular wilt pathogens (Rutherford et al., 2009). Initial entry is considered to be primarily

Figure 5 Blue-black staining under the bark of an infected tree © CABI.

through roots and lower stem, facilitated by the presence of wound sites as a consequence of management practices (slashing) or from livestock grazing in the coffee fields. Once inside the plant, the fungus colonises the vascular system and surrounding tissues. The observed blue-black staining in the wood is probably due to changes in phenolic metabolism and may constitute a host response (Beckman, 1987)

Perithecia are commonly observed on the wood at the base of infected coffee in the field and are primarily seen as sexual reproductive structures but may also be a means of survival in the absence of chlamydospores (Flood and Brayford, 1997). Ascospores (sexual spores) are produced within the perithecia. Profuse production of conidia (macro and micro conidia) was noticed on mycelium at the base of stems (Fraselle, 1950). All these spore types could disseminate the pathogen and cause infection-conidial suspensions, have been routinely used to infect plants for experimentation. A putative life cycle of the CWD pathogen has been proposed by H. Maraite based on earlier drawings (Rutherford et al., 2009), but more study of the life cycle is needed.

Studies on the spatial and temporal development of CWD (Rutherford and Flood, 2005; Musoli et al., 2008) revealed that plants from a range of susceptible clones showed increased infection of less than 3% to 45% over a 2 year period, with an average of six plants developing external symptoms each month. These studies indicated that initial foci were randomly distributed and these foci enlarged over time but in no particular direction. Infection of adjacent trees was observed, suggesting root-to-root contact or short-distance dispersal of fungal material (such as via rain splash) as being involved in disease spread, together with on-farm practices such as through wounds made at the base of the bushes during weeding. Transmission did not appear to be linked to wind direction. Musoli et al. (2008) further reported that in the early stages of plot infection, the effective range of influence of an infected tree was up to two trees away, but this rose to three trees at the end of a 6-year observation. This would suggest that once a tree exhibits disease, all trees within a three-tree radius (10m) should be grubbed out and destroyed.

6 Historical management recommendations

When CWD first emerged as a significant constraint to coffee production in the mid-20th century, management options considered were a combination of sanitation and breeding for resistance, although some chemical and biological options were also investigated.

Cultural practices included frequent inspection of the crop, along with uprooting and burning of infected material (*in situ*). Removal of coffee bushes ahead of the infection (to reduce spread between plantations) was considered effective in Cote d'Ivoire; gaps of a few hundred metres were considered enough to confine the disease (Deassus, 1954).

Several authors reported varietal differences in resistance to the pathogen, and suggested the use of resistant varieties as a means of control (Fraselle, 1950; Bouriquet, 1959; Porteres, 1959). Cultivars of *Coffea canephora* (notably robusta) then formed the basis of many of the West African breeding programmes in the mid-20th century. Resistance mechanisms were unknown but preliminary studies indicated that caffeine, thought to inhibit the pathogen, was detected in higher concentrations in tissues of *C. canephora* than in *C. liberica* (Rabechault, 1954) and a higher content of chlorogenic acid in the wood of resistant material (variety Robusta) was also reported (Bouriquet, 1959). Similarly, van der Graaff and Pieters (1978) reported that coffee lines of *Coffea arabica* in Ethiopia also differed widely in resistance to *F. xylarioides* and considered that these differences provided an excellent opportunity to control the disease with resistant varieties. Pieters and van der Graaff (1980) reported two methods of screening for resistance - a seedling test which involved wounding seedlings with a knife dipped in spore suspension and a conidial germination test conducted directly on the bark of the tree. Both tests correlated significantly with field scores and with each other, and provided the basis for a screening programme to be used in a more extensive programme involving selection for resistance to Coffee Berry Disease (*Colletotrichum kahawae*) and *F. xylarioides*. van der Graaff and Pieters (1978) suggested that resistance in *C. arabica* was likely qualitative in nature with no evidence of vertical resistance. However, Zadoks and van Leur (1983) re-analysed the data from Pieters and van der Graaff (1980) and demonstrated a significant cultivar-isolate interaction indicating that vertical resistance was involved. Girma et al. (2009b) further corroborated the work of Zadoks and van Leur (1983). The nature of the resistance has considerable implications for breeding, with a more qualitative resistance being likely to be more sustainable while vertical resistance (involvement of major genes) could be more easily overcome by evolution of more virulent races of the pathogen.

Chemical control was also investigated and spraying the soil surface with 2.5% copper (II) sulphate, was advocated as an effective control measure (Saccas, 1956), while Gaudy (1956) reported that spraying the affected coffee bushes with copper oxychloride was effective in controlling the disease. However, the immense costs of using this approach as well as its environmental impact meant that the chemical approach was rarely used.

With regard to biocontrol, Rabechault (1954) isolated four actinomycetes, one bacterium and *Corticium*, *Marasmius* and *Trichoderma* spp., all of which were inhibitory to the pathogen, but this avenue of research does not seem to have been much pursued and there are no biological control agents currently available. Consequently, sanitation and screening/breeding for resistance were prioritised for the management of the disease. An international conference held in 1956 made recommendations for a management programme for CWD to include (1) systematic sanitation over vast areas where affected coffee plants were uprooted and destroyed; (2) where possible, relocation of coffee production to new locations; and (3) replanting with more resistant germplasm. Implementation of these recommendations reduced the impact of CWD. However, it should be noted that much of this early work in the mid-20th century involved a plantation model of coffee production where large-scale, systematic sanitation (over several hundred metres ahead of the disease front) was possible. Decisions were made by relatively few people and there was an ample supply of labour.

7 Re-emergence

Successful implementation of the recommendations above appeared to reduce the impact of the disease and literature produced during the 1970s and 1980s referred to CWD as being a minor disease, of little importance to arabica and robusta coffee production (Flood, 2009). Yet in the remote areas of DRC, the disease continued to be observed around the abandoned plantations in the north-east of the country (Kalonji-Mbuyi et al., 2009). As early as the 1970s, farmers in Aketi (76 km from Isiro) had observed the disease in abandoned plantations and during a survey conducted from 1974 to1975, a number INERA (Institut National pour l'Etude et la Recherche Agronomique) fields around Yangambi were also reported as affected. Reports continued throughout the 1980s around the town of Isiro, where abandoned plantations appeared to be the focus for the disease.

In 1995, CABI's plant health diagnostic laboratory isolated *F.xylarioides* from samples of robusta coffee from the Beni and Rutchuru areas in eastern DRC. The International Coffee Organization (ICO) facilitated contact between Office Zairois du Café (OZACAF) and CABI in March 1996 and requested that a plant pathologist be sent to conduct an independent assessment of the situation in country; this was the start of my own personal involvement with CWD.

I travelled to the North East of DRC in July 1996 and conducted surveys of plantations and smallholder farms as well as interviewing farmers about the disease they called 'coffee AIDS'. Growers wanted information on the management of the problem. Unfortunately, the only recommendations available was the same as in the 1950s – sanitation coupled with long-term breeding programmes. On some plantations, sanitation was being conducted, but the disease was spreading faster than the workers could undertake the work (Fig. 6). On one plantation I visited, 80% of the production area had already been destroyed. For smallholders, uprooting of infected coffee bushes *in situ* using just hand tools was impossible (Flood, 1996).

I undertook a similar visit to Uganda in 1997 as similar symptoms had been seen on coffee in Uganda. From 1994, samples from the districts in Western Uganda had been received by the Coffee Research Institute (CORI) and the pathogen isolated (Hakiza et al., 2009). Several surveys had already been undertaken in Uganda and the pathogen confirmed. The disease was also confirmed in a CORI experimental plot at Kituza in Mukono (Hakiza et al., 2009). After 1993, the incidence and severity of CWD continued to rise in Uganda and yields began to decline. In the worse affected districts. 40-50% of coffee fields

Figure 6 Affected plantation in Zaire in 1996. Gaps in the foreground indicate losses © Julie Flood, CABI.

were completely destroyed or abandoned. An estimate by the Ugandan Coffee Development Authority (UCDA) in 1999 indicated that 14.5 million robusta coffee trees (4.8% of the coffee crop) were destroyed country-wide. Yield loss at a district level indicated that CWD caused 77% loss in yield of robusta coffee at the national level (Hakiza et al., 2009)

Initially, in Uganda, there was some success with sanitation at the experimental farm but the approach proved unsustainable due to the high labour cost and the spread of inoculum from coffee in the adjacent forest introducing fresh inoculum. Initially, only robusta coffee was affected in Uganda, but Rutherford et al. (2009) also reported that the disease had been observed on excelsa coffee held at a coffee germplasm conservation site; arabica coffee was free of symptoms in Uganda.

I confirmed the presence of the pathogen (Flood, 1997) as reported by the national scientists and recommended regional collaboration as a means of managing this disease. A proposal was developed as a fully integrated programme of activities addressing different aspects of the disease and its management. Five interrelated and interdependent projects were developed, each dealing with specific components. The collective name of these projects was the Regional Coffee Wilt Programme (RCWP).

8 Case study: Regional Coffee Wilt Programme (RCWP)

The Regional Coffee Wilt Programme (RCWP) (2000–2007) was co-ordinated by CABI, the co-ordination being funded by the CFC.

Activities of the RCWP included biological and socio-economic surveys in project participating countries (DRC, Uganda, Ethiopia, Tanzania, Rwanda, Cote d'Ivoire and Cameroon) to ascertain the extent of the disease and its impact on livelihoods. These surveys were funded by the EU through the Association for Strengthening Agricultural Research in Eastern and Central Africa/African Coffee Research Network. The RCWP also funded research on durable resistance (through EU-INCO-DEV), epidemiology and pathogen variation (funded by UK Department for International Development (Crop Protection Programme) and EU INCO-DEV). There was considerable focus too on extensionist and farmer training and the raising awareness of the disease to facilitate surveillance and limit spread. This component was also funded by the CFC. For further details of the interrelations between the projects see Phiri and Baker (2009).

Summaries of the findings are outlined in the following sections.

8.1 Biological surveys in country

Major findings from the surveys undertaken, indicated that CWD was present in four of the participating countries - in DRC, Uganda, Tanzania on robusta coffee and in Ethiopia on arabica coffee. The disease appeared to be absent from other surveyed countries (Rwanda, Cote d'Ivoire and Cameroon). In Ethiopia, several surveys undertaken prior to the RCWP had indicated the presence of CWD in the major coffee-growing regions (Girma, 1997). From the surveys undertaken under RCWP, it seemed that in Ethiopia, CWD was affected by the farming system with lower rates of infection in forest and semi-forest coffee but much higher rates in garden and plantation coffee (Phiri and Baker, 2009). These observations could be linked to the degree of intervention, with greater opportunity for the spread of the pathogen through agronomic practices. Phiri and Baker (2009) reported that the incidence of CWD under plantation conditions ranged from 77% in Chira to 10.8% in Gomma with a mean of 31%+/− 18% so there was considerable variation in effect. These authors also commented that the strains affecting arabica coffee in Ethiopia may be less aggressive than the strains attacking robusta; Ethiopia had continued to increase national production (Phiri and Baker, 2009). Nevertheless, Girma et al. (2009a) reported many coffee-producing areas had over 50% of infected trees on-farm. For example, the disease was particularly high in garden coffee in the quality coffee production zones of Sidama and Gedeo (with an incidence of over 90% and severity of 25%).

In DRC, the surveys revealed that the disease was present in North Kivu, Oriental and Equateur provinces but absent in provinces of Bandundu, Bas Congo, Kasai Occidental, Kasai Orientale and South Kivu (Kalonji-Mbuyi et al., 2009). The highest incidence of the disease was in Uganda, where 90% of farms surveyed had CWD and an average severity of 45%. In Tanzania, CWD was only detected in the Kagera region that borders Uganda, but the incidence and severity were increasing in that region (Kilambo et al., 2009).

8.2 Socio-economic surveys in country

Socio economic surveys were also undertaken under the RCWP, and the results suggested that in many of the participating countries, coffee was in a poor state with poor soil fertility and old tree stock (>59% over 15 years old). None of the farm characteristics, agronomic practices or environmental characters had a consistent effect on CWD occurrence across Uganda, Ethiopia and Tanzania. However, many farmers weeded using hand tools - machete and hoes - which can wound the trees, aiding the invasion of the pathogen. Interestingly, Tanzanian farmers conducted more mulching and intercropping together with less pruning and slashing (all practices believed to lessen the spread and impact of the disease). This may have contributed to the overall lower incidence of CWD seen in Tanzania (Phiri and Baker, 2009), but the outbreak was confined to one area in that country. In contrast, Ugandan farmers used lower levels of fertilization and highest level of pruning and did little weed slashing, but still the incidence was high in the country. Farmers' awareness of CWD varied widely from country to country, with 90% of farmers being aware of CWD in Uganda but only 17% in Ethiopia and 15% in Tanzania.

Farmers found it difficult to understand the nature of the disease and this was reflected in their use of diseased coffee wood - often dragging it through their stands of living trees, for use as firewood or selling it as firewood to others. Many used diseased stems as supports for other crops (tomatoes and beans) as well as using it for fencing, or for house and animal shed construction. These actions undoubtedly contributed to the spread of the disease. Initially, farmers tried to up-root and burn the affected bushes (47%) or conducted pruning (26%), although this was not practiced in Ethiopia. Cutting down the trees and burning the above ground materials had also been attempted (12%) whilst some farmers tried removing the bark of infected trees (7%). The use of various "concoctions" such as ash, cow urine had also been attempted. Yet by the time of the surveys, 80% of farmers interviewed had given up trying to manage CWD because they considered the activities to be ineffective. Some farmers (6%) also indicated that management was too labour intensive - they could not undertake measures themselves and hiring labour was too expensive. Farmers surveyed obtained management information from extension staff (37%) or from other farmers (22%), with 19% receiving information from the radio, newspapers, pamphlets and direct contact with researchers. As farmers could not undertake management options, many farmers adopted coping strategies (Phiri and Baker, 2009) which included diversification of crop and activities such as brewing and trade or opening new coffee fields, etc. Many farmers in Uganda had abandoned their fields when the disease was spreading quickly in the country (early 1990s–early 2000s), while in Ethiopia, farmers replanted in a new locality (77%) or replaced with new coffee bushes (22%). There was less diversification in Ethiopia, which again may suggest a less aggressive strain or there was a greater possibility to relocate their coffee production to new areas.

Phiri and Baker (2009) concluded it was difficult to summarise across the countries, but they suggested a combination of education and motivation lead some farmers to make sensible agronomic choices. For details of the history, spread, incidence and management of CWD in each participating country, see chapters in Flood (2009).

8.3 Studies of the pathogen

Also as part of the RCWP, studies of pathogen variability were undertaken.

The presence of a sexual stage (*Gibberella xylarioides*) implies some degrees of variability, but relatively little was known about this. With the emergence of the pathogen as a major constraint to coffee in Africa, coupled with the development of molecular methodologies, there was an opportunity to examine the pathogen. A broad range of primarily PCR-based approaches, microsatellite loci, AFLPs, RAPDs, ISSRs, nuclear housekeeping genes, IGS and ITS regions were used. Several hundred isolates of the pathogen were examined (Rutherford et al., 2009) and although overall variability within the pathogen appeared to be limited, two major distinct populations were detected (Adugna et al., 2005; Tshilenge-Djim et al., 2004). The first were those isolates obtained from robusta coffee affected in DRC, Uganda and Tanzania (since the re-emergence of CWD), while the second group of isolates was obtained from arabica coffee in Ethiopia. Genetical variability among isolates recovered historically from robusta and excelsa coffee from CAR, Guinea and Côte d'Ivoire indicated they were all different to the two populations responsible for the disease in the 1990s (Rutherford et al., 2009). Yet, isolates from robusta coffee in the late 20th-century epidemic were identical to those in DRC in 1960, which suggested that CWD had never been completely eradicated in DRC after the mid-20th-century outbreak and the disease has spread from DRC to neighbouring countries (Lepoint, 2006; Phiri and Baker, 2009). Unfortunately, the original isolate from Steyaert (1948) is lost, so we can never know how that links to the later isolates.

Mating trials between the isolates from the different hosts (arabica, robusta and excelsa) suggested three biological species. A fourth group was suggested as a group of incompatible strains. Lepoint et al. (2005) and Lepoint (2006) proposed that there could be four different species, namely, *Gibberella abyssiniae* (responsible for the Ethiopian outbreaks *on arabica); G. congoensis* (responsible for epidemic in DRC, Uganda and Tanzania on robusta); *Fusarium quineensis*, an asexual form responsible for the mid-century outbreaks in Guinea, Côte d'Ivoire and possibly CAR; and *Gibberella xylarioides*, originally described as a disease on *Coffee excelsa*. Investigating the host-pathogen interactions, Girma et al. (2009b) reported host specialization with isolates from robusta causing disease on robusta but not on arabica and vice versa. This linked to previous field reports of the disease affecting robusta in Uganda but not arabica in that country (Flood, 1997). Yet one strain (DSMZ62457) caused some mortality in seedlings of robusta, arabica and excelsa, and Girma et al. (2009b) suggested that further work was required to confirm host specificity.

8.4 Host variation and screening for resistance

Significant variations were detected between wild and cultivated populations of robusta coffee in Uganda. Ugandan genotypes were sufficiently different from the Congolese and Guinean types to represent a genetically distinct diversity within the *C canephora* genome. It was suggested (Phiri and Baker, 2009) that these valuable resources need to be planted at secure conservation locations as a matter of urgency because of problems with CWD and with rapid deforestation. Implementation of *in situ* conservation would also help sensitise local communities to ensure farmers multiply, conserve and utilise the plants in their own fields to maintain genetic diversity (Musoli et al., 2006).

In addition to conservation of valuable genetic resources, screening robusta coffee for resistance was also undertaken in Uganda, DRC and Tanzania. Various methods were evaluated in Uganda to establish a reliable, cost-effective method of evaluating coffee germplasm to CWD. Two methods were trialled: the root dip method, where coffee roots were dipped in a spore suspension of the pathogen for 20 minutes, and, the stem nicking method, involving scraping the stem of coffee seedlings with a scalpel (with inoculum). Standard concentrations and exposure times established that spore concentrations as low as 13 spores per ml were sufficient to cause seedling death (Phiri and Baker, 2009). Wild robusta from the Kalangala and Itwara forests showed a high level of resistance to CWD. In Uganda, there was a considerable variation in the susceptibility of field-grown clones to the disease, while in Tanzania, all commercial clones that had been previously released were susceptible to CWD (using similar screening methods). In Tanzania, the Maruka germplasm collection seemed to have high levels of resistance, but more field work is needed to confirm this. Several genotypes showed resistance in screening trials in DRC, with mortality rates lower than in Uganda and Tanzania. Studies in Uganda highlighted that CWD resistance was heritable, but its transmission from parent to progeny is only about 33%, suggesting that commercial CWD-resistant robusta varieties should be cloned. In Uganda, a major breeding programme involving screening for resistance as well as agronomic traits was initiated. Seven candidate clones were released to farmers in 2010 with the intention of further evaluating these clones under farmer conditions as well as agronomic traits such as cup quality and yield.

Arabica coffee was screened in Ethiopia and the stem nicking method at the cotyledon stage was found to be the most effective way to screen for resistance (Girma et al., 2009b). In trials, Catimor lines J19 and 7440 were resistant, but Caturra Rojo and cultivar 24./85 were more susceptible (Girma et al., 2009a). Having tested arabica cultivars from several countries including Kenya, Girma et al. (2009b) concluded arabica strains of CWD would pose a serious threat to *C arabica* in other countries.

Host-pathogen interaction studies in Uganda (Musoli et al., 2009) indicated that although there was host specialisation, it was not clear cut; one isolate (CAB007) collected from Arabica in Ethiopia induced some initial symptoms on *C. canephora*, but it was not fatal. This may be due to the inoculation method used. Further, several isolates from *C. arabica* and *C. canephora* induced symptoms in *C. liberica*, and Musoli et al. (2009) suggested that further studies are required to confirm host specificity.

8.5 Transmission pathways and management

Transmission from contaminated wood to adjacent un-infected seedlings was confirmed in screen house trials (Phiri and Baker, 2009). This confirmed that leaving infected wood near uninfected trees in the field is a significant pathway and underlines the concern about use of contaminated wood for fences, firewood and stakes. CWD transmission from contaminated soil to healthy seedlings was also confirmed. Infectivity in soil lasted at least 3 months but then declined. Consequently, a fallow period could be a useful management strategy. A fallow of 1 year was advised before replanting to avoid re-infection. Despite proving that contaminated soil could be an inoculum source, it is difficult to isolate the pathogen from soil, which would be the critical test to assess the pathogen's presence. The pathogen was not isolated from insect pests of coffee (including coffee berry borer) or from bees or termites on affected farms (Rutherford and Flood, 2005) A wide-scale survey of 6 coffee growing districts in Uganda (with a high incidence of CWD) revealed that of 270 plant samples (105 species of crops and weeds) the pathogen was not recovered including from roots (Kangire et al., 2002). Yet Serani et al. (2007) reported isolation of the pathogen from banana roots, and these isolates induced symptoms in inoculated coffee plants. This raises concerns about the use of bananas as an intercrop and a replacement crop for coffee. Artificial inoculation of a wide range of crops and weeds (from coffee fields) with the pathogen is needed. There was evidence that wounds made with a machete previously used to cut infected coffee stems could transmit the pathogen to uninfected trees, but Phiri and Baker (2009) advised further work using more replication and more comparison of the positioning of, for example, wounds near the tree base and pruning wounds in branches in order to have some definite conclusions.

8.6 Environmental influence on infection

With regard to favourable environmental conditions, Musoli et al. (2008) reported no apparent effect of temperature on CWD development but temperatures do not fluctuate markedly in Uganda (Phiri and Baker, 2009). Some negative correlation between the number of coffee plants showing symptoms for the first time and severity of rainfall at least in first 18 month of assessment and the number of symptomatic plants increased during periods of low rainfall that followed periods of high rainfall. This pattern could be linked to death of roots during the wet season, allowing more infection to occur while symptoms become more apparent in the dry season when vascular occlusion (pathogenic activity) and drought combine to ensure insufficient water is being distributed internally. It is likely that with extended dry seasons predicted in future climatic scenarios that CWD could have a more severe impact.

8.7 Agronomic treatments of CWD

A major part of the RCWP was the initiation of on-farm participatory trials in Ethiopia, DRC, Tanzania and Uganda. Through preliminary workshops, farmers were consulted about the treatments to be applied and some of their suggestions were incorporated. The trials were then evaluated over a number of years. Treatments included mulching and herbicides to reduce slashing, use of fungicide stem paints to be applied to the base of trees, fungicides sprays and applications of ash. On farm trials and on station trials were set up. Workshops were held to discuss the results and any promising options were incorporated into Farmer Field School curricula. Copper-based fungicides used as stem paint applications and herbicide sprays showed promise but were too expensive for many producers. Interestingly, although effective in DRC and Uganda, stem paints were not so effective in Ethiopia. Stems were painted with copper oxychloride, which is contact fungicide but *F xylarioides* is a systemic pathogen so the reduction in incidence could be linked to the indirect effect for example of sealing of wounds (Phiri and Baker, 2009) or an effect on the pathogen's development under the bark or production of perithecia at the tree base. Reducing the use of machetes, that is, reducing the chances of wounding trees had a big effect on reducing CWD incidence. As herbicides were expensive for farmers, slashing with machetes was advocated together with hand weeding around the base of the trees to reduce the incidence of wounds at the base of the trees. This approach was very effective especially in Ethiopia. The application of mulches to the base of the trees was also beneficial in Ethiopia and could have been due to mulches being a weed suppressant but also may have influenced the nutrient status of the plants.

8.8 Recommendations to reduce CWD

From the results of the trials and investigations, recommendations stressing an integrated management approach were made (Phiri and Baker, 2009). This approach included regular monitoring to detect infections early and management options that included regular cleaning of tools following use on trees with symptoms of CWD (using household disinfectant or heating); minimising inadvertent wounding made whilst slashing weeds; uprooting and burning of infected trees *in situ*; removal of coffee tree stumps; leaving a fallow period of at least 3 months but, preferably, planting an intercrop and leaving for a year before replanting coffee; replanting with more resistant coffee material (if available); discouraging the use of diseased firewood and if essential (due to fire wood shortage) cutting the short lengths *in situ and* placing them in a sack rather than dragging the diseased wood through areas with unaffected trees.

Apart from on-farm activities, regional and national distribution of coffee material may constitute a means of transmission, so it was recommended that distribution systems of coffee materials from nurseries are investigated further. Quarantine was also strongly advocated by Phiri and Baker (2009). Monitoring and surveillance needs to be maintained to reduce the possible spread to other coffee-growing areas of Africa and to other coffee-growing areas worldwide.

Phiri and Baker (2009) stressed that the re-emergence of CWD in Central Africa should be an example of what happens when there is little preparedness for disease epidemics and when governments do not act quickly enough to contain the disease and reduce its impact. There should be lessons learned from this disease - the need for improved preparedness in order to achieve greater sustainability in coffee production in Africa. Chemical control is difficult for this pathogen, but painting cut stems and the base of infected plants with copper fungicides did seem to help reduce incidence, and for valuable germplasm (e.g. in germplasm collections), benomyl applied as root drenching (once every month) may be effective (Phiri and Baker, 2009). Nevertheless, wide-scale use of fungicides is not advocated due to cost to smallholders and to environmental concerns. One of the major concerns about the destructive ability of this pathogen was that it attacks all *Coffea* species including wild relatives, which normally could be a likely source of resistance to be used in breeding programmes, so conservation of germplasm is essential. Breeding for resistance remains an ideal strategy for managing this disease especially for smallholders. In-country breeding programmes in Ethiopia, Uganda and Tanzania were ongoing at the time of the report (Phiri and Baker, 2009), but this is likely to be a long-term approach which will need significant investment. It is also important that countries share any resistant germplasm.

This integrated approach would use a combination of cultural, mechanical, physical, chemical and biological control in a systematic way. However, for this approach to be effective, a greater awareness of the impact of CWD was needed. Consequently, a major effort (through the RCWP) was made to train a range of stakeholders in CWD-affected countries and also to raise awareness of the threat of CWD in Africa and beyond. Over 2500 extensionists were trained as trainers and facilitators under the RCWP. In addition to the technical training of CWD identification and management, FFS members learned better husbandry practices for coffee which often resulted in improved productivity. It is estimated that more than 1 million farmers were trained by extensionists between 2002 and 2007, including a substantial number of women farmers.

An estimated total of 487 700 items of information concerning CWD were produced under the RCWP and were disseminated to coffee-growing regions of Ethiopia, Uganda, Tanzania, DRC and Rwanda. Printed materials were produced in 13 languages and over 250 radio broadcasts were made in 5 countries as well as television programmes reaching tens of millions or people (Negussie et al., 2009).

9 Future trends in research

Although the RCWP had a significant impact on understanding more about the pathogen, the disease, its spread and management, there remain some interesting avenues for future research. This includes a greater understanding of the proposed host specificity. Buddie et al. (2015) confirmed the complexities of the CWD pathogen already highlighted by previous researchers (Lepoint et al., 2005; Lepoint, 2006; Girma et al., 2009b; Rutherford et al., 2009) regarding the presence of multiple taxa (or biological species) within *G. xylarioides sensu lato*. Buddie et al. (2015) emphasized the significant and consistent differences observed in IGS amplicon size between C. *arabica* and C. *canephora* strains, which would be of considerable importance to plant pathologists and epidemiologists monitoring development and spread of CWD as well as for breeders trying to derive resistant germplasm, and for national and local authorities providing management advice and technical support to farmers. Buddie et al. (2015) further tested the species concepts in this complex using additional markers and the extent of similarity across all *G. xylarioides* strains examined, and implied that the population is largely homogenous but with clear variation, allowing discrimination of strains from *C. arabica*, C. *canephora* or from *C. excelsa*. These conclusions were supported by the largely nuclear ISSR analysis, and may suggest recent divergence of the two established host-specific groups, but this does need further study.

Also, the strains are heterothallic, and perithecia have been reported to occur frequently on coffee trees in the field (Lepoint et al., 2005; Hakiza et al., 2009), so the lack of genetic variation revealed in a sexually reproducing population is surprising, suggesting that other factors may be involved, which needs further research. Members of the *G. fujikuroi* complex are known to include strains with various 'spore killer' genes that, depending on mating combination, can result in perithecia containing only one genetic type of ascospore or perithecia with aborted ascospores (van der Gaag et al., 2000). If this were the mechanism for the CWD pathogen, it would have major implications for selection in, and management of, a largely clonal pathogen reliant on development and dispersal primarily through mycelial growth and conidia production. It would give coffee breeders more confidence in trying to identify or develop resistance to currently recognised forms of the pathogen. Further work is needed on this aspect. One output of the RCWP was the establishment of CWD isolate collections at CABI (UK), in CIRAD (France) and in UCL (Belgium), so further studies can be conducted.

Another area for future research involves understanding more about the underlying resistance mechanisms in this H-P interaction, which would aid breeding programmes, but little has been published on this. Breeding programmes for cultivars of both arabica and robusta were ongoing in the first decade of the 21st century (Flood, 2009) and it would be useful to understand more about the progress of these breeding programmes including adoption by farmers of any new planting material. For a smallholder production system, the use of resistant material provides a more environmentally sustainable management option. Kilambo et al. (2012) working at the Tanzania Coffee Research Institute reported some success with the screening of 875 breeding lines of *C. canephora* (artificially inoculated with pathogen using the root dip method) and 201 lines were found to be resistant to CWD. In 2006, these 201 genotypes were planted in clonal mother garden to raise planting materials for field evaluations. Production and cup taste of the 875 lines were also assessed between 2001 and 2008. Six CWD-resistant robusta lines that were also resistant to leaf rust with productivity ranging from 0.5 kg to 3.0 kg of clean coffee per tree and cup taste described as 'clean/smooth cup of natural robusta' were selected. These lines were planted in multi-location evaluation trials in CWD hot spot areas in the Kagera region of Tanzania. Eighteen months later the varieties were considered resistant, and Kilambo et al. (2012) reported that the six breeding lines were expected to be released for commercial use. It would be helpful to have updates of progress with trials and with the proposed commercial release. The availability of such resistant material to researchers would also allow investigation of resistance mechanisms.

Understanding more about alternative hosts for this pathogen would also be useful in determining a field management strategy. The pathogen has been isolated from tomatoes (Onesirosan and Fatunla, 1976) and from cotton (Pizzinatto and Menten (1991), so potentially, members of the *Malvaceae* and *Solanaceae* found in coffee fields (weeds) could act as alternative hosts and should be tested by artificial inoculation. Similarly, the pathogen was isolated from bananas (Serani et al., 2007), indicating that *Musa* species could act as an alternative host. The reduction of alternative hosts in the coffee gardens would reduce inoculum as part of an integrated approach.

Biological control has not really been exploited for this pathogen and given the vascular nature of the pathogen could be difficult due to the specific ecological niche, that is, the vascular system to which these pathogens have adapted. However, many investigations have reported success for different forms of the species of *Fusarium oxysporum* (Alabouvette et al., 2009), including the use of non-pathogenic fusaria to compete for infection sites on the roots of host plants (to reduce inoculum prior to entry into the plant). Nevertheless, these authors also noted that, despite an increasing number of papers dealing with biological control, there are very few products available especially in the field situation and more research into the field aspects of biological control are needed. Mulaw et al. (2013) have reported the successful antagonism of endophytic *Trichoderma* to *F. xylarioides*, but the interactions were only studied in *vitro*. More research on these interactions *in vivo* are needed including glasshouse studies, but ultimately, trials are required under conditions in the farmers' fields. We need to determine a practical approach for producers to be used as part of any ongoing integrated management of the disease. It was also suggested by Phiri et al. (2009) that a biocontrol approach could be further trailed using antagonistic microbes as sealants for wounds following rejuvenation (stumping) of coffee trees.

An alternative approach to using a specific antagonist as a putative biocontrol would be to understand more of the ecology of the pathogen in the soil with a view to increasing the diversity and activity of a wide range of antagonistic microbes so as to decrease the pathogen inoculum. Such approaches could include the use of mulches (a practical approach for the farmers) to encourage and support a more diverse microbiome. In recent years, the study of disease organisms in crops including coffee has been greatly facilitated through the development of next-generation sequencing technologies. This includes comparative genomics through the analysis of whole genome sequences but also the more in-depth analysis of microbial communities in the rhizosphere (and phyllosphere) through metagenomics and transcriptomics approaches. The latter have been particularly useful in developing our understanding of the microbiome and the role it plays in the functioning of the phytobiome. These approaches have broader implications for sustainable coffee management and the development of biological solutions to reduce losses and enhance yield.

Phiri and Baker (2009) previously stressed that the re-emergence of CWD in Central Africa was an example of what happens when there is little preparedness for disease epidemics and when governments do not act quickly enough to contain the disease and reduce its impact. Although not a future research trend as such, improved biosecurity and preparedness are needed going forward. On-going monitoring of this disease (and other coffee diseases) is required and regular surveys should be undertaken across the African continent to detect new pathogens, new strains and new antagonists as well as establishing protocols to deal with these constraints. In addition, other coffee-producing countries in Africa and beyond need to be aware of the threat of CWD. In trials, arabica cultivars from several countries including Kenya were susceptible to the arabica strain of the pathogen (Girma et al., 2009b), so raising awareness of the threat from CWD remains critical for sustainable coffee production globally.

10 Where to look for further information

1 Bieysse, D. (2007). Development of a long-term strategy based on genetic resistance and agro-ecological approaches against coffee

wilt disease in Africa. Final Technical Report INCO-DEV Contract ICA4-CT-2001-100006. Montpellier, France: CIRAD-AMIS.
2 CABI (2020). *Gibberella xylarioides*. In: *Crop Protection Compendium*. Wallingford, UK: CAB International. www.cabi.org/cpc.
3 Flood, J. (2009). *Coffee Wilt Disease*. Wallingford, UK: CAB International, 199 pp.
4 Phiri, N. and Baker P. (2009). A synthesis of the work of the Regional Coffee Wilt Programme 2000-2007. Coffee Wilt Disease in Africa. Wallingford, UK: CABI, 233 pp.
5 Lepoint, P. C. E. (2006). Speciation within the African coffee wilt pathogen. PhD Thesis. Belgium Universite Catholique de Louvain, Belgium.

11 References

Abrego, L. (1965). Coffee trees killed by stem pitting disease. *Nota Fitopatologica Bot. Inf. Int. Salvador. Tropical Abstracts* 21, 550.

Adugna, G., Hindorf, H., Steiner, U., Nirenberg, H. I., Dehne, H. W. and Schellander, K. (2005). Genetic diversity in the coffee wilt pathogen (*Gibberella xylarioides*) populations: differentiation by host specialization and RAPD analysis. *Zeitschrift fur Pflanzenkrankheiten und Pflanzenschutz* 112, 134-145.

Alabouvette, C., Olivain, C., Migheli, Q. and Steinberg, C. (2009). Microbiological control of soil-borne phytopathogenic fungi with special emphasis on wilt-inducing Fusarium oxysporum. *New Phytologist* 184(3), 529-544. DOI 10.1111/j.1469-8137.2009.03014.x.

Anon (1989). *A New Disease of Coffee in South Africa? Information Bulletin of Citrus and Subtropical Fruit Research Institute* (vol. 205), p 13.

Beckman, C. H. (1987). *The Nature of Wilt Diseases of Plants*. St Paul, MN: The American Phytopathology Society. ISBN:0-89054-074-8, p 2.

Bouriquet, G. (1959). Plant diseases and pests in some African territories. *FAO Plant Protection Bulletin* 7, 61-63.

Buddie, A. G., Crozier, J., Rutherford, M. A., Flood, J. and Bridge, P. D. (2015). Population development within the coffee wilt pathogen *Gibberella xylarioides* reflects host-related divergence. *European Journal of Plant Pathology* 142(2), 291-304. DOI 10.1007/s10658-015-0613-z.

CABI (2020). *Gibberella xylarioides*. In: *Crop Protection Compendium*. Wallingford, UK: CAB International. Available at: www.cabi.org/cpc.

Chiarappa, L. (1969). International assistance in plant pathology in developing countries with particular reference to FAO programmes. *FAO Plant Protection Bulletin* 17, 7-8.

Deassus, E. (1954), La tracheomycosis de Cafeier. *Bull. Sci. Minist.Colon. Sect. Agron. Trop.*, 5, 345-348.

Figueres, R. (1940). *Sur une maladie tres grave du cafeier en Oubangui*. Paris: Rapport. Miniestere des colonies.

Flood, J. (1996). A Study of Tracheomycosis or Vascular Wilt Disease of Coffee in Zaire. Report Presented to Zairean Coffee Organization (OZACAF) and the International Coffee Organization (ICO). 13pp.

Flood, J. (1997). Tracheomycosis or Vascular Wilt Disease of Coffee in Uganda. Report Presented to the Ugandan Coffee Development Authority (UCDA) and the International Coffee Organization (ICO). 12pp.

Flood, J. (2009). *Coffee Wilt Disease*. Wallingford, UK: CAB International, 199 pp.

Flood, J. and Brayford, D. (1997). Re-emergence of Fusarium wilt of coffee in East and Central Africa. *Proceedings of the 17th International Conference on Coffee Science*, July, 1997. Paris, France: Association Scientifique Internationale du Cafe (ASIC), pp 621-628.

Fraselle, J. (1950). Observations preliminaires sur une tracheomycosis de *Coffea robusta*. *Bulletin Agricole du Congo Belge* 41, 361-372.

Fraselle, J. V., Vallaeys, G. and De Knop, O. (1953). La lutte contre la tracheomycose du cafeier a Yangambi et le probleme que pose actuellement cette maladie au Congo Belge. *Bulletin Agricole du Congo Belge* 116, 373-394.

Gaudy, M. R. (1956). Contribution du techniques, scientifique ou development de l'agriculture en Afrique Occidentale Francaise. *Journal of the West African Science Association* 2, 172-197.

Girma, A. (1997). Status and Economic Importance of Fusarium Wilt Disease of Arabica Coffee in Ethiopia. *First Regional Workshop on Coffee wilt Disease (Tracheomycosis)*. Kampala, Uganda, pp 53-61.

Girma, A., Hulluka, M. and Hindorf, H. (2001). Incidence of tracheomycosis, *Gibberella xylarioides* (*Fusarium xylarioides*), on Arabica coffee in Ethiopia. *Zeitschrift fu^umlaut~r Pflanzenkrankheiten und Pflanzenschutz* 108(2), 136-142.

Girma, A., Million, A., Hindorf, H., Arega, Z., Teferi, D. and Jefuka, C. (2009a). Coffee wilt disease in Ethiopia. In: Flood, J. (Ed), *Coffee Wilt Disease*. Wallingford, UK: CAB International, pp 50-68.

Girma, A., Bieysse, D. and Musoli, P. (2009b). Host pathogen interactions in the *Coffea-Gibberella xylarioides* pathosystem. In: Flood, J. (Ed), *Coffee Wilt Disease*. Wallingford, UK: CAB International, pp 120-136.

Hakiza, G. J., Kyetere, T. D., Musoli, P., Wetala, P., Njuki, J., Kucel, P., Aluka, P., Kangire, A. and Ogwang, J. (2009). Coffee wilt disease in Uganda. In: Flood, J. (Ed), *Coffee Wilt Disease*. Wallingford, UK: CAB International, pp 28-49.

Heim, R. and Saccas, A. (1950). La tracheomycose des *Coffea* et robusta des plantations de l'Oubangui-Chari. *Revue de Mycologia* 15 (Suppl. Colonial), 97.

Kalonji-Mbuyi, A., Tshilenge Dijm, P. and Saiba, N. T. (2009). Coffee wilt disease in the Democratic Republic of the Congo. In: Flood, J. (Ed), *Coffee Wilt Disease*. Wallingford, UK: CAB International, pp 7-27.

Kangire, A., Kytere, D., Hakiza, G., Warren, H., Erbaugh, M. and Kabol, C. (2002). Determination of Alternative Hosts for *F. xylarioides*, Causal Agent of Coffee Wilt Disease (CWD) in Uganda. Internal Report. Uganda: Coffee Research Institute.

Kilambo, D. L., Ng'Homa, N. M., Teri, J. M. and Masumbuko, L. (2009). Status of coffee wilt disease in Tanzania. In: Flood, J. (Ed.), *Coffee Wilt Disease*. Wallingford, UK: CAB International, pp 69-82.

Kilambo, D. L., Ng'Homa, N. M., Mtenga, D. J., Teri, J. M. and Phiri, N. (2012). Progress in breeding for resistance to Coffee Wilt Disease (Tracheomycosis) in Tanzania. *World Research Journal of Agricultural & Biosystems Engineering* 1(1), 12-16. ISSN:2319-3913 & E-ISSN: 2319-3921.

Kranz, J. and Mogk, M. (1973). *Gibberella xylarioides* Heim et Saccas on arabica coffee in Ethiopia. *Journal of Phytopathology* 78(4), 365-366.

Lejeune, P. (1958). Rapport Au Governement Imperial D'Ethiopie sur le Production Cafeiere. Rome: FAO /58/3/1881.

Lepoint, P. C. E. (2006). Speciation within the African coffee wilt pathogen. PhD Thesis. Belgium: Louvain. Universite Catholique de Louvain.

Lepoint, P. C. E., Munaut, F. T. J. and Maraite, H. M. M. (2005). *Gibberella xylarioides sensu lato* from *Coffea canephora*: a new mating population in the *Gibberella fujikuroi* species complex. *Applied and Environmental Microbiology* 71(12), 8466–8471.

Mulaw, T. B., Druzhinina, I. S., Kubicek, C. and Atanasova, L. (2013). Novel endophytic *Trichoderma* spp. Isolated from healthy *Coffea arabica* roots are capable of controlling coffee tracheomycosis. *Diversity* 5(4), 750–766. DOI 10.3390/d5040750.

Musoli, P. C., Aluka, P., Cubry, P., Dufour, M., de Bellis, F., Nakendo, S., Nabaggala, A., Ogwang, J., Kyetere, D., Leroy, T., Bieysse, D. and Charrier, A. (2006). Fighting against CWD: Uganda: Uganda wild *C. canephora* genetic diversity and usefulness. *Proceedings of the 21st International Conference on Coffee Science*. Montpelier, France: Association Scientifique Internationale du Cafe (ASIC), pp 1268–1276.

Musoli, P. C., Pinard, F., Charrier, A., Kangire, A., ten Hoopen, G. M., Kabole, C., Ogwang, J., Bieysse, D. and Cilas, C. (2008). Spatial and temporal analysis of coffee wilt disease caused by *Fusarium xylarioides* in *Coffea canephora. European Journal of Plant Pathology* 122(4), 451–460.

Musoli, P., Girma, A., Hakiza, G., Kangire, A., Pinard, F., Agwanda, C. and Bieysse, D. (2009). Breeding for resistance against coffee wilt disease. In: Flood, J. (Ed), *Coffee Wilt Disease*. Wallingford, UK: CAB International, pp155–175.

Negussie, E., Kimani, M., Girma, A., Phiri, N. and Teshome, D. (2009). Extension approaches and information dissemination for coffee wilt disease management in Africa: experiences from Ethiopia. In: Flood, J. (Ed), *Coffee Wilt Disease*. Wallingford, UK: CAB International, pp 176–195.

Onesirosan, P. T. and Fatunla, T. (1976). Fungal fruit rots of tomatoes in Southern Nigeria. *Journal of Horticultural Science* 51(4), 473–479.

OZACAF (1995). Project de lutte contre la tracheomycose du Cafeier au Zaire. Report for ICO on Coffee Wilt in Zaire. Kinshasa: OXACAF.

Phiri, N. and Baker, P. (2009). *Coffee Wilt in Africa: Final Technical Report of the Regional Coffee Wilt Programme (2000-2007).*Wallingford, UK: CABI Publishing, 233 pp.

Phiri, N., Kimani, M., Negussie, E., Simons, S. and Oduor, G. (2009). Management of coffee wilt disease. In: Flood, J. (Ed), *Coffee Wilt Disease*. Wallingford, UK: CAB International, pp 137–154.

Pieters, R. and van der Graaff, N. A. (1980). Resistance to *Gibberella xylarioides* in *Coffea arabica*: evaluation of screening methods and evidence for the horizontal nature of the resistance. *Netherlands Journal of Plant Pathology* 86(1), 37–43.

Pizzinatto, M. A. and Menten, J. O. M. (1991). Pathogenicity of eight *Fusarium* species isolated from seeds to cotton seedlings. *Summa Phytopathologica* 17, 124–134.

Porteres, R. (1959). Valeur agronomique des Cafeiers des types Kouilou et robusta cultivars en Cote d'Ivoire. *Cafe, Cacoa. Therapie* 3, 3–13.

Rabechault, H. (1954). Sur quelques facteurs de resistance du Cafier a la Tracheomycose. *Bull. Sci. Minist. Colon. Sect. Agron.Trop.* 5, 292–295.

Rutherford, M., Bieysse, D., Lepoint, P. and Maraite, H. M. M. (2009). Biology, taxonomy and epidemiology of the coffee wilt pathogen *Gibberella xylarioides sensu lato*. In: Flood, J. (Ed), *Coffee Wilt Disease*. Wallingford, UK: CAB International, pp 99–119.

Rutherford, M. and Flood, J. (2005). Epidemiology and Variability of *Gibberella xylarioides*, the Coffee Wilt Pathogen. UK Department for International Development (DFID) Crop Protection Programme Project R8188. Final Technical Report. Egham. UK: CABI Publishing.

Saccas, A. M. (1951). La tracheomycose (carbunculariose) des *Coffee excelsa, neo-arnoldiana et robusta* en Oubangui- Chari. *Agronomique Tropicale* 6, 453-506.

Saccas, A. M. (1956). Recherches experimentales sur la Tracheomycose des Cafeiers en Oubangui-Chari. *Agronomie Tropicale Nogent* 11, 7-38.

Serani, S., Taligoola, H. K. and Hakiza, G. J. (2007). An investigation into *Fusarium* spp. associated with coffee and banana plants as potential pathogens of robusta coffee. *African Journal of Ecology* 45(s1), 91-95.

Stewart, R. B. (1957). Some diseases occurring in Kaffa province, Ethiopia. In: *Imperial Ethiopian College of Agriculture and Mechanical Arts*. Ethiopia: Alemay, pp 15-16.

Steyaert, R. L. (1948). Contribution a l'etude des parasites des vegetaux du Congo Belge. *Bulletin de la Societe Royale de Botanique du Belque* 80, 11-58.

Tshilenge-Djim, P., Munaut, F., Kalonji-Mbuyi, A. and Maraite, H. (2004). Caracterisation des *Fusarium* spp associees au deperissement du cafeier robusta en Republique Democratique du Congo. *Parasitica* 60, 19-34.

van der Gaag, M., Debets, A. J. M., Oosterhof, J., Slakhorst, M., Thijssen, J. A. G. M. and Hoekstra, R. F. (2000). Spore-killing meiotic drive factors in a natural population of the fungus *Podospora anserina*. *Genetics* 156(2), 593-605.

van der Graaff, N. A. and Pieters, R. (1978). Resistance levels in *Coffea arabica* to *Gibberella xylarioides* and distribution pattern of the disease. *Netherlands Journal of Plant Pathology* 84(4), 117-120.

Zadoks, J. C. and van Leur, J. A. (1983). Durable resistance and host-pathogen-environment interaction. In: Lambert, L., Waller, J. M. and van der Graaff, N. A. (Eds) *Durable Resistance in Crops*. New York: Plenum Press, pp 125-140.

Chapter 12

Integrated management of nematodes of coffee

Regina M. D. G. Carneiro and Marcilene F. A. dos Santos, Embrapa Recursos Genética e Biotecnologia, Brazil

1 Introduction

Coffee is a widely used global commodity that is threatened by climate change. Increasing temperatures and rainfall variability are predicted in the Americas, Africa and Asia. By 2050, some countries in the Americas could see a reduction of 40–50% in the land area that is suitable to grow coffee. As the air temperature increases, there will be less and less suitable altitudes at which coffee can grow. Rainfall will become more variable. It is known that coffee is very sensitive to temperature, rainfall and duration of sunshine, suggesting that any substantial climatic change will have a significant effect on coffee growth and yields (Groenen, 2018).

These climatic changes may also exacerbate the pests and diseases affecting these regions that have already caused millions of dollars in damage and losses to the coffee industry. Among the factors limiting coffee growing and production, plant parasitic nematodes, soil-borne pests and soil diseases are capable of parasitizing the plant's root system during practically the entire crop cycle in the field. Roots are fundamental plant organs responsible, among other things, for water, nutrient and mineral absorption, as well as producing substances vital to plant physiology. Successful coffee plant growth depends on maintaining an adequate balance in growth and function between roots

http://dx.doi.org/10.19103/AS.2021.0096.27

and branches. With the predicted changes in climate, the development of an efficient root system better adapted to changing soil and environmental conditions is crucial to maintaining and enhancing crop productivity. Root traits associated with improved adaptation to rising temperatures are increasingly being analyzed to obtain more suitable crop varieties (Calleja-Cabrera et al., 2020).

Soil-borne nematodes, pests and diseases depend on the host plant for their development and reproduction. Parasitism of root systems by nematodes, pests and diseases compromises plant development. Pests (such as nematodes) and diseases are known to be able to adapt to changes in abiotic conditions. The emergence of desiccation and heat-tolerant races of pests, together with increased potential susceptibility of agricultural crops to plant parasite nematodes, will be new challenges for plant breeders and growers (Gaur, 2011). There have been many studies on the vast number of nematode species, fungi and insects capable of parasitizing coffee roots. However, the extent of pest pressure can differ greatly from country to country and even from region to region in the same country. These differences and their effects depend on factors such as edaphoclimatic conditions in different regions, cultivars planted, cultural practices and existing pest population levels.

Problems affecting coffee plant roots arising from parasitism by soil nematodes, pests and diseases are often difficult to detect. In general, when diseased coffee trees are observed, a thorough survey of their root systems should be carried out first. Monitoring through the collection of root and soil samples for laboratory analysis is an important strategy in pest management as it makes it possible to identify and quantify the numbers of a particular pest. Based on information from monitoring, it is possible to select suitable management tactics within an integrated pest management (IPM) program.

In this and an accompanying chapter, we review the main soil-based nematodes, insect pests and fungal diseases affecting coffee plants. In each case, the chapter discusses symptoms from pest attack, causal agents, epidemiology and management. In general, we have placed more emphasis on prophylactic measures and resistant varieties with respect to nematodes and diseases. This is because chemical and biological control measures are generally not very efficient in controlling nematodes and soil diseases and are often highly polluting, as has been seen in the case of some commercially banned nematicides.

The first sections in this chapter discuss nematodes. Numerous genera and species of nematodes have been associated with coffee worldwide, including some that are responsible for significant damage or plant mortality, resulting in important economic losses for coffee farmers and local economies. In Africa, the continent where coffee originated, less damage to coffee by nematodes has been identified, though this is probably due to lower monitoring (Danny

Coyne, personal information). In contrast, coffee suffers drastic damage from nematodes in both Latin America and Asia, where global coffee production is now concentrated. In Latin America, RKNs (*Meloidogyne* spp.) constitute the major nematode threat, while in Asia, lesion nematodes (*Pratylenchus* spp. and *Radopholus* spp.) are the most damaging and widely distributed on coffee (Villain et al., 2018; Trinh et al., 2009a; Wiryadiputra and Tran, 2008).

2 The root-knot nematode (*Meloidogyne* spp.)

Root-knot nematodes (RKNs) are globally the most widely distributed nematodes and probably the most harmful species affecting coffee, especially in Latin America (Villain, et al., 2018). To date 18 *Meloidogyne* species have been described and reported as parasitizing coffee (Le et al., 2019; Trinh et al., 2019; Villain et al., 2018; Carneiro and Cofcewicz, 2008). These can be separated into three categories (Monteiro et al., 2016; Carneiro and Cofcewicz, 2008):

1 major damaging species (widespread), namely *M. exigua*, *M. paranaensis* and *M. incognita*;
2 minor species (less widespread), although some are highly pathogenic on coffee, namely *M. africana* (=*M. decalineata*, *M. oteifae*), *M. arabicida*, *M. arenaria*, *M. coffeicola*, *M. daklakensis*, *M. enterolobii* (=*M. mayaguensis*), *M. hapla*, *M. izalcoensis*, *M. kikuyensis*, *M. lopezi*, *M. megadora* and *M. moensis*; and
3 misidentified species that do not parasitize coffee, namely *M. javanica*, *M. inornata* and *M. konaensis*, which are no longer considered coffee parasites.

Due to the previous misidentification caused by overlapping morphological features and lack of enzymatic and molecular markers, the distribution and relative importance of RKN species is well known in Brazil, but remains uncertain in other countries in Latin America, Asia and Africa (Villain et al, 2018; Janssen et al., 2016, 2017; Carneiro and Cofcewicz, 2008). *M. exigua* and *M. paranaensis* have wide American distribution, while *M. incognita* is particularly widely spread. These species also appear to be present over most of the altitudinal and latitudinal range of coffee cultivation in Latin America (López-Lima et al., 2015).

All esterase phenotypes of RKN detected on coffee are available in Carneiro and Cofcewicz (2008) and Villain et al. (2013), excluding the recently described species *M. lopezi* (Humphreys-Pereira et al., 2014), *M. daklakensis* and *M. moensis* (described without esterase phenotypes). More recently, the esterase phenotype of *M. megadora* (Maleita et al., 2016) and *M. africana* (=*M. decalineata* and *M. oteifae*) have been characterized (Janssen et al,

2017). Based on studies on *Meloidogyne* found in coffee, isozyme analysis also remains a useful and efficient methodology for RKN identification, when females are available.

The molecular sequence-characterized amplified region (SCAR) markers developed by Randig et al. (2002) have proved to be the best tool to identify all *Meloidogyne exigua* populations from coffee and rubber trees (Muniz et al., 2008). Species-specific SCAR molecular markers have been developed for most RKN species parasitizing coffee in Latin America: *M. exigua*, *M. incognita, M. paranaensis*, *M. enterolobii*, *M. arenaria*, *M. arabicida* and *M. izalcoensis* (Correa et al., 2013; Randig et al., 2002, 2004). Use of SCAR markers in multiplex-PCR provides a reliable tool for monitoring complex RKN species communities present on coffee plantations and for detecting species at low densities, due to the high sensitivity of this technique (Correa et al., 2013; Randig et al., 2002, 2004). Other advantages of this technique are that RKN DNA can be extracted directly from the eggs or J2s or females extracted from the roots or soil and amplified with PCR. This technique enables RKN detection using any life stage of the nematode, compared with electrophoresis analysis, which requires females.

For newly described species, such as *M. lopezi*, or for species present only in Africa and Vietnam, species-specific markers have not been developed. Recent exploration of mitochondrial coding genome analysis of tropical *Meloidogyne* species, however, has provided strong support for haplotype-based diagnostics, offering a seemingly durable *Meloidogyne* diagnostic technique (Janssen et al., 2016; Pagan et al., 2015). However, it remains prudent to use both isoenzyme electrophoresis analysis and molecular markers to conduct RKN species identification from field surveys because there is a possibility of misidentification using SCAR markers, especially for cryptic species (Santos et al., 2018; Villain et al., 2018).

3 *Meloidogyne exigua*

3.1 Distribution

M. exigua is the most widely distributed RKN, affecting Arabica coffee in the Caribbean, Central and South America, with a presence in all major coffee-growing countries from Honduras to Paraná state in Brazil (Villain et al., 2013; Carneiro and Cofcewicz, 2008; Campos and Villain, 2005). *M. exigua* has been detected in primary natural forest areas: the Atlantic forest in Brazil, Rio de Janeiro State (Lima et al., 2005) and the Amazon forest in the northern region of Mato Grosso State (Silva et al., 2006) in Brazil, as well as in Costa Rica (López and Vilchez, 1991). All these observations indicate that *M. exigua* is a native species of Latin America and probably has wide original distribution. However, *M.*

exigua, together with *M. coffeicola*, has also recently been identified in Vietnam (Trinh et al., 2013) and, surprisingly, has been found on *C. canephora*. This is very improbable since neither species parasites *C. canephora,* and *M. coffeicola* is a rare species even in Brazil (its country of origin). The characterization of these two species (Trinh et al., 2013) was based on the perineal region, which is highly susceptible to incorrect identification (Carneiro and Cofcewicz, 2008).

3.2 Symptoms

M. exigua generally causes numerous small, rounded or elongated galls, mostly on new roots. Each gall usually contains several females and egg masses, which are produced in the cortex and generally not exposed. The galls are initially yellow to brown and turn dark brown as the root ages (Fig. 1a). Necrotic areas

Figure 1 Symptoms of plant parasitic nematodes on *Coffea arabica*. (a) Roots infected with *Meloidogyne exigua* showing rounded galls. (b) Coffee seedling infected with *M. exigua*. (c) Roots of young plants infected with *M. paranaensis*, showing elongated galls. (d, e) Roots of old plants infected with *M. paranaensis* showing necrosis, peeling, cracking and corky root symptoms. (f, g) Roots with mixed populations of *M. exigua* and *M. paranaemsis* showing symptoms of both species. (h) Root infected with *M. izalcoensis* and stained with Phloxin B, showing a large number of external egg masses and necrosis in the root extremities. (i) Root infected with *Pratylenchus coffeae* showing symptoms of necrosis. Photographs courtesy: Rui Gomes Carneiro, Sônia M. L. Salgado, Fátima Muniz, Inorbert M. Lima and Luc Villain.

can appear on old galls and may be aggravated by secondary infections caused by fungi and bacteria. Root symptoms should not be used to identify species, only to indicate evidence of RKN infection (Villain et al., 2013; Muniz et al., 2009). *M. exigua* alters the physiological state of the coffee plant by affecting water and nutrient translocation (Rezende et al., 2013; Barbosa et al., 2010, 2004). Galls caused by *M. exigua* are mainly distributed on the new, whitish roots and remain intact for a long period without cracking or splitting the roots. Necrosis on older galled roots is somewhat compensated by the high rate of production of feeder (fine) roots. Damage to the plants, therefore, depends mainly on the physiological state of the plant (Villain et al., 2018).

Significant production losses due to *M. exigua* have been reported even from commercial coffee plantations that use modern and advanced technologies, correct fertilizer use and IPM practices (Barbosa et al., 2004). Rezende et al. (2013) observed a 23% reduction in productivity among susceptible cultivars, compared to resistant ones. In Costa Rica, production losses caused by *M. exigua* on susceptible *C. arabica* cv. Caturra were assessed as 8–16%, with no detection of foliar nutrient level alteration, indicating a likely impact of the nematode mostly on water uptake (Rojas and Salazar, 2014). Similarly, young coffee plants were the most susceptible to *M. exigua* during the dry season, with reduced growth, defoliation and some mortality observed (Villain et al., 2018; Muniz et al., 2009; Barbosa et al., 2004).

3.3 Diversity and diagnosis

Muniz et al. (2008) and Carneiro et al. (2000) characterized three races of *M. exigua* by their capacity to reproduce on different hosts:

- race 1 (coffee and pepper);
- race 2 (coffee, pepper and tomato); and
- race 3 (only rubber tree).

According to Muniz et al. (2008), phylogenetic analysis showed high intraspecific genetic diversity, a polymorphism of 25.9–59.6% among the 16 *M. exigua* populations studied, which all clustered closely together with 100% bootstrap support. Four esterase phenotypes (Est E1, E2, E2a and E3) and three malate dehydrogenase (Mdh) phenotypes (N1, N1a and N2) have been described for *M. exigua* populations, but only band E1 is common to all variants and a good marker for the species; however, these enzymatic phenotypes do not separate *M. exigua* races (Muniz et al., 2008).

Although the Est E1 phenotype is species-specific, the reaction resolution often does not appear in the gel due to the low concentration of esterase enzyme in *M. exigua*, making the molecular marker the most accurate technique

(Randig et al., 2002). One population of *M. exigua* from Rio de Janeiro State has been described as highly virulent on *C. arabica* cv. IAPAR 59 (Sarchimor), bearing the resistance gene to *M. exigua*, *Mex-1* (Muniz et al., 2009).

3.4 Life cycle and management

The biology of *M. exigua* has been studied on coffee plants at the embryonic and postembryonic stages of the nematode (D'Arc de Lima and Ferraz, 1985a,b). The whole life cycle of *M. exigua* was found to take 32–42 days at 25–30°C on a susceptible cultivar. Histopathology of susceptible coffee roots infected by *M. exigua* revealed swollen root tips and axes; nematode egg masses were visible on root surfaces only in small root galls, whereas they were commonly embedded in the root tissues (Di Vito et al., 2000).

On coffee, *M. exigua* is almost exclusively a parasite of *C. arabica*, since most *C. canephora* germplasm shows complete resistance to this species, controlled by a major gene, *Mex-1* ('IAPAR 59'), which was transferred to most introgressed cultivars of Catimor and Sarchimor varietal families (Noir et al., 2003). No resistant accessions have been found in *C. arabica*, nor in wild coffee trees collected in Ethiopia (Bertrand and Anthony, 2008). On the other hand, several resistant accessions have been identified in *C. canephora* and in some progenies of interspecific hybrids (*C. arabica* X *C. canephora*) (Bertrand and Anthony, 2008; Gonçalves et al., 1998; Bertrand et al., 1997, 1998). Hypersensitive-like reactions and induced and constitutive defense mechanisms have been reported in IAPAR 59 cultivar (*C. arabica* x *C. canephora*) bearing the *Mex-1* gene (Anthony et al., 2005).

M. exigua is not a polyphagous RKN and can be eradicated from the soil by one-year rotation with non-host crops (cotton, soybean, maize, bean, sorghum, *Crotalaria spectabilis*, *Stilozobium atterrinum* and *Panicum maximum*) and can also be managed through nematicides or organic matter application (Campos and Silva, 2008).

Many nematicides are effective in decreasing nematode densities of *M. exigua* up to 3 months after application in plants with typical root galls, after which population densities may increase on treated plants (Campos and Silva, 2008). Almeida et al. (2015) observed efficient control of *M. exigua* in plantations up to 90 days after application, using dosages of fluensulfone at ≥2.0 l/ha (Nimitz®, ADAMA). The study did not continue longer than the 90-day period, and further studies are necessary to assess the duration of effect as well as to elucidate the mode of action and determine whether fluensulfone is a 'true' nematicide affecting nematode metabolism prior to death (Kearn et al., 2014). *M. exigua*-infected coffee treated for five consecutive years with this nematicides produced 30.9% higher yields than non-treated coffee. However, the nematicide does not eradicate the nematode (Villain et al., 2018).

Nematicides can reduce or nearly cancel nematode multiplication in the field for up to 90 days, as has been seen with the organophosphorus cadusafos (Rugby®, FMC) applied at 200 g/L against *M. exigua* (Campos and Silva, 2008). The application of granular nematicides needs to be incorporated into the soil underneath the foliage, toward the stem, which requires specialized equipment. Rugby® also needs to be applied following litter clearance, which again requires an adapted soil cleaning device with the application machine, adding to cost. The timing of application can also be important, such as for granular products that require water to liberate activity and therefore need to be applied at the beginning of the rainy season (Campos and Silva, 2008; Villain et al, 2018). However, there is consensus that more environmentally sensitive, biologically based options for nematode management should be explored and exploited, building on the biodiversity and natural mechanisms of agroecosystems (Bertrand et al., 2016; Villain et al., 2018).

This is especially important in relation to biological control used in organic coffee, which prohibits chemical use and commands a higher market price. The bacterium *Pasteuria penetrans* has been investigated in relation to biological control, having been found in coffee fields in Colombia (Baeza-Aragón, 1978) and in Brazil (Sharma and Lordello, 1992). In Brazil, in Minas Gerais state, 21-65% of *M. exigua* J2 in one coffee field were naturally infected by *P. penetrans* (Maximiniano et al., 2001). This nematode-parasitic bacterium is resistant to heat, drought and pesticides (Campos et al., 1998). However, this bacterium is not currently commercially available (Carneiro et al., 2020).

In Cuba, the fungus *Pochonia chlamydosporia* (=*Verticillium chlamydosporium*), isolated from coffee plantations, has shown potential as a biological control agent for RKNs affecting coffee (Hidalgo-Diaz et al., 2000). In Brazil, predators and egg parasitic fungi have also been isolated from coffee plantations (Ribeiro and Campos, 1993; Naves and Campos, 1991). The efficacy of *Arthrobotrys conoides*, *Arthrobotrys musiformes*, *Purpureocillium lilacinum* (=*Paecillomyces lilacinus*) and *P. chlamydosporia* for the control of *M. exigua* in coffee has been shown by Campos and Campos (1997), but studies in field conditions on the effective use of nematophagous fungi have been neglected in recent years (Carneiro et al., 2020).

4 *Meloidogyne paranaensis* and *M. incognita*

4.1 Distribution

M. paranaensis was first described on a coffee farm located in Paraná state, Brazil (Carneiro et al., 1996). This species is restricted to the Americas, including Brazil, Peru, Guatemala, Hawaii (Campos and Villain, 2005; Villain et al., 2018) and Mexico (López-Lima et al., 2015). This fragmented distribution could be related to early coffee seedling introductions but also linked to banana introductions in Central America since banana is also a good host for *M. paranaensis* (López-Lima et al., 2015; Villain et al., 2013).

The range of *M. paranaensis* is constantly expanding, at least inside the countries where it is already present, via propagation material or soil transport. There have been many reports of *M. paranaensis* in Brazil, for example, in Paraná state (Andressa C. Z. Machado, personal information), in São Paulo state and some localities of Minas Gerais state (Castro and Campos, 2004; Carneiro et al., 2005a,b; Salgado et al., 2015), in Espírito Santo state on *C. canephora* crop (Barros et al., 2011, 2014) and in Goiás state (Silva et al., 2009). In Guatemala, where *M. paranaensis* seems to be the predominant RKN species (Villain et al., 2013), corky root symptoms similar to those observed in cases of coffee root parasitism by *M. paranaensis* (Fig. 1d, e) were already being observed at the beginning of the twentieth century (Alvarado, 1935), suggesting a long presence (Villain et al., 2018). In Brazil, this species was initially identified as *M. incognita* race 5 until it was fully and correctly described (Carneiro et al., 1996).

The distribution of *M. incognita* on coffee remains uncertain, since numerous previous reports of putative *M. incognita* populations could likely have been misidentified using perineal patterns that do not discriminate between species such as *M. paranaensis*, *M. incognita* and *M. izalcoensis* (Carneiro and Cofcewicz, 2008; Villain et al., 2013). Currently, the only country in Latin America where a wide distribution of *M. incognita* on coffee may be confirmed is Brazil, because of numerous reliable identifications based on esterase phenotypes and species-specific sequence-characterized amplified regions (SCAR markers) (Correa et al, 2013; Randig et al., 2002). The presence of *M. incognita* has been confirmed in Nicaragua (Herrera et al., 2011b) and Costa Rica (Villain et al., 2013). However, in both countries, *M. incognita* seems to have restricted distribution on coffee (both *C. arabica* and *C. canephora*, Villain, et al., 2018). In Vietnam, *M. incognita* is widely distributed on *C. canephora* in the Western Highlands and has also been detected on *C. arabica* in Dak Lak province (Trinh et al., 2013). But identifications based only on the perineal regions are not conclusive, especially in the case of *M. incognita*, whose pattern is not species-specific (Carneiro and Cofcewicz, 2008).

4.2 Symptoms

The symptoms caused by *M. paranaensis* and *M. incognita* are very similar and difficult to separate under field or laboratory conditions. Both nematodes cause peeling and cracking of the root surface due to hypertrophy of infected tissues, in addition to disruptions of the cortex and epidermis caused by protruding egg masses breaking the root surface. Sometimes, root swellings rather than galls are observed on lateral roots. These cortical tissue disruptions become necrotic, leading to root death (Fig. 1 c-g). The infected plants show symptoms of foliar chlorosis and nutritional deficiency, defoliation, general decline and reduced growth (Fig. 2a-e). Progressive decay of the plants sometimes leads to plant death for both *C. arabica* and *C. canephora* (Fig. 2b and e) (Villain et al., 2013, 2018; López-Lima et al., 2015; Barros et al., 2011). In new coffee plantations, where abiotic stresses are intense, due to low altitude with high

Figure 2 Symptoms of plant parasitic nematodes in *Coffea arabica*. (a, b) *Meloidogyne paranaensis* infecting adult plants in Astorga, PR, Brazil: severe decay (defoliation, death, low production). (c, d) Replanted coffee in area infected with *M. paranaensis*, young plants showing symptoms of chlorosis, reduced development and defoliation (Piuí, MG, Brazil). (e) Dead plants infected with *M. incognita* in Monte Carmelo, MG, Brazil. (f) Symptoms caused by *Pratylenchus coffeae* in an infested area in Veraccruz, Mexico, chlorosis and growth reduction. Photographs courtesy: Rui Gomes Carneiro, Sônia M. L. Salgado, Fátima Muniz, Inorbert M.Lima and Luc Villain.

temperatures for *C. arabica*, extended drought, low fertility and/or low pH soils, high plant mortality rates are observed before the infected plants begin to produce in the second or third year (Fig. 2c) (Villain et al., 2002, 2018; López-Lima et al., 2015).

In the case of *M. paranaensis* and *M. incognita* parasitism, the development of females close to the surface of the roots, and rupturing of the cortex, which leads to emergence of egg masses (unlike with *M. exigua*), probably favor infection by secondary pathogens such as *Fusarium* leading to cracking and corky root symptoms (Fig. 1e). Under controlled conditions, the inoculation of both *M. arabicida* and *F. oxysporum* resulted in corky root symptoms on *C. arabica* cvs. Caturra and Catuai, while *M. arabicida* alone only caused gall formation, without corky root symptoms (Bertrand et al., 2000). In fields planted with cultivars that are susceptible and resistant to *M. arabicida*, only the susceptible ones developed corky root symptoms (Rivillas et al., 2015; Bertrand et al., 2000). Unfortunately, there have been no studies of nematode–fungus interactions in relation to *M. paranaensis* and *M. incognita,* which also present corky root symptoms in field conditions in Brazil and Guatemala. In Veracruz state, Mexico, López-Lima et al. (2020) observed the presence of *F. oxysporum* associated with *M. paranaensis* in internal tissues of sampled coffee corky roots (Fig. 1 d, e) (Villain et al., 2018).

Plant parasitic nematodes are estimated to cause between 10 and 20% coffee losses worldwide (Abd-Elgawad and Askary, 2015). However, for perennial crops, estimating economic losses is difficult, since infection occurs over a long period and directly affects not only production but also the cost of replanting following plant death, especially in relation to *M. paranaensis* and *M. incognita*. Over the years, the relative importance of the major coffee RKN species has shifted in Brazil. From 1975 to 1990, it was assumed that *M. incognita* (incorrectly identified) affected the best coffee-growing areas, especially in north Paraná and west São Paulo states, resulting in the complete destruction of entire plantations; this forced farmers to switch crops and led Lordello (1984) to declare *M. incognita* as 'a disaster pathogen becoming the worst enemy of coffee in some areas of São Paulo state'.

However, some of the damage attributed to *M. incognita* was due to *M. paranaensis*, which was previously identified as *M. incognita* race 5. More recently, up to 50% of the decline and dieback of coffee trees, along with yield suppression, has been associated with *M. paranaensis* in Brazil (Moens et al., 2009). This nematode has recently spread quickly in Minas Gerais state (Salgado et al., 2015; Castro et al., 2008). In the traditional Arabica coffee-growing region of Alta Paulista in north São Paulo state, replanted Arabica cultivars must now be grafted on to *C. canephora* cv. Apoatã rootstock, which is resistant to *M. incognita* and *M. paranaensis* and immune to *M. exigua*, in order to continue cropping coffee in this region (Villain et al., 2018).

In Mexico, the coffee corky root disease associated with *M. paranaensis* is now considered a severe threat to coffee production, causing up to 35% of annual *C. arabica* plant losses in the two highest coffee-producing states, Veracruz and Chiapas (López-Lima et al., 2015). Although Mexico has an important *C. canephora* cv. Robusta industry for soluble (instant) coffee production, *M. paranaensis* has not been reported on this crop, even though some Robusta cropping areas are situated close to Arabica sites. In Guatemala, like Brazil, no statistics are available on economic losses caused by *M. paranaensis* on coffee (the principal coffee RKN species in Guatemala), although this species is responsible for the entire destruction of coffee plantations in the south-western region of Sierra Madre. As in Brazil, many coffee growers have had to replant *C. arabica* cultivars grafted on to resistant rootstock *C. canephora* cv. Nemaya, provided by the National Association of Coffee (Villain et al., 2013, 2008, 2018). *M. paranaensis* has also been observed causing serious damage to commercial *C. canephora* plantations, with progressive decay beginning with chlorosis and production loss. On pruning the decayed plants, most fail to regenerate and eventually die (Villain et al., 2013, 2018).

4.3 Diversity and diagnosis

Perineal patterns are sometimes helpful for the identification of *Meloidogyne* species but do not differentiate *M. incognita* from *M. paranaensis*, which is possible via esterase phenotypes using electrophoresis (Carneiro and Cofcewicz, 2008; Carneiro et al., 2000).

4.4 Life cycle and management strategies

The biology of *M. paranaensis* and *M. incognita* in susceptible cultivars is very similar and has been studied in detail at a temperature of 25–30°C. At 2 days after inoculation (DAI), numerous second-stage juveniles (J2s) were observed in the subapical meristem of roots. At 4 DAI, J2s were present in the root cortex, migrating toward the vascular cylinder, causing cell wall damage. At 4–8 DAI, numerous J2s were observed within the vascular cylinder, and feeding J2s were visible inside the vascular cylinder at 8 DAI, close to giant cells. At 16–32 DAI, numerous J3s/J4s had established feeding sites with 10–12 well-defined giant cells associated with each nematode. Giant cells were hypertrophied, oval-shaped, vacuolated and with dense cytoplasm containing some nuclei. At this time, enlargement of the vascular cylinder region took place, and small galls were apparent. Giant cells displayed thickened cell walls. At 32 DAI, adult females were observed associated with severe injuries to the surrounding cells and disruption of the root cortex. At 45 DAI, egg masses were observed inside and at root surface, completing one life cycle (Lima et al., 2015).

M. paranaensis and *M. incognita* management in a perennial crop like coffee is more difficult than in annual crops, due to difficulties in reducing nematode infestations after they have become established. The long-term condition of perennial crops makes rotation schemes difficult to implement, since these two species are polyphagous. With perennial crops, nematodes that survive control practices have time to recover and build up to destructive levels. Moreover, old roots or plants remaining in the field, weeds or other crop hosts provide sources of nematode re-infestations, negating the beneficial effects of control practices (Campos and Villain, 2005).

The first measure to avoid these two highly pathogenic RKNs is prevention by restricting the dissemination of nematodes through infected seedlings or soil (Salgado and Rezende, 2010). Phytosanitary inspections of commercial coffee seedling nurseries often remain weak or nonexistent but can make a big difference, such as in Minas Gerais, São Paulo and Paraná states, where coffee seedlings are certified by law for the absence of *Meloidogyne* spp. In Brazil, the Agriculture Department has implemented regulations for the production and commercialization of coffee seedlings since 2012, with regard to *Meloidogyne* spp., but this now needs to be revised to include *Pratylenchus* spp.

Agricultural tools should be washed thoroughly before moving to a new field. Water runoff from infested plots should be prevented by trenches and canals, and dense hedges planted to restrict movement of infected material between fields (Dhanam et al., 2011). Problems arise, however, where coffee growers produce their own seedlings without taking nematode control into account (Fig. 1 b) (Villain et al., 2018). Seedlings infected with RKNs should be destroyed (burned) and should under no circumstances be planted into new fields (Villain et al., 2018; Campos and Villain, 2005).

The application of nematicides on coffee infected with *M. incognita* or *M. paranaensis* has not been recommended, due to the destruction of large parts of the root system by the nematode and inefficiency of chemical control. For other *Meloidogyne* species causing similar symptoms to *M. coffeicola* and *M. arabicida*, the use of nematicides as a control measure is not recommended (Campos and Silva, 2008; Villain et al., 2018).

The use of genetic resistance is a fundamental strategy in the control of *Meloidogyne paranaensis* and *M. incognita*. Breeders have been assessing resistance against RKNs in *C. canephora* germplasm in relation to controlling these two nematodes in *C. arabica* by grafting. Breeding for rootstocks has traditionally focused on *C. canephora*, *C. congensis* and *C. dewevrei* germplasm as a rich source of resistance to nematodes and other pathogens due to their vigor and abundant root systems (Gonçalves and Silvarolla, 2001). However, resistance genes found in wild or semi-wild lines of *C. arabica* from Ethiopia or Yemen could also be used in coffee rootstock breeding programs for implementing interspecific hybridizations with resistant diploid *Coffea* spp. lines. For example, the interspecific hybrid 'Arabusta' (*C. canephora* × *C. arabica*), in addition to nematode resistance, could provide better adaptability than *C. canephora* to the cooler coffee-growing areas at altitude, in addition to its strong vigor (Berthaud, 1978a,b; Capot, 1972; Villain et al., 2018).

More recently, selection of resistance to RKNs has also been implemented for *C. canephora* itself, especially for cv. Conilon in Brazil, against *M. incognita* and *M. paranaensis* (Lima et al., 2015). The widespread distribution and the aggressive nature of these two RKN species in the west of São Paulo forced Brazilian researchers to rethink their options for efficient control measures other than chemicals. The introduction of *C. canephora* cv. 2258 from the CATIE germplasm collection, Turrialba, Costa Rica, provided strong resistance to *M. exigua* and resistance and/or tolerance to races 1, 2 and 3 of *M. incognita* (Fazuoli, 1986), as well as to *M. paranaensis* (Fazuoli et al., 2002; Santos et al., 2018).

The high level of resistance of cv. 2258, initially of 70%, was further improved through selection from fields highly infested with *M. incognita/M. paranaensis*. The resulting improved line, cv. Apoatã, is widely used as a rootstock in Brazil (Fazuoli et al., 2002). *C. arabica* cv. Mundo Novo grafted on to *C. canephora* cv. Apoatã has shown a yield 3.6 times greater than non-grafted plants grown (Villain et al., 2018). Although Apoatã rootstock is particularly promising, it has a segregation of about 12% for susceptible plants, as was proved by Santos et al. (2018). Grafting on to cv. Apoatã rootstock has effectively become the only control method to enable sustainable cultivation of *C. arabica* in areas infested with *M. incognita* and/or *M. paranaensis* in Brazil. The Agronomic Institute of Campinas (IAC) has developed other resistant *C. canephora* rootstocks, with a larger genetic base composed of four clones, which will soon be launched, with the aim of replacing Apoatã rootstock (Oliveiro Guerreiro, pers. information) and with resistance to *M. exigua*, *M. paranaensis* and *M. incognita* (Gonçalves et al., 2021).

In Turrialba, Costa Rica, among numerous *C. canephora* clones of the coffee germplasm of CATIE, just two clones (from one of which Apoatã originated) demonstrated high levels of resistance to the main RKN species detected in Central America: *M. paranaensis*, *M. izalcoensis*, *M. arabicida* and *M. exigua* (Bertrand and Anthony, 2008). By multiplying these two clones *in vitro* and then crossing them in seed-producing parcels (crossed pollination), another resistant rootstock cultivar, with a low segregation rate, was created and named cv. Nemaya (Bertrand and Anthony, 2008; Villain et al., 2018).

More recently, the drought-tolerant *C. canephora* cv. Conilon clone 14 rootstock (released as ES8161 Goytacá by INCAPER, ES) (Lima, 2016 and Inorbet Lima, pers. information) was identified with multiple resistance to *M. paranaensis*, to races 1 and 3 of *M. incognita* and to *M. exigua* (Lima et al., 2015). Since *C. canephora* is now widely affected by these aggressive RKNs, this clone can now be used as a rootstock for *C. canephora* as well as for *C. arabica* in infested areas, as well as being a potential source of *Meloidogyne* resistance genes (Lima et al., 2015; Lima, 2016). Early and late defense mechanisms, such as hypersensitive-like reactions, have been observed in resistant *Coffea* germplasm, for example, *C. canephora* cv. Conilon, clone 14 (Lima et al., 2015), while late flavonoid accumulation in giant cells was observed without evidence of hypersensitive reaction in *C. canephora* cv. Nemaya and one *C. arabica* Ethiopian wild accession (Villain et al., 2015).

Resistance to *M. paranaensis* and *M. incognita* has also been found in many Brazilian coffee lines, for example, *C. arabica* cv. IPR-100 (genes from *C. liberica*) with a segregation of 29% (Peres et al., 2017; Santos et al., 2018), IPR 106 (genes from *C. canephora* - Icatú) (Shigueoka et al., 2016). Progenies from crosses between Catuai and Amphilo cultivars have also provided resistance against *M. paranaensis* and *M. incognita* in greenhouse conditions (Peres et al., 2017) and had an excellent performance in an area infested by *M. paranaensis* in Piumhi, MG, and are now being validated in other regions of Brazil. These materials, originating from crosses with Amphilo, were released as two varieties - MG Vereda and MG Graiçara - that have excellent agronomic traits and resistance to *M. paranaensis* in greenhouse and field conditions (Salgado et al., 2021).

Several semi-wild Ethiopian accessions of *C. arabica* from the Centro Agronómico Tropical de Investigación y Enseñanza (CATIE) collection showed high levels of resistance to *M. paranaensis* populations from Guatemala and Brazil (Anzueto et al., 2001; Boisseau et al., 2009). Some of these wild Ethiopian accessions with resistance to *M. paranaensis* have been used to create F1 hybrids, in particular by crossing with Sarchimor lines resistant to *H. vastatrix* (van der Vossen et al., 2015). However, seven of these F1 hybrids were phenotyped as highly susceptible to *M. paranaensis*, under controlled inoculations (Villain et al., 2015, 2018; Xalapa, Veracruz, Mexico, unpublished). Since these hybrids possess many useful characteristics, such as high yield, excellent quality, adaptation to shade and resistance to other diseases, these lines can be grafted on to resistant rootstocks such as cv. Nemaya for use in areas infested by *M. paranaensis* (Villain et al, 2018).

There are several reports of biological control agents or bionematicides to manage *Meloydogyne* spp. on coffee, such as *Pasteuria penetrans*, *Pochonia clamydosporia* and *Purpuriocilim lilacinum*, among others (Carneiro et al., 2020). However, none of these microorganisms can be considered effective in controlling RKN of coffee, due to the root destruction caused by these two species, *M. paranaensis* and *M. incognita*.

Carneiro and Carneiro (1982) screened 29 crops for rotation in *M. incognita*/*M. paranaensis*-infested coffee fields and found that only *Arachis hypogaea* and *Ricinus communis* were immune, while *Mucuna pruriens* var. *utilis* (=*Stizolobium deeringianum*) and *Crotalaria spectabilis* were poor hosts to this nematode. There was no penetration of J2 or gall/egg mass formation of *M. paranaensis* and *M. incognita* races 1, 2, 3 and 4 in *Arachis pintoi* roots following inoculation, and this could provide a potential rotation crop (Santiago et al., 2001).

Carneiro et al. (2013) have studied the effect of integrated pest management methods using crop rotation, genetic resistance and biological control with the bacterium *Pasteuria penetrans*, in experiments conducted in clay soil, heavily infested with *M. paranaensis*. The authors used two coffee varieties, IAPAR 59 (susceptible) and Icatú H4782-7-925 (moderately resistant), treated or not with *P. penetrans* in areas managed previously or not with crop rotations (black oat, peanut and grey mucuna, for 12 months). The most efficient treatment during the first 3 years was coffee cv. Icatú after 1 year of plant succession. At 4 years after coffee planting, populations of nematodes had equalized in managed and unmanaged (crop succession) areas planted with two coffee cultivars. No effect of the bacterium was observed, and *P. penetrans* disappeared from the field; the endospores probably remained attached to the clay soil particles.

Rotation with non-hosts of *M. paranaensis* and *M. incognita* remains a useful strategy to reduce initial densities in infested areas before planting coffee grafted on resistant 'Apoatã' or 'Nemaya', since neither approach on its own was 100% effective in reducing nematode populations (Villain et al., 2018). Intercropping velvet beans between rows of coffee and incorporating them at the flowering stage protects the coffee from cold wind and improves soil texture, organic matter and nutrients, reducing the damage caused by *M. incognita* and *M. paranaensis* in sandy soils depleted in organic matter (Fazuoli et al., 2002).

5 Minor species of *Meloidogyne*

5.1 Distribution

In Africa, where coffee is a major cash crop in a number of countries, there have been surprisingly few RKN problems affecting coffee, possibly due to limited surveys providing scant information on nematode distribution (Villain et al., 2018). From the available data, the redescribed *M. africana* corresponds to *M. decalineata* and *M. oteifae* (Janssen et al., 2017). This species was detected in Kenya and Tanzania on *C. arabica* and in the Democratic Republic of Congo in *C. canephora*.

M. megadora appears to have a quite restricted distribution. First found in coffee plants in Angola, it is the most important nematode pathogen affecting coffee plants in the Democratic Republic of São Tomé and Príncipe (Maleita et al., 2016) and in Uganda (Whitehead, 1969a,b). *M. megadora* is also recorded on some species of coffee (*C. arabica*, *C. canephora*, *C. congensis* and *C. eugenioides*) in Angola and Uganda (Whitehead, 1968, 1969a,b). *M. kikuyensis* was found in the Kilimanjaro area (Swai, 1981) and Kenya (Campos and Villain, 2005), but Whitehead (1969a,b) considered coffee to be a highly resistant plant to both *M. javanica* and *M. kikuyensis*, probably a non-host of these two species.

However, it would appear there has been some confusion over the identity of some species in Africa. From a limited number of field samples in Kenya, Tanzania and Uganda, several species have been misidentified in the past. *M. hapla*, *M. incognita* and *M. izalcoensis* have been identified, including the occurrence of at least one undescribed species, indicating a potentially diverse and expanding diversity of *Meloidogyne* species on coffee in East Africa, its center of origin (Villain et al., 2018).

Two newly described species of RKN from coffee have also been added to the list of minor species. One is *M. izalcoensis*, identified in the region of the Izalco volcano, El Salvador (Carneiro et al., 2005a), and recently detected in Brazil (Stefanelo, 2019) and in Africa (Kenya and Tanzania, Jorge-Junior et al., 2016). The second is *M. lopezi*, detected in a southern region of Costa Rica and closely related to *M. izalcoensis*, *M. arabicida* and *M. paranaensis* (Humphreys-Pereira et al., 2014). *M. izalcoensis* is present over a range of altitudes (430–1100 m), covering most of the altitudinal range where *C. arabica* grows in Central America, although the species seems to have a relatively small area of distribution in the case of coffee, limited to the region where it was first described (Villain et al., 2013). However, we do not know its distribution in Africa and Brazil. These two species, *M. izalcoensis* and *M. lopezi* (esterase phenotypes: Est S4 and Est S2, respectively), were also collected on different farms in the same area in El Salvador (Hernández et al., 2004). They were very closely related in DNA analysis (Carneiro et al., 2004). Recently, the phenotype Est SA2 was described as *M. lopezi* (Humphreys-Pereira et al., 2014) from coffee trees in southern Costa Rica, but the distribution of this species is not known.

M. arabicida, also described from Costa Rica (López and Salazar, 1989), appears to have a distribution restricted to this country only (Villain et al., 2013). *M. arenaria* has been recovered from coffee in Jamaica, El Salvador and Guatemala (Campos and Villain, 2005; Villain et al., 2013). *M. morocciensis* was detected on coffee in El Salvador but with an esterase phenotype Est A3, previously considered as belonging to *M. arenaria* (Villain et al., 2013), but which should, together with an Mdh N1 phenotype, be attributed to *M. morocciensis* (Carneiro et al., 2016). *M. hapla* has been detected in a few coffee plantations in Latin America and Africa, mostly at higher altitudes in coffee-growing regions, in accordance with its general distribution in temperate climate regions (Villain et al., 2013). Whitehead (1969a) and Bridge (1984) suggested that coffee is not a good host for this species, or for *M. javanica*, and these results were confirmed by Carneiro and Cofcewicz (2008). These authors confirm the presence of *M.*

javanica in several areas planted with coffee in Brazil but never found in coffee roots.

M. inornata, described by Lordello (1956) from Brazil and redescribed by Carneiro et al. (2008), also occurs in coffee in Guatemala (Schieber and Sosa, 1960). However, recent studies have shown that this species does not parasitize coffee, and detection in Guatemala must be a misidentification of *M. paranaensis*, since the perineal region was used as a diagnostic feature to differentiate these two species (Carneiro and Cofcewicz, 2008).

M. enterolobii was first detected on coffee in Cuba (Sampedro et al., 1989) and then in Guatemala and Costa Rica (Hernández et al., 2004; Villain et al., 2013). This species is widespread in the tropics and subtropics on various plant hosts, especially guava, and is now considered a major emerging RKN (Castagnone-Sereno, 2012). On coffee, its limited occurrence in low densities would indicate that coffee is not a suitable host for this species (Villain et al., 2013, 2018), as observed in greenhouse experiments (Muniz et al., 2009). To further support this, *M. enterolobii* was recovered from banana plants in a coffee plantation in Mexico, but attempts to reproduce this population on *C. arabica* failed (Villain et al., 2015, 2018, unpublished).

M. coffeicola is also considered a minor species category, since its distribution has become limited in Brazil over the past decade, where it was described by Lordello and Zamith (1960). This reduced distribution appears to be related to the fact that this species mostly affects older coffee trees (Carneiro and Cofcewicz, 2008). However, *M. coffeicola* has been detected in Vietnam at one locality of the Western Highlands in combination with *M. exigua* on *C. canephora* (Trinh et al., 2013). This also constitutes the first reliable record of these two species outside Brazil (Villain et al, 2018). Recently, the species *M. daklakensis* and *M. moensis* have been described in Vietnam, but very little is known about their distribution. As the esterase patterns and molecular markers have not been characterized, precise identification in other areas will be very difficult to achieve (Trinh et al., 2019; Le et al., 2019).

We have removed *M. konaensis* from the list of coffee RKNs because this species does not parasitize coffee plants. For *M. konaensis*, three esterase phenotypes (Est) were detected by Sipes et al. (2005). A mixture of three species, *M. paranaensis* (Est F1 = P1), *M. incognita* (Est I1) and *M. konaensis* (Est K3) (Carneiro et al., 2000, 2004), are commonly found in the coffee crop. *M. konaensis* has been described by Eisenback et al. (1994) in tomato plants, although the profile of the esterase enzyme was not clearly characterized. Among the three species detected by Sipes et al. (2005), only the Est F1 (= P1) population parasitizes the coffee plant, that is, *M. paranaensis*. Recently, Monteiro et al. (2016) showed the difference in morphological, enzymatic and molecular features between *M. konaensis* and *M. paranaensis*, demonstrating once again that *M. konaensis* does not parasitize coffee plants.

5.2 Diagnosis

Considering the 12 valid minor species of *Meloidogyne* described from coffee, 9 have a species-specific enzyme esterase profile: *M. africana* (Est AF2, Janssen et al., 2017), *M. arabicida* (Est Ar 2), *M. arenaria* (A2), *M. coffeicola* (Est C2), *M. enterolobii* (Est En2 = M2), *M. hapla* (Est H1), *M. izalcoensis* (Est I4) (Carneiro and Cofcewicz, 2008), *M. lopezi* (Est L2 = SA2, Humphreys-Pereira et al., 2014; Carneiro et al., 2004) and *M. megadora* (Est Me3, Maleita et al., 2016). The other three species: *M. kikuyensis* (Eisenback and Vieira, 2020), *M. daklakensis* (Trinh et al., 2019) and *M. moensis* (Le et al., 2019) have been characterized without esterase phenotypes. Additionally, there are SCAR markers to identify *M. arabicida*, *M. izalcoensis* and *M. enterolobii* (Correa et al, 2013).

5.3 Symptoms, economic importance and management

Meloidogyne africana, recently redescribed by Janssen et al. (2016), usually causes small- to moderate-sized galls from 1 mm to 5 mm in diameter, mainly at the root tips, from where numerous rootlets are produced in reaction. It is reported as causing poor growth in both *C. arabica* and *C. canephora* nursery seedlings, as well as affecting mature coffee trees (Campos and Villain, 2005; Villain et al., 2018). The importance and distribution of this species are not known (Danny Coyne, personal information).

M. arabicida causes corky root symptoms similar to those caused by *M. paranaensis* and *M. incognita*, with numerous swellings evolving into extensive development of corky tissues on the taproot up to the collar, and on the primary and secondary roots. In the field, infected plants show a progressive decline with leaf chlorosis and leaf fall, followed by flower and fruit fall, leading to plant death within 2-4 years after planting (Villain et al., 2018). Under controlled conditions, only the inoculation of both *M. arabicida* and *F. oxysporum* resulted in corky root symptoms on *C. arabica* cvs. Caturra and Catuai, while *M. arabicida* alone caused only gall formation, without corky root symptoms. In fields planted with cultivars that are susceptible and resistant to *M. arabicida*, only the susceptible developed the corky root symptoms. The predisposition of the host to *M. arabicida* consequently plays a major role in this complex etiology (Bertrand et al., 2000; Rivillas et al., 2015; Villain et al., 2018).

The outbreak of *M. coffeicola* in Paraná state in Brazil resulted in the death of numerous coffee trees and had a major economic impact (Lordello and Zamith, 1960). However, *M. coffeicola* has not been detected in Paraná state in recent surveys carried out on more than 500 analyzed samples (Andressa Machado, personal information). *M. coffeicola* causes peeling and cracking of roots in old coffee plants but does not produce galls. Females are found in older root tissue, especially on the taproot, and egg masses are predominantly external, emerging from cracks (Villain et al., 2018).

M. izalcoensis causes severe root destruction, frequently killing coffee trees in El Salvador (Carneiro et al., 2005a,b). Coffee roots infected with *M. izalcoensis* produce relatively small galls with necrotic tissues at the ends of the roots. Egg masses are produced outside the roots in large quantities (Fig. 1 H). Each gall is characterized by a single female exposing an egg mass on the outside of the gall. There is no cracking of the root surface, as typically caused by *M. paranaensis* or *M. incognita*. The root symptoms caused by *M. izalcoensis* on coffee are completely different from those caused by *M. paranaensis*, *M. incognita*, *M. arabicida* and *M. exigua*, but identical to *M. lopezi* (Humphreys-Pereira et al., 2014). *M. izalcoensis* and *M. lopezi* cause very similar symptoms: small galls on the fine roots, each containing a single female, with the egg mass exposed outside the gall tissues. These two species (Est Sa4 and Sa2) were previously studied by Hernández et al., 2004 and Carneiro et al., 2004. Both occurred in El Salvador, in nearby coffee areas, and were considered variants of the same species on this occasion, before they had been properly identified.

M. arenaria causes large root swellings without corky formation across the whole root system, including the taproot up to the collar, and with dramatic induction of adventitious roots. Similar symptoms have been observed with *M. morocciensis* on *C. arabica* in Guatemala. On the very few infected *C. arabica* trees observed in Guatemala and Costa Rica, large galls (larger than those produced by *M. exigua*) and some swellings without cortex cracking have been observed (Villain et al., 2013, 2018). *M. hapla* causes light root galling and swellings, different from the symptoms of other species occurring in Tanzania (Bridge, 1984).

Although there is no recent information available in Africa on the actual yield losses caused by nematodes, yield losses of trees severely infected with the African coffee root-knot nematodes (RKN) were estimated by Bridge (1984) to be in the region of 20% under optimum conditions, extending to 100% at worst. Nematode damage also causes premature fruit drop, twig dieback and defoliation, nutrient deficiency and stunted growth. (Villain et al., 2018). Although *M. arabicida* distribution is restricted to a small area in Costa Rica and has not yet been detected elsewhere, this parasite presents a threat to coffee-growing regions. Its economic impact can be so severe that it results in the abandonment of coffee plantations (Anon., 1989).

Considering there is little practical knowledge about *Meloidogyne* species from Africa, it is very difficult to establish management strategies. However, the assessment of *Coffea* species, crosses and selections against RKNs in Tanzania by Bridge (1984) indicated some resistance and suggested that grafting on to resistant rootstocks could be useful. Reports from the Kenya Coffee Research Station, cited by Whitehead (1968), also indicated some resistance to *Meloidogyne* sp. in *Coffea carrisoi*, Conuga or Congusta coffee (hybrid of *C. congensis* and *C. canephora*) and some lines of *C. congensis* in Angola. Bertrand et al. (2002) found that resistance to the corky root disease complex (which is caused by *M. arabicida* and *F. oxysporum* and has affected *C. arabica* in Costa Rica), was heritable and that genetic resistance to *M. arabicida* was an effective strategy against the complex. By using *C. canephora* rootstocks, it was possible to reduce mortality substantially in the field and to reduce by half the number of plants with corky root symptoms. *C. canephora* rootstock cv. Nemaya is resistant to *M. arabicida* and *M. izalcoensis* (Bertrand and Anthony, 2008). There has been no assessment of coffee resistance against *M. lopezi* given its recent characterization (Villain et al., 2018).

6 Root lesion nematodes (*Pratylenchus* spp.)

6.1 General aspects

Root lesion nematodes (RLNs) are considered major parasites of Arabica and Robusta coffee in many coffee-producing countries or regions (Villain, 2008; Wiryadiputra and Tran, 2008; Rivillas et al., 2015; Villain et al, 2018). Among the 12 described species, *P. brachyurus* and *P. coffeae* are the most widespread in the tropics (Inomoto and Oliveira, 2008). However, in São Paulo, Brazil, P. *brachyurus* is more widely distributed, while *P. coffeae* occurs in higher densities, causing greater damage where it occurs (Kubo et al., 2004). *Pratylenchus* spp. lay their eggs in the root cortical tissue, as well as on the surface of roots. All stages feed on cortical tissue, and extended periods of feeding result in cell death and necrotic lesions, hence their name (Moens and Perry, 2009; Villain et al., 2018).

6.2 Pratylenchus coffeae

This species is described in coffee from Java Island (Zimmermann, 1898; Sher and Allen, 1953), and it is the most widespread and damaging RLN affecting coffee worldwide (Campos and Villain, 2005; Handoo et al., 2008; Inomoto and Oliveira, 2008). *P. coffeae* is a bisexual species that reproduces by obligatory amphimixis, so males are frequent. In some countries, it has long been considered a major threat for *C. arabica*, such as in Guatemala and El Salvador (Villain et al., 2018) or in India (Dhanam and Sreedharan, 2008), but also for *C. canephora*, especially in Indonesia and Vietnam (Wiryadiputra and Tran, 2008). Indeed, in Vietnam, *P. coffeae* is the dominant plant parasitic nematode species on both Robusta and Arabica coffee, occurring in all coffee-growing areas (Tran, 2002). *P. coffeae,* which has a wide distribution, may have the same geographical origin as banana and plantain, that is, the Pacific islands and neighboring Asian countries, from which it may have been spread through the transfer of infected plant material (Bridge et al., 1997; Villain et al., 2018).

6.3 Pratylenchus brachyurus and other species

P. brachyurus is a monosexual species (males absent or rare) that reproduces by mitotic parthenogenesis. This species is present in many coffee-growing regions in Brazil and appears to be the most widely distributed RLN in Brazil, especially in São Paulo and Minas Gerais (Inomoto and Oliveira, 2008). It was also reported on coffee in Vietnam (Nguyen and Nguyen, 2001). *P. pratensis* has been reported from one locality in south India by Somasekhar, cited by Whitehead (1969b), and *P. loosi* from Sri Lanka by Hutchinson, cited by Whitehead (1968). *P. goodeyi* occurs frequently on coffee in Tanzania (Bridge, 1984). *P. vulnus* was detected on coffee in Minas Gerais state, Brazil (Monteiro et al., 2001; Ferraz, 1980). *P. zeae* has been found only in soil samples associated with graminaceous weeds in coffee plantations, so its status as a coffee parasite is uncertain (Inomoto and Oliveira, 2008; Villain et al., 2018).

Due to difficulties in differentiating *Pratylenchus* spp., morphologically, many populations collected from coffee from many countries remain misidentified or unidentified, as is the case across Central America and Brazil (Campos and Villain, 2005; Villain et al., 2018). This is especially the case for the complex of amphimictic species morphologically similar to *P. coffeae*, such as *P. gutierrezi*, *P. panamaensis*, *P. pseudocoffeae* (not yet recorded in coffee) and likely undescribed species (Campos and Villain, 2005; Zamora-Araya et al., 2016; Villain et al., 2018). Additional studies are needed to determine the diversity of *Pratylenchus* spp. parasitizing coffee through an integrated taxonomy approach, involving classic morphology, molecular tools and pathogenicity assessment of the detected taxa (Villain et al., 2018).

6.4 Symptoms and losses caused by Pratylenchus spp.

The same atypical aerial symptoms mentioned for RKN infections are also observed for coffee RLN infections: stunting, chlorosis, leaf fall and progressive decay (Fig. 2e), with intensity varying according to the nematode species, the *Coffea* species, cultivar and age, and the agroecological conditions. RLNs cause root symptoms that are not always easy to recognize, with atypical necrosis observed mainly on fine roots, due to RLNs feeding on the cortical parenchyma cells. With the more aggressive RLN species, this necrotic process of absorbing fine roots leads to rotting and destruction of the entire root system (Fig 1i) and the death of infected plants. Due to the unspecific aerial and root symptoms, the occurrence of RLNs often remains unnoticed, with soil fertility issues or other soil-borne diseases blamed (Villain et al., 2002, 2018; Villain, 2008).

In Guatemala, an RLN population, presumed to be *P. panamaensis* according to Zamora-Araya et al. (2016), proved highly pathogenic on *C. arabica* under controlled conditions and in the field (Villain et al., 2002; Villain, 2000), affecting both the quantity and quality of coffee produced. The size and yield of coffee beans were strongly correlated negatively with RLN root densities (Villain et al., 2008; Villain, 2000), with average yields of 6 t/ha in the least infested plots down to 0.5 t/ha in the most infested plots. Bean size followed a similar trend, while high rates of plant death were also observed in the most infested plots, with up to 76% mortality 4 years after planting, with densities of 125 nematodes/g root. Consequently, particular attention is required for this species, which is highly pathogenic and likely has a wide distribution (Villain et al., 2018).

In Brazil, *P. coffeae* populations collected from coffee caused root destruction of *C. arabica* seedlings following reinoculation (Inomoto et al., 2007; Kubo et al., 2002). Severe root necrosis in the cortex, and even in the collar area of the stem, was observed, leading to severe decay of trees and death in up to 70% of the plantation. *P. coffeae* is the most destructive nematode on *C. arabica* in south India (Palanichamy, 1973). In Indonesia, *P. coffeae* is reported as highly destructive, causing production losses between 29% and 79% on *C. canephora* cv. Robusta plantations (Toruan-Mathius et al., 1995). In Vietnam, *P. coffeae* is the only RLN species reported from Robusta coffee in the Western Highlands (Tran, 2002), where it is considered among the main causes of leaf yellowing disease (Villain et al., 2018; Vu et al., 2014; Le et al., 2013; Trinh et al., 2009a).

P. brachyurus reduces plant and root growth and results in shedding of leaves and root darkening (Kubo et al., 2004). Infection by *P. goodeyi*, *P. loosi*, *P. pratensis* and *P. zeae* on coffee is not clearly defined. *P. brachyurus* and *P. zeae* have been reported from the rhizosphere of *C. arabica* fields previously planted with sugarcane in Hawaii, but both species tended to disappear 3 years after planting, indicating that coffee is a poor host (Villain et al., 2018; Schenk and Schmitt, 1992).

6.5 Management of *Pratylenchus* spp.

The most important measure to manage RLNs is preventative action to stop dissemination via nursery seedlings or soil. Effective control of the highly destructive *P. coffeae* requires the integration of a range of physical, cultural, chemical, biological control measures together with the use of plant resistance (Villain, 2008). Control of *P. coffeae* by chemical or physical means has been found to have limited efficacy.

Planting in suspected infested areas should be conducted with Robusta (*C. canephora*) or Robusta-Arabica grafted plants (Dhanam et al., 2011). *C. canephora* genotypes, studied by Dhanam and Sreedharan (2008), were considered the most resistant to *P. coffeae*, followed by *C. liberica*, while *C.*

arabica genotypes were considered highly susceptible. Variable resistance to *Pratylenchus* populations from Guatemala was observed among *C. canephora* progenies (Villain et al., 2004), in accordance with the substantial genetic variability observed within this species (Cubry et al., 2008; Musoli et al., 2009). The two reciprocal crossings of the two-parent clones of the rootstock *C. canephora* cv. Nemaya, which is also resistant to *Meliodogyne* spp. from Central America (see above), were the progenies most resistant to *Pratylenchus* spp.

Pre- and post-infective resistance factors have been observed on *C. canephora* cv. Nemaya (Villain, 2000; Villain et al., 2004). At the pre-infective stage, early penetration studies showed that *C. canephora* cv. Nemaya was unattractive to *Pratylenchus* spp., compared with the high attractiveness of *C. arabica* cv. Catuai. No histological structure likely to prevent or hinder penetration by nematodes was detected. In the post-infective stage, the poor reproduction of *Pratylenchus* spp. observed on *C. canephora* may be related to the presence of many polyphenols, especially in cv. Nemaya roots. Indeed, a strong correlation between resistance to *P. coffeae* with the total polyphenol content was observed in roots among numerous *C. canephora* clones in Indonesia (Toruan-Mathius et al., 1995). In addition, cambium lignifications, close to nematode lesions in the cortical parenchyma, were observed on cv. Nemaya roots, indicating inducible resistance mechanisms (Villain et al, 2004, 2018). If the main *Pratylenchus* resistance factors are linked to phenolic metabolism, it is to be hoped that resistance is not very specific (Dalmasso et al., 1992) and would thus provide the plant with an acceptable level of resistance to *Pratylenchus* species and/or pathotypes.

Schieber and Grullon (1969) also suggested the use of *C. canephora* var. Robusta in Guatemala as a source of resistance for rootstocks in grafted plants. Since grafting on to *C. canephora* cv. Robusta at the cotyledon stage was perfected by Reyna (1968), this practice has become increasingly common for controlling nematodes in Guatemala, Costa Rica and other countries. Grafting on common *C. canephora* rootstock provides efficient control of *Pratylenchus* spp., keeping densities at very low levels, even without chemical control (Villain et al., 2000, 2004). Grafting on to *C. canephora* also has the advantage of maintaining both physical and chemical bean qualities, as well as the organoleptic beverage quality of *C. arabica* (Villain et al., 2001, 2018).

6.6 Other nematode parasites of coffee

There are other genera and species of plant parasitic nematodes like *Radopholus* spp. and *Rotylenchus* spp. For these species, there are no accurate studies of damage and management. They will not be commented on in this chapter. For more information, consult Villain et al. (2018).

7 Summary and future trends

Coffee is affected by a wide diversity of plant parasitic nematodes, insects and soil-based fungi that may occur simultaneously or otherwise in different regions, countries or plantations. The control of nematodes is difficult and their eradication from the infested area is practically impossible. The main strategy for the management in coffee areas is to prevent the spread of infected coffee seedlings and contamination with infested soils, water and agricultural machines. Except for some examples of nursery inspections, official measures are generally too limited to prevent the dissemination of infected coffee seedlings. Indeed, in many countries, we observe increasing distribution ranges of the most pathogenic nematodes such as RKNs and RLNs. This is undoubtedly related to the increase in global temperatures that boosts the reproduction of nematodes with an increase in the number of cycles per year.

The use of germplasm resistant to nematodes is a key pillar for integrated nematode management in coffee, whether directly by planting resistant cultivars or by using susceptible cultivars grafted on to resistant rootstocks. The choice of resistant germplasm depends mostly on climatic conditions and/or targeted organoleptic quality. Coffee breeding programs must take into account the complexity of nematode communities present in coffee fields. The use of coffee germplasm that does not offer resistance to all the damaging nematodes could be counter-productive. For example, in Costa Rica, planting cv. Sarchimor resistant to *M. exigua* but not to *Pratylenchus* resulted in substantially decreased densities of *M. exigua* but significantly higher RLN populations, which could be more damaging than *M. exigua* (Bertrand et al., 1998; Villain et al., 2018). Another important aspect to consider is that resistance to nematodes in coffee, such as against *Pratylenchus*, is quantitative or incomplete (Villain et al., 2018).

Although reliable tools (molecular and biochemical) are increasingly available for accurate species-specific diagnosis of plant parasitic nematodes, there remains too little information on the diversity of nematodes parasitizing coffee in each country and/or area. Agriculture departments and/or coffee institutions need to promote monitoring studies to create the basic knowledge necessary to implement appropriate strategies for plant nematode management before and after planting coffee. It is especially important to have accurate information on plant nematode diversity so as to implement the adequate use of resistance, crop rotation or other strategies. However, to ensure the durable management of selected resistances, particularly in perennial crops, complementary non-chemical control measures must be developed to reduce field inoculum pressures (Villain et al., 2018). Biological control must become an increasingly adopted strategy against coffee parasitic nematodes (Carneiro et al., 2020). Durable use of resistance must be planned with appropriate coffee farming practices, such as appropriate fertilizer and soil management practices, shade in some geographical regions and, more globally, all practices that contribute to conferring an optimum physiological status on the plant, especially mitigating abiotic stress effects generated by climate change. For all these reasons, cropping under agroforestry systems seems to be the most sustainable cropping option for most coffee-growing regions and likely the most resilient way to mitigate both the effects of pests and climate change (Villain et al., 2018).

8 References

Abd-Elgawad, M. M. M. and Askary, T. H. (2015). Impact of phytonematodes on agriculture economy. In: Askary, T. H. and Martinelli, P. R. P. (Eds), *Biocontrol Agents of Phytonematodes*, 3–49. National Research Centre, Giza, Egypt.

Almeida, V. F. and Campos, V. P. (1991). Reprodutividade de *Meloidogyne exigua* em plantas antagonistas e em culturas de interesse econômico. *Nematologia Brasileira* 15: 24–29.

Almeida, A. A., Santiago, D. C., Benetti, E., Fudo, C. H., Boss, A. L. and Takachi, M. T. (2015). Eficiência do nematicida fluensulfone no controle de Meloidogyne exigua do cafeeiro. Proceedings of the 32nd Brazilian Congress of Nematology. Sociedade Brasileira de Nematologia, Londrina, Paraná, Brazil, pp.102–103.

Alvarado, J. A. (1935). *Tratado de Caficultura Practica*, Tipografia Nacional, Guatemala.

Anon (1974). India Coffee Board, 33rd *Annual Report*, 1972–1973.

Anon (1989). Corchosis del cafeto, una amenaza a la agricultura. *Boletim Informativo* 13: 1–3.

Anthony, F., Topart, P., Martinez, A., Silva, M. and Nicole, M. (2005). Hypersensitive-like reaction conferred by the *Mex-1* resistance gene against *Meloidogyne exigua* in coffee. *Plant Pathology* 54(4): 476–482.

Anzueto, F., Bertrand, B., Sarah, J. L., Eskes, A. B. and Decazy, B. (2001). Resistance to *Meloidogyne incognita* in Ethiopian *Coffea arabica* accessions. *Euphytica* 118(1): 1–8.

Araya, T. Z., Padilla, W. P., Archidona-Yuste, A., Cantalapiedra-Navarrete, C., Liébanas, G., Palomares-Rius, J. E. and Castillo, P. (2016). Root-lesion nematodes of the genus *Pratylenchus* (Nematoda: Pratylenchidae) from Costa Rica with molecular identification of *P. gutierrezi* and *P. panamaensis* topotypes. *European Journal of Plant Pathology* 145(4): 973–998.

Baeza-Aragón, C. A. (1978). Parasitismo de *Bacillus penetrans* en *Meloidogyne exigua* estabelecido em *Coffea arábica*. *Cenicafé* 29: 94–97.

Barbosa, D. H. S. G., Souza, R. M. and Vieira, H. D. (2010). Field assessment of coffee (*Coffea arabica* L.) cultivars in *Meloidogyne exigua* infested or free fields in Rio de Janeiro State, Brazil. *Crop Protection* 29(2): 175–177.

Barbosa, D. H. S. G., Vieira, H. D., Souza, R. M., Viana, A. P. and Silva, C. P. (2004). Field estimates of coffee yield losses and damage threshold by *Meloidogyne exigua*. *Nematologia Brasileira* 28: 49–54.

Barros, A. F., Oliveira, R. D. L., Lima, I. M., Coutinho, R. R., Ferreira, A. O. and Costa, A. (2014). Root-knot nematodes, a growing problem for Conilon coffee in Espírito Santo state, Brazil. *Crop Protection* 55: 74–79.

Barros, A. F., Oliveira, R. D. L., Zambolim, L., Ferreira, A. O. and Coutinho, R. R. (2011). *Meloidogyne paranaensis* attacking coffee trees in Espirito Santo State, Brazil. *Australasian Plant Disease Notes* 6(1): 43–45.

Berthaud, J. (1978a). L'hybridation interspécifique entre *Coffea arabica* L. et *Coffea Canephora* Pierre. Obtention et comparaison des hybrids triploids, Arabusta et hexaploïdes Premièe partie. *Café Cacao Thé* 22: 3-12.

Berthaud, J. (1978b). L'hybridation interspécifique entre *Coffea arabica* L. et *Coffea Canephora* Pierre. Obtention et comparaison des hybrids triploids, Arabusta et hexaploïdes. Deuxième partie. *Café Cacao Thé* 22: 87-112.

Bertrand, B., Aguilar, G., Bompard, E., Rafinon, A. and Anthony, F. (1997). Comportement agronomique et résistance aux principaux déprédateurs des lignées de Sarchimors et Catimors au Costa Rica. *Plante, Recherche, Développement* 4: 312-321.

Bertrand, B. and Anthony, F. (2008). Genetics of resistance to root-knot nematodes (*Meloidogyne* spp.) and breeding. In: Souza, R. M. (Ed.), *Plant-Parasitic Nematodes of Coffee*, 165-190. Springer, Dordrecht, Netherlands.

Bertrand, B., Cilas, C., Herve, G., Anthony, F., Etienne, H. and Villain, L. (1998). Relations entre les populations de deux espèces de nématodes, *Meloidogyne exigua* et *Pratylenchus* sp. In: Plante and Recherche (Eds), *Développement* (vol. 5), 279-286. les racines de *Coffea arabica* au Costa Rica, Dans.

Bertrand, B., Marraccini, P., Villain, L., Breitler, J. C. and Etienne, H. (2016). Healthy tropical plants to miti- gate the impact of climate change - As exemplified in coffee (d. o. Chapitre, Trans.). In: Torquebiau, E. (Ed.), *Climate Change and Agriculture Worldwide*, 83-95. Springer, Heidelberg, Germany.

Bertrand, B., Nuez, C. and Sarah, J.-L. (2000). Disease complex in coffee involving *Meloidogyne arabicida* and *Fusarium oxysporum*. *Plant Pathology* 49(3): 383-388.

Bertrand, B., Ramirez, G., Topart, P. and Anthony, F. (2002). Resistance of cultivated coffee (*Coffea arabica* and *C. canephora*) trees to corky-root caused by *Meloidogyne arabicida* and *Fusarium oxysporum*, under controlled and field conditions. *Crop Protection* 21(9): 713-719.

Boisseau, M., Aribi, J., Rodríguez de Sousa, F. Rd, Carneiro, R. M. D. G. and Anthony, F. (2009). Resistance to *Meloidogyne paranaensis* in wild *Coffea arabica*. *Tropical Plant Pathology* 34(1): 38-41.

Bridge, J. (1984). Coffee nematode survey of Tanzania. *Report on a Visit to Examine Plant Parasitic Nematodes of Coffee in Tanzania*. Commonwealth Institute of Parasitology, UK, 22p.

Bridge, J., Fogain, R. and Speijer, P. (1997). *Parasites et Ravageurs des Musa: fiche Technique no. 2. Les nématodes parasites des bananiers. P. Coffeae (Zimmermann, 1898) Filip. & Schu. Stek., 1941; P. goodeyi Sher & Allen, 1953. INIBAP, Parc Scientifique Agropolis II*. Montpellier, France.

Calleja-Cabrera, J., Boter, M., Oñate-Sanchez, L. and Pernas, M. (2020). Root growth adaptation to climate change in crops. *Frontiers in Plant Science* 11: 544.

Campos, H. D. and Campos, V. P. (1997). Efeito da época e forma de aplicação dos fungos *Arthrobotrys conoides, A. musiformis, Paecilomyces lilacinus* e *Verticillium chlamydosporium* no controle de *Meloidogyne exigua* do cafeeiro. *Fitopatologia Brasileira* 22: 361-365.

Campos, V. P. and Silva, J. R. C. (2008). Management of *Meloidogyne* spp. in coffee plantations. In: Souza, R. M. (Ed.), *Plant-Parasitic Nematodes of Coffee*, 149-164. Springer, Houten, the Netherlands.

Campos, V. P., Souza, J. T. and Souza, R. M. (1998). Controle de fitonematóides por meio de bactérias. In: Luz, W. C. (Ed.), *Revisão Anual de Patologia de Plantas* 6: 285-327. Sociedade Brasileira de Nematologia, Brasília- DF.

Campos, V. P. and E., Gomes, A. M., Correa, V. R. and Mattos, V. (2005). Coffee and cocoa. In: Luc, M., Sikora, R. A. and Bridge, J. (Eds), *Plant Parasitic Nematodes in Subtropical and Tropical Agriculture* (2nd edn), 529–580. CAB International, Wallingford, UK.

Cannell, M. G. R. (1971). Seasonal patterns of growth and developement of arabica coffee in Kenya. *Kenya Coffee* 36: 68–74.

Capot, J. (1972). L'amélioration du caféier en Côte d'Ivoire. Les hybrides 'Arabusta'. *Café Cacao Thé* 16: 3–18.

Carneiro, R. G. and Carneiro, R. M. D. G. (1982). Seleção preliminar de plantas para rotação de culturas em áreas infestadas por *Meloidogyne incognita* nos anos de 1979 e 1980. *Nematologia Brasileira* 6: 141–148.

Carneiro, R. G., Sousa, M. G., Moita, A. W., Correa, V. R. and Carneiro, R. M. D. G. (2013). Manejo integrado de *Meloidogyne paranaensis*, utilizando sucessão de culturas, resistência genética e controle biológico, Boletim de Pesquisa e Desenvolvimento 296. Embrapa, Brasília, DF, 23p.

Carneiro, R. M. D. G., Carneiro, R. G., Abrantes, I. M. O., Santos, M. S. N. A. and Almeida, M. R. A. (1996). *Meloidogyne paranaensis* n.sp. (Nemata: Meloidogynidae), a root- knot nematode parasitizing coffee in Brazil. *Journal of Nematology* 28(2): 177–189.

Carneiro, R. M. D. G. and Cofcewicz, E. (2008). Taxonomy of coffee-parasitic root-knot nematodes, *Meloidogyne* spp. In: Souza, R. M. (Ed.), *Plant-Parasitic Nematodes of Coffee*, 87–122. Springer, Houten, Netherlands.

Carneiro, R.M.D.G, Almeida, M.R.A., Gomes, A.C.M.M. and Hernández, A. (2005a). *Meloidogyne izalcoensis* n.sp. (Nematoda: Meloidogynidae), a root-knot nematode parasitizing coffee in El Salvador. *Nematology* 7(6): 819–832.

Carneiro, R. M. D. G., Randig, O., Almeida, M. R. A. and Gonçalves, W. (2005b). Identificação e caracterização de espécies de *meloidogyne* em cafeeiro nos estados de São Paulo e Minas Gerais através dos fenótipos de esterase e SCAR-Multiplex-PCR. *Nematologia Brasileira* 29: 233–241.

Carneiro, R. M. D. G., Mendes, Md, Almeida, M. R. A., dos Santos, M. F. A., Gomes, A. C. M. M. and Karssen, G. (2008). Additional information on *Meloidogyne inornata* Lordello, 1956 (Tylenchida: Meloidogynidae) and its characterisation as a valid species. *Nematology* 10(1): 123–136.

Carneiro, R. M. D. G., Monteiro, T. S. A., Eckstein, B. and Freitas, L. G. (2020). Controle de nematoides fitoparasitas. In: Fontes, G. and Valadares-Inglis, M. C. (Eds), *Controle Biológico de Pragas da Agricultura, E.M.*, 371–411. Embrapa Editora, Brasília, Brazil.

Carneiro, R. M. D. G., Monteiro, J. M. S., Silva, U. C. and Gomes, G. (2016). Gênero *meloidogyne*: Diagnose através de eletroforese de isoenzimas e marcadores SCAR. In: Oliveira, C. M. G., Santos, M. A. and Castro, L. H. S. (Eds), *Diagnose de Fitonematoides*, 47–70. Millennium Editora, Campinas, São Paulo, Brazil.

Carneiro, R. M. D. G., Quénéhervé, P. and Almeida, M. R. (2000). Enzyme phenotypes of *Meloidogyne* spp. populations. *Nematology* 2(6): 645–654.

Carneiro, R. M. D. G., Tigano, M., Randig, O., Almeida, M. R. and Sarah, J. L. (2004). Identification and genetic diversity of *Meloidogyne* spp. (Tylenchida: Meloidogynidae) on coffee from Brazil, Central America and Hawaii. *Nematology* 6(2): 287–298.

Castagnone-Sereno, P. (2012). *Meloidogyne enterolobii* (= *M. mayaguensis*): Profile of an emerging, highly pathogenic, root-knot nematode species. *Nematology* 14(2): 133–138.

Castagnone-Sereno, P., Danchin, E. G. J., Perfus-Barbeoch, L. and Abad, P. (2013). Diversity and evolution of root-knot nematodes, genus *meloidogyne*: New insights from the genomic era. *Annual Review of Phytopathology* 51: 203–220.

Castro, J. M. C. and Campos, V. P. (2004). Ocorrência de *Meloidogyne paranaensis* em cafeeiro na região do sul de Minas Gerais. *Summa Phytopathologica* 30: 507.

Castro, J. M. C., Campos, V. P., Pozza, E. A., Naves, R. L., Andrade Júnior, W. C., Dutra, M. R., Coimbra, C. M. and Silva, J. R. C. (2008). Levantamento de fitonematóides em cafezais do sul de Minas Gerais. *Nematologia Brasileira* 32: 56-64.

Correa, V. R., dos Santos, M. F. A., Almeida, M. R. A., Peixoto, J. R., Castagnone-Sereno, P. and Carneiro, R. M. D. G. (2013). Species-specific DNA markers for identification of two root-knot nematodes of coffee: *Meloidogyne arabicida* and *M. izalcoensis*. *European Journal of Plant Pathology* 137(2): 305-313.

Cubry, P., Musoli, P., Legnaté, H., Pot, D., Bellis, F. B. D., Poncet, V., Anthony, F., Dufour, M. and Leroy, T. (2008). Diversity in coffee assessed with SSR markers: Structure of the genus *Coffea* and perspectives for breeding. *Genome* 51(1): 50-63.

D'Arc de Lima, R. and Ferraz, S. (1985a). Biologia de *Meloidogyne exigua*. I. Desenvolvimento embriogênico e efeito da temperatura na embriogênese. *Revista Ceres* 32: 339-348.

D'Arc de Lima, R. and Ferraz, S. (1985b). Biologia de *Meloidogyne exigua*. II. Desenvolvimento pos-embriogênico em cafeeiro 'Mundo Novo'. *Revista Ceres* 32: 349-361.

Dalmasso, A., Castagnone-Sereno, P. and Abad, P. (1992). Seminar: Tolerance and resistance of plants to nematodes - Knowledge, needs and prospects. *Nematologica* 38: 466-472.

Dhanam, M., Kumar, P. K. V. and Jayarama (2011). Management of nematode infestation in coffee. *Indian Coffee* 75: 27-28.

Dhanam, M. and Sreedharan, K. (2008). India. In: Souza, R. M. (Ed.), *Plant-Parasitic Nematodes of Coffee*, 293-303. Springer, Houten, Netherlands.

Di Vito, M., Crozzoli, R. and Vovlas, N. (2000). Pathogenicity of *Meloidogyne exigua* on coffee (*Coffea arabica* L.) in pots. *Nematropica* 30: 55-61.

Eisenback, J. D., Bernard, E. C. and Schmitt, D. P. (1994). Description of the Kona coffee root-knot nematode, *meloidogyne konaensis* n. sp. *Journal of Nematology* 26(4): 363-374.

Eisenback, J. D. and Vieira, P. (2020). Additional notes on the morphology and molecular data of the Kikuyu root-knot nematode, *Meloidogyne kikuyensis* (Nematoda: Meloidogynidae). *Journal of Nematology* 52: 1-13.

Fazuoli, L. C. (1986). Genética e melhoramento do cafeeiro. In: Rena, A. B., Malavolta, E., Rocha, M. and Yamada, T. (Eds), *Cultura do cafeeiro, fatores que afetam a produtividade*, 87-113. Associação Brasileira para Pesquisa da Potassa e do Fosfato, Piracicaba, SP, Brazil.

Fazuoli, L. C., Medina Filho, H. P., Gonçalves, W., Guerreiro Filho, O. and Silvarolla, M. B. (2002). Melhoramento do cafeeiro: Variedades tipo arabica obtidas no Instituto Agronomico de Campinas. In: Zambolim, L. (Ed.), *O Estado da Arte de Tecnologias na Produção de Café*, 162-215. Editora UFV, Viçosa, MG, Brazil.

Ferraz, S. (1980). Reconhecimento das espécies de fitonematóides presente nos solos do estado de Minas Gerais. *Experientia* 26: 255-328.

Fornazier, M. J., Martins, D. S., Fanton, C. J. and Benassi, V. L. R. M. (2017). Manejo de Pragas do café Conilon. In: Ferrão, R. G., Fonseca, A. F. A., Ferrão, M. A. G. and De Muner, L. H. (Eds), *Café Conilon* (2nd edn), 399-433. Incaper, Vitória, ES.

Gaur, H. S. (2011). The implications of global climate change on plant-parasitic nematodes and nematology. *Proceedings of the National Symposium on Nematodes: A*

Challenge under Changing Climate and Agricultural Practices, Thiruvananthpuram, Kerala, India, November 16-18.

Gonçalves, W., Andrade, V. T., dos Reis Fatobene, B. J., de Brito Caixeta, L., Padilha, L., Oliveira, C. M. G., Rosa, J. M. O., Rodrigues, L. M. R. and Guerreiro Filho, O. (2021). Selection strategy of a *Coffea canephora* rootstock with simultaneous nematode resistance to *Meloidogyne exigua*, *M. incognita* and *M. paranaensis*. *European Journal of Plant Pathology* 160(1): 81-95.

Gonçalves, W. and Silvarolla, M. B. (2001). Nematóides parasitos do cafeeiro. In: Zambolim, L. (Ed.), *Tecnologias de Produção de Café com Qualidade*, 199-267. Editora Universitária, UFV, Viçosa, MG, Brazil.

Gonçalves, W., Silvarolla, M. B. and Lima, M. M. A. (1998). Estratégia visando a implementação de manejo integrado de nematoides parasitos do cafeeiro. *Informe Agropecuário* 19: 36-47.

Groenen, D. (2018). The effects of climate change on the pests and diseases of coffee crops in Mesoamerica. *Journal of Climatology and Weather Forecasting* 6(3): 3. doi: 10.4172/2332-2594.1000239.

Handoo, Z. A., Carta, L. K. and Skantar, A. M. (2008). Taxonomy, morphology and phylogenetics of coffee- associated root-lesion nematodes, *Pratylenchus* spp. In: Souza, R. M. (Ed.), *Plant-Parasitic Nematodes of Coffee*, 29-50. Springer, Dordrecht, Netherlands.

Hernández, A., Fargette, M. and Sarah, A. J. (2004). Characterisation of *Meloidogyne* spp. (Tylenchida: Meloidogynidae) from coffee plantations in Central America and Brazil. *Nematology* 6(2): 193-204.

Herrera, I., Bryngelsson, T., Monzon, A. and Geleta, M. (2011). Identification of coffee root-knot nematodes based on perineal pattern, SCAR markers and nuclear ribosomal DNA sequences. *Nematologia Mediterranea* 39: 101-110.

Hidalgo-Diaz, L., Bourne, J. M., Kerry, B. R. and Rodríguez, M. G. (2000). Nematophagous *Verticillium* spp. in soils infested with *Meloidogyne* spp. in Cuba: Isolation and screening. *International Journal of Pest Management* 46(4): 277-284.

Humphreys-Pereira, D. A., Flores-Chaves, L., Gómez, M., Salazar, L., Gómez-Alpízar, L. and Elling, A. A. (2014). *Meloidogyne lopezi* n.sp. (Nematoda: Meloidogynidae), a new root-knot nematode associated with coffee (L.) in Costa Rica, its diagnosis and phylogenetic relationship with other coffee-parasitising *Meloidogyne* species. *Nematology* 16: 643-661.

Inomoto, M., Kubo, R., Silva, R., Oliveira, C., Tomazini, M. and Mazzafera, P. (2007). Pathogenicity of two Pratylenchus coffeae populations from Brazil on coffee plants. *Nematology* 9: 853-858.

Inomoto, M. M. and Oliveira, C. M. G. (2008). Coffee associated *Pratylenchus* spp. - Ecology and interactions with plants. In: Souza, R. M. (Ed.), *Plant-Parasitic Nematodes of Coffee*, 51-64. Springer, Houten, Netherlands.

Janssen, T., Karssen, G., Topalovic,O., Coyne, D.. and Bert, W. (2017). Integrative taxonomy of root-knot nematodes reveals multiple independent origins of mitotic parthenogenesis. *PLoS ONE* 12(3): e0172190.

Janssen, T., Karssen, G., Verhaeven, M., Coyne, D. and Bert, W. (2016). Mitochondrial coding genome analysis of tropical root-knot nematodes (*meloidogyne*) supports haplotype based diagnostics and reveals evidence of recent reticulate evolution. *Scientific Reports* 6: 22591.

Jorge-Junior, A. S., Cares, J. E., Mattos, V. S., Coyne, D., dos Santos, M. F. A. and Carneiro, R. M. D. G. (2016). First report of *Meloidogyne izalcoensis* (Nematoda: Meloidogynidae) on coffee, cabbage, and other crops in Africa. *Plant Disease* 100(10): 2173-2173.

Kearn, J., Ludlow, E., Dillon, J., O'Connor, V.and Holden-Dye, L. (2014). Fluensulfone is a nematicide with a mode of action distinct from anticholinesterases and macrocyclic lactones. *Pesticide Biochemistry and Physiology* 109: 44-57.

Kubo, R. K., Oliveiro, C. M. G., Antedomênico, S. R., Monteiro, A. R., Ferraz, L. C. C. B. and Inomoto, M. M. (2004). Ocorrência de nematoides do gênero *Pratylenchus* em cafezais do Estado de São Paulo. *Nematologia Brasileira* 28: 159-165.

Kubo, R. K., Silva, R. A., Tomazini, M. D., Oliveira, C. M. G., Mazzafera, P. and Inomoto, M. M. (2002). Patogenicidade de *Pratylenchus coffeae* em plântulas de cafeeiro cv. Mundo Novo. *Fitopatologia Brasileira* 28(1): 41-48.

Le, D. K., Le, Q. K., Dao, T. H., Tran, T. H., Nguyen, N. C. and Trinh, P. Q. (2013). Plant parasitic nematodes associated with coffee and black pepper in the several major growing areas in the Central Highlands. *Journal of Plant Protection* 6: 19-24.

Le, T. M. L., Nguyen, T. D., Nguyen, H. T., Liebanas, G., Nguyen, T. A. D. and Trinh, Q. P. (2019). A new root-knot nematode, *Meloidogyne moensi* n. sp. (Nematoda: Meloidogynidae), parasitizing robusta coffee from Western Highlands, Vietnam. *Helminthologia* 56(3): 229-246.

Lima, E. A., Furlanetto, C., Nicole, M., Gomes, A. C. M. M., Almeida, M. R. A., Jorge-Junior, A., Correa, V. R., Salgado, S. M., Ferrão, M. A. G. and Carneiro, R. M. D. G. (2015). The multi-resistant reaction of drought-tolerant coffee 'Conilon Clone 14' to *Meloidogyne* spp. and late hypersensitive-like response in *Coffea canephora*. *Phytopathology* 105(6): 805-814.

Lima, I. N. (2016). Crescimento vegetativo, trocas gasosas e produtividade de clones de cafeeiros Conilon suscetíveis e resistentes a *Meloidogyne incognita* e *M. paranaensis*. Tese de Doutorado. UFV, Viçosa, MG. 2, 61p.

Lima, I. N., Souza, R. M., Silva, C. P. and Carneiro, R. M. D. G. (2005). *Meloidogyne* spp. from preserved areas of Atlantic forest in the state of Rio de Janeiro, Brazil. *Nematologia Brasileira* 29: 31-38.

López, R. and Salazar, L. (1989). *Meloidogyne arabicida* sp. n. (Nemata: *Heteroderidae*) nativo de Costa Rica. Un nuevo y severo patógeno del cafeto. *Turrialba* 39: 313-323.

López, R. and Vilchez, H. (1991). Two new hosts of the coffee root knot nematode. *Meloidogyne exigua in Costa Rica, Agronomica Costarricence* 15: 163-166.

López-Lima, D., Carrion, G., Sanchez-Nava, P., Desgarennes, D. and Villain, L. (2020). Fungal diversity and *Fusarium oxysporum* pathogenicity associated with coffee corky-root disease in Mexico. *Revista de la Facultad de Ciencias Agrarias. Universidad Nacional de Cuyo* 52(1): 276-292.

López-Lima, D., Sánchez-Nava, P., Carrion, G., Espinosa de los Monteros, A. and Villan, L. (2015). Corky root symptoms for coffee in central Veracruz are linked to the root-knot nematode *Meloidogyne paranaensis*, a new report for Mexico. *European Journal of Plant Pathology* 141(3): 623-629.

Lordello, L. G. E. (1956). *Meloidogyne inornata* sp. n. a serious pest of soybean in the state of São Paulo, Brazil (Nematoda, Heteroderidae). *Revista Brasileira de Biológia* 16: 65-70.

Lordello, L. G. E. (1984). Nematoides das Plantas Cultivadas. *Editora Nobel, São Paulo, Brasil.*

Lordello, L. G. E. and Zamith, A. P. L. (1960). *Meloidogyne coffeicola* sp.n., a pest of coffee trees in the state of Paraná, Brazil (Nematoda, Heteroderidae). *Revista Brasileira de Biologia* 20: 375-379.

Maleita, C. M., de Almeida, A. M. S. F., Vovlas, N. and Abrantes, I. (2016). Morphological, biometrical, bio-chemical, and molecular characterization of the coffee root-knot nematode *Meloidogyne megadora*. *Plant Disease* 100(8): 1725-1734.

Maximiniano, C., Campos, V. P., Souza, R. M. and Almeida, A. R. (2001). Flutuação populacional de *Meloidogyne exigua* em cafezal naturalmente infestado por *Pasteuria penetrans*. *Nematologia Brasileira* 21: 63-69.

Moens, M. and Perry, R. N. (2009). Migratory plant endoparasitic nematodes: A group rich in contrasts and divergence. *Annual Review of Phytopathology* 47: 313-332.

Moens, M., Perry, R. N. and Starr, J. L. (2009). *Meloidogyne* species - A diverse group of novel and important plant parasites. In: Perry, R. N., Moens, M. and Starr, J. L. (Eds), *Root-Knot Nematodes*, 225-248. CAB International, Wallingford, UK.

Monteiro, A. R., Antedomênico, S. R., Ferraz, L. C. C. B., Inomoto, M. M., Kubo, R. K. and Oliveira, C. M. G. (2001). Primeira ocorrência de *Pratylenchus vulnus* Allen & Jensen, 1951, em cafeeiro. In: *Congresso Brasileiro de Nematologia XXIII 88*. Marilia, SP, Brazil.

Monteiro, J. S., Cares, J. E., Gomes, A. M., Correa, V. R., Mattos, V. S., Santos, M. A., Almeida, M. R. A., Santos, C. G., Castagnone-Sereno, P. and Carneiro, R. D. G. (2016). First report of, and additional information on, *Meloidogyne konaensis* (Nematoda: Meloidogyninae) parasitising various crops in Brazil. *Nematology* 18(7): 831-844.

Muniz, MdF. S., Campos, V. P., Moita, A. W., Gonçalves, W., Almeida, M. R. A., Sousa, F. Rd and Carneiro, R. M. D. G. (2009). Reaction of coffee genotypes to different populations of *Meloidogyne* spp.: Detection of a naturally virulent *M. exigua* population. *Tropical Plant Pathology* 34(6): 370-378.

Muniz, MdF., Carneiro, R., Almeida, M. R., Campos, V. P., Castagnone-Sereno, P. and e Castro, J. MdC. (2008). Diversity of *Meloidogyne exigua* (Tylenchida: Meloidogynidae) populations from coffee and rubber tree. *Nematology* 10(6): 897-910.

Musoli, P., Cubry, P., Aluka, P., Billot, C., Dufour, M., Belllis, F., Pot, D., Bieysse, D., Charrier, A. and Leroy, T. (2009). Genetic differentiation of wild and cultivated populations: Diversity of *Coffea Canephora* Pierre in Uganda. *Genome* 52(7): 634-646.

Naves, R. L. and Campos, V. P. (1991). Ocorrência de fungos predadores de nematoides no sul de Minas Gerais e estudo da capacidade predatória e crescimento *in vitro* de alguns de seus isolados. *Nematologia Brasileira* 15: 152-162.

Nguyen, N. C. and Nguyen, V. T. (2001). Plant parasitic nematodes associated with coffee crop in some North and Tay Nguyen Highland provinces of Vietnam. In: *Selected Works on Ecological and Bio-Resource Studies (1996-2000)*, 188-195. Hanoi Agriculture Publishing House, Hanoi, Vietnam.

Noir, S., Anthony, F., Bertrand, B., Combes, M.-C. and Lashermes, P. (2003). Identification of a major gene (Mex-1) from *Coffee canephora* conferring resistance to *Meloidogyne exigua*. *Plant Pathology* 52(1): 97-103.

Oliveira, C., Mazzafera, P., Silva, R., Kubo, R., Inomoto, M. and Tomazini, M. (2007). Pathogenicity of two *Pratylenchus coffeae* populations from Brazil on coffee plants. *Nematology* 9(6): 853-858.

Pagan, C., Coyne, D., Carneiro, R., Kariuki, G., Luambano, N., Affokpon, A. and Williamson, V. M. (2015). Mitochondrial haplotype-based identification of ethanol-preserved root-knot nematodes from Africa. *Phytopathology* 105(3): 350-357.

Palanichamy, L. (1973). Nematode problems of coffee in India. *Indian Coffee* 37: 99-100.

Peres, A. C. J., Salgado, S. L., Correa, V. R., Santos, M. A., Mattos, V. S., Monteiro, J. S. M. S. and Carneiro, R. D. G. (2017). Resistance of *Coffea arabica* genotypes against *Meloidogyne paranaensis* and *M. incognita* under controlled and field conditions. *Nematology* 19(5): 617-626.

Randig, O., Bongiovanni, M., Carneiro, R. M. D. G. and Castagnone-Sereno, P. (2002). Genetic diversity of root-knot nematodes from Brazil and development of SCAR markers specific for the coffee-damaging species. *Genome* 45(5): 862-870.

Randig, O., Carneiro, R. M. D. G. and Castagnone-Sereno, P. (2004). Identification of Brazilian coffee-damaging species of *meloidogyne* using SCAR-coffee markers in multiplex-PCR. *Nematologia Brasileira* 28: 1-10.

Reyna, E. H. (1968). La técnica de injerto hipocotiledonar del cafeto para el control de nematodos. *Café (Lima)* 17: 5-11.

Rezende, R. M., Salgado, S. M. L., Renzende, J. C., Carvalho, G. R., Pereira, A. A., Lima, R. R. and Ferreira, A. D. (2013). Resistência de progênies de *Coffea arabica* em área infestada por *Meloidogyne exigua*. *Nematropica* 43: 233-240.

Ribeiro, R. C. F. and Campos, V. P. (1993). Isolamento e identificação e efeito da temperatura no crescimento *'in vitro'* de fungos parasitos de ovos de *Meloidogyne* spp. no sul de Minas Gerais. *Nematologia Brasileira* 17: 132-138.

Rivillas, C. A., Villain, L. and Bertrand, B. (2015). Diseases caused by nematodes. In: Gaitán, A. L., Cristancho, M. A., Castro-Caicedo, B. L., Rivillas, C. A. and Cadena-Gomez, G. (Eds), *Compendium of Coffee Diseases and Pests*, 37-42. The American Phytopathological Society, St Paul, MN.

Rojas, M. and Salazar, L. (2014). Loss caused by *Meloidogyne exigua* on the Caturra variety under controlled conditions in Costa Rica. In: *Proceedings of the 25th International Conference on Coffee Science*, ASIC, Bussigny, Switzerland.

Salgado, S. M. L., Guimarães, N. R. M. B., Botelho, C. E., Tassone, G. A. T., Marcelo, A. L., Souza, S. R., Oliveira, R. D. L. and Ferreira, D. F. (2015). *Meloidogyne paranaensis* e *Meloidogyne exigua* em lavouras caffeiras na região sul de Minas Gerais. *Coffee Science* 10: 475-481.

Salgado, S. M. L. and Rezende, J. C. (2010). Manejo de fitonematoides em cafeeiro. In: P. R. Reis and R. L. Cunha. (Eds.), *Café Arabica: do plantio à colheita*, 757-804. Epamig, Lavras, MG, Brazil.

Salgado, S. M. L. S., Fatobene, B. J. R., Terra, W. C. and Carneiro, R. M. D. G. (2021). Nematoides parasitas do cafeeiro: Aspectos gerais e controle. *EPAMIG Boletim Técnico* 111: 56.

Sampedro, J. J., Perez, J., Fowler, R. V., Fernandez, E., Gandarila, H. A. O., Lorenzo, E., Basterrechea, J. G. and O'Connor, B. (1989). Nematodos parasitos associados al cultivo del cafeto em Cuba. *Ciencia y Técnica em la Agricultura*. Prot. Plant 12:59-71.

Santiago, D. C., Homochin, M., Krzyzanowski, A. A., Carvalho, S. and Fonseca, I. C. B. (2001). Efeito antagônico de *Arachis pintoi* sobre *Meloidogyne paranaensis* e *Meloidogyne incognita* (Kofoid & White), 1919. *Chitwood* 1949, raças 1: 2, 3 e 4. Anais Congresso Brasileiro de Nematologia, Marília, SP (vol. XXIII), 150p.

Santos, M. F. A., Correa, V. R., Peixoto, J. R., Mattos, V. S., Silva, J. G. P., Moita, A. W., Salgado, S. M. L., Castagnone-Sereno, P. and Carneiro, R. M. D. G. (2018). Genetic variability of *Meloidogyne paranaensis* populations and their aggressiveness to susceptible coffee genotypes. *Plant Pathology* 67(1): 193-201.

Schenk, S. and Schmitt, P. (1992). Survey of nematodes on coffee in Hawaii. *Supplement to the Journal of Nematology* 24(4): 771-775.

Schieber, E. and Grullon, L. (1969). El Problema de nematodos que atacam el café (*Coffea arabica* L.) en la Republica Dominicana. *Turrialba* 19: 513–517.

Schieber, E. and Sosa, O. N. (1960). Nematodes on coffee in Guatemala. *Plant Disease Reporter* 44: 722–723.

Sharma, R. D. and Lordello, R. R. A. (1992). Occurrence of Pasteuria penetrans in coffee plantations infested by Meloidogyne exigua in the State of Sao Paulo, Brazil. Proceedings XXV Congr. Bras. Fitopatol 183.

Sher, S. A. and Allen, M. W. (1953). Revision of the genus *Pratylenchus* (Nematoda: Tylenchidae). *University of California Publications in Zoology* 57: 441–470.

Shigueoka, L. H., Sera, G. H., Sera, T., Fonseca, I. C. D. B., Andreazi, E., Carvalho, F. G., Carducci, F. C. and Ito, D. S. (2016). Reaction of arabica coffee progenies derivative from Icatu to *Meloidogyne paranaensis*. *Bragantia* 75: 193–198.

Silva, MdC., Várzea, V., Guerra-Guimarães, L., Azinheira, H. G., Fernandez, D., Petitot, A. S., Bertrand, B., Lashermes, P. and Nicole, M. (2006). Coffee resistance to the main diseases: Leaf rust and coffee berry disease. *Brazilian Journal of Plant Physiology* 18(1): 119–147.

Silva, R. V., Oliveira, R. D. L. and Zambolim, L. (2009). Primeiro relato da ocorrência de *Meloidogyne paranaensis* em Cafeeiro no Estado de Goiás. *Nematologia Brasileira* 33: 187–190.

Sipes, B. S., Schmitt, D. P., Xu, K. and Serracin, M. (2005). Esterase polymorphism in Meloidogyne konaensis. *Journal of Nematology* 37(4): 438–443.

Stefanelo, D. R., Santos, M. F. A. , Mattos, V. S. , Braghini, M. T., Mendonça, J. S. F., Cares, J. E. and Carneiro, R. M. D. G. (2018). *Meloidogyne izalcoensis* parasitizing coffee in Minas Gerais state: the first record in Brazil. *Tropical Plant Pathology* 45(2): 209–212.

Swai, I. S. (1981). Root knot nematodes, *Meloidogyne* species. In: Carter, C., and Sasser, J. N. (Eds), *Proceedings of the* Third Research Planning *Conference on Root-Knot Nematodes, Meloidogyne* spp. (Vols. 28–30). Department of Plant Pathology, North Carolina State University, Ibadan Nigeria, Raleigh, NC.

Toruan-Mathius, N., Pancoro, A., Sudarmadji, D., Mawardi, S. and Hutabarat, T. (1995). Root characteristics and molecular polymorphisms associated with resistance to Pratylenchus coffeae in robusta coffee. *Menara Perkebunan* 63: 43–51.

Tran, K. L. (2002). Research on the causal agents of Robusta coffee yellow rot root symptoms in Dak Lak Province and assessment of control potential. PhD Thesis, Hanoi Agriculture University, Vietnam.

Trinh, P. Q., De la Pena, E., Nguyen, C. N., Nguyen, H. X. and Moens, M. (2009). Plant-parasitic nematodes associated with coffee in Vietnam. *Russian Journal of Nematology* 17: 73–82.

Trinh, P. Q., Pham, K. D. and Nguyen, N. C. (2013). Emerging *Meloidogyne* species (root-knot nematodes) threats to coffee in the Western Highlands in Vietnam. In: *Proceedings of the Second VAST-KAST Work- Shop on Biodiversity and Bio-Active Compounds*, 313–319.

Trinh, Q. P., Le, T. M. L., Nguyen, T. D., Nguyen, H. T., Liebanas, G. and Nguyen, T. A. D. (2019). *Meloidogyne daklakensis* n. sp. (Nematoda: Meloidogynidae), a new root-knot nematode associated with robusta coffee (*Coffea Canephora* Pierre ex A. Froehner) in the Western Highlands, Vietnam. *Journal of Helminthology* 93(2): 242–254.

van der Vossen, H., Bertrand, B. and Charrier, A. (2015). Next generation variety development for sustainable production of arabica coffee (*Coffea arabica* L.): A review. *Euphytica* 204(2): 243–256.

Villain, L. (2000). Caractérisation et bioécologie du complexe parasitaire du genre *Pratylenchus* (Nemata: Pratylenchidae) présent sur caféiers (*Coffea* spp.) au Guatemala. PhD Thesis. ENSAR, Rennes, France.

Villain, L. (2008). Economic importance, epidemiology and management of *Pratylenchus* sp. in coffee plantations. In: Souza, R. M. (Ed.), *Plant-Parasitic Nematodes of Coffee*, 65–84. Springer, Houten, Netherlands.

Villain, L., Anzueto, F. and Sarah, J. L. (2004). Resistance to root-lesion nematodes on *Coffea canephora*. In: Cook, R. and Hunt, D. J. (Eds), *Proceedings of the Fourth International Congress of Nematology, Nematology Monographs and Perspectives* (vol. 2), 289–302. Brill, Leiden, Netherlands.

Villain, L., Baujard, P., Anzueto, F., Hernández, A. and Sarah, J. L. (2002). Integrated protection of coffee plantings in Central America against nematodes. *Plante Recherche Développement (Special Issue: Research and Coffee Growing)*, 118–133.

Villain, L., Hernández, A. and Anzueto, F. (2008). Central America. In: Souza, R. M. (Ed.), *Plant-Parasitic Nematodes of Coffee*, 261–275. Springer, Houten, Netherlands.

Villain, L., Martinez, D., Mondolot-Cosson, L., La Fisca, P., Nicole, M. and Campa, C. (2015). Histological, histochemical and phytochemical characterization of *Coffea* spp. resistance to the root-knot nematode *Meloidogyne paranaensis*. In: *International Conference on Coffee Science 25*, 42–46. ASIC, Bussigny, Switzerland.

Villain, L., Molina, A., Sierra, S., Decazy, B., Anzueto, F., Bernard, G., Jean-Jacques, P. and Jean-Louis, S. (2001). Evaluation of grafting and nematicide treatments for the management of a root-lesion nematode, *Pratylenchus* sp. in *Coffea arabica* L. plantations in Guatemala. In: *International Conference on Coffee Science* (vol. XIX). ASIC, Paris.

Villain, L., Molina, A., Sierra, S., Decazy, B. and Sarah, J. L. (2000). Effect of grafting and nematicide treaments on damage by root lesion nematodes (*Pratylenchus* spp.) to *Coffea arabica* L. in Guatemala. *Nematropica* 30: 87–100.

Villain, L., Salgado, S. M. L. and Trinh, P. Q. (2018). Nematode parasites of coffee and cocoa. In: Sikora, R. A., Coyne, D., Hallmann, J. and Timper, P. (Eds), *Plant Parasitic Nematodes in Subtropical and Tropical Agriculture* (3rd edn.), 536–583. CAB International, Wallingford, UK.

Villain, L., Sarah, J. L., Hernández, A., Bertrand, B., Anthony, F., Lashermes, P., Anzueto, F., Figueroa, P. and Carneiro, R. M. D. G. (2013). Diversity of root-knot nematodes parasitizing coffee in Central America. *Nematropica* 43: 194–206.

Vu, A. T., Trinh, Q. P. and Nguyen, V. T. (2014). Plant nematodes associated to replanting coffee in basalt soil and their correlation with yellow leaf symptom in Gia Lai. *Journal of Agriculture and Agricultural Development* 23: 36–43.

Whitehead, A. G. (1968). Nematodea. In: Le Pelley, R. H. (Ed.), *Pests of Coffee*, 407–422. Longmans, Green and Co. Ltd, London.

Whitehead, A. G. (1969a). The distribution of root-knot nematodes *Meloidogyne* spp. in tropical Africa. *Nematologica* 15(3): 315–333.

Whitehead, A. G. (1969b). Nematodes attacking coffee, tea and cocoa and their control. In: Peachey, J. E. (Ed.), *Nematodes of Tropical Crops: Technical Communication* (vol. 40), 238–250. Commonwealth Bureaux of Helminthology, St Albans, UK.

Wiryadiputra, S. and Tran, L. K. (2008). Indonesia and Vietnam. In: Souza, R. M. (Ed.), *Plant-Parasitic Nematodes of Coffee*, 277–292. Springer, Dordrecht, the Netherlands.

Zimmermann, A. (1898). De nematoden der koffiewortels. Batavia (=Jakarta, Indonesia) and S'Gravenhage, La Hague, pays bas, G. Kolf & Co., coll., Mededeelingen uit's lands pantentuin 27: 67.

Chapter 13

Integrated management of soil-borne insect and fungal pests of coffee

Cesar J. Fanton and Renan B. Queiroz, Instituto Capixaba de Pesquisa, Assistência Técnica e Extensão Rural (INCAPER), Brazil; and Laércio Zambolim, Universidade Federal de Viçosa, Brazil

1 Introduction

Like other major crops, coffee (*Coffea arabica* and *Coffea canephora*) is threatened by climate change. Increasing temperatures and rainfall variability are predicted in the Americas, Africa and Asia. Since coffee is very sensitive to temperature, rainfall and available sunshine, climate change threatens to have a significant effect on coffee development and yields (Groenen, 2018). As a result, by 2050, some countries in the Americas could see a reduction

http://dx.doi.org/10.19103/AS.2021.0096.32

of 40–50% in the land area that is suitable to grow coffee. Climate change potentially both induces stress in plants and, through warmer and more humid conditions, promotes pests and diseases. Stressed plants may then have lower resistance to pest and disease attacks, increasing the incidence and severity of damage to plants at the expense of growth, yields and quality.

Plant parasitic nematodes, soil-borne pests and soil diseases are capable of parasitizing the plant's root system during practically the entire crop cycle in the field from the nursery stage through to harvest. Given the key importance of a healthy root system to plant health and growth, damage from these pests can be substantial. This chapter reviews key research on soil-inhabiting insect pests such as termites, ants, mealybugs, root flies and cicadas. In each case, it looks at topics such as pest life cycle, damage caused and strategies for pest management. The chapter also covers soil-borne fungal diseases, rhizoctoniosis, rosellinia root rot, Ceratocystis canker stain and coffee wilt disease, looking at symptoms, causal agents, epidemiology and disease management. An accompanying chapter reviews soil-borne nematode pests of coffee.

Coffee-associated insects that have a subterranean habit can be separated into two groups:

1. those that directly attack the roots, causing injuries on coffee plants by mechanical damage or by sucking sap;
2. those that have a mainly subterranean life cycle but attack the aerial part of the plant.

Although pests that cause obvious and visible damage to aerial parts such as seeds are clearly important for crop quality and yield, pests predominantly attacking roots may also cause substantial losses even if the damage is less immediately visible. Generally, an increase in the population of these pests is favoured by a combination of local conditions, notably soil conditions, linked with climatic factors such as higher or lower rainfall or temperature, and the absence of natural enemies, such as parasitoids, predators and entomopathogens. Le Pelley (1968) and Fornazier et al. (2017a,b) list coffee-associated pests and the parts of the plant they attack.

2 Asian subterranean termite (*Coptotermes gestroi*) (Isoptera: Rhinotermitidae)

2.1 Overview

Termites are generalist insects which can occasionally attack Robusta and Arabica coffee plants, although it is not common to see a resulting economic loss. Termite damage is more frequent in young coffee plants. Commonly known as 'Asian subterranean termite,' the species *Coptotermes gestroi* has a worldwide distribution in tropical and subtropical regions (Evans et al., 2013). A study carried out in Florida (USA) showed a high dispersion of this species between 1990 and 2015 (Chouvenc et al., 2016a,b). Since subterranean termites can forage over long distances, temperature and humidity are important factors in this dispersion (Bordereau and Pasteels, 2010).

2.2 Life cycle

Termite colonies are formed after selected males and females take wing to become the future kings and queens of new colonies. A colony is divided into castes: the royal couple is responsible for multiplication, the soldiers defend the colony and the workers are responsible for food collecting and nest cleaning (Constantino, 2002). The generation of new individuals is continuous, so the colony's longevity can extend indefinitely since kings and queens can be replaced by other individuals. The subterranean termite does not form nests above the soil surface as mound termites do. Instead, nests comprise extensive subterranean chambers which make it difficult to verify their presence in coffee plantations (Chouvenc et al., 2016a,b).

2.3 Damage caused

Colony workers are the termites responsible for damage to coffee plants, mainly to seedlings. During foraging, workers attack the coffee plant's main stem and cause bark removal. This damage can interrupt the young plant's development, and more severe attacks can remove all the stem bark which then kills the plant. They can also feed on seedling roots up to 2 years old. Older coffee crops are less affected, perhaps due to the availability of cellulosic material as a food source. This includes remains of branches from pruning carried out on the coffee plants (mainly Robusta coffee) which serve as food for termites and are available in sufficient quantity (Constantino, 2002; Harris, 1961).

2.4 Management

The symptoms of termite attack can be confused with nutritional deficiency, since affected plants show reduced development, leaf yellowing and then leaf fall (Junqueira et al., 2008). It is necessary to remove affected plants to examine the root system, as well as look for activity and presence of termites in the soil. A monitoring station which detects movement of termites over soil has been developed. When termites are detected, the station is replaced by slow-acting baits containing hexaflumuron, a physiological insecticide that acts as a chitin synthesis inhibitor (Su and Scheffrahn, 1998).

3 Leaf-cutting ants (*Atta* sp. and *Acromyrmex* spp.) (Hymenoptera: Formicidae)

3.1 Overview

Leaf-cutting ants, both those of the genera *Atta* and *Acromyrmex*, are generalist social insects which attack a great diversity of plant species (Schowalter and Ring, 2017). They have worldwide distribution (Della Lucia et al., 2014). Ants use coffee leaves (*C. arabica* and *C. canephora*) as a substrate for the production of two fungal genera (*Leucoagaricus* and *Leucocoprinus*) on which they feed (Shik et al., 2016; Mueller et al., 2001).

The genus *Atta* is larger and has three pairs of pronotum spines. Colonies forage over long distances and build subterranean nests with many chambers, removing a large amount of soil which makes it easier to locate nests. The genus *Acromyrmex* is smaller and has four pairs of pronotum spines. They forage over shorter distances and build smaller nests with one or more subterranean chambers which are more difficult to find (Fisher and Cover, 2007).

3.2 Life cycle

Leaf-cutting ant colonies are formed during nuptial flights in which several males fertilize queens. Nests are organized in chambers from a single to a few (*Acromyrmex*), up to several dozen (*Atta*) (Moser, 2006). The colony individuals are divided into castes and with a division of labour characteristic of social insects. The queen is responsible for reproduction and soldiers defend the nest. The workers carry out foraging; they cut and transport leaves used as a substrate for fungus cultivation that serves as food for the colony. Finally, the gardeners are responsible for growing fungus and cleaning the nest (Fisher and Cover, 2007). A leaf-cutting ant colony's longevity is determined by the queen's reproductive capacity. The colony will become extinct when the queen ends her reproductive life.

3.3 Damage caused

Leaf-cutting ants cause damage to Conilon and Arabica coffee by defoliation (Fanton and Queiroz, 2020) (Fig. 1). This defoliation can stimulate the plant to use extra energy to replace the removed leaves and thus decreases its normal

Figure 1 Leaves of Conilon coffee (*Coffea canephora*) after attack by leaf-cutting ants. Photo: Renan Queiroz.

growth. This slows the development of seedlings and young plants. In newly transplanted seedlings, high infestations can cause plant death, requiring replanting which increases production costs.

3.4 Management

A key first step in management is prior inspection of a plot before planting and, where activity is detected, use of insecticidal baits. It is important to prevent insecticide granules from becoming wet since this makes them unattractive to ants who will then not carry them to the nest. Chemical control can also be carried out by blowing powder insecticides into nest entrance holes. The use of cultural control methods, using bait plants as intercrops, has reduced losses caused by leaf-cutting ants (Montoya-Lerma et al., 2012). In a study carried out in Costa Rica with Arabica coffee, the ant species *Atta cephalotes* preferred other plants, such as *Erythrina poeppigiana*, reducing the infestation by up to 40% (Varón et al., 2007).

4 Coffee root mealybug (*Dysmicoccus texensis*) (Hemiptera: Pseudococcidae)

4.1 Overview

Dysmicoccus texensis occurs in several countries in the neotropics (Central America, the Caribbean and South America) (Fornazier et al., 2017a,b). The species is polyphagous, feeding on 36 plant species, including both Arabica and Robusta coffee plants (García Morales et al., 2016),. This species shelters on coffee plant roots, resulting in the production of the *Bornetina* fungus. This fungus is associated with honeydew excreted by the mealybug which forms waxy scales around the roots, commonly called 'crypts' or 'popcorn,' within

Figure 2 Conilon coffee roots (*Coffea canephora*) with 'crypts' and coffee root mealybug, *Dysmicoccus texensis*. Photo: Renan Queiroz.

which mealybugs take shelter (Santa-Cecília et al., 2005; Willians and Granara De Willink, 1992) (Fig. 2).

4.2 Life cycle

The entire life cycle takes place inside the crypts, where the mealybug females, eggs and nymphs are found. Only the male adult has wings. Females have an oval body, pink colouring and can reach up to 25 mm in length. They are covered by white wax and have 34 waxy filaments arranged around the body. Coffee root mealybug has three nymph stages over 21 to 49 days. It also has a total lifespan of 80 to 120 days, can generate up to 253 offspring by parthenogenetic reproduction and has an oviposition period of 52–87 days (Santa-Cecília et al., 2000, 2002, 2005).

4.3 Damage caused

Damage from the *D. texensis* mealybug is due to continuous suction of sap from roots and injection of toxins. When the population is high, mealybug infestation can kill coffee plants (Mibey, 1997). Infested coffee plants do not grow well and not respond to irrigation and fertilization.

4.4 Management

The main control measure against coffee root mealybug is insecticides. Nakano et al. (2001) recommends control should be undertaken even in minor outbreaks since the pest can quickly disperse throughout the coffee crop. Once the initial infestation is identified, both symptomatic plants and asymptomatic neighbouring plants should be treated. Neonicotinoid insecticides on arabica coffee using a drench application have shown good efficiency in controlling *D. texensis* (Souza et al., 2007a,b). Natural biological control agents include predatory ladybugs (Coleoptera: Coccinellidae) and *Verticilium lecanii* fungus. However, their effectiveness has not been determined. Studies using entomopathogenic nematodes from *Heterorhabditis* sp. have efficacy of up to 65% in field tests (Alves et al., 2009).

5 Coffee mealybug (*Planococcus* sp.) (Hemiptera: Pseudococcidae)

5.1 Overview

Several species of the genus *Planococcus* are economically important pests in coffee crops (*C. arabica* and *C. canephora*) (Le Pelley, 1968). These insects have worldwide distribution (Ben-Dov, 1994; Cox, 1989). In Brazil, the most common species in *C. canephora* is *P. citri* (Fornazier, 2016). As well as attacking roots, mealybugs can attack aerials of plants, including sucking sap from the small fruits of coffee. Dispersion depends on climatic conditions and crop management.

5.2 Life cycle

Adults, eggs and nymphs form colonies at the base of coffee fruits. They form white waxy masses which protect the colony and prevent the insecticide. Females have an oval-shaped body and are 3–5 mm long with dark yellow colouring. Females have 18 waxy filaments arranged on each side of the body and 2 larger filaments in the terminal part of the body. The oviposition period is approximately 90 days, and each female can generate about 400 offspring (Costa et al., 2016). The eggs are placed inside the waxy secretion mass that the females form at the base of fruits. After 10–20 days, depending on the climatic conditions, nymphs appear and, after about 10 days, reach adulthood (Fornazier et al., 2017a,b; Santa-Cecília et al., 2002).

5.3 Damage caused

The level of economic damage caused to coffee by the coffee mealybug has not been determined. Insects attack small fruits of Robusta coffee plants by sucking sap and causing them to fall, reducing plant productivity (Fornazier, 2016) (Fig. 3). *C. canephora* plants have a mechanism for adjusting the number of fruits that they can produce. This natural drop occurs when the rapid expansion phase of the fruits begins. Sometimes this natural process of the Robusta coffee is confused with fruit fall caused by coffee mealybug.

5.4 Management

High populations of the coffee mealybug in *C. canephora* have been observed since 2005 in Espírito Santo state, Brazil. Damage has been linked to excessive

Figure 3 Conilon coffee fruits totally damaged and small fruits with *Planococcus citri* nymphs. Photo: Renan Queiroz.

use of herbicides for weed control (personal communication). If some weeds are left, they provide alternative hosts for the mealybug which reduces the infestation of coffee plants. The first rains at the end of the dry season of the year, around September, stimulate the migration of the coffee mealybug population from the roots to aerial parts (Fornazier, 2016). Spraying with insecticides is recommended at this point since insects are not yet protected by a waxy powdery secretion.

6 Cicadas (*Quesada gigas, Dorisiana drewseni, D. viridis, Fidiciana pronoe, Carineta fasciculata, C. spoliata, C. matura*) (Hemiptera: Cicadidae)

6.1 Overview

Adult cicadas are free-living insects laying eggs on the plants on which they have grown. Several species are associated mainly with Arabica coffee (Martinelli, 2004). *Quesada gigas* is the most important coffee pest (Andrade et al., 2020). Subterranean nymphs develop from eggs and feed on coffee roots (Decaro Junior et al., 2012). Cicadas are more frequently associated with sandy textured soils, though they can also occur in clay-textured soils. In Brazil, Arabica coffee planting has increased in areas of the Cerrado Biome, mainly in Minas Gerais state, resulting in increased frequency of this pest (Souza et al., 2007a,b).

6.2 Life cycle

Cicada adult females have a dark green to brown colouring with spots on the transparent wings. After mating, the cicadas lay their eggs on the branches, bark and trunk of the coffee plants (Decaro Junior et al., 2012). When the nymphs hatch, they move to the soil, looking for a place to adhere to coffee roots. After attaching to the roots, they start sucking sap continuously (Maccagnan and Martinelli, 2004). After feeding for about a year on the roots, the nymphal stage is complete and then they leave the roots and come out of the soil. Nymphs' emergence is associated with circular holes. The nymphs then climb up the stem where they settle and remain immobile. The adults then emerge and leave the exuvia attached to the plants. This exuvia is another indication of the pest's presence in the crop. Adult emergence varies from species to species but extends from September to February, which is the rainy season in Brazil. The longevity of *Q. gigas* in the field has been estimated at 49 days (Andrade et al., 2020).

6.3 Damage caused

Cicadas prefer old coffee plantations from 6 to 10 years. The continuous sucking of sap by the nymphs weakens plants, reducing their productive capacity. The plant gradually wastes, a process which is accelerated during the dry season. There can be both severe attacks, with more than 400 nymphs per plant, or a series of smaller-scale attacks, both of which can kill plants. The leaves of the attacked plants become chlorotic and yellowish, with an early and extensive fall, which also affects flower buds, flowers and new fruits. These symptoms can be confused with nutritional deficiency (Souza et al., 2007a,b). However, intense attacks can be identified by the presence of exuvia adhered to the coffee plants as well as circular exit holes made by the nymphs.

6.4 Management

Early detection is essential. Many coffee plantations have sampling plans to both detect pests and establish damage thresholds requiring pest management. Once thresholds are reached, specific insecticides for cicada control should be applied. Systemic granulated insecticides are most efficient and should be applied during the rainy season (Souza et al., 2007a,b).

7 Coffee root fly (*Chiromyza vittata*) (Diptera: Stratiomyidae)

7.1 Overview

Chiromyza vittata species was first described in 1986 and has been recorded in Arabica plantations (*C. arabica*) in several states in Brazil (Matiello et al., 1990). It also causes minor infestations in Robusta (*C. canephora*) (Matiello et al., 1999). The larvae attack the coffee roots which, when perforated, can cause plant death (D'Antonio, 1991).

7.2 Life cycle

The insect cycle, from egg to adult, occurs over a period of about 1 year, with adults emerging from February to March until May–June. Eggs are laid in fissures in stem bark and near to the soil. The larvae are subterranean and can vary from 2 mm to 25 mm in length given overlapping generations (about 5 per year). Most larvae are found at depths of up to 30–40 cm, but they are highly mobile and can affect the entire root system. Their colouring varies from light brown at the initial instars to dark brown at the end instars. They have a strongly segmented body (Pujol-Luz and Vieira, 2000). Their presence is indicated by the large number of pupal exuvia on the ground when the adults emerge.

7.3 Damage caused

Coffee root fly larvae perforate the roots and then feed on decomposing organic matter from the damage caused. This results in a reduction in the plant's ability to absorb water and nutrients. The initial symptoms of attack are a drastic loss in growth associated with severe defoliation and dryness of the plant's branches. High larval populations can result in plant death (Matiello et al., 1998).

7.4 Management

Chemical control with granulated insecticides applied to the soil can be effective. The imidacloprid insecticide (from the neonicotinoid chemical group) has shown up to 80% efficacy against *C. vitatta* larvae (Souza and Reis, 2000). Natural enemies include the parasitoid *Monomachus fuscator* (Hymenoptera: Monomachidae) which emerges at the same time as coffee root fly adults. However, no evaluation of its effectiveness has been undertaken.

8 White grubs (*Phyllophaga* spp.) (Coleoptera: Scarabaeidae)

8.1 Overview

The genus *Phyllophaga* belongs to the Scarabaeidae family which includes several herbivorous species that can cause economic damage to crops. The larvae live in soil where they feed on plant roots. Infestations have been reported in coffee plant nurseries and plantations in America and Asia (Waller et al., 2007).

8.2 Life cycle

Phyllophaga adults have a reddish-brown colour with a length up to approximately 2 cm. Like all individuals in the Scarabaeidae family, the larvae have a curved body in a 'C' shape with three pairs of legs and visible brown hair (Fig. 4). The eggs are spherical, around 2.3 mm in diameter and have an incubation period of around 15 days.

8.3 Damage caused

Larvae feed on coffee plants' roots, especially during the rainy season. Damage from larvae is higher when plants are still seedlings. Larval feeding on roots results in plants turning yellow and losing vigour.

8.4 Management

Physical, chemical and biological methods can be used in white grub management. Insecticides applied to the soil in planting lines or on seedlings have shown good efficacy, mainly at the beginning of the rainy season, when the larvae are still in the first instar and are more sensitive to insecticides (Oliveira and Frizzas, 2017). Physical techniques such as ploughing and harrowing of soil can expose larvae to natural predators. Biological control using entomopathogenic nematodes (Girón-Pablo et al., 2015), such as *Heterorhabditis bacteriophora* and *Steinernema carpocapsae* (Liesch and Williamson, 2010), and entomopathogenic fungus native isolates of *Beauveria bassiana* and *Metarhizium anisopliae* has shown good efficacy (Rivera, 2014).

Figure 4 Larvae of *Phyllophaga* sp. with characteristic body in a 'C' shape. Photo: Renan Queiroz.

9 Soil-borne fungi: rhizoctoniosis

9.1 Overview

Rhizoctoniosis is caused by the fungus *Rhizoctonia solani* and is a serious disease in nurseries and seedlings. Seedlings can be attacked about a year after planting, delaying normal development and even causing death. The disease causes significant damage to coffee seedlings as well as planting failures. The fungus does not survive in soil which is poor in organic matter content. Seeds treated with specific fungicides can resist fungal attack. *R. solani* rarely attacks mature coffee trees (Zambolim et al., 1997, 1999, 2000).

9.2 Symptoms

Pre-emergence attacks cause seedlings to die before they reach the soil surface. However, the most common infection occurs post-emergence. The stem is attacked in the neck region, causing brown lesions which can reach 1-3 cm in length. The disease tends to rotate around the stem with strangulated necrotic regions due to rotting of the bark caused by penetration of mycelium into the tissues (Fig. 5). Sap circulation is interrupted, causing the above-ground part of the seedling to wilt and die. When the lesion reaches the wood, it can cause the stem to break. The pathogen extracts sugars and pectin using pectinolytic enzymes. This causes tissue death and maceration as the hyphae penetrate the host's tissues. The fungus cannot attack cellulose and lignin so fibrous plants remain upright despite being dead.

With high humidity, the mycelium of the fungus develops a brown or grey colour. The lower part of the stem develops a light brown, sometimes reddish constricted area which affects the functioning of vascular bundles, restricting transport of water to cotyledonal and first true leaves (Gaitán, 2003). The typical

Figure 5 Symptoms of damping-off caused by *Rhizoctonia solani* in coffee seedlings before and after the germination of the seeds. Photo: G. Hoyos, Cenicafé.

symptoms of the disease are shown in Fig. 5. The disease manifests itself up to a year after planting, with lesions appearing on or just above the soil surface. The affected area covers 1-2 cm or more of the stem. At the upper limit of the lesion, scar tissue is formed. Swelling comes from accumulation of descending sap (Gaitán, 2003).

9.3 Causal agent

Damping-off or 'Mal del Talluelo' is linked to a very heterogeneous complex of fungi, particularly *Rhizoctonia solani* Kühn. This fungus inhabits soil and can live in crop debris from one year to the next in the form of sclerotia (Mora, 1992). In the case of *C. canephora*, rhizoctoniosis is also caused by *Rhizoctonia bataticola*. The asexual phase of the fungus is classified as follows (Gaitán et al., 2015a,b):

- Kingdom - Fungi
- Phylum - Basidiomycota
- Class - Hyphomycetes
- Subclass - Incertais sedis
- Order - Agonomycetales
- Family - Agonomycetaceae
- Genus - Rhizoctonia
- Genus and Species - *R. solani*

The sexual phase of the fungus is classified as follows (Donk, 1956):

- Kingdom - Fungi
- Phylum - Basidiomycota
- Class - Agaricomycetes
- Subclass - Incertae sedis
- Order - Cantharellales
- Family - Ceratobasidiaceae
- Genus - Thanatephorus
- Genus and Species - *Thanatephorus cucumeris*

R. solani is divided into three groups, according to the number of nuclei per cell:

1. multinucleated Rhizoctonia with three or more nuclei per cell with wide hyphae of 6-10 μm; the teleomorph is *Thanatephorus cucumeris* Donk., which produces basidia with four apical sterigmas, forming labile, ovoid and hyaline basidiospores;

2 binucleate Rhizoctonia, whose sexual status is *Ceratobasidium Rogers*, and the hyphae have an average width of 4–7 μm;
3 the last group represents the final state of Waitea Warcoup and Talbot, with a multinucleated mycelium; these include *R. oryzae* and *R. zeae*.

9.4 Epidemiology

The fungus survives from one season to another in soil by means of sclerotia, either as mycelia colonizing crop residues or as a saprophyte (Galindo et al., 1983). Rain is the main agent in the dissemination of sclerotia or mycelial fragments from the soil to the first leaves of the plant. The pathogen needs a film of water to cause infection, and when the leaves dry, the growth of the mycelium is stopped. Incubation occurs ideally at 18–24°C, with a range from 16°C to 30°C, taking 36–72 h. Very humid and poorly drained soil favours the development of the disease (Mora, 1992).

Infection of nursery seedlings and plants in the field is promoted by excess moisture, high nitrogen content and shade. Attacks are severe during spring and summer due to the abundance of rain and optimal temperatures between 25°C and 28°C. Deep planting favours the pathogen, especially when the pit is filled prematurely. Sowing seeds treated with fungicides in washed sand without organic matter provides a less favourable environment for the development of the fungus.

The disease appears during sowing and in the nursery when substrate contaminated with the pathogen is used. Soils from old coffee plantations and forests rich in organic matter usually contain the fungal inoculum. Infection of seedlings in the growth phase occurs by hyphae or by the germination of sclerotia that colonize the hypocotyl region (Flentje et al., 1963; Dodman et al., 1968). High humidity (above 90%) and temperatures of 18–28°C are favourable to the attack of the pathogen in sowing and nursery conditions. In these conditions, the inoculum of the fungus spreads when sprinkler irrigation is used. Soils contaminated with the fungus are also responsible for the spread of the pathogen to nurseries. Forest soil used to prepare nurseries for the growth of coffee seedlings is the main source of contamination. Light promotes the initiation of the perfect state of the fungus, promoting the formation and release of basidia and basiodiospores. The fungus then needs a film of water, permanent dew and little light to infect the plants.

9.5 Disease management

There are no varieties of coffee resistant to the disease. Disease control must be preventive using cultural practices (Zambolim et al., 1997, 1999; Solano-Bonilla and Brenes-Chacón, 2012; Cordero, 2013). The main measures are as follows:

- Avoid terraces originating from forest or undergrowth rich in organic matter to plant seedlings, and do not use sand rich in organic matter for seedlings in nurseries and sowing.
- Organic compounds, when used in the seedling substrate, must be well decomposed and free of the pathogen.
- Use artificial substrate with appropriate fertilization since it is free of pathogens.Spray the substrates used for seed germination and seedling growth with appropriate fungicides.
- Treat the seeds with protectant and/or systemic fungicides before sowing.
- As another option for seedlings, use soil originating from the B horizon without organic matter, mix it with washed sand in a 3:1 ratio and fertilizers.
- Use plastic bags, which must measure at least 22.0 × 11.0 cm, to grow seedlings.
- Spray the substrate before growing coffee seedlings with the appropriate fungicides.
- In contaminated nurseries, eliminate not only the bags of seedlings with symptoms but also those around the diseased seedlings and spray the rest with appropriate fungicides.
- During the seedling growth phase, spray with specific fungicides against *R. solani*.
- Choose the appropriate place to install the nursery, avoiding humid, poorly drained and shady places, and control irrigation and shading of the nursery.

10 Rosellinia root rot

10.1 Overview

Rosellinia root rot has been observed for many years in most coffee regions in the world. The disease is caused by *Rosellinia bunodes*. The disease is difficult to eradicate once established. It usually occurs in new coffee plantations, planted in recently cleared and ridged land, especially where there are decaying tree trunks and remains of old forests. The disease is restricted to soil rich in decomposing organic matter. There are no data on the damage that the disease causes to coffee plantations.

The disease was first diagnosed in coffee by Berkeley and Broome in Ceylon, now Sri Lanka, in 1870. Zimmerman identified *R. bunodes* from roots of coffee trees in Java in 1902. The pathogens that attack coffee are:

- *R. bunodes*, causing black root rot in Central America (Costa Rica, Honduras, Guatemala and El Salvador) and also in India, Brazil and Colombia (Caicedo, 2003);

- *R. pepo* causing star-shaped rot in South America (Fernández and López, 1964);
- *R. arcuata* Petch., the causal agent of black root in India.

In Colombia, the presence of these pathogens dates back to the 1930s, with incidence increasing as coffee crops were established in forest areas on soils with abundant organic matter, or cocoa or cassava was previously grown. Losses from *R. bunodes* infection of up to 50% have been reported for a coffee plantation established in an area with surviving cassava roots (Castro, 1999a,b). All varieties of coffee are susceptible to the disease.

10.2 Symptoms

External symptoms are similar to those caused by *Ceratocystis fimbriata*. The first symptoms in above-ground parts of the plant appear in coffee aged 3–5 years. Symptoms include yellowing, withering and falling of leaves, death of branches and then coffee plants. The primary site of infection is at the root collar area, where the bark is superficially cracked. The mycelial mass can be found as black dots or short lines embedded in the wood, when caused by *R. bunodes*, or as a white stellate mycelial growth under the bark of affected roots, when caused by *R. pepo* (Gaitán et al., 2015a,b) (Fig. 6). Flowers may fall before fertilization. Coffee cherries are small and malformed, some remaining green and others not forming a seed. Leaves fall off, leaving bare branches which dry quickly, followed by the death of the coffee tree.

Reliable diagnosis of the disease requires examination of the root system. The roots located immediately below the surface of the soil show typical

Figure 6 Symptoms of white rot caused by *Rosellinia pepo* in coffee plantations. Photo: A. L. Gaitán et al. (2015a,b).

symptoms of blackening and decaying bark. Whitish and branched filaments, the rhizomorphs of the pathogen appear on root surfaces (Fig. 6). Subsequently, the rhizomorphs become dark, presenting a series of nodes along their branches. Under the bark, thick black layers are formed which appear as streaks when the bark is cracked. Transversal cuts in the roots show branches as black punctuations while longitudinal cuts show black lines (Fernández and López, 1964).

10.3 Aetiology

The genus *Rosellinia* was originally described by De Notaris in 1844. *Rosellinia* belongs to the class Ascomycetes, subclass Hemiascomycetes, Order Sphaeriales and family Xylariaceae. The species identified in coffee roots in Colombia are *R. bunodes* and *R. pepo* (López, 1965; Gaitán et al., 2015a,b). The fungi have two forms (Bermudez and Carranza-Morse, 1992):

1 the ascosporic (teleomorphic) *Rosellinia* spp.
2 the rarer conidial (anamorph) *Dematophora* sp.

Asci are hyaline, cylindrical, 9–12 × 250–350 μm, long-stalked and unitunicate with eight ascospores. Ascospores are ellipsoidal, without cellular appendages. Several dark reddish to brown synnemata arise from a common enlarged base, are 1.5 mm × 40.0–140.0 μm long and are composed of septate hyphae measuring 3–4 μm in diameter. Conidiophores are formed at the top of synnemata, up to 70 μm long, frequently di- or trichotomously branched and light brown. Conidia are one-celled, simple ellipsoidal or ovoid, hyaline to pale brown and smooth. The *R. bunodes* conidia are 3.0–4.0 × 6.0–7.0 μm, and the *R. pepo* conidia are 2.0–3.0 × 5.0–9.0 μm (Gaitán et al., 2015a,b). The genus *Dematophora* forms a white mycelium, with dilations next to the septa in the old hyphae. *Rosellinia* is characterized by the production of rhizomorphs on surfaces.

10.4 Epidemiology

Both *R. bunodes* and *R. pepo* are soil-borne, facultative saprophytes that become pathogenic on dead stumps or logs on newly cleared land, forming rhizomorphs that spread into a circle of trees around the infection centre via root contact (López, 1965). The pathogens start to colonize with a thin, brilliant white mycelium that invades the larger roots, changing progressively until it turns grey. *R. bunodes* mycelium then forms black masses on the wood and black stripes and dots under the bark. *R. pepo* produces dense grey, cottony mycelium on roots and fan- or star-shaped mycelium under the bark. After the substrate has been consumed, synnemata emerge from the root or from infested soil. Infected seedlings develop the disease after 9 days while plantlets show symptoms after 20 days. The mycelium initially develops on the root surface and then penetrates the cells until it reaches conducting vessels and the medulla. Black mycelia colonize the conducting vessels, forming the black stripes and dots in the roots that characterize the disease (Gaitán et al., 2015a,b).

Rosellinia spp. can survive saprophytically for several years on rotted roots in soil. Infected soil, cassava or other crop debris is a rich medium for fungal growth. *R. bunodes* and *R. pepo* also develop in residues from shade trees, especially Inga spp., spreading by root contact to neighbouring trees (Castro, 1999a,b). Optimum conditions for the development of *Rosellinia* spp. include soils rich in organic matter, temperatures of 20-28°C, soil moisture of 70-80% and pH of 4.0-7.0. Extremely well-drained or flooded soils limit pathogen growth.

The penetration of the rhizomorphs in healthy plants usually takes place in damaged roots located close to the surface of the soil. Rhizomorphs grow on and under the bark; secondary branches spread in different directions and penetrate cells. This results in the destruction of protective tissues and conductive vessels which kills roots. Conidia and perithecia can also appear. Coffee trees typically die in rows which gradually widen as the fungus spreads. High temperatures, high precipitation and little exposure to the sun favour the development of the fungus. Excess humidity does not favour fungal growth (Fernández and López, 1964). Direct sunshine and very high temperatures inhibit the growth of the pathogen (Castro and Esquivel, 1991). *R. pepo* infection occurs in a similar way to *R. bunodes*. The pathogen condenses its mycelium at various points along the roots, forming grey-white fans or stars that turn dark as they age (Realpe, 2001). Both fungal species can be disseminated by root contact, soil transport or by shoes worn by farm staff. Cacao, cassava, forest trees, fruit trees, plantains, pineapple and tomato are hosts of the two pathogens (Caicedo, 2003).

10.5 Disease management

There is no known resistance to this disease in coffee varieties. The management of the disease must be preventive. Plants with symptoms must be removed from the field. Care should be taken to avoid spreading the inoculum (contaminated soil and roots) in the growing area. All affected trees and their neighbours should therefore be uprooted and destroyed, and care should be taken to remove fragments of diseased roots in the soil. Immediate replanting must be avoided. A period of at least 1 year should be observed before planting coffee seedlings in a contaminated field (Castro, 1999a,b). The soil contaminated by the pathogen should be exposed to sunlight for at least 6 months to reduce the amount of inoculum in the soil (Aranzazu et al., 1999). Forest soil should not be used for seedlings; stumps, roots and tree trunks should be removed after logging. For coffee seedlings, the use of artificial substrate is recommended since it is free of pathogens.

11 Ceratocystis canker stain

11.1 Overview

Canker stain is caused by *C. fimbriata sensu lato*. The disease was first reported in 1900 in Indonesia on the island of Java. The disease subsequently appeared in Colombia in 1932 (Caicedo, 2003; Gaitan et al., 2015a,b). Canker stain is now widely found in Central and South America, as well as in Indian coffee-growing areas. The disease is present in Ecuador, Colombia, Guyana, Peru, Costa Rica, Mexico, Guatemala, the Dominican Republic, Trinidad, Jamaica, Haiti, the Philippines, Sri Lanka, Fiji and New Guinea (Castro, 1999a,b). Malaguti (1956) refers to the death of more than a million cacao trees from the disease in Venezuela. The fungus can cause heavy losses in adult plants, poorly nourished plantations and especially in poorly drained soils. It has been found to cause losses of 20–50% in coffee plantations in mountainous areas of South America (Caicedo, 2003). All varieties of coffee are susceptible to the disease (Castro, 1999a,b).

11.2 Symptoms

The fungus causes both canker stain and wilt diseases and can affect plants of any age. The first symptom of canker stain is the yellowing of mature leaves which show dark green areas between the ribs. Leaves then wilt, take on a reddish, uniform colouration, curl slightly on their vertical axis and dry in 1–2 months. Foliar wilt appears 3 months to a year later and then spreads to the whole coffee plantation. Diseased plants die over a period of a few weeks (Fig 7a–c). The fungus occurs in vascular tissues. The affected plant produces tylose that obstructs the xylem-carrying vessels and prevents water from reaching the various parts of the plant, causing water deficiency.

The primary site of infection, as seen by vascular streaking, is the plant's collar or other parts of the stem where dark sunken lesions can appear. When the bark is removed, darkly coloured lesions with a streaked pattern can be found in the infected wood and can also extend into the roots. In advanced stages of the disease, lesions circle trunks, resulting in tree death. Common sites of infection are the pruning wounds made when plants are rejuvenated by stem coppicing.

Another disease symptom is dried and sunken areas in roots, stems and branches that show perforations (Fig. 8). These are caused by wood-boring

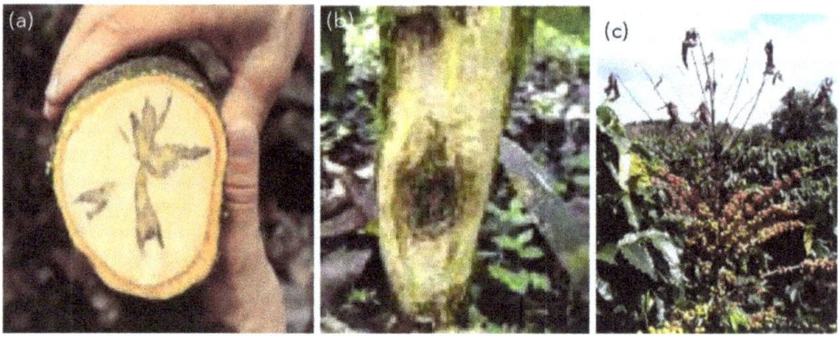

Figure 7 Ceratocystis canker stain: (a) internal symptoms that resemble decorative figures of dark coffee colour to light; (b) external symptoms on plant stain; (c) symptoms of total wilting and dried branches affected by the fungus. Photos : A. C. Alfenas et al. (2009).

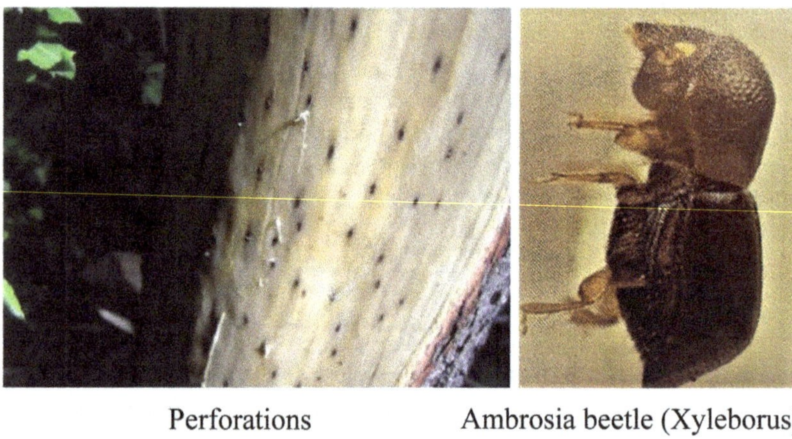

Figure 8 Perforations performed by the insect *Xylosandrus* (Xyleborus) *morigerus* (Bland) in avocado plant, from where the galleries and the development of sexual and asexual structures of the fungus are produced. Photos: C. B. L. Castro (1999a,b).

ambrosia beetles (Coleoptera - Curculionidae - Scolytinae) which have a symbiotic relationship with the canker stain fungus. Ambrosia beetles attack weakened, dying or dead trees, causing pinhole defects and dark staining in the outer wood. Galleries are formed in the sapwood or heartwood. Adults introduce and feed on fungi rather than the wood itself. The most obvious sign of an ambrosia beetle attack is the fine, white dust caused by boring that accumulates at the base of the tree and in the bark crevices. Adult beetles bore straight into the tree, creating perfectly round, small-diameter holes. If the bark is removed, the entrance points of adult ambrosia beetles and galleries are distinctive and are often surrounded by a dark brown or black fungal stain. Ambrosia beetles are small, roughly 2 mm long, elongated and cylindrical insects. These beetles have a symbiotic relationship with the fungi on which the larvae feed.

Perforations are shown up by sawdust and gummy exudates accumulated at the foot or trunks of plants (Fig. 8). On the bark, multiple bluish-brown spots are observed that grow in a cortical and medullary direction, intersected by galleries, where the different stages of the insects develop. In Florida, the red bay leaf ambrosia beetle (*Xyleborus glabratus*) is the main vector of the fungus (*Raffaelea lauricola*) that causes laurel wilt. In coffee cultivation, *Xyloborus* sp. disseminates the fungus *C. fimbriata* through the galleries (spots) where the mycelium of the fungus is found in the asexual and sexual phase. These spots can appear from the base of the trunk to the area of the fork and branches; when uncovering the root area, necrotic areas with perforations of the insect can also be observed.

11.3 Causal agent

The genus *Ceratocystis* was first described in 1890. *Ophiostoma ulmi*, the sexual stage, has caused severe epidemics on *Ulmus* spp. in Europe and in the United States (Caicedo, 2003). The asexual stage (anamorph) of *Ceratocystis* is Chalara and Thielaviopsis, and Ophiostoma for the genus *Graphium*.

The fungus *Ceratocystis* spp. belongs to the class of ascomycetes causing canker stain and wilt diseases, especially in fruit, forest and ornamental trees. Structures of the sexual states of *Ceratocystis* are perithecia with globose bases and a diameter of 138–185 μm, giving rise to long necks of 408–627 μm, typically terminating in distinct ostiolar hyphae. Asci are evanescent and seldom seen in the case of *C. fimbriata* Ellis & Halst. Ascospores are hat shaped, 2.4–2.7 × 5.7–6.5 μm and develop in sticky masses at the apices of ascomatal necks (Upadhyay, H.P 1981). The asexual state of these fungi is in the genus Thielaviopsis, typified by distinct tubular conidiophores and rectangular endoconidia that are 2.4 × 19.4 μm and produced in chains. These fungi also produce distinct, dark-coloured chlamydospores that facilitate survival in the soil.

The causal agent of canker stain of coffee is broadly referred to as the sexual stage (teleomorph) of *C. fimbriata sensu lato*. This species forms resistance structures, known as chlamydospores between affected tissues. Recent studies of the pathogen in Colombia have shown that the isolates reside in two distinct phylogenetic clades other than *C. fimbriata sensu lato*. Isolates in these clades have been given the names *C. colombiana* sp. nov. and *C. papillata* sp. nov. It is not known whether the same *Ceratocystis* species in the *C. fimbriata sensu lato* species complex cause canker stain of coffee elsewhere in the world (Gaitan et al., 2015a,b).

11.4 Epidemiology

The fungus *C. fimbriata* is considered as a facultative saprophyte, present in all types of soils and in altitudes from 800 m to 2000 m. The fungus can live in the soil in its dormant stage and in debris as macroconidia. Infected soil is one of the major sources of inoculum for *Ceratocystis* spp. causing canker stain of coffee. These fungi can easily be isolated from the soil using freshly cut coffee twigs which then become infected (Gaitan et al., 2015a,b). Dissemination is by wind, water, movement soil, people and equipment. The fungus penetrates plant tissue via wounds during cultural practices such as pruning (Castro, 1999a,b). Infected coffee trees are usually randomly scattered in the fields, without any clear pattern of distribution. Lesions develop at the base of the stem of infected plants.

The disease is classified as monocyclic. *Ceratocystis* spp. in the *C. fimbriata sensu lato* species complex have close associations with causal insects such as picnic beetles and flies that feed on wounded plant tissue. Insects of the Scolytidae family are natural disseminators of the disease. *Ceratocystis* spp. produce fruity odours that are attractive to these insects and, when the fungi are present on the surface of the lesions, their spores attach to insects that carry them to fresh wounds. When feeding on a suitable host, ambrosia beetles excavate a complex gallery system. The fungi sprout and grow in the gallery walls, producing nutritious spores eaten by adults and larvae. The excretion of uric acid by insects is used by the fungi.

Dissemination depends on types of cultural practices, the population of disseminating insects and the climatic conditions of the region. There have been no detailed studies of insect associations with *Ceratocystis* spp. causing canker stain of coffee, but it is most likely that these fungi are transmitted by insects (Gaitan et al., 2015a,b). Propagules of *C. fimbriata* are disseminated by wind, floods and irrigation, as well as by the movement of materials, agricultural equipment, animals and workers.

Canker stain is reported in most coffee-growing regions of the world. The fungus has the ability to survive as a saprophyte in the rainforest biome, from where it infects coffee plantations. The fungus spreads from soil to the plant mainly via rain splashing. It penetrates plant tissue via wounds made by cultural practices such as tilling, weeding and pruning, or via galleries made by insects. The fungus is present in most high-altitude coffee plantations, especially with poor drainage and old plants.

Favourable conditions for growth include rain, high humidity, cold temperatures at night (18°C) and warm temperatures during the day (28°C). The disease is dispersed by conidia, mycelium and ascospores, which can survive for several years in the soil and in wood debris colonized by the fungus. It is common to have infection rates of 10% even in well-managed coffee plantations which, in a population of 5000 plants per hectare, can represent a significant impact on yields. Outbreaks can spread very quickly via insect vectors.

11.5 Disease management

The most important strategy to manage canker stain is to avoid any form of injury to the tree trunk. A longer-term strategy is to plant resistant varieties. *C. canephora* is reported as resistant. In *C. arabica*, just one variety, *C. arabica* var. Bourbon, is known to be resistant to *C. fimbriata sensu lato*. Lineages resistant to Ceratocystis canker stain have been bred from either *C. arabica* var. Bourbon or *C. canephora*, crossed with *C. arabica* var. Caturra (Castro and Cortina, 2009).

The fungus invades the host via vectors such as infected tillage tools or galleries produced by ambrosial insects penetrating the host. The best way to avoid the establishment of this fungus is by prevention. It is important to avoid causing unnecessary wounds to the tree and to apply fungicide as a preventive measure to any cuts in trunks or branches. It is also recommended to disinfect pruning and weeding tools several times during the working day. Working tools can be disinfected with products such as Formol, diluted in 2% water or sodium hypochlorite in commercial doses. Preventive control is via benzimidazole, chlorothalonil and triazole fungicides. However, no fungicide can eliminate the disease once inside plant tissues. Fungicides should not be applied to the soil (Castro and Montoya, 1994). Pastes or species of paints based on cupric fungicides can also be used as disinfectants and to protect pruned plants. All these products should be used as part of an integrated disease management programme. Diseased plants should be removed from the field. After the eradication of diseased plants, a new seedling can be planted immediately. Preventive biological control is possible using various species of the fungus *Trichoderma* spp., which can be used in the production of seedlings and in the establishment of a new plantation. *Trichoderma* spp. is also a biostimulant promoting beneficial activity in the rhizosphere of coffee plants.

Other cultural practices to reduce the incidence of disease are pruning above 40 cm and preventing particles of organic matter containing the fungus from coming into contact with the wood. If the top of a plant is infected, it is recommended to cut the stem 15 cm below the lesion. When pruning branches near the soil line, it is recommended to cut them to at least 2 cm from the orthotropic stem without causing direct damage to the main stem (Castro, 1998).

It is advisable to remove all the pruned material from the plantation and incinerate it. Plants with characteristic symptoms of the disease, such as irreversible yellowing, must be removed from the plantation; the hole left by the diseased plant must be solarized for a period of 4-8 weeks by placing transparent plastic over it. A fungicide can then be added combined with biocontrol measures.

12 Coffee wilt disease (tracheomycosis)

12.1 Overview

Coffee wilt disease (CWD) or tracheomycosis is one of the most important coffee diseases in Africa, mainly on *C. canephora* var. *robusta*. The disease was first reported on *Coffea liberica* var. *dewevrei* in 1927 in the Central African Republic (Figueres, 1940; Felix, 1954; Saccas, 1956). In the mid-1900s, an epidemic of CWD was detected in the Central African Republic, Côte d'Ivore and Democratic Republic of Congo, which caused great damage to *C. liberica* and *C. canephora*. The disease advanced to west and central Africa (Guillemat, 1946). From the 1950s, it extended to the Ivory Coast, Congo, Ethiopia, Uganda, Nigeria, Sudan, Zimbabwe, Tanzania and Rwanda, attacking species of *C. canephora* and *C. arabica* (Stewart, 1957; Pieters and Van der Graaff, 1980). In the 1970s, epidemics attacked new plantations of *C. canephora* in the Democratic Republic of Congo as well as Uganda and Tanzania. The disease then declined in severity with the adoption of sanitary practices and improvement programmes but began to increase again from 1984 (Flood, 1996; Girma, 1977; Girma and Hindorf, 2001; Flood et al., 2001). Outside Africa, there are no reports of CWD. The highest impact on coffee production has been in the Democratic Republic of Congo and Uganda. In Uganda, 90% of *C. canephora* farms were affected and annual losses were estimated at US$3 million. In Ethiopia, losses varied from 44% in Gera province to 61% in Bebeka province, though some producers lost up to 100% (Girma, 1977).

12.2 Symptoms

The symptoms of the disease are similar to those caused by *R. bunodes*, *R. pepo* and *C. fimbriata* though CWD causes higher and more rapid plant mortality. CWD can manifest at any stage of plant development. Leaves of young plants develop brown necrotic lesions often along the veins and margins. Leaves then dry, shrivel and fall 3-4 days after the first symptoms. Lower leaves tend to be the first to display symptoms. In adult plants, it is very common to observe chlorosis, wilting, curling and drying of leaves. Dark brown discolouration of the leaf veins may be apparent. Berries of the diseased plants become red and ripen prematurely. Younger branches develop a dark brown-black necrosis that may also be laterally restricted. Sometimes the symptoms extend to the entire tree.

The key symptom of CWD upon removal of bark is blue-black staining of vascular tissues. Discolouration is usually more pronounced towards the stem base, but it may extend from below the soil level to the apex of the trees and may spiral along stems and branches. Along the trunk, the bark may become swollen with spiral cracks. Small black perithecia ascomata form under wet conditions. The perithecia may be visible on the surface of the bark and often in the cracks. Perithecia can be found also on the dead and decaying parts of affected plants. The plants can die between 3 and 15 months after the first visible symptoms, though this can be as little as 1-2 months in some cases (Govindarajan and Subramanian, 1968). Older and poorly managed trees appear to be more susceptible to the disease (Figs. 9-11).

12.3 Causal agent

CWD is caused by *Gibberella xylarioides* (anamorph *Fusarium xylarioides*; syn *F. oxysporum* forma *xylarioides*). Fungal colonies cultivated in the lab are initially

Figure 9 Symptoms of the fungus *Gibberela xylariodes* (= *Fusarium xylariodes*) in coffee plants, on the stem and the effect on branches, leaves and grains. Photos: M. A. Rutherford, 2006. Symposium, Fusarium-induced diseases of tropical perennial crop.

Figure 10 Symptoms of the fungus *Gibberela xylariodes* (= *Fusarium xylariodes*) in coffee plants, on the stem and the effect on branches, leaves and grains. Photos: M. A. Rutherford, 2006. Symposium, Fusarium-induced diseases of tropical perennial crop.

Figure 11 Coffee grower in Africa inspecting coffee plants affected by the disease. https://www.monitor.co.ug/uganda/magazines/farming/climate-change-threatens-coffee-production-1537246.

pale beige with sparse white mycelium. Purple discolouration may develop with age and be accompanied by dark bluish black, discrete stromata. Microconidia and macroconidia are produced in slimy masses, as short, conidiotenous cells or as vegetative mycelium. Microconidia are unicellular, allantoid, curved and 2.5–3 × 5–10 μm. Macroconidia, which tend to be less abundant, are fusoid and falcate, with two to three septa and measure 4–5 × 20–25 μm. Chlamydospores are oval to globose, smooth or rough and 8–10 × 10–15 μm. Perithecia are globose, measure 180–300 × 200–400 μm, have a flattened base and are embedded, singly or in groups, in dark purple stromata. In the absence of chlamydospores, perithecia may provide a way of survival under adverse conditions. Asci are cylindrical and thin walled, have short pedicels and measure 7–9.5 × 90–110 μm. Asci contain eight monostichous, hyaline to straw-coloured ascospores and are fusoid, having one to three septa, are finely roughened and measure 4.5–6 × 12–14.5 μm.

12.4 Infection process

Conidia and ascospores of *G. xylarioides* are transported over long distances by wind, rainwater splashes as well as by human or animal activity (Jacques, 1954). The fungus penetrates the trunk via injuries caused by cultural practices such as pruning or by insects (Siddiqui and Corbett, 1968). The mycelium develops inter- and intracellularly and progressively invades the xylem. The plant reacts to infection by obstructing the water conduction system through the production of tyloses. The incubation period of the fungus in plant tissues is influenced by age and environmental conditions. Appearance of initial symptoms varies from a few days to 4–6 months. The first symptoms indicate general obstruction of the vascular leaves caused by both the fungus structures and the plant's defence reaction. The plant dies after 2 weeks (Heim and Saccas, 1950).

12.5 Epidemiology

The life cycle of *G. xylarioides* is not well known. The fungus is believed to be a natural inhabitant of the soil; however, the fungus does not live long in the soil as it rarely produces resting structures known as chlamydospores (Van der Graaff and Pieters, 1978). *G. xylarioides* is considered a soil-borne fungus that induces symptoms in a manner similar to other vascular wilt pathogens. Ascospores produced in diseased plants or on dead branches and stems could act as survival structures in the soil. The fungus can also live on plant residues or on alternative hosts.

The fungus penetrates roots and lower stems through wound sites. The fungus colonizes the vascular tissue of the xylem, resulting in blockage of vessels and toxin production, which leads to the symptoms such as wilting and discolouration. Microconidia, macroconidia and ascospores are probably dispersed by air, water and human activity. Soil and plant debris, particularly wood, are important sources of inoculum and disease spread. Wood cuttings used for vegetative propagation are also a source. Germination of conidia is favoured at a temperature of 30°C and humid conditions. Shade and marginal areas of low fertility favour disease spread (Siddiqui, 1965). The disease is more severe in old and under-fertilized plants. Spores can germinate in moist environments without the presence of liquid water (Krantz and Mogk, 1973). Transmissibility of the fungus via seeds still needs confirmation.

12.6 Management of the disease

The most important method of disease control is resistant varieties. *C. canephora* was initially the basis for resistance breeding programmes. In Ethiopia, different degrees of resistance were found between *C. arabica* L. semi-wild accessions (Van der Graaff and Pieters, 1978). Those authors suggested that resistance is quantitative in nature in *C. arabica*. There is no evidence of qualitative resistance. The mechanisms of resistance in coffee plants remain unclear. In 2001, researchers from Jima Station in Ethiopia demonstrated that coffee varieties resistant to CBD are at some level resistant to CWD (Girma and Hindorf, 2001).

In the absence of resistant varieties, the most efficient measure to prevent the disease is to exclude the pathogen from disease-free areas. This may be achieved by strict quarantine measures to control the distribution of seeds, vegetative cuttings, clones, seedlings and plantlets, as well as soil from affected areas. New plantations should be established with disease-free material. Other measures to minimize the severity of the disease in the field include maintaining a vigorous and well-managed crop, minimizing physical damage to coffee trees to reduce the risk of infection and delay subsequent disease development. It is very important to avoid wounding the stem with cultural measures such as pruning. Treating pruning tools with disinfectant products before use and between pruning is also essential. Cultural measures are also vital when the disease is already present, such as burning diseased and adjacent plants even if they have no symptoms of the disease. It is recommended to leave a 300 m strip free of plants around the site from where the sick plants were burned. Coffee may be replanted with resistant cultivars (if available) or an alternative crop or left fallow for at least 1 year. The planting of non-host plants is recommended to reduce the population of the fungus in the soil.

13 Summary and future trends

Coffee is affected by a wide diversity of insects and soil-borne fungi in many different regions, countries or plantations. Future prospects for the management of soil-borne pest insects associated with coffee will depend less on the selection and release of new varieties from plant breeding, since current coffee (Arabica and Robusta) varieties do not show much resistance to soil insect pests. Further research is needed on the soil and climate characteristics associated with different scales of pest infestation. A promising avenue is the identification of novel natural biological control agents. The need for more sustainability in coffee farming involves reducing the pesticide use for pest control, with a consequent decrease in environmental and farm contamination.

The two most important soil-borne diseases of coffee are Ceratocystis canker stain (*C. fimbriata*) and CWD (*G. xylarioides*). The main method of controlling these diseases is by using resistant varieties. For diseases such as rhizoctoniosis (*R. solani*) and rosellinia root rot (*R. bunodes*), there are no current sources of resistance in coffee germplasm collections. Rosellinia root rot is not a problem in Brazil where coffee plantations grow under sunlight but is a problem in Central America where coffee pants are grown under shade trees. A major issue is the enforcement and improvement of quarantine measures to prevent the spread of diseases between countries and between plantations.

14 References

Alfenas, A. C., Zauza, E. A. V., Maffia, R. G. and Assis, T. F. (2009). *Clonagem e doenças do eucalipto*. 2. ed. Viçosa: Editora UFV, 500.

Alvarado, J. A. (1935). *Tratado de Caficultura Practica, Tipografia Nacional*. Guatemala.

Alves, V. S., Junior, A. M., Santa-Cecilia, L. V. C., Rohde, C. and Tramontin, M. A. (2009). Tests for control of coffee root mealybug with Heterorhabditis. *Revista Brasileira de Entomologia* 53(1): 139–143.

Andrade, S. C., Rossi, G. D. and Martinelli, N. M. (2020). Dispersion pattern of giant cicada (Hemiptera: Cicadidae) in a Brazilian coffee plantation. *Environmental Entomology* 49(5): 1019–1025.

Anon. (1974). *India Coffee Board. 33rd Annual Report, 1972–1973.*

Aranzazu, H. F., Cárdenas, L. J., Mujica, J. J. and Goméz, Q. R. (1999). Manejo de las llagas radicales (*Rosellinia* sp.). *Boletin de Sanidad Vegetal* 25, 35.

Barba, D. C. (1961). *Estudio morfológico y pruebas de patogenicidade de varias cepas de Ceratocystes fimbriata EL. & HALST. Costa Rica Noviembre.* Instituto Interamericano de Ciencias Agrícolas de la OEA Centro Tropical de Investigación Y Enseñanza para Graduados, Turrialba, 59.

Ben-Dov, Y. (1994). *A Systematic Catalogue of the Mealybugs of the World*. Intercept Ltd, Andover, UK, 397–400.

Bermudez, M. and Carranza-Morse, J. (1992). Estado anamorfico de *Rosellinia bunodes* (Derk. & Dr.) Sacc. y *Rosellinia pepo* Pat. (Ascomycotina: Xylariaceae). *Revista de Biología Tropical* 40: 43–46.

Bianchini, P. C. L. (1955). Las llagas del café en Costa Rica. *Revista de Biología Tropical* 3(2): 203–235.

Bordereau, C. and Pasteels, J. M. (2010). Pheromones and chemical ecology of dispersal and foraging in termites. In: *Biology of Termites: A Modern Synthesis*. Springer Netherlands, 279–320.

Caicedo, B. L. C. (2003). Lagas radicales. *Rosellinia bunodes* Berk. y Br. Sacc. y *Rosellinia pepo* Pat. In: Vallejo, et al. (Eds), *Enfermedades del Cafeto en Colombia. Cenicafé*, 100–106.

Calleja-Cabrera, J., Boter, M., Oñate-Sanchez, L. and Pernas, M. (2020). Root growth adaptation to climate change in crops. *Frontiers in Plant Science* 11: 544.

Cannell, M. G. R. (1971). Seasonal patterns of growth and development of arabica coffee in Kenya. *Kenya Coffee* 36: 68–74.

Castro, C. B. L. (1995). Antagonismo de algunos aislamientos de *Trichoderma koningii*, originarios del suelo colombiano contra *Rosellinia bunodes*, *Sclerotinia sclerotiorum* y *Phytium ultimum*. *Fitopatología Colombiana* 19(2): 7–17.

Castro, C. B. L. (1998). Incidencia de la Llaga Macana *Ceratocystes fimbriata* (Ellis, Halst) Hunt. En: la práctica de poda de ramas bajeras del café. *Avances Tecnólógicos Ceneicafe* 250: 8.

Castro, C. B. L. (1999a). Las llagas del cafeto. *Avances Técnicos Cenicafé* 268: 1–8.

Castro, C. B. L. (1999b). Las plagas del cafeto. Chinchiná, Colombia. *Cencafé (avances Técniicos)* 268: 1–8.

Castro, C. B. L. and Chaves, C. B. (1999). Evaluación del daño económico ocasionado por la Macana del cafeto *Ceratocystes fimbriata*, (Ellis, Halts) Hunt. In: *XX Congreso Nacional de Fitopatología y Ciencias Afines*, 65.

Castro, C. B. L. and Cortina, H. A. (2009). Evaluación de resistencia a *Ceratocystes fimbriata* Ell. Halst. Hunt. P*rogenies F5 de café Bourbon resistente x Caturra* 60(2): 115–125.

Castro, C. B. L. and Esquivel, R. V. H. (1991). Las Lagas Radicularis del cafeto. *Avances Técnicos Cenicafé* 163: 1–4.

Castro, C. B. L. and Montoya, R. E. (1994). Evaluación de fungicidas para el control de *Ceratocystis fimbriata* en café. *Cenicafé* 45(4): 137-153.

Castro, C. B. L. and Montoya, R. E. C. (1997). El zoqueo de cafetales y su relación con la infección por Llaga Macana. *Avance Tecnico* 240: 8.

Chouvenc, T., Li, H. F., Austin, J., Bordereau, C., Bourguignon, T., Cameron, S. L., Cancello, E. M., Constantino, R., Costa-leonardo, A. M., Eggleton, P., Evans, T. A., Forschler, B., Grace, J. K., Husseneder, C., Křeček, J., Lee, C., Lee, T., Lo, N., Messenger, M., Mullins, A., Robert, A., Roisin, Y., Scheffrahn, R. H., Sillam-dussès, D., Šobotník, J., Szalanski, A., Takematsu, Y., Vargo, E. L., Yamada, A., Yoshimura, T. and Su, N. (2016a). Revisiting *Coptotermes* (Isoptera: Rhinotermitidae): A global taxonomic roadmap for species validity and distribution of an economically important subterranean termite genus. *Systematic Entomology* 41(2): 299-306.

Chouvenc, T., Scheffrahn, R. H. and Su, N. Y. (2016b). Establishment and spread of two invasive subterranean termite species (*C. formosanus* and *C. gestroi*; Isoptera; Rhinotermitidae) in metropolitan southeastern Florida (1990-2015). *Florida Entomologist* 99(2): 187-191.

Constantino, R. (2002). The pest termites of South America: Taxonomy, distribution and status. *Journal of Applied Entomology* 126(7-8): 355-365.

Cordero, C. M. (2013). Cómo controlar los hongos que causan el Mal del Talluelo? *Cuadernillo* 2: 1-16.

Costa, M. B., Souza, B., Prado, E. and Santa-Cecília, L. V. C. (2016). Tabela de vida de fertilidade de *Planococcus citri* (Risso) e *Planococcus minor* (Maskell) (Hemiptera: Pseudococcidae) em cafeeiro. *Coffee Science* 11(2): 204-210.

Cox, J. M. (1989). The mealybug genus *Planococcus* (Homoptera: Pseudococcidae). *Bulletin of the British Museum (Natural History), Entomology* 58(1): 1-78.

D'Antonio, A. M. (1991). A descoberta da mosca das raízes. *Bayer Correio Agrícola* 2: 8-9.

Decaro Junior, S. T., Martinelli, N. M., Maccagnan, D. H. B. and Ribeiro, E. S. D. B. P. (2012). Oviposition of *Quesada gigas* (Hemiptera: Cicadidae) in coffee plants, *Ver. Colomb. Entomology* 38: 1-5.

Della Lucia, T. M., Gandra, L. C. and Guedes, R. N. (2014). Managing leaf-cutting ants: Peculiarities, trends and challenges. *Pest Management Science* 70(1): 14-23.

Dodman, R. L., Barker, L. R. and Walker, J. C. (1968). Modes of penetration by different isolates of *Rhizoctonia solani*. *Phytopathology* 58: 31-33.

Donk, M. (1956). Notes on resupinate hymenomycetes. *Reinwardtia* 3: 363-379.

Duque, O. H., Castro, L. B. y Montoya, R. E. C. (2003). Importancia económica de la Llaga Macana del Cafeto. *Avances Tecnicos Cenicafé* 314: 4.

Evans, T. A., Forschler, B. T. and Grace, J. K. (2013). Biology of invasive termites: A worldwide review. *Annual Review of Entomology* 58: 455-474.

Fanton, C. J. and Queiroz, R. B. (2020). Manejo de Pragas do cafeeiro Conilon. *Informe Agropecuário* 41(309): 41-52.

Felix, J. A. (1954). La Carbunculariose. *Sci. Minist. Colon. Sect. Agron. Trop.* 5: 296-344.

Fernández, B. O. and López, D. S. (1964). Lagas radiculares negras (*Rosellinia bunodes*) y estrelada (*Rosellinia pepo*) del cafeto. I. Patogenicidad e influencia de la clase de inóculo en la infección. *Avances Tecnicos Cenicafé* 15: 126-144.

Figueres, R. (1940). Sur une maladie très grave du caféier en Oubangui. Rapport. Ministère des colonies, Paris, France.

Fisher, B. L. and Cover, S. P. (2007). *Ants of North America: A Guide to Genera*. University of California Press, Berkeley, p. 216. https://doi.org/10.1525/9780520934559.

Flentje, N. T., Dodman, R. L. and Kerr, A. (1963). The mechanism of host penetration by *Thanatephorus cucumeris*. *Australian Journal of Biological Sciences* 16(4): 784-799.

Flood, J. A. (1996). A study of Tracheomycosis or vascular wilt disease of coffee in Zaire. Report. *Presented to Zairean Coffee Organization, Zaire, 13*.

Flood, J. L., Gil, L. F. and Waller, J. M. (2001). Coffee diseases: A clear and a present danger. In: Baker, P. S. (Ed.), *Coffee Futures*. CABI-FEDERACAFÉ-USAID-ICO, Chinchiná, Colombia, 82-93.

Fornazier, M. J. (2016). Bioecologia, dano e controle de *Planococcus citri* (Risso) (Hemiptera: Pseudococcidae) em *Coffea Canephora* Pierre ex Froehner (Rubiaceae). *Tese doutorado*. Universidade Federal de Viçosa, 91.

Fornazier, M. J., Martins, D. S., Fanton, C. J. and Benassi, V. L. R. M. (2017a). Manejo de Pragas do café Conilon. In: Ferrão, R. G., Fonseca, A. F. A., Ferrão, M. A. G. and De Muner, L. H. (Eds), *Café Conilon* (2nd edn.). Incaper, Vitória, ES, 399-433.

Fornazier, M. J., Martins, D. S., Willink, M. C. G., Pirovani, V. D., Ferreira, P. S. F. and Zanuncio, J. C. (2017b). Scale insects (Hemiptera: Coccoidea) associated with arabica coffee and geographical distribution in the Neotropical region. *Anais da Academia Brasileira de Ciências* 89(4): 3083-3092.

Gaitán, A. L. B. (2003). *Volcamiento o Mal del Tallito Rhizoctonia solani Kuhn, Enfermedades del cafeto en Colombia*, 85-90.

Gaitán, A. L. B., Cristancho, M. A., Caicedo, B. L. C., Rivillas, C. A. and Gómez, G. C. (2015a). *Compendium of Coffee Diseases and Pests*. The American Phytopathological Society, 79.

Gaitán, A. L., Cristancho, M. A., Caicedo, B. L. C., Rivillas, C. A. and Gómez, G. C. (2015b). *Rosellinia Root Rot*. APS Press, St. Paul, MN, 14-15.

Galindo, J. J., Abawi, G. S., Thurston, H. D. and Galvez, G. E. (1983). Source of inoculum and development of bean web blight in Costa Rica. *Plant Disease* 67(9): 1016-1021.

Gallo, D., Nakano, O., Silveira Neto, S., Carvalho, R. P. L., Baptista, G. C., et al. (2002). *Entomologia Agrícola*. Fealq, Piracicaba, SP, 920.

García Morales, M. et al. (2016). ScaleNet: a literature-based model of scale-insect biology. Available at: http://scalenet.info.

Girma, A. and Hindorf, H. (2001). Recent investigation on coffee tracheomycosis, Gibberella xylarioides (*Fusarium xylarioides*) in Ethiopia. *Indian Coffee* 32(9): 270-291.

Girma, A. S. (1977). Coffee wilt disease in Ethiopia. International Workshop of Tracheomycosis, Kampala.

Girón-Pablo, S., Ruiz-Vega, J., Pérez-Pacheco, R., Aquino-Bolaños, T. and Martínez-Martínez, L. (2015). Biological control of *Phyllophaga vetula* (Horn) with entomopathogenic nematodes in various formulations and moisture conditions. *Southwestern Entomologist* 40(3): 511-517.

Govindarajan, T. and Subramanian, S. (1968). Fusarium wilt in India. *Indian Coffee Journal* 32(9): 270-291.

Groenen, D. (2018). The effects of climate change on the pests and diseases of coffee crops in Mesoamerica. *Journal of Climatology and Weather Forecasting* 6(3): 1-5.

Guillemat, J. (1946). Quelques observarions sur la trachéomycose du *Coffea excelsa*. *Revue de Botanique Appliquée et d'Agriculture Tropicale* 26: 542–550.
Harris, W. V. (1961). *Termites: Their Recognition and Control*. New York, 187.
Heim, R. and Saccas, A. M. (1950). La tracheomicose des Coffee et robusta des plantations de l'Oubangui-Chari. *Revue de Mycologia* 15(97).
Jacques, F. A. (1954). La Carbunculariose. *Boletin Science* 5: 296–344.
Junqueira, L. K., Diehl, E. and Berti Filho, E. (2008). Termites in eucalyptus forest plantations and forest remnants: An ecological approach. *Bioikos* 22: 3–14.
Krantz, J. and Mogk, M. (1973). *Gibberella xylarioides* Heim et SAccas on arábica coffee in Ethiopia. *Journal of Phytopathology* 78(4): 365–366.
Le Pelley, R. H. (1968). *Pests of Coffee*. Longmans, London, 590.
Liesch, P. J. and Williamson, R. C. (2010). Evaluation of chemical controls and entomopathogenic nematodes for control of *Phyllophaga* white grubs in a fraser fir production field. *Journal of Economic Entomology* 103(6): 1979–1987. https://doi.org/10.1603/EC10176.
López, D. D. S. (1965). *Estudios sobre La Laga Negra radicular del cafeto causada por R. bunodes Berk. Y Sacc.* Universidad de Caldas, Manizales.
Martinelli, N. M. (2004). Cigarras associadas ao cafeeiro. In: Salvadori, J. R., Ávila, C. J. and Silva, M. T. B. (Eds), *Pragas de solo no Brasil*. Espiral Comércio de Livros, Passo Fundo, 517–541.
Matiello, J. B., Coelho, C. and Baeta-Neves, A. M. (1999). Ocorrência da mosca das raízes (*Chyromiza* spp.) em cafezais conilon. *Congresso Brasileiro* de *Pesquisas Cafeeiras* 25: 17–18.
Matiello, J. B., D'Antonio, A. M., Garcia, A. W. R., Fernandes, D. R., Almeida, S. R. and Paulini, A. E. (1990). Ocorrência da mosca das raízes (*Chyromiza* spp.) nas regiões cafeeiras do Brasil. *Congresso Brasileiro* de *Pesquisas Cafeeiras* 16: 81–82.
Matiello, J. B., Miguel, A. E., Silva, W. L., Almeida, S. R., Ferreira, R. A. and Resende, A. R. (1998). Nível de dano econômico da mosca das raízes (*Chyromiza*) no cafeeiro. *Congresso Brasileiro* de *Pesquisas Cafeeiras* 24: 36–37.
Mibey, R. K. (1997). Sooty moulds. In: Ben-Dov, Y. and Hodgson, C. J. (Eds), *Soft Scale Insects: Their Biology, Natural Enemies and Control*. Elsevier, Amsterdam, 275–290.
Moller, W. J., Devay, J. E. and Backman, P. A. (1969). Effect of some ecological factor on *Ceratocystes fimbriata* canker on stone fruits. *Phytopathology* 59: 938-942.
Montoya-Lerma, J., Giraldo-Echeverri, C., Armbrecht, I., Farji-Brener, A. and Calle, Z. (2012). Leaf-cutting ants revisited: Towards rational management and control. *International Journal of Pest Management* 58(3): 225–247.
Mora, B. (1992). Epidemiology and agronomical management of Rhizoctonia web blight. PhD thesis. University of Florida, USA.
Mora, B. B. (2012). Epidemiology and agronomical management of Rhizoctonia. Web Blight in common bean. PhD Thesis. University of Florida, 145.
Mora, J. R. (2001). Control biológico de la pudrición radicular por *Fusarium oxysporum* en semilleros de café usando endomicorriza y *Trichoderma harzianum*. Tesis. Zamorano Carrera de Ciencia y Producción Agropecuaria, 25.
Moser, J. C. (2006). Complete excavation and mapping of a Texas leafcutting ant nest. *Annals of the Entomological Society of America* 99(5): 891–897.
Mueller, U. G., Schultz, T. R., Currie, C. R., Adams, R. M. M. and Malloch, D. (2001). The origin of the attine ant fungus mutualism. *The Quarterly Review of Biology* 76(2): 169–197.

Musoli, P., Cubry, P., Aluka, P., Billot, C., Dufour, M., Belllis, F., Pot, D., Bieysse, D., Charrier, A. and Leroy, T. (2009). Genetic differentiation of wild and cultivated populations: Diversity of *Coffea Canephora* Pierre in Uganda. *Genome* 52(7): 634-646.

Nakano, O., Romano, F. C. B. and Pessini, M. M. O. (2001). *Pragas de Solo*. ESALQ/USP, Piracicaba, SP, 213.

Oliveira, C. M. and Frizzas, M. R. (2017). How climate influences the biology and behaviour of *Phyllophaga capillata* (Coleoptera: Melolonthidae) in the Brazilian Cerrado. *Austral Entomology* 58: 336-345. https://doi.org/10.1111/aen.12309.

Palanichamy, L. (1973). Nematode problems of coffee in India. *Indian Coffee* 37: 99-100.

Parra, J. R. P., Batista, G. C. and Zucchi, R. A. (1992). Pragas do Cafeeiro. In: *FEALQ*. Curso de Entomologia Aplicada à Agricultura. FEALQ, Piracicaba, SP, 355-386.

Pieters, R. and Van der Graaff, N. A. (1980). Resistance to Gibberella xylaroides in Coffea arabica: Evaluation of screening methods and evidence for the horizontal nature of resistance. *Netherlands Journal Plant Pathology* 86: 37-43.

Pujol-Luz, J. R. and Vieira, F. D. (2000). A larva de Chyromyza vittata Wiedemann (Diptera: Stratiomyidae). *Anais da Sociedade Entomológica do Brasil* 29(1): 49-55.

Realpe, Q. C. E. (2001). Estudios biológicos de *Rosellinia pepo* Pat. causante de la llaga estrellada en la macadamia. Tesis, *Maestría en Fitopatología*. Universidad de Caldas, Manizales, 88.

Rivera, M. (2014). Evaluación de cuatro productos para el control biológico de la gallina ciega (*Phyllophaga* spp.) en los viveros de café de la finca Colombia; Antigua Guatemala, Sacatepéquez. Bachelor's Thesis. Universidad Rafael Landívar, Guatemala, 79. http://biblio3.url.edu.gt/Tesario/2014/06/17/Rivera-Mynor.pdf.

Rivillas, C. A., Villain, L. and Bertrand, B. (2015). Diseases caused by nematodes. In: Gaitán, A. L., Cristancho, M. A., Castro-Caicedo, B. L., Rivillas, C. A. and Cadena-Gomez, G. (Eds), *Compendium of Coffee Diseases and Pests*. The American Phytopathological Society, St Paul, MN, 37-42.

Rutherford, M. A. (2006). Current knowledge of coffee wilt disease, a major constraint of coffee production in Africa. *Phytopathology* 96(6): 663-666.

Saccas, A. M. (1956). Recherches experimentales sur la tracheomycose des caffeiers em Oubanghui-Chari. *Agronomic Tropical* 11(1): 7-38.

Santa-Cecília, L. V. C., Reis, P. R. and Souza, J. C. (2002). Sobre a nomenclatura das espécies de cochonilhas farinhentas do cafeeiro nos estados de Minas Gerais e Espírito Santo. *Neotropical Entomology* 31(2): 333-334.

Santa-Cecília, L. V. C., Souza, B., Prado, E., Souza, J. C. and Fornazier, M. J. (2005). Cochonilhas-farinhentas em cafeeiros: Reconhecimento e controle. *Circular Técnica* 189. EPAMIG, Lavras.

Santa-Cecília, L. V. C., Souza, J. C. and Reis, P. R. (2000). Novas constatações da cochonilha-da-raiz *Dysmicoccus cryptus* em lavouras de café no sul de Minas. *Em Minas Gerais. Circular técnica* 130. EPAMIG-CTSM, Lavras, 2.

Schowalter, T. D. and Ring, D. R. (2017). Biology and management of the Texas Leafcutting ant (Hymenoptera: Formicidae). *Journal of Integrated Pest Management* 8: 1-8.

Shik, J. Z., Gomez, E. B., Kooij, P. W., Santos, J. C., Wcislo, W. T. and Boomsma, J. J. (2016). Nutrition mediates the expression of cultivar-farmer conflict in a fungus-growing ant. *Proceedings of the National Academy of Sciences of the United States of America* 113(36): 10121-10126.

Siddiqui, M. A. (1965). A note on coffee bark disease in East Africa. *Kenya Coffee* 30(356): 345-346.

Siddiqui, M. A. and Corbett, D. C. M. (1968). Coffee bark diseases in Malawi. II. *Transactions of the British Mycological Society* 51(1): 129–135.

Silva, MdC., Várzea, V., Guerra-Guimarães, L., Azinheira, H. G., Fernandez, D., Petitot, A. S., Bertrand, B., Lashermes, P. and Nicole, M. (2006). Coffee resistance to the main diseases: Leaf rust and coffee berry disease. *Brazilian Journal of Plant Physiology* 18(1): 119–147.

Solano-Bonilla, M. and Brenes-Chacón, D. (2012). Evaluación de métodos de curación de sustratos para la prevención del Mal de Talluelo. *Revista Forestal Mesoamericana Kurú* 9(22): 63-65.

Souza, J. C. and Reis, P. R. (2000). Eficiência de inseticidas no controle da mosca-da-raiz do cafeeiro, *Chiromyza vittata* (Diptera: Stratiomyidae). *Simpósio de Pesquisa dos Cafés do Brasil* 2: 1490.

Souza, J. C., Reis, P. R., Ribeiro, J. A., Santa-Cecília, L. V. C. and Silva, R. A. (2007a). Controle químico da cochonilha-da-raiz, *Dysmicoccus texensis* (linsley, 1900) em cafeeiro (*Coffea arabica* L.). *Coffee Science* 2(1): 29–37.

Souza, J. C., Reis, P. R. and Silva, R. A. (2007b). Cigarras-do-cafeeiro em Minas Gerais: Histórico, reconhecimento, biologia, prejuízos e controle. *Boletim Técnico* 80 (2nd edn.). EPAMIG, Belo Horizonte, 48.

Stewart, R. B. (1957). *Some Diseases Occurring in Kaffa Province*. Imperial Ethiopian College of Agriculture and Mechanical Arts, Ethiopia, Alemaya, 15–16.

Su, N. Y. and Scheffrahn, R. H. (1998). A review of subterranean termite control practices and prospects for integrated pest management programmes. *Integrated Pest Management Reviews* 3(1): 1–13.

Tran, K. L. (2002). Research on the causal agents of robusta coffee yellowing and rot roots symptoms in Dak Lak province and assessment of the control potential. PhD Thesis. Hanoi Agriculture University, Hanoi.

University of Florida and IFAS. (2014). Escarabajos ambrosiales (Coleoptera: Curculionidae: Scolytidae. TREC. *Tropical Fruit Entomology*. http://trec.ifas.ufl.edu /tropicalentomology/factsheets/ ambrosia_beetles_es.shtml.

Updhyay, H. P. (1981). *A Monograph of Ceratocystis and Ceratocystiopsis*. University of Georgia Press, Athens, GA.

Van der Graaff, N. A. and Pieters, R. (1978). Resistance levels of *Coffea arabica* to *Gibberella xylaroides* and distributuion patterns of the disease. *Journal of Plant Pathology* 84: 117–120.

Van Der Vossen, H., Bertrand, B. and Charrier, A. (2015). Next generation variety development for sustainable production of arabica coffee (*Coffea arabica* L.): A review. *Euphytica* 204(2): 243–256.

Varón, E. H., Eigenbrode, S. D., Bosque-Pérez, N. A. and Hilje, L. (2007). Effect of farm diversity on harvesting of coffee leaves by the leaf-cutting ant *Atta cephalotes*. *Agricultural and Forest Entomology* 9(1): 47–55.

Waller, J. M., Bigger, M. and Hillocks, R. J. (2007). *Coffee Pests, Diseases and Their Management*. Crop Protection Programme. CAB International, Wallingford, UK, 434.

Williams, D. J. and Granara De Willink, M. C. (1992). *Mealybugs of Central and South América*. CABI Publishing, Wallingford, 629.

Wiryadiputra, S. and Tran, L. K. (2008). Indonesia and Vietnam. In: Souza, R. M. (Ed.), *Plant-Parasitic Nematodes of Coffee*. Springer, Dordrecht, The Netherlands, 277-292.

Zambolim, L. and Vale, F. X. R. (2000). Perdas na produtividade e qualidade do cafeeiro causadas por doenças bióticas e abióticas. In: Zambolim, L. (Ed.), *Café: Produtividade, qualidade e sustentabilidade*. UFV, Viçosa, MG, 239–261.

Zambolim, L., Vale, F. X. R., Pereira, A. A. and Chaves, G. M. (1997). Café (*Coffea arabica* L.): controle de doenças - doenças causadas por fungos, bactérias e vírus, in Vale. F.X.R. and Zambolim, L. (Eds.), *Controle de doenças de plantas: grandes culturas 1*. UFV, Viçosa, MG, 83–140.

Zambolim, L., Vale, F. X. R., Pereira, A. and Chaves, G. M. (1999). Manejo integrado das doenças do cafeeiro. In: Zambolim, L. (Ed.), *Encontro sobre produção de café com qualidade I*. UFV, Viçosa, MG, 134–215.

Chapter 14

Integrated weed management in coffee production

Cláudio Pagotto Ronchi, Federal University of Viçosa, Brazil

1 Introduction

Coffee plants grow perennially, with a production cycle of up to 30 years. Inter-row spacing and coffee plant density may change significantly among coffee-growing areas around the world, and even among areas in the same country. In Brazil, for example, coffee cropping systems consist mainly of large-spacing inter-rows (2.5–4.0 m), with an optimal planting density of about 5000 trees ha^{-1} (DaMatta et al., 2010; Matiello et al., 2016). In Central America (Costa Rica), a narrowed inter-row spacing (~2.0 m) is used, but keeping the same plant density (Ramírez, 2009) as in Brazil. In both cases, however, a high soil exposure is observed.

In addition, coffee plants show a very low initial growth rate (DaMatta et al., 2010) that also impairs soil covering (Ronchi et al., 2001; Borges et al., 2019). Thus, mainly during the juvenile phase (up to 2 years in fields), the coffee crop is highly sensitive to weed species competition (Ronchi and Silva, 2006; Araújo et al., 2012). This results in a remarkable decrease in coffee growth and yield, and weed control is one of the largest field management tasks, which can entail high costs (Alcântara and Ferreira, 2000a; Ramírez, 2009; Ronchi et al., 2014). Weed control thus is required at all steps of coffee production (Ronchi and Silva, 2018).

http://dx.doi.org/10.19103/AS.2021.0096.28

Weed control starts at an early stage when the land for coffee nurseries and both the soil and manure required to produce coffee seedlings have been identified. During the nursery period, coffee seedlings must also be kept weed-free. In the field, the area where the coffee plants will be established must be totally, or at least partially, cleared of weeds and natural vegetation, depending on the production systems. During the 2 years after transplanting, an appropriate weed control system is crucial for proper crop establishment, minimizing any possible harmful effects on the coffee environment (e.g. high soil exposure). Finally, during the life of a productive plant, including the pruned ones, weeds should be managed using appropriate (or sustainable) weed control practices in order to preserve yearly crop yields and shoot growth, as well as to ensure that coffee managing practices, like fertilizer and pesticide applications, and coffee harvesting are efficiently applied to the crop (Ronchi and Silva, 2018).

Coffee is grown worldwide in different production systems that have a direct impact on weed dynamics and, hence, its control strategy. For example, shading plantations with low soil disturbance is used in Central America (Aguilar et al., 2003; Ramírez, 2009) or in Africa (Deribe, 2018), whereas, in Brazil, coffee is grown under full sunlight with constant soil disturbance, especially during harvesting (Matiello et al., 2016). In Brazil, the world's largest coffee producer, there are different primary coffee production regions, each utilizing particular cultural practices for crop management (Matiello et al., 2016). Therefore, adopted integrated weed management (IWM) practices will vary between coffee farms worldwide depending on the local characteristics. Whatever the case, this chapter deals with preventive, cultural, cover crops, mechanical (hand weeding, soil cultivation, and mowing), physical (electrical discharge), and biological (animals as grazers) methods for weed management in coffee plantations.

2 Weed interference: yield losses and economic impact

It is well known that weeds affect the coffee crop in several ways during its life cycle. For example, it has been shown that young coffee plants suffer from competition with different weed species under both controlled conditions (Ronchi and Silva, 2006; Ronchi et al., 2007; Fialho et al., 2010, 2011; Carvalho et al., 2013; Borges et al., 2019) and in field experiments (Lemes et al., 2010; Araújo et al., 2012; Magalhães et al., 2012).

The reduction in plant growth correlated with decreasing photosynthetic efficiency (Matos et al., 2013) and nutrient accumulation by both the coffee plant's shoot (Ronchi et al., 2003; Carvalho et al., 2013) and root (Ronchi et al., 2007) systems. These mentioned studies also showed that the effect of weed competition on coffee plants was strongly dependent on both the weed species and its density and on the coffee plant's age after transplanting. Therefore, an

IWM in coffee must consider the characteristics of weed species individually, as well as their high potential for nutrient recycling.

The impaired crop growth due to weed competition soon after transplanting in the field will certainly cause irreversible losses in crop yield, as demonstrated by Lemes et al. (2010). These authors reported coffee yield loss of approximately 40% during the first three harvest cycles as a consequence of inadequate weed control practices that led to weed-crop competition after transplanting. In fact, yield losses due to weed competition varied from 24% (Moraima García et al., 2000) to 92% (Lemes et al., 2010) in different adult commercial coffee crops.

Weeds can also hinder coffee harvesting and, indirectly, the bean quality. In this case, especially in some Brazilian regions where mechanized harvesting methods are used, like the Cerrado Mineiro Region, weed control in the pre-harvest period has the main objective of facilitating the harvesting and ensuring high efficiency of the sweeping process of ground coffee (Matiello et al., 2016; Ronchi, 2020). Furthermore, the control over the weeds may contribute to the reduction of moisture on the soil surface (due to organic residues), capable of causing the fermentation of grains fallen on the soil (Matiello et al., 2016).

Currently, with great concern over the attack of the coffee berry borer and its significant damage to the yield, type, and quality of the coffee harvested (among other effects), high efficiency is sought in collecting coffee during sweeping so that there are no remaining grains in which the pest can multiply in the off-season (cultural control). In this sense, the careful management of weeds in the pre-harvest starts to play a major role in reducing the coffee berry borer attack in the next harvest (Ronchi, 2020).

There are several weed species that occur in coffee plantations worldwide (Aguilar et al., 2003; Sánchez and Gamboa, 2004; Gómez, 2005; Silva et al., 2006; Maciel et al., 2010; Partelli et al., 2010; Ferreira et al., 2011; Araújo et al., 2012; Moreira et al., 2013), and they will not be nominated here. The fact is that the weed species' biomass and/or composition significantly differ among coffee sites, depending on several factors, including the coffee cropping system - monocrop or intercropped (Concenço et al., 2014), use of shade trees and the level of shading (Silva et al., 2006), crop age, season, ground cover management (Aguilar et al., 2003), and the use of cover crops (Partelli et al., 2010; Moreira et al., 2013). Therefore, potential weed species-coffee interactions are high and diverse around the coffee-producing countries.

For example, while the species of the genus *Commelina* are efficiently used as ground cover in shaded coffee plantations in Central America (Ramírez, 2009), in Brazil, these weed species strongly compete with coffee plants, especially for water and nutrients, in addition to impairing coffee harvesting. Therefore, they need to be controlled (Ronchi et al., 2001). Also, in Brazil, in addition to *Commelina* spp., other weed species have been of major importance in all coffee growing areas of either Arabica or Robusta coffee: *Conyza* spp., *Digitaria insularis*, *Ipomoea* spp., and *Urochloa* spp., most of them due to their tolerance or even resistance to herbicides (Ronchi and Silva, 2018).

Weed control practices are directly related to the economic sustainability of coffee production (Ronchi and Silva, 2018). Among the several factors that affect coffee crop profitability, the coffee yield (measured as 60 kg bags of green coffee per hectare) is of major importance, since profitability tends to increase with coffee yield. Similarly, reductions in production costs tend to increase profits. Therefore, when weeds (and weed control) are considered in the production system, they may decrease crop profitability by affecting both the coffee yield, due to the resource competition that impairs plant growth, and the production costs, because the coffee grower must spend money to control them (Ronchi et al., 2014).

Thus, it can be deduced that the practices applied to control weeds in coffee must be properly balanced in order to reduce costs of weed control and to efficiently prevent weed competition to ensure high crop yield (Ronchi et al., 2014). For example, in a typical Brazilian Arabica coffee farm, where all the management practices including weed control are mechanized, the costs of weed control are about 5–8% of the total input costs per year. Such expenditures consist of herbicides, diesel oil, labor, and equipment (springers, mowers, etc.) maintenance. Although weed control in coffee is small, it should never be neglected; otherwise, the losses in crop yield due to weed competition would undoubtedly reduce the profitability of the crop (Ronchi et al., 2014).

3 Young coffee plantations: the most weed-sensitive crop phase

Young coffee plants are highly sensitive to weed competition soon after transplanting in the field (Ronchi et al., 2003; 2007; Ronchi and Silva, 2006; Fialho et al., 2010, 2011, 2012; Lemes et al., 2010; Araújo et al., 2012; Magalhães et al., 2012; Borges et al., 2019), with sensitivity decreasing with age (Aguilar et al., 2003; Fialho et al., 2010). Thus, attention must be paid to weed control in young coffee plantations, since this crop stage is considered a critical period for weed control in coffee, irrespective of the region where the crop is grown (Aguilar et al., 2003; Sánchez and Gamboa, 2004; Sarno et al., 2004; Ramírez, 2009; Araújo et al., 2012).

The weeds, irrespective of species' types, are only harmful to young coffee plants when they grow near the coffee rows. For example, competition trials with young pot-grown coffee plants show that the distance between weeds and coffee plants affects coffee growth: the degree of weed competition increases as they get closer to the crop (Marcolini et al., 2009). Such negative interference occurs because proximity to neighboring plants leads to resource limitations for coffee plants, especially light and nutrients (Radosevich et al., 1996). This is because the main part of the root system of an Arabica coffee tree is generally concentrated in the first 0.2 m layer from the soil surface and is distributed near the coffee trunk (Ronchi et al., 2015), which aggravates the competition effect on coffee. Therefore, to preserve the initial growth of coffee plants, it is necessary to ensure that the area around them is kept bare, as evidenced in a study by Sarno et al. (2004) in Indonesia.

To prevent weed competition, farmers sometimes decide to eliminate all weeds from the crop fields, even in young and widely spaced (up to 3.8 m between rows) coffee plantations. However, such a drastic agronomic practice is not recommended, since it requires high inputs of energy for mechanical operations and, mainly, because soil exposure after weeding increases the likelihood of soil erosion, an unsustainable practice (Ronchi and Silva, 2018). Nonetheless, coffee is grown worldwide mainly on steep slopes where soil erosion is a severe problem (Sarno et al., 2004). In fact, several studies have demonstrated that natural soil coverage or the introduction of cover crops (discussed below) between coffee rows is effective for improving soil and crop characteristics (Aguilar et al., 2003; Shivaprasad et al., 2005; Alcântara et al., 2009), especially in hilly areas (Sarno et al., 2004).

Adequate weed control in young coffee crops, and other perennial crops, is achieved by eliminating weed species only within a strip on both sides of the coffee row during the 2 years after transplanting. Weeds in the inter-rows may be managed by mower operations, and even by cover crops. Managed this way, coffee can grow without weed competition, and producers avoid unnecessary weeding. However, the width of the weed control strip is usually randomly defined by growers without any criteria (Araújo et al., 2012).

Generally, coffee genotype and crown architecture, plant spacing, soil texture, climatic conditions, and weed species density and diversity are factors that affect the width of a weed control strip. The age of the coffee plant also affects it (Araújo et al., 2012), since the diameter of the coffee plant's canopy and the root system progressively increases after transplantation (DaMatta et al., 2010; Matiello et al., 2016; Araújo et al., 2012; Ronchi et al., 2015).

Even though coffee producers commonly keep young coffee rows permanently weeding, information regarding the criteria for weed control strip establishment in young coffee plantations is scarce in the published literature, which suggests that little importance has been given to this subject. In Brazil, the adequate width of the weed control strip at each side of the coffee rows to prevent young coffee plants from weed competition ranges between 0.6 m (Dias et al., 2008) and 1.0 m (Souza et al., 2006).

However, establishing a fixed weed control range irrespective of the coffee plant's age may not be an effective strategy, as can be depicted from the work of Lemes et al. (2010). Indeed, Araújo et al. (2012) clearly demonstrated that integrated management of *Urochloa decumbens* in young coffee crops must focus on weed control only in a minimum range along the coffee rows, which must increase with the coffee plant's age (Fig. 1).

The weed strip control values reported by Araújo et al. (2012) for each crop age, however, are not standard fixed values for coffee plantations. Depending on both the occurring weed species and the adopted IWM methods applied in the inter-rows, as well as on several other crop aspects, weed control strips may increase or decrease. Consequently, each farm must be analyzed separately to properly define the best weed control strip.

Over the past few decades, some general agronomical recommendations to prevent weed competition focused on controlling weeds only below the projected coffee plant canopy (Sánchez, 1991; Ronchi et al., 2001). Araújo

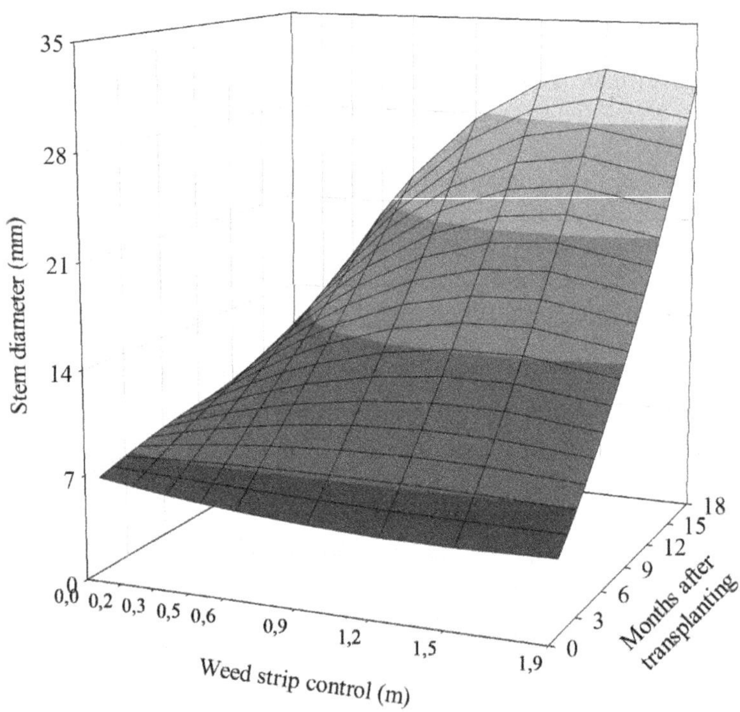

Figure 1 Coffee plant growth (represented by stem diameter) as a function of both the weed control strip and coffee plants' age in months after transplanting (MAT). In the first period after transplanting (within 2-3 MAT), weed competition exerts no or very little negative effect on plant growth; as the coffee plant ages, its growth is drastically reduced when weed control practices are not employed. However, adequate coffee growth is ensured at approximately 18 MAT if a minimum weed strip control is applied in the coffee rows. As the coffee plant ages, weed control strips must be progressively wider to prevent weed competition and to enable the coffee plants to grow properly. Source: This figure is adapted from Araújo et al. (2012).

et al. (2012) showed, however, that weed control must extend beyond the projected coffee plant canopy, which many Brazilian coffee growers are already applying. Recent research (Pedrosa, 2013) has also shown that by controlling *U. decumbens* (that was sowed in the coffee inter-rows as ground cover) in a 0.5-m control strip beyond the projected coffee plant canopy, nutrient (nitrogen) competition was reduced to only 1%, ensuring maximum coffee growth and yield.

Recommending the control of weeds beyond the projected coffee plant canopy does not mean that they should be eliminated in the entire inter-row spacing. As shown in Fig. 1 (for a detailed discussion see Araújo et al. (2012)), maximum coffee plant growth for each coffee plant age was reached at a weed control strip smaller than the largest one that was possible; above such a critical weed control strip value, coffee growth was slightly reduced. This suggests that weed elimination in the inter-rows, not the weed-coffee competition in the coffee rows, may be the reason for such a negative effect.

Therefore, maintaining natural weed coverage in the center of coffee inter-rows may favor edaphoclimatic conditions for the improvement of coffee plant growth (Ronchi and Silva, 2018). In fact, there are many benefits of covering the inter-rows of coffee plantations, as will be discussed later. Moreover, the cost of weed control may be diminished if it is applied only to the minimum weed control strip in the coffee rows.

4 Integrated weed management

According to Harker and O'Donovan (2013), integrated weed management (IWM) is a holistic approach to weed management that integrates different methods of weed control to provide the crop with an advantage over the weeds. In addition to considering all plant growth factors, IWM allows for efficient utilization of environmental resources by the crops and provides increased yields and higher quality products (e.g. no defensive residues in foods). Moreover, IWM aims to ensure production under environmental, economic, and social sustainability, increasing or at least maintaining crop yields. For these reasons, IWM practices are being increasingly used in many crops worldwide (Bajwa, 2014; Harker and O'Donovan, 2013).

IWM in coffee is based on the rational combination of different weed control practices (Ronchi and Silva, 2018) discussed herein. Thus, as stated by Bajwa (2014) and Harker and O'Donovan (2013), for general crops, no weed control practice is used in isolation in coffee plantations. For example, IWM associates herbicide use with other weed control practices, especially cultural or mechanical practices (Ronchi, 2020), to ensure better conditions for coffee growth and development. If the IWM practices are precisely established, the negative effects of weeds on coffee plants (e.g. crop yield reduction) and the environmental impacts of such practices are both diminished. Moreover, a reduction in the production costs, due to lesser use of herbicides and labor, is also observed (Silva and Silva, 2013).

An IWM program must consider several aspects that include: potential benefits and harmful effects of each weed species, coffee crop characteristics, climatic conditions, costs of weed control, impacts of each practice on the environment, coffee crop and human health, labor supply, the availability of farm implements and machinery, critical periods of coffee-weed competition, and the expected effectiveness of the practice on controlling weeds (Ronchi and Silva, 2018). Moreover, an IWM must consider the possibility of diversification in the coffee farms (e.g. grazing with ovines – Sánchez and Chacón (2000) – as will be discussed below). Other excellent examples of IWM in coffee have been reported by Sarno et al. (2004), Gómez (2005), Silva et al. (2006), Araújo et al. (2012), Aguilar et al. (2003), and Ronchi (2020).

Ronchi and Silva (2018) recently made a holistic analysis of all the scientific work published in the last decades and showed that the sustainability of weed management practices in coffee plantations has improved in the last decades worldwide. It is in agreement with Harker and O'Donovan (2013), who concluded that considerable evidence exists that research on nonherbicidal weed management strategies as well as strategies that integrate other weed management systems with herbicide use has increased in the last decades for all crops worldwide.

According to Ronchi and Silva (2018), coffee farmers, researchers, and technicians have changed from simply targeting to eradicate the weeds *per se* to the concept of sustainable and integrated weed management. Complete eradication of natural weed vegetation from coffee farms is now rarely used or recommended (with the exception of those areas where the harvesting process is intensively mechanized). Quite the opposite, special attention is paid to several advantages of maintaining and managing natural weed vegetation in coffee farms, whether or not the crop is on a steep slope of a hilly area or on a gently sloping area. This is also true for Robusta coffee (see Section 6). It must be emphasized that harvesting machines capable of gathering fallen coffee fruits from the soil (just below the coffee plant canopy, in the coffee row) without disturbing the coffee inter-rows vegetation have already been developed in Brazil.

Thus, weed vegetation is eliminated only from areas where, and when, it is necessary, reducing soil disturbance and erosion, as well as costs of weed control. Moreover, Ronchi and Silva (2018) pointed out that the introduction of cover plants in coffee inter-rows has become a successful and common weed control practice in many coffee farms worldwide. Selective weed control has also been applied in some areas based on the ecological characteristics of each weed species, reducing their harmful effects on coffee plants while preserving their ground cover ability. Most likely, these practices have resulted in reduced labor and energy inputs in coffee plantations.

Although the sustainability of weed control has increased, coffee growers are challenged to choose the weed control practices that are most suitable for each field situation. Moreover, the growers must know how to apply each practice correctly to ensure satisfactory weed control without causing harmful effects on the environment or damaging crops (Ronchi and Silva, 2018). For example, inappropriate technologies for herbicide application can lead to herbicide spray drift, which affects the coffee plant, as shown in previous studies on glyphosate (França et al., 2013; Carvalho et al., 2014). Moreover, glyphosate drift occurring during the coffee fruit development stage can result in the accumulation of herbicide residues in the harvested fruits, hampering coffee commercialization. Like chemical, even mechanical weed control practices can be harmful to the coffee plant if used in an unsustainable way, as will be discussed in the following sections.

5 Weed control methods

In this section, the main weed control methods in coffee plantations will be described, focusing on those particularly important to improve environmental sustainability and to reduce reliance on herbicides. Although there are some advantages to chemical weed control, if it is not correctly executed, weed competition will not be reduced, the coffee plants may become toxic, and it may result in health and environmental risks, such as soil contamination and natural selection of tolerant or resistant weed species (Ronchi et al., 2001, 2005).

Because of repeated applications in the same coffee area each year and over the years (considering coffee is perennial, and crop rotations are not possible), weed populations that have glyphosate tolerance such as Benghal dayflower (*Commelina benhalensis* and *C. diffusa* - Santos et al., 2002) and Broadleaf buttonweed (*Spermacoce latifolia*) have spread throughout coffee crops, particularly in Brazil. Moreover, some weed biotypes of Hairy fleabane (*Conyza* spp.) and Sourgrass (*Digitaria insularis*) have shown high levels of glyphosate resistance (Christoffoleti et al., 2008; de Carvalho et al., 2011).

Therefore, the use of chemical weed control in coffee plantations has reduced effectiveness in these cases. Thus, in order to control both the tolerant and resistant weed biotypes, as well as to prevent new biotype selection, integration of different weed control practices must be used. Readers are encouraged to read Santos et al. (2014) and Oliveira and Brighenti (2018), since they bring a detailed description of several sustainable methods to control weeds without using herbicides. In the case of Santos et al. (2014), they explore sustainable weed control specifically in coffee plantations. In any case, Ronchi and Silva (2018) resumed how chemical weed control in coffee plantations can be safely used, aiming for environmental sustainability.

5.1 Preventive

Similar to cultural methods, preventive measures for weed suppression are less expensive and also not harmful to the coffee crop. There are few, but relatively important, preventive measures that should be applied in coffee production systems either to restrain the entrance or to diminish weed seeds spreading over the coffee plantations.

The land where the coffee nursery will be established should be carefully selected. The nursery site should be free from hard-to-control weeds such as *Cyperus rotundus* and *Cynodon dactylon* or other weed species that are asexually propagated. Moreover, considering that most coffee seedlings are produced in polyethylene bags filled with a substrate composed of a mixture of soil and manure (clod-seedling), close attention should be paid to the origin of these substrate components. Although there are several physical and chemical methods potentially suitable for substrate sterilization (Miranda et al., 2007), it is not a common practice under field conditions.

Therefore, the removal of topsoil layers (which represent the soil weed-seed bank – Roberts, 1981) before soil withdrawal and the use of completely decomposed manure are important preventive measures that will drastically reduce the occurrence of weeds at the nursery stage (Ronchi et al., 2001). Other weed control practices such as hand weeding can be used during the nursery stage to prevent the introduction of new weed species in new coffee areas.

In field-grown coffee plantations, preventive weed control practices basically include: (i) cleaning manure for any soil amendments; (ii) maintenance of weed-free farm roads by mowing operations that prevent soil erosion while keeping the soil covered; (iii) cleaning machinery during or after any mechanized operation in the coffee farm; (iv) removing any new weed infestation before they become too dense; and (v) controlling weed species up to the flowering stage to prevent weed seeds spreading over the area, or to avoid increasing the soil weed-seed bank (Ronchi et al., 2014).

Weed species such as *Bidens pilosa* and *Cenchrus echinatus*, for example, are commonly present in coffee areas, and their seeds are easily dispersed by animals or even by farm workers. Moreover, as mentioned earlier, in some important Brazilian coffee growing areas, mechanized coffee harvesting has contributed to the spread of *Ipomoea* spp. seeds over the cropland (Matiello and Santinato, 2016). By controlling this weed species during the early development stages or by cleaning harvesting machines frequently, the problem may diminish.

5.2 Cultural

Cultural control is one of the least costly measures and plays an essential role in any modern IWM. It relies on the design and manipulation of cropping systems to reduce weed pressure. It is achieved through either reducing weed emergence or increasing crop competitive ability, or both, and can only be precisely established if the ecological characteristics of the weeds and crops are known (Mohler, 1996). In this context, some cultural weed control methods in coffee plantations are discussed in the following, with the exception of crop rotations, which cannot be applied in coffee, since this crop is perennial.

Although coffee crop seedlings have an initial size advantage over new-germinated weeds, coffee plants show a very low initial growth rate and low soil covering capacity as compared to annual crops (Borges et al., 2019). Therefore, high (or increasing) planting densities do not contribute to weed suppression during the critical period following transplanting. However, as the crop ages, they become effective in suppressing new weed growth compared to the conventional coffee planting density (Matiello et al., 2016; Ronchi and Silva, 2018). Such a positive effect on early weed suppression is almost the same as that induced by shading in Central American coffee plantations (Aguilar et al., 2003; Ramírez, 2009).

As coffee and shade trees age, the weed species composition also changes (a reduction in the species less adapted to shade) and a reduction in weed biomass is commonly observed, which, together, significantly reduces the costs of weed control (Soto-Pinto et al., 2002; Aguilar et al., 2003; Silva et al., 2006; Ricci et al., 2008; Concenço et al., 2014). Weed suppression occurs because closer coffee plant canopies (as coffee ages and/or in high planting density) and over-story cover crop trees restrict light for weed growth, and probably also increase soil nutrient competition (Soto-Pinto et al., 2002; Aguilar et al., 2003). Thus, for cover crops, effective high planting density and shading trees must be established for each coffee plantation.

The knowledge of cultivars with a greater capacity to tolerate the presence of weeds can contribute to the selection of more competitive genotypes as a crop management practice for newly planted crops (Medeiros et al., 2016). This is particularly important in areas where weed competition is very high or where chemical and even mechanical methods of control are restricted (Ronchi and Silva, 2018), for example, in the cases of organic coffee, high planting density, and steep slope in a hilly area.

The varied ability of different cultivars to compete with plants has been studied in several crops, such as wheat (Lamego et al., 2013), soybeans (Bianchi et al., 2006; Fleck et al., 2007), barley (Galon et al., 2011), rice (Balbinot Junior et al., 2003), and eucalyptus clones (Medeiros et al., 2016), with the latter being a perennial plant similar to coffee. However, for the coffee crop, only a single work seems to have been recently published worldwide.

Borges et al. (2019) explore the response of Arabica coffee cultivars to competition from beggarticks (*B. pilosa*). According to these authors, the choice of a coffee cultivar that has, besides the traditional requirements usually taken into account (e.g. plant vigor; productive potential; fruit and grain quality; and tolerance to diseases, pests, and abiotic stresses), higher tolerance to weed competition in the early stage of crop plantation could represent a low-cost strategy for weed management and thus could be used as a criteria for selecting coffee cultivars.

However, the 14 Arabica coffee cultivars tested by Borges et al. (2019) were equally susceptible to beggarticks competition, whether they were established at either low or high levels of infestation, in the second and third month after transplanting. It appeared that morphophysiological differences detected among cultivars may not necessarily assure different competitive abilities with perennial fast-growing weeds, such as the beggarticks. In fact, they found that, up to flowering, the beggarticks plants reached a height 2.78 times higher than the coffee seedlings, either by the rapid growth of the weed or by the slow growth of the coffee tree after transplanting, or by both (Ronchi and Silva, 2018).

Other important cultural practices to improve a crop's competitive ability include the production of high-quality seedlings with a vigorous root system, transplanting only fully acclimated seedlings, establishing an equilibrated plant nutrition program, and controlling pest infestations and plant diseases (Ronchi and Silva, 2018). One practice that has been successfully used in coffee plantations, which minimizes weed-crop competition for nutrients, is fertilizer distribution using the drip fertigation system set up in the coffee row, near the trunk under the coffee plant canopy. Considering that roots of coffee plants, not weeds, are concentrated in this soil region (Ronchi et al., 2015), nutrient uptake by the crop is improved. Long-term weed-crop competition trials conducted in modern Brazilian coffee plantations using such rigorous agronomic practices revealed a high tolerance of coffee plants to strong weed pressure (Ronchi C.P., unpublished data).

5.3 Cover crops

As emphasized in the previous sections, if weed vegetation is maintained at a sufficient distance from the coffee row (to prevent resource competition), there is no need to eliminate the natural vegetation of the entire area (except during the harvesting period in some Brazilian coffee areas). Moreover, cover crops or green manure crops can be successfully intercropped with coffee. Field studies carried out in different experimental conditions worldwide have demonstrated the high importance of green manure crops, mulching, or even the maintenance of natural vegetation in the coffee inter-rows for both crop and environment improvement (Aguilar et al., 2003; Sarno et al., 2004; Shivaprasad et al., 2005; Pais et al., 2011; Araújo et al., 2012; Pedrosa, 2013; Siqueira et al., 2014a; Martins et al., 2015).

The benefits of cover crops revealed in those studies include: prevention or reduction of soil exposure and erosion; improvement of physical, chemical, and biological soil characteristics; maintenance of coffee yield potential; conserving soil moisture; improvement of soil organic carbon content; and weed suppression associated with allelopathic and physical effects or with direct resources competition (Ronchi and Silva, 2018). For example, Martins et al. (2015) found that the plots subjected to *M. deeringiana* cover crop in the inter-rows showed more than 90% reduction in weed density, which was attributed to the chemical composition of the plant species, thus possibly leading to allelopathic effects of this cover crop, as also reported by Rosa et al. (2013). Thus, besides weed suppression, the cover crop must provide other agroecosystem services (Lemessa and Wakjira, 2015).

The other important effect of such sustainable practices is the reduction in soil temperature variation, preventing high soil temperatures from occurring. This is particularly important for coffee plants since their root systems are physiologically highly sensitive to high soil temperature (Franco, 1958). In Central American countries, selective ground cover management is carefully used in shaded coffee plantations to control weeds in patches, leaving uncontrolled species that are considered suitable to protect the soil from erosion and compaction and to suppress more aggressive weed growth (Aguilar et al., 2003; Ramírez, 2009).

However, the weed control methods and cover crops do not always alter soil characteristics. For example, Martins et al. (2015; 2016a) assessed the effects of long-term different weed control methods and cover crops between coffee rows on the chemical soil characteristics, soil organic carbon and humic substances content, soil organic matter quality, and weed diversity in Brazilian coffee plantations. They found that neither the different cover crops (*Arachis hypogeae*, *Mucuna deeringiana*, and natural vegetation - no weed control) nor the weed control methods (hand weeding, portable mechanical mower, and herbicide application) altered the humic and fulvic acid carbon content at superficial soil layers (0–20 cm), or the organic matter humification degree. In addition, little or no changes were detected in chemical soil attributes among weed control treatments. However, the topsoil organic carbon content was up to 26% higher in the plots with no weed control as compared to other weed control methods, even those possessing cover crops (Martins et al., 2015). Such an increase in topsoil organic carbon content was especially observed when mowed residues from coffee shrub pruning were left on the soil surface (Martins et al., 2016b).

Plant residues from other crops (if they are available on the farm without additional costs), from the coffee plants (leaves and stems), or from shade trees, especially after pruning operations, can be used as mulches. Actually, different types of organic materials, including coffee residues such as coffee pulp, husk (Minassa, 2014), and grounds (Yamane et al., 2014), have the potential to be used for controlling weeds through top dressing applications. For example, Yamane et al. (2014) recently demonstrated that top dressing application of coffee grounds at 16 kg m^{-2} resulted in significant weed control for half a year.

Such inhibition was a result of an allelopathic effect due to the presence of caffeine, tannins, and polyphenols in the coffee grounds (Pandey et al., 2000). Regardless of the type of organic materials, their possible inhibitory effects on coffee plant growth and yield should also be evaluated before adding them to the coffee plantations to control weeds. Moreover, the studies mentioned above showed that adequate weed control is reached only if an optimum amount of top-dressing organic material is used; otherwise, it may stimulate weed growth.

In this sense, both natural and introduced vegetation need to be properly established and managed per local crop characteristics, ensuring the weed control practices benefit the environment without affecting the coffee plants. For example, some cover crops such as *Mucuna aterrima* (Fialho et al., 2011) and *U. decumbens* (Araújo et al., 2012) or possible inter-planted annual crops such as *Gossypium hirsutum, Arachis hypogaea, Ricinus communis,* and *Zea mays* (Paulo et al., 2004) or *Phaseolus vulgaris* (Carvalho et al., 2007) may strongly affect coffee growth or yield if they are not appropriately established or managed.

Another important example occurs in several important Arabica coffee growing regions of Brazil, where coffee plantations are commonly established on land previously used for grazing. Since *U. decumbens* (Surinam grass) and other species of the genus *Urochloa* are the most used herbage in Brazil, it has become an important weed in some coffee crop areas (Dias et al., 2004; Souza et al., 2006; Araújo et al., 2012). In addition, *Urochloa* spp. has been widely and successfully used as ground cover (by sowing them in the coffee inter-rows) to add biomass to the soil surface for soil moisture conservation and nutrient recycling (Alcântara and Silva, 2010; Pedrosa, 2013).

During the first months of coffee crop establishment in the field (and even in adult coffee crops), the occurrence of high temperatures and abundant rainfall, associated with high irradiance levels, collectively promote the growth of *U. decumbens*, since this weed species shows a C4 photosynthetic metabolism. In contrast to the slow coffee plant growth just after transplanting (DaMatta et al., 2010), weed species such as *U. decumbens* quickly develop a large leaf area and height (a dense canopy), allowing them to better compete for light.

This suggests that even before reaching pre-flowering stages, when slashing (mower) operations are commonly recommended for weed control in the coffee inter-rows, weed shoots can shade both the lower and medium sections of the coffee plant canopy if they are growing near the crop line. This can interrupt the photosynthetic performance of the coffee plants, which also affects the carbohydrate supply for crop growth (Ronchi and Silva, 2018).

In fact, the use of ground cover vegetation leads to a larger turnover of organic matter and nutrients, which is associated with increased soil water content and improves long-term soil fertility (Aguilar et al., 2003) and coffee plant nutrition (Pedrosa, 2013). Although such practices may increase the crop yield, care must be taken to control Surinam grass so that it does not grow too close or too tall so as to compete with the coffee plants and permanently impair their growth and yield, as demonstrated by Dias et al. (2004), Souza et al. (2006), and Araújo et al. (2012) (see Fig. 1).

Therefore, a lot of attention must be paid to the cover crops' species, density, planting-spacing, root system and crown architecture, and to the frequency of pruning or mowing/slashing operations, as well as to the coffee crop's characteristics and climatic conditions. In some cases, it may not be recommended. For example, the use of cover crops, although effective also in controlling weeds, is not common in unshaded Arabica coffee because of excessive competition for moisture (Deribe, 2018). Anyway, all these aspects must be properly considered by any IWM aiming at the sustainability of either the coffee-producing system or integrated crop management that uses cover crops as an approach to suppressing weeds (Lemessa and Wakjira, 2015).

5.4 Mechanical

5.4.1 Hand weeding

Hand weed control methods are of major importance in coffee farms, although they are slow and labor-intensive. During nursery stages, if preventive measures fail or selective herbicides are not used, weed plants that emerge in the clods must be removed during the period of seedling formation and growth (Ronchi et al., 2001). Within 2 years after transplantation in the field, several hand hoeing (with heavy-blended hoe) operations are used to establish and maintain an adequate weed control strip along coffee rows, although selective herbicides can also be judiciously applied, mainly in large coffee areas.

Even in those coffee farms where selective pre-emergence herbicide application is the major method for weed control in the coffee rows, at least one hand hoeing operation is accomplished 2-3 weeks after transplanting prior to herbicide application to smooth the ground surface. In coffee inter-rows, other weed control methods are recommended (e.g. mower operations).

Although hand weeding is a traditional weed control method applied in coffee plantations, an assay conducted from transplanting to 30-month coffee crop in the field (Campos, 2020) has recently shown that hand-weed control may impair the growth of young coffee plants. According to Campos (2020), decreases in coffee plant height, stem, and crown diameter and the number of plagiotropic branches were observed at 7–8 months after transplanting in hand-weed treatment (three operations) as compared to traditional pre-emergent herbicides.

It happens basically due to two reasons. First, the top-soil layer near the coffee plant stem is removed; hence the plants fall over, especially if wind breaks were not previously established (which is highly recommended), and also the coffee plants have their shallow root system exposed to the harmful environmental conditions (e.g. soil drying and high temperature). Second, the fragile coffee stems are damaged directly by the hoe. In fact, a major limitation of this method, according to Deribe (2018), is that it is not very effective in dealing with weeds growing close to the crop and may result in physical injuries on coffee stems and lateral roots. Together, it allows us to conclude that, in spite of being a simple weed control method, hand weeding must be consciously accomplished, particularly in young coffee plantations.

5.4.2 Soil cultivation

The use of weed control practices comprising soil cultivation, especially those applied in the coffee inter-rows (such as the use of chisel plow, disk harrow, rotary hoe, and field cultivator), have drastically decreased in the last decades in coffee farms (Ronchi and Silva, 2018). This is in contrast to an increase in mower operations (as mentioned above). Nowadays, many coffee growers are conscious that soil disturbance must be reduced to prevent soil erosion and to ensure sustainable weed management. Although soil cultivation may have limitations (e.g. it may not be efficient under high soil moisture content and for controlling asexually propagated weed species), and that its use may be partially restricted to both sloping and high planting density or intercropped areas, it is still successfully used.

In Brazil, for example, different types of disk-harrow implements have been developed and used to till the soil along young coffee rows, before the coffee root system is fully developed. Moreover, in the coffee fields where sweeping and coffee gathering operations are necessary for the recovery of fallen fruit due to natural fall and mechanical harvesting, a brush shredder is commonly used to smooth the soil surface and to crush organic residues during the pre-harvesting period (Borba et al., 2016). In countries or coffee fields where there is no need to gather fallen coffee fruits (because they are completely harvested directly from coffee trees instead), such mechanic soil cultivation operations are not necessary.

Several important long-term field studies (Alcântara and Ferreira, 2000b; Alcântara et al., 2007, 2009; Araújo-Junior et al., 2011a,b; Pais et al., 2011, 2013; Siqueira et al., 2014a,b; Pires et al., 2017a,b) have considered the impacts of mechanical (and many other) weed control methods, including rotary hoe, disk harrow, and brush shredder on the soil's physical and chemical properties, as well as their effect on hydraulic attributes. The results of those studies were highly diverse and sometimes conflicting, showing large variations over the years and depending on both the field situations and sampled soil layers, which limits any generalization. A brief analysis of those works shows that local soil conditions and weed-crop interactions must be known to precisely define the best soil cultivation method required for each weed control situation, thereby preserving physical and chemical soil attributes. This is the basis of integrated weed control practices for sustainable agriculture.

5.4.3 Mowing

Mowing is one of the most widely used weed control practices in coffee plantations, especially in Brazil, to manage either the natural weed vegetation or introduced cover crops in the coffee inter-rows. It is mainly used during the warm and wet season, when they (weeds or cover crops) grow very fast. Mowing can be performed by hand-held or mechanized equipment. Nowadays another tractor equipment that has been used in the same way as a mower is the brush shredder. It is important to mow the weeds when they have developed maximum biomass, but before the seeds are viable. In addition, a regular mowing frequency that would prevent excessive shading to the bottom of the coffee plant canopy should be used.

In addition to providing mulching, improving nutrient cycling (discussed in the following sections), and controlling soil erosion, successive mowing operations allow a constant weed regrowth and, hence, promote the maintenance of soil organic carbon stocks (Cogo et al., 2013). Compared to other mechanical weed control methods, such as the use of a disk harrow, the brush-cutter maintains a better state of soil particle aggregation (Siqueira et al., 2014a) and a lower soil bulk density (Siqueira et al., 2014b). In some coffee farms, mowing operations are commonly and successfully applied in alternating coffee inter-rows to conserve the populations of the natural enemies of insect pests.

A recent study comparing three inter-row weed control practices in a Brazilian coffee plantation for almost two following years (Campos, 2020), revealed that either the brush shredder or the mower (which are operations that slash the weeds a few centimeters above the soil level) respectively recycled an average of 41% and 35% more nutrients as compared to chemical weed control with glyphosate (Table 1). In the case of the brush shredder, not the mower, an increase of 29% was also observed in dry matter production of the weed vegetation, which was composed of the main species *Alternanthera tenella*, *Amaranthus deflexus*, *A. hybridus*, *A. retroflexus*, *C. benghalensis*, *Eleusine indica*, *Galinsoga parviflora*, *Richardia brasiliensis*, *Sida rhombifolia*, *Spermacoce latifolia*, *U. decumbens*, and *U. plantaginea*. It can be concluded that weed control practices that allow constant weed regrowth in the coffee inter-rows, in comparison to desiccation with non-selective herbicides, may represent a well-thought-out IWM strategy to increase cropping-system sustainability

Some coffee growers prefer to use the brush shredder instead of the mower believing that the former promotes weed control for a longer time, due to a slower weed regrowth rate. However, it seems not to be true, as demonstrated by Campos (2020). In this case, he showed that, during almost one and half year, the same number of operations were required by either the brush shredder or the mower. Actually, the mower has several advantages over the brush shredder: mowed weed residues are greater in size, increasing the time required for mineralization, which, thus, protects the soil for a longer time; it allows mowed residues to be directed to the coffee rows, where the coffee root system is mainly located; the operation speed is almost twice as fast as that with the brush shredder, spends less diesel, and requires less traction power; finally, the mower does not exert any pressure on the soil surface, as may occur by using the brush shredder (A. J. Ribeiro, personal communication). In any case, more scientific researches are required to prove these practical agronomical observations.

5.5 Physical method: electrical discharge

This method of weed control has strongly evolved in Brazil since the first drafts, more than 30 years ago (Almeida, 1988). Nowadays it is largely used in soybeans (Brighenti and Brighenti, 2009) but also in sugarcane, citrus, eucalyptus, vineyards, non-agricultural lands (like urban areas, railways, airports, highways, etc.), and, of course, in coffee plantations (Oliveira and Brighenti, 2018; Oliveira and Coutinho Filho, 2018; Zasso Brazil, 2020).

Since it is a new method for weed control in coffee, little or no scientific data is yet available. In any case, Oliveira and Coutinho Filho (2018) mentioned that the electrical discharge promoted a non-selective control (as high as 95%) of weeds present in the coffee crop inter-row, particularly *Commelina* spp., which is one of the most important weed species in Brazilian coffee farms. Moreover, they observed no negative effect on the coffee plants. The equipment mentioned here was developed by the company Zasso Brazil Indústria e Comércio Ltda. and the equipment, as well as several technical information, can be found on Zasso Brazil's website (Zasso Brazil, 2020).

Table 1 Weed shoot dry matter and nutrient contents recycled by three different weed control practices applied in a 3- to 4-year-old *Coffea arabica* crop in the Cerrado Mineiro Region in Brazil

Weed control practices	Weed dry matter (t/ha/year)	Nutrients								
		N	P	K	Ca	Mg	Fe	Zn	Mn	Cu
		(kg/ha/year)					(g/ha/year)			
Glyphosate	4.59	73.7	5.9	263.4	33.6	29.2	1020.1	69.3	78.9	56.5
Brush shredder	5.94 (+29%)	116.4 (+58)	7.4 (+25)	368.9 (+40)	46.1 (+37)	42.2 (+45)	1555.0 (+52)	100.7 (+45)	129.7 (+64)	55.9 (0)
Mower	4.63 (+1%)	105.8 (+44)	6.9 (+17)	332.0 (+26)	39.8 (+18)	37.0 (+27)	1397.6 (+37)	101.6 (+47)	118.1 (+50)	84.6 (+50)

Note: The numbers between parentheses represent the percentual increases of either brush shredder or mower practices in relation to glyphosate. Source: Extracted and adapted from Campos (2020).

According to the manufacturer, the equipment works plugged into the tractor battery and uses high-voltage electrodes installed in an application bar. They perform high-voltage discharges directed through the leaves to the root system, then to the soil, and return to the passive electrode. The deepest weed roots are reached. The death of the plant occurs immediately because electricity destroys the compartmentalization of vascular systems (xylem and phloem), causing irreparable lethal damage. Thus, no regrowth is observed as occurs with mower or brush shredder operations. The weed species are desiccated in a few days, so the results can be quickly attained. Since there is no need to refuel with water, as occurs with chemical weed control, the operation is very efficient. Moreover, the equipment used in coffee plantations is very versatile. Its mechanical arms work in a hydraulic way, being designed to touch the weeds even under the bottom of the coffee plants without reaching the crop.

This weed control method has gained importance in Brazilian coffee crop areas, because it allows for the efficient control of glyphosate-resistant or glyphosate-tolerant weeds (*Commelina* spp., *Conyza bonariensis*, *Digitaria insularis*, *Eleusine indica*, and *Lolium multiflorum*). Moreover, it is a non-invasive technology, which makes a significant contribution to preserving and promoting agricultural practices of soil conversation. It does not leave toxic residues in the soil. Also, unlike mechanical weeding, there is no movement of the soil or risk of erosion, disturbance of the structure is avoided, and inactive seeds are protected. It has an organic and safe certificate for coffee plantations, and, according to Zasso Brazil (2020), it is completely safe for the coffee plants, without affecting coffee growth and yield, or even the coffee root system.

5.6 Biological control: animals as grazers

Biological control of weeds consists of using a biological agent for weed suppression. Even though recent researches have identified species of fungus presenting high potential as a biological agent for controlling *Commelina benghalensis* (Ferreira et al., 2018) and *C. diffusa* (Boyette et al., 2015), which are some of the worst weeds in coffee plantations in Brazil, biological weed control in coffee areas worldwide resumes to the use of animals, basically ovines (Matiello et al., 1980; Sánchez and Chacón, 2000, 2001; García et al., 2014; Dávila-Solarte et al., 2019). Chickens have also been mentioned in the past (Prado Filho and Lima, 1981) as a biological agent for controlling some weed species in small coffee farms in Brazil, but nowadays such use is negligible. Moreover, in Brazil, the world's largest coffee producer, the use of ovines is also unusual.

On the contrary, in Central America, grazing with ovines is a good biological method for weed control in coffee plantations (Sánchez and

Chacón, 2000, 2001; García et al., 2014; Dávila-Solarte et al., 2019). Actually, it is not used for weed control *per se*, but instead, it seems to be more of an alternative for diversification in coffee farms. Even though the costs of weed control either by ovines or by traditional methods are similar, the former has additional environmental, social, and economic advantages, since it provides meat, manure, and other profitable benefits (Sánchez and Chacón, 2000; Dávila-Solarte et al., 2019). The fact is that, by using ovines, the costs of weed control with traditional weeding methods in coffee plantations are reduced, since weed infestation (shoot fresh matter) is also reduced.

The use of ovines is possible because they do not feed on coffee leaves and new stems, and even less on older coffee crops. However, one must be careful because, despite weed control level increases with increasing grazing stoking rate, injuries to the coffee plants (defoliation, broken branches) also significantly rise under high grazing stoking rates. For example, Sánchez and Chacón (2000) showed that under 4.33 sheep unit - SU (one SU = 30 kg live weight) - the weed control level was 46% and the damage to coffee plants was 1.1% (after 5 days of grazing); however, under 9.16 SU, the weed control level raised to 69%, but the damage to the coffee plants was 11.1%. The height of the coffee plant stems is also reduced under high stoking grazing rates as compared to the low ones. In any case, damage to the coffee plants by animals may not ever occur, since García et al. (2014) ensure that the sheep can feed on several weed species without causing any damage to the coffee plants.

In parallel to the negative effects that grazing with ovines can cause on the coffee plants, attention must be paid to how grazing affects weed-dynamic in coffee plantations. According to Sánchez and Chacón (2001), it does not change weed frequency and diversity, regardless of whether the grazing stoking rate was low (4.33 SU) or high (9.16 SU) during 5 days. It probably would be different if, instead, an over-grazing for long periods had been accomplished (Sánchez and Chacón, 2001).

In fact, a long-term experiment conducted in an integrated crop-livestock system (ICLS) in Brazil, even though not being in a coffee plantation nor with sheep, showed that decreasing the grazing intensity (sward grazing height of 40 cm as compared to sward grazing height of 10 cm) reduced the number of weed species, the density of emerged weed seedlings, and the density of the weed-seed bank (Schuster et al., 2016). Thus, integrated weed management strategies should consider minimizing grazing intensities in an ICLS. If such a concept is applied to grazed areas with ovines in coffee plantations, it may cause a significant reduction (or at least no increase) in some undesirable weed species (those rejected by the sheep), as compared to those selected by the animals (see the following discussion).

One disadvantage that must be considered in using ovines is that they do not feed equally on all weed species. Actually, animals select those fodders (weed species) of better nutritional quality (García et al., 2014). Thus, weed infestation is only partially controlled, which suggests that such a biological weed control method needs to be integrated with other practices to reach an adequate level of weed control (or efficiently suppress weed competition). In any case, the costs of weed control will diminish since only the reminiscent weeds (the rejected ones) need to be posteriorly controlled (Sánchez and Chacón, 2000).

Since sheep ingest various types of seeds when grazing mature plants or seeds scattered on the soil surface (Michael et al., 2006; Schoenbaum et al., 2009), they could be an important vector for dispersing weed seeds, particularly hard-coated seeds (e.g. *Amaranthus retrofexus*, *Cuscuta campestris*), which remain viable in the animal's digestive system for more than 3 days (Haidar et al., 2010). Thus, according to these authors, sheep should be kept in pens for about 4 days before being released into coffee plantations to reduce the risk of weed infestation and spread, and their fresh manure should not be applied on arable lands. It is a preventive weed control method. However, sheep grazing is an effective weed management method against grasses and soft-seeded weeds (Haidar et al., 2010), which corroborates the finds of Matiello et al. (1980), in Brazil.

6 Main aspects of weed control in Robusta coffee

Even though *C. canephora* (Robusta coffee) has large economic importance worldwide, particularly in Africa and Asia, almost all published scientific articles that investigated weed control in coffee in the 21st century have considered weed control in *C. arabica* (Ronchi and Silva, 2018). Nonetheless, several weed control practices applied to Arabica coffee can also be recommended for Robusta coffee. Thus, in the following text, it will be briefly discussed only some major aspects of integrated weed control in Robusta coffee, focusing on the sustainable practices used in Brazil, where the variety Conilon (also known as 'Kouillou') is grown. This is practically the sole cultivated variety of *C. canephora* in Brazil, but with significant importance in the world's coffee bean production (DaMatta et al., 2010). In any case, a detailed discussion concerning weed control in Conilon coffee can be found in Ronchi et al. (2019).

As in *C. arabica*, Conilon coffee is also grown under relatively large spacing, generally 3.0 × 1.0 m. This leads, especially in the implantation phase, to a low degree of soil cover. Conilon coffee also presents slow growth right after the transplanting (DaMatta et al., 2010; Ronchi and DaMatta, 2019). Consequently, the annual weeds that occur in the planting rows will easily present rapid growth in height in comparison to that of the coffee tree, reducing the light absorption capacity of the latter. It must be emphasized that, opposite to Arabica coffee, in regions where the Conilon coffee is grown at predominate high temperatures, which, together with the presence of crop irrigation or even during the rainy season of the year, intensify the weed germination, growth, and development, aggravating their effect on the culture. Thus, weeds must also be controlled in Robusta crops.

Conilon coffee, however, has some positive growth characteristics that ensure a more efficient cultural weed control as compared to *C. arabica*. Even though the spacings between rows are still large (2.0–3.0 m), they are not so large as in traditional *C. arabica* growing areas (that are greater than 3.0 m, usually 3.5–4.0 m). Moreover, Conilon coffee, like other varieties of Robusta coffee, has a specific crown architecture formed by multiple flexible orthotropic stems. In addition, in highly modern Conilon crops (Fonseca et al., 2019), the coffee plants are stimulated very early in the cycle (between 3 and 6 months after transplanting), by using an appropriate pruning technique, to produce several orthotropic stems (three to five), which are allowed to grow for production along several years before they are renewed.

Considering what was mentioned above, this means that very early in the cycle the coffee plants start to impair light intensity reaching the ground, which in turn naturally reduces weed infestation. Thus, as the coffee plants age, competition from weeds is drastically reduced (much more than in Arabica coffee), as is the energy input for controlling.

In the first years in the field, when competition from weeds is strong, but the light availability in the inter-rows is not yet limiting to the growth of cover crops (Baligar et al., 2006), they can be used to reduce weed infestation, as successfully showed by Partelli et al. (2010) on organic Conilon coffee plantations. These authors grew *Pennisetum glaucum* and the legume plants *Canavalia ensiformis*, *Mucuna deeringiana*, and *Cajanus cajan* between the coffee rows. Cover crops promote modifications to the succession dynamics of weeds: they reduce the weed species diversity as compared to control treatment (with no cover crops). Moreover, they do not interfere with the development of the coffee trees, besides helping weed control due to its fast soil covering.

The use of chemical weed control in Conilon coffee, particularly that of ghyphosate, has been reduced due to several reasons. As in Arabica crop areas, weed species showing tolerance to herbicides, particularly to glyphosate (like *Commelina* spp., *Conyza* spp., and *Digitaria insularis*) have spread over Conilon coffee plantations (Ronchi et al., 2019). Thus, farmers have to look for alternative weed control methods instead of chemicals. In this case, mower operations have predominated in coffee plantations, even in those located in hilly areas. In addition to efficiently controlling such weed species in either coffee rows or inter-rows, nowadays several portable back-carried motorized equipment are available, which show high operational mowing performance with low costs.

Finally, glyphosate drift occurring during the coffee fruit development stage can result in the accumulation of herbicide residues in the harvested fruits, hampering coffee commercialization. Such cases of contamination have recently been detected in Conilon coffee beans. Thus, the Brazilian coffee producers have been challenged to improve weed control in Conilon coffee areas, and the adoption of rigorous integrated weed control strategies is a decision that will not be looked back upon.

7 Summary and future trends

There is no doubt that coffee growth and yield both may be strongly affected by weed interference. Weeds can also impair the coffee-harvesting process. Thus, their control by using integrated managing practices, despite increasing the production costs, is of capital importance aiming for crop yield and profitability. However, the best-integrated weed control practices in coffee farms vary significantly between different regions, considering that coffee is grown in many different production systems worldwide. Moreover, two different coffee species (*C. arabica* and *C. canephora* Pierre) are of major economic and social importance to the producing countries. Hence, the weed dynamics, as well as the IWM, will change accordingly, even though some weed control practices may be effective for both.

Substantial advances have been made in the last decades to understand the long-term effects of weed control practices on both the environment and the coffee crop worldwide. For instance, adequate weed control is achieved by eliminating weed species only within a strip on both sides of the coffee rows, especially in the weed-sensitive younger crops. Weed vegetation in the inter-rows can be managed by mower operations or by cover crops, which are sustainable practices providing other agroecosystem services (e.g. nutrient recycling) besides weed suppression. Moreover, selective weed control is crucial in some field situations and, thus, the beneficial effects of each weed species must also be considered.

Unfortunately, coffee genotypes seem to be all equally highly sensitive to weed competition just after transplantation, because they grow slowly as compared to high-aggressive annual weeds. Thus, exploring higher tolerance of a particular coffee cultivar to weed competition may not work as a good cultural weed control method in an IWM in a coffee plantation. On the contrary, grazing with ovines in Central America is an alternative for diversification in the coffee farms: besides suppressing weeds, it reduces weed control costs, since it provides other profitable benefits.

Finally, electrical discharge, particularly in mechanized Brazilian coffee growing areas, is emerging as a great agroecological practice for the long-term sustainability of coffee production. Besides causing no damage to the coffee plants, it is a non-selective weed control method in both coffee rows and inter-rows. Moreover, no soil disturbance is observed as much as in hand hoeing practices. It is a fast tool for controlling weeds in certified or organic coffee farms, including those glyphosate-resistant or glyphosate-tolerant weeds. Thus, it contributes to reducing the reliance on herbicides. In any case, it lacks further scientific validation.

8 Where to look for further information

Readers are encouraged to access Oliveira and Brighenti (2018) at: https://www.embrapa.br/busca-de-publicacoes/-/publicacao/1103281/controle-de-plantas-daninhas-metodos-fisico-mecanico-cultural-biologico-e-alelopatia to download the book published by EMBRAPA (Empresa Brasileira de Pesquisa Agropecuária). This book provides a detailed discussion of several sustainable weed control methods without using herbicides that can be applied on any cultivated crop, including the coffee crop. Readers who are interested in either organic or agroecological agriculture will probably find it very useful.

Moreover, the publication entitled 'Agroecological Weed Management in the Coffee Crop' is available for download at: https://www.embrapa.br/busca-de-publicacoes/-/publicacao/1027715/manejo-agroecologico-de-plantas-daninhas-da-cultura-do-cafe. In this publication, readers can see alternative weed control methods particularly described for use in the coffee crop.

9 References

Aguilar, V., Staver, C. and Milberg, P. (2003). Weed vegetation response to chemical and manual selective ground cover management in a shaded coffee plantation. *Weed Res.* 43(1): 68–75.

Alcântara, E. N. and Ferreira, M. M. (2000a). Efeito de diferentes métodos de controle de plantas daninhas sobre a produção de cafeeiros instalados em Latossolo Roxo distrófico. *Ciênc Agrotecnol.* 24: 54–61.

Alcântara, E. N. and Ferreira, M. M. (2000b). Efeitos de métodos de controle de plantas daninhas na cultura do cafeeiro (*Coffea arabica* L.) sobre a qualidade física do solo. *Rev. Bras. Cienc. Solo* 24(4): 711–721.

Alcântara, E. N. and Silva, R. A. (2010). Manejo do mato em cafezais. In Reis, P. R. and Cunha, R. L. (Eds.), *Café arábica do plantio à colheita*. Lavras, Brasil: Epamig, U. R. SM: 519–572.

Alcântara, E. N., Nóbrega, J. C. A. and Ferreira, M. M. (2007). Métodos de controle de plantas invasoras na cultura do cafeeiro (*Coffea arabica* L.) e componentes da acidez do solo. *Rev. Bras. Cienc. Solo* 31(6): 1525–1533.

Alcântara, E. N., Nóbrega, J. C. A. and Ferreira, M. M. (2009). Métodos de controle de plantas daninhas no cafeeiro afetam os atributos químicos do solo. *Ciênc Rural* 39(3): 749–757.

Almeida, F. M. (1988). *O uso de descarga elétrica no controle de plantas daninhas. Dissertação (Mestrado em Agronomia).* Botucatu: Faculdade de Ciências Agronômicas, Universidade Estadual Paulista Júlio de Mesquita Filho.

Araújo, F. C., Ronchi, C. P., Almeida, W. L., Silva, M. A. A., Magalhães, C. E. O. and Good-God, P. I. V. (2012). Optimizing the width of strip weeding in arabica coffee in relation to crop age. *Planta Daninha* 30(1): 129–138.

Araujo-Junior, C. F., Dias Junior, M. S., Guimarães, P. T. G. and Alcântara, E. N. (2011a). Sistema poroso e capacidade de retenção de água em Latossolo submetido a diferentes manejos de plantas invasoras em uma lavoura cafeeira. *Planta Daninha* 29(3): 499–513.

Araujo-Junior, C. F., Guimarães, P. T. G., Dias Junior, M. S., Alcântara, E. N. and Mendes, A. D. R. (2011b). Alterações nos atributos químicos de um Latossolo pelo manejo de plantas invasoras em cafeeiros. *Rev. Bras. Cienc. Solo* 35(6): 2207–2217.

Bajwa, A. A. (2014). Sustainable weed management in conservation agriculture. *Crop Prot.* 65: 105–113.

Balbinot Junior, A. A., Fleck, N. G., Barbosa Neto, J. F. and Rizzardi, M. A. (2003). Características de plantas de arroz e a habilidade competitiva com plantas daninhas. *Planta Daninha* 21(2): 165–174.

Baligar, V. C., Fageria, N. K., Paiva, A. Q., Silveira, A., Pomella, A. W. V. and Machado, R. C. R. (2006). Light intensity effects on growth and micronutrient uptake by tropical legume cover crops. *J. Plant Nutr.* 29(11): 1959–1974.

Bianchi, M. A., Fleck, N. G. and Federizzi, L. C. (2006). Características de plantas de soja que conferem habilidade competitiva com plantas daninhas. *Bragantia* 65(4): 623–632.

Borba, M. A. P., Tavares, T. O., Oliveira, B. R., Santos, A. F. and Silva, R. P. (2016). Perdas na varrição e recolhimento mecanizado do café em quatro manejos do solo. In *Proceedings of the 45th Congresso Brasileiro de Engenharia Agrícola, Florianópolis, Brasil*, Small Business Exporters Association.

Borges, L. G. M., Miranda, F. R., Borges, A. M., Silva, J. R. O., Campos, A. A. V. and Ronchi, C. P. (2019). Response of arabica coffee cultivars to competition from beggarticks. *Planta Daninha* 37: e019213814.

Boyette, C. D., Hoagland, R. E. and Stetina, K. C. (2015). Biological control of spreading dayflower (*Commelina diffusa*) with the fungal pathogen *Phoma commelinicola*. *Agronomy* 5(4): 519–536.

Brighenti, A. M. and Brighenti, D. M. (2009). Controle de plantas daninhas em cultivos orgânicos de soja por meio de descarga elétrica. *Cienc Rural* 39(8): 2315–2319.

Campos, A. A. V. (2020). *Seletividade de herbicidas e controle de plantas daninhas em lavouras de café arabica*. Viçosa, Minas Gerais: Universidade Federal de Viçosa.

Carvalho, A. J., Andrade, M. J. B. and Guimarães, R. J. (2007). Sistemas de produção de feijão intercalado com cafeeiro adensado recém-plantado. *Ciênc. agrotec* 31(1): 133–139.

Carvalho, F. P., Souza, B. P., França, A. C., Ferreira, E. A., Franco, M. H. R., Kasuya, M. C. M. and Ferreira, F. A. (2014). Glyphosate drift affects arbuscular mycorrhizal association in coffee. *Planta Daninha* 32(4): 783–789.

Carvalho, L. B., Alves, P. L. C. A. and Bianco, S. (2013). Sourgrass densities affecting the initial growth and macronutrient content of coffee plants. *Planta Daninha* 31(1): 109–115.

Christoffoleti, P. J., Gali, A. J. B., Carvalho, S. J. P., Moreira, M. S., Nicolai, M., Foloni, L. L., Martins, B. A. B. and Ribeiro, D. N. (2008). Glyphosate sustainability in South American cropping systems. *Pest Manag. Sci.* 64(4): 422–427.

Cogo, F. D., Araujo-Junior, C. F., Zinn, Y. L., Dias Junior, M. D. S., Alcântara, E. N. and Guimarães, P. T. G. (2013). Estoques de carbono orgânico do solo em cafezais sob diferentes sistemas de controle de plantas invasoras. *Sem. Ci. Agr.* 34(3): 1089-1098.

Concenço, G., Motta, I. S., Correia, I. V. T., Santos, S. A., Mariani, A., Marques, R. F., Palharini, W. G. and Alves, M. E. S. (2014). Infestation of weed species in monocrop coffee or intercropped with banana, under agroecological system. *Planta Daninha* 32(4): 665-674.

Carvalho, L. B., Cruz-Hipolito, H., González-Torralva, F., da Costa Aguiar Alves, P. L., Christoffoleti, P. J. and De Prado, R. (2011). Detection of Sourgrass (*Digitaria insularis*) biotypes resistant to glyphosate in Brazil. *Weed Sci.* 59(2): 171-176.

DaMatta, F. M., Ronchi, C. P., Maestri, M. and Barros, R. S. (2010). Coffee: Environment and crop physiology. In DaMatta, F. M. (Ed.), *Ecophysiology of Tropical Tree Crops*. New York: Nova Science Publishers: 181-216.

Dávila-Solarte, P., Sanginés-García, L., Amezcua, T. and Solano, L. (2019). Productive performance and economic evaluation of sheep grazing on weeds in coffee plantations compared to pastures with or without supplementation. *Agroforest. Syst.* 93(1): 175-183.

Deribe, H. (2018). Review on effect of weed on coffee quality yield and its control measures in southwestern Ethiopia. *Int. J. Res. Stud. Agric. Sci.* 4(10): 7-16.

Dias, G. F. S. D., Alves, P. L. C. A. and Dias, T. C. S. (2004). *Urochloa decumbens* suppresses the initial growth of *Coffea arabica*. *Sci. Agric.* 61: 579-583.

Dias, T. C. S., Alves, P. L. C. and Lemes, L. N. (2008). Faixas de controle de plantas daninhas e seus reflexos na produção do cafeeiro. *Científica* 36: 81-85.

Ferreira, B. W., Alves, J. A., Miranda, B. E. C. and Barreto, R. W. (2018). Fungi on *Commelina benghalensis* from Brazil, with notes on potential for weed biological control. *Trop. Plant Pathol.* 43: 21-35.

Ferreira, E. A., França, A. C., Carvalho, R. F., Santos, J. B., Silva, D. V. and Santos, E. A. (2011). Avaliação fitossociológica da comunidade infestante em áreas de transição para o café orgânico. *Planta Daninha* 29(3): 565-576.

Fialho, C. M. T., França, A. C., Tironi, S. P., Ronchi, C. P. and Silva, A. A. (2011). Interferência de plantas daninhas sobre o crescimento inicial de *Coffea arabica*. *Planta Daninha* 29(1): 137-147.

Fialho, C. M. T., Silva, A. A., Faria, A. T., Torres, L. G., Rocha, P. R. R. and Santos, J. B. (2012). Teor foliar de nutrientes em plantas daninhas e de café cultivadas em competição. *Planta Daninha* 30(1): 65-73.

Fialho, C. M. T., Silva, G. R., Freitas, M. A. M., França, A. C., Melo, C. A. D. and Silva, A. A. (2010). Competição de plantas daninhas com a cultura do café em duas épocas de infestação. *Planta Daninha* 28: 969-978.

Fleck, N. G., Lamego, F. P., Schaedler, C. E. and Ferreira, F. B. (2007). Resposta de cultivares de soja à competição com cultivar simuladora da infestação de plantas concorrentes. *Sci. Agr. Paran* 8(3): 213-218.

Fonseca, A. F. A., Verdin, A. C., Ronchi, C. P., Volph, P. S., Lani, J. A., Guarçoni, M. A., Ferrão, M. A. G. and Ferrão, R. G. (2019). Management of conilon coffee cultivation: Planting, spacing, pruning and pinching. In Ferrão, R. G., Fonseca, A. F. A., Ferrão, M. A. G. and DeMuner, L. H. (Eds.), *Conilon Coffee Updated and Expanded*: The Coffea *Canephora Produced in Brazil* (3rd ed.). Brasil: Incaper, Espírito Santo: 326-339.

França, A. C., Carvalho, F. P., Fialho, C. M. T., D'Antonino, L., Silva, A. A., Santos, J. B. and Ferreira, L. R. (2013). Deriva simulada do glyphosate em cultivares de café Acaiá e Catucaí. *Planta Daninha* 31(2): 443-451.

Franco, C. M. (1958). *Influence of Temperature on Growth of Coffee Plant*. Bulletin 16. New York: Irish Business and Employers Confederation Research Institute.

Galon, L., Tironi, S. P., Rocha, P. R. R., Concenço, G., Silva, A. F., Vargas, L., Silva, A. A., Ferreira, E. A., Minella, E., Soares, E. R. and Ferreira, F. A. (2011). Habilidade competitiva de cultivares de cevada convivendo com azevém. *Planta Daninha* 29(4): 771-781.

García, L. S., Solarte, P. D., Solano, L. and Pérez-Gil, R. F. (2014). Arvenses de cafetal: Identificacion, evaluacion quimica y comportamiento etologico de ovinos en pastoreo. *Ecos. Recursos Agropec.* 1(3): 249-260.

Gómez, R. (2005). Effecto del control de malezas com paraquat y glifosato sobre la erosión y perdida de nutrimentos del suelo em cafeto. *Agron. Mesoamericana* 16: 55-87.

Haidar, M. A., Gharib, C. and Sleiman, F. T. (2010). Survival of weed seeds subjected to sheep rumen digestion. *Weed Res.* 50(5): 467-471.

Harker, K. N. and O'Donovan, J. T. (2013). Recent weed control, weed management, and integrated weed management. *Weed Technol.* 27(1): 1-11.

Lamego, F. P., Ruchel, Q., Kaspary, T. E., Gallon, M., Basso, C. J. and Santi, A. L. (2013). Habilidade competitiva de cultivares de trigo com plantas daninhas. *Planta Daninha* 31(3): 521-531.

Lemes, L. N., Carvalho, L. B., Souza, M. C. and Alves, P. L. C. A. (2010). Weed interference on coffee fruit production during a four-year investigation after planting. *Afr. J. Agric. Res.* 5: 1128-1143.

Lemessa, F. and Wakjira, M. (2015). Cover crops as a means of ecological weed management in agroecosystems. *J. Crop Sci. Biotechnol.* 18(2): 123-135.

Maciel, C. D. G., Poletine, J. P., Neto, A. Md, Guerra, N. and Justiniano, W. (2010). Levantamento fitossociológico de plantas daninhas em cafezal orgânico. *Bragantia* 69(3): 631-636.

Magalhães, C. E. O., Ronchi, C. P., Ruas, R. A. A., Silva, M. A. A., Araújo, F. C. and Almeida, W. L. (2012). Seletividade e controle de plantas daninhas com oxyfluorfen e sulfentrazone na implantação de lavoura de café. *Planta Daninha* 30(3): 607-616.

Marcolini, L. W., Alves, P. L. C. A., Dias, T. C. S. and Parreira, M. C. (2009). Effect of the density and of the distance of *Brachiaria decumbens* staff on the initial growth of *Coffea arabica* L. seedligns. *Coffee Sci.* 4: 11-15.

Martins, B. H., Araujo-Junior, C. F., Miyazawa, M. and Vieira, K. M. (2016a). Humic substances and its distribution in coffee crop under cover crops and weed control methods. *Sci. Agric.* 73(4): 371-378.

Martins, B. H., Araujo-Junior, C. F., Miyazawa, C. F., Vieira, K. M., Hamanaka, C. A. and Silva, A. S. (2016b). Weed control methods and coffee shrub residue effects on carbon stocks in a Latosol under conservation management practices. *Agron. Sci. Biotechnol.* 2: 66-78.

Martins, B. H., Araujo-Junior, C. F., Miyazawa, M., Vieira, K. M. and Milori, D. M. B. P. (2015). Soil organic matter quality and weed diversity in coffee plantation area submitted to weed control and cover crops management. *Soil Till. Res.* 153: 169-174.

Matiello, J. B., Carvalho, F., Shalders, C. C. and Rasga, A. R. D. (1980). Uso de ovinos para controle do mato em cafezais. In *Congresso Brasileiro de Pesquisas Cafeeiras, IBC/GERCA*: 444-446.

Matiello, J. B. and Santinato, F. (2016). Corda-de-viola avança nos cafezais. In *Folha Técnica 318*. Varginha, Brasil: Fundação Procafé.

Matiello, J. B., Santinato, R., Almeida, S. R. and Garcia, A. L. (2016). *Cultura de café no Brasil: Manual de recomendações, ed. 2015*. São Paulo, Brasil: Futurama Editora.

Matos, C. C., Fialho, C. M. T., Ferreira, E. A., Silva, D. V., Silva, A. A., Santos, J. B., França, A. C. and Galon, L. (2013). Características fisiológicas do cafeeiro em competição com plantas daninhas. *Biosci. J.* 29: 1111-1119.

Medeiros, W. N., Melo, C. A. D., Tiburcio, R. A. S., Silva, G. S., Machado, A. F. L., Tuffi Santos, L. D. and Ferreira, F. A. (2016). Crescimento inicial e concentração de nutrientes em clones de *Eucalyptus urophylla* x *Eucalyptus grandis* sob interferência de plantas daninhas. *Cienc. Flor* 26: 147-157.

Michael, P. J., Steadman, K. J., Plummer, J. A. and Vercoe, P. (2006). Sheep rumen digestion and transmission of weedy Malva parviflora seeds. *Aust. J. Exp. Agric.* 46(10): 1251-1256.

Minassa, E. M. C. (2014). *Efeito alelopático da palha de café (Coffea Canephora* L. *e Coffea arabica* L.*) sobre plantas cultivadas e espontâneas*. Thesis, Universidade Estadual do Norte Fluminense Darcy Ribeiro, Campos dos Goytacazes, Rio de Janeiro.

Miranda, G. R. B., Guimarães, R. J., Campos, V. P., Botrel, E. P., Almeida, G. R. R. and Gonzalez, R. G. (2007). Métodos alternativos de desinfestação de plantas invasoras em substratos para formação de mudas de cafeeiro (*Coffea arabica* L.). *Coffee Sci.* 2: 168-174.

Mohler, C. L. (1996). Ecological bases for the cultural control of annual weeds. *J. Prod. Agric.* 9(4): 468-474.

Moraima Garcia, S., Canizares, A., Salcedo, F. and Guillen, L. (2000). A contribution to determine critical levels of weed interference in coffee crops of Monagas state, Venezuela. *Bioagro* 12: 63-70.

Moreira, G. M., Oliveira, R. M., Barrella, T. P., Fontanétti, A., Santos, R. H. S. and Ferreira, F. A. M. (2013). Fitossociologia de plantas daninhas do cafezal consorciado com leguminosas. *Planta Daninha* 31(2): 329-340.

Oliveira, M. F. and Brighenti, A. M. (2018). *Controle de plantas daninhas: Métodos físico, mecânico, cultural, biológico e alelopatia*. Brasília-DF, Brasil: Embrapa.

Oliveira, M. F. and Coutinho Filho, S. A. (2018). Eletrocussão de plantas - capina elétrica. In *Circular técnica 242*. Sete Lagoas, Brasil: Embrapa.

Pais, P. S. M., Junior, D. M. S., Dias, A. C., Iori, P., Guimarães, P. T. G. and Santos, G. A. (2013). Load-bearing capacity of a red-yellow Latosol cultivated with coffee plants subjected to different weed managements. *Cienc. Agrotecnol.* 37: 145-151.

Pais, P. S. M., Junior, D. M. S., Santos, G. Ad, Dias, A. C., Guimarães, P. T. G. and Alcântara, E. N. (2011). Compactação causada pelo manejo de plantas invasoras em Latossolo Vermelho-amarelo cultivado com cafeeiros. *Rev. Bras. Cienc. Solo* 35(6): 1949-1957.

Pandey, A., Soccol, C. R., Nigam, P., Brand, D., Mohan, R. and Roussos, S. (2000). Biotechnological potential of coffee pulp and coffee husk for bioprocesses. *Biochem. Eng. J.* 6(2): 153-162.

Partelli, F. L., Vieira, H. D., Freitas, S. D. P. and Espindola, J. A. A. (2010). Aspectos fitossociológicos e manejo de plantas espontâneas utilizando espécies de cobertura em cafeeiro Conilon orgânico. *Sem. Ci. Agr.* 31(3): 605-618.

Paulo, E. M., Berton, R. S., Cavichioli, J. C. and Kasai, F. S. (2004). Comportamento do cafeeiro Apoatã em consórcio com culturas anuais. *Bragantia* 63(2): 275-281.

Pedrosa, A. W. (2013). *Eficiência da adubação nitrogenada no consórcio entre cafeeiro e Brachiaria brizantha*. Thesis, Universidade de São Paulo, Piracicaba, São Paulo.

Pires, L. F., Araujo-Junior, C. F., Auler, A. C., Dias, N. M. P., Dias Junior, M. S. and Alcântara, E. N. (2017a). Soil physico-hydrical properties changes induced by weed control methods in coffee plantation. *Agric. Ecosyst. Environ.* 246: 261-268.

Pires, L. F., Araujo-Junior, C. F., Piccolomini Dias, N. M., Dias Junior, M. D. S. and Alcântara, E. N. (2017b). Weed control methods effect on the hydraulic attributes of a Latosol. *Acta Sci. Agron.* 39(1): 119-128.

Prado Filho, H. P. A. and Lima, D. B. (1981). Uso de galinhas no controle de plantas daninhas em cafezal. In *Congresso Brasileiro De Pesquisas Cafeeiras*, IBC/GERCA.

Radosevich, S. R., Holt, J. and Ghersa, C. (1996). Physiological aspects of competition. In Radosevich, S. R., Holt, J. and Ghersa, C. (Eds.), *Weed Ecology: Implication for Managements*. New York: John Wiley and Sons, EUA: 217-301.

Ramírez, J. E. (2009). *Hacia la caficultura sostenible*. San José: confederación unitaria de trabajadores C. R. ICAFE.

Ricci, M. S. F., Filho, E. M. and Costa, J. R. (2008). Diversidade da comunidade de plantas invasoras em sistemas agroflorestais com café em Turrialba, Costa Rica. *Pesq. agropec. bras.* 43(7): 825-834.

Roberts, H. A. (1981). Seed banks in the soil v 6. In *Advances in Applied Biology*. Cambridge: Academic Press.

Ronchi, C. P. (2020). Association of chemical and mechanical weed control methods during the pre-harvest of coffee crops. *Planta Daninha* 38: e020216110.

Ronchi, C. P., Carvalho, F. P. and Silva, A. A. (2019). Weed integrated management. In Ferrão, R. G., Fonseca, A. F. A., Ferrão, M. A. G. and DeMuner, L. H. (Eds.), *Conilon Coffee Updated and Expanded: The Coffea Canephora Produced in Brazil* (3rd ed.). Brasil: Incaper, Espírito Santo: 472-491.

Ronchi, C. P. and DaMatta, F. M. (2019). Physiological aspects of conilon coffee. In Ferrão, R. G., Fonseca, A. F. A., Ferrão, M. A. G. and DeMuner, L. H. (Eds.), *Conilon Coffee Updated and Expanded*: The Coffea *Canephora Produced in Brazil* (3rd ed.). Brasil: Incaper, Espírito Santo: 110-143.

Ronchi, C. P., Ferreira, R. T. and Silva, M. A. A. (2014). Manejo de plantas daninhas na cultura do café. In Monquero, P. A. (Ed.), *Manejo de plantas daninhas nas culturas agrícolas*. São Carlos, Brasil: RiMa Editora: 132-154.

Ronchi, C. P. and Silva, A. A. (2006). Effects of weed species competition on the growth of young coffee plants. *Planta Daninha* 24(3): 415-423.

Ronchi, C. P. and Silva, A. A. (2018). Sustainable weed control in coffee. In Korres, N., Burgos, N. R. and Duke, S. O. (Eds.), *Weed Control: Sustainability, Hazards and Risks in Cropping Systems Worldwide*. Boca Raton: Science Publishers: 425-441.

Ronchi, C. P., Silva, A. A. and Ferreira, L. R. (2001). *Manejo de plantas daninhas em lavouras de café*. Viçosa, Brasil: Suprema Gráfica e Editora.

Ronchi, C. P., Silva, A. A., Terra, A. A., Miranda, G. V. and Ferreira, L. R. (2005). Effect of 2,4-dichlorophenoxyacetic acid applied as a herbicide on fruit shedding and coffee yield. *Weed Res.* 45(1): 41-47.

Ronchi, C. P., Souza Junior, J. M., Almeida, W. L., Souza, D. S., Silva, N. O., Oliveira, L. B., Guerra, A. M. N. M. and Ferreira, P. A. (2015). Morfologia radicular de cultivares de café arábica submetidas a diferentes arranjos espaciais. *Pesq. agropec. bras.* 50(3): 187-195.

Ronchi, C. P., Terra, A. A. and Silva, A. A. (2007). Growth and nutrient concentration in coffee root system under weed species competition. *Planta Daninha* 25(4): 679-687.

Ronchi, C. P., Terra, A. A., Silva, A. A. and Ferreira, L. R. (2003). Acúmulo de nutrientes pelo cafeeiro sob interferência de plantas daninhas. *Planta Daninha* 21(2): 219-227.

Rosa, D. M., Nóbrega, L. H. P., Mauli, M. M., Lima, G. P. and Boller, W. (2013). Weeds suppression and agronomic characteristics of maize crop under leguminous crop residues in no-tillage system. *Trop. Subtrop. Agroecosyst.* 16: 455-463.

Sánchez, F. L. E. (1991). Lãs malezas em los cafetales: Recomendaciones prácticas para su control. *Fonaiap Divulga* 9: 18-27.

Sánchez, F. L. E. and Chacón, L. C. (2000). Control de malezas en café usando ovinos. *Rev. Fac. Agron.* 7: 424-433.

Sánchez, F. L. E. and Chacón, L. C. (2001). Cambios en la vegetación de un agrossistema cafetalero por efecto del pastoreo con ovinos. *Rev. Fac. Agron.* 18: 73-81.

Sánchez, F. L. E. and Gamboa, E. (2004). Control de malezas con herbicidas y métodos mecánicos en plantaciones jóvenes de café. *Bioagro* 16: 1-4.

Santos, I. C., Ferreira, F. A., Silva, A. A., Miranda, G. V. and Santos, L. D. T. (2002). Eficiência do 2,4-D aplicado isoladamente e em mistura com glyphosate no controle da trapoeraba. *Planta Daninha* 20(2): 299-309.

Santos, J. C. F., Costa, R. S. C., Leônidas, F. C., Mendes, A. M. and Rodrigues, V. G. S. (2014). Manejo agroecológico de plantas daninhas da cultura do café. *Embrapa Rondônia Doc.* 159.https://ainfo.cnptia.embrapa.br/digital/bitstream/item/132238/1/Doc-159-cafe.pdf.

Sarno, Lumbanraja, J., Adachi, T., Oki, Y., Senge, M. and Watanabe, A. (2004), Effect of weed management in coffee plantation on soil chemical properties. *Nutr. Cycling. Agroecosyst.* 69(1): 1-4.

Schoenbaum, I., Kigel, J., Barkai, D. and Landau, S. (2009). Weed infestation of wheat fields by sheep grazing stubble in the Mediterranean semi-arid region. *Crop Pasture Sci.* 60(7): 675-683.

Schuster, M. Z., Pelissari, A., Moraes, A., Kent Harrison, S. K., Mark Sulc, R. M., Lustosa, S. B. C., Anghinoni, I. and Carvalho, P. C. F. (2016). Grazing intensities affect weed seedling emergence and the seed bank in an integrated crop-livestock system. *Agricult. Ecosyst. Environ.* 232: 232-239.

Shivaprasad, P., Biradar, I. B., Salakinkop, S. R., Raghuramulu, Y., D'Souza, M. V., Hariyappa, N., Hareesh, S. B., Murthy, M. A. and Jayarama (2005). Influence of soil cultivation methods in young coffee on soil moisture, weed suppression and organic matter. *J. Coffee Res* 33: 1-14.

Silva, A. A. and Silva, J. F. (2013). *Tópicos em manejo de plantas daninhas*. Viçosa, Brasil: UFV.

Silva, S. O., Matsumoto, S. N., Bebé, F. V. and José, A. R. S. (2006). Diversidade e frequência de plantas daninhas em associações entre cafeeiros e grevíleas. *Coffee Sci.* 1: 126-134.

Siqueira, S. H. R., Ferreira, M. M., Alcântara, E. N. and Carvalho, R. C. S. (2014a). Agregação de um Latossolo Vermelho-amarelo submetido a métodos de controle de plantas invasoras na cultura do café. *Rev. Bras. Cienc. Solo* 38(4): 1128-1134.

Siqueira, S. H. R., Ferreira, M. M., Alcântara, E. N., Silva, B. M. and Silva, R. C. (2014b). Water retention and S index of an oxisol subjected to weed control methods in a coffee crop. *Cienc. Agrotecnol.* 38: 471-479.

Soto-Pinto, L., Perfecto, I. and Caballero-Nieto, J. (2002). Shade over coffee: Its effects on berry borer, leaf rust and spontaneous herbs in Chiapas, México. *Agrofor. Syst.* 55(1): 37-45.

Souza, L. S., Losasso, P. H. L., Oshiiwa, M., Garcia, R. R. and Goes Filho, L. A. (2006). Efeitos das faixas de controle do capim-braquiária (*Brachiaria decumbens*) no desenvolvimento inicial e na produtividade do cafeeiro (*Coffea arabica*). *Planta Daninha* 24(4): 715–720.

Yamane, K., Kono, M., Fukunaga, T., Iwai, K., Sekine, R., Watanabe, Y. and Iijima, M. (2014). Field evaluation of coffee grounds application for crop growth enhancement, weed control, and soil improvement. *Plant Prod. Sci.* 17(1): 93-102.

Zasso Brazil. (2020). *Electroherb™: The Non-chemical Crop Protection Alternative - Advanced Power Electrophysics for Clean Weed Control*. Available at: www.zasso .com (accessed 7 September 2020).

Index